D0578324

Applied Regression
Including Computing and Graphics

Applied Regression Including Computing and Graphics

R. DENNIS COOK

SANFORD WEISBERG
The University of Minnesota
St. Paul, Minnesota

A Wiley-Interscience Publication

JOHN WILEY & SONS, INC.

New York · Chichester · Weinheim · Brisbane · Singapore · Toronto

Copyright © 1999 by John Wiley & Sons, Inc. All rights reserved.

Published simultaneously in Canada.

For ordering and customer service, call 1-800-CALL WILEY.

Library of Congress Cataloging-in-Publication Data:

Cook, R. Dennis.
    Applied regression including computing and graphics / R. Dennis Cook, Sanford Weisberg.
       p.   cm. — (Wiley series in probability and statistics. Texts and references section)
    "A Wiley-Interscience publication."
    Includes bibliographical references and index.
    ISBN 0-471-31711-X (alk. paper)
    1. Regression analysis.   I. Weisberg, Sanford, 1947– .
II. Title.   III. Series.
QA278.2.C6617   1999
519.5'36—dc21   99-17200
                                                      CIP

Printed in the United States of America

10 9 8 7 6 5 4 3 2 1

This book is dedicated to the students
who turn words and graphs into ideas and discoveries.

Contents

Preface

This textbook is about regression, the study of how a response variable depends on one or more predictors. Regression is one of the fundamental tools of data analysis. This book is intended for anyone interested in applying regression to data. The mathematical level required is intentionally low; prerequisite is only one semester of basic statistical methods.

The main features of the book are as follows:

- Emphasis is placed on *seeing* results through graphs. We present many easily used graphical methods, most based on simple two-dimensional scatterplots, that provide analysts with more insight into their data than would have been possible otherwise, including a deeper appreciation for interpretation.
- We provide user-friendly computer software called *Arc* that lets the reader immediately apply the ideas we present, both to the examples in the book and to their own data. Many of the methods we use are impossible or at best difficult to implement in standard statistical software packages.
- This book is a *complete* textbook on applied regression suitable for a one semester course. We start with very basic ideas, progressing from standard ideas for linear models integrated with newer graphical approaches, to regression graphics and more complex models like generalized linear models.
- The book includes over 300 figures. The reader following along on the computer can reproduce almost all of them.
- Most of the examples and homework problems are based on real data. All of the data sets used in the book are included with *Arc*.
- A companion Internet site includes the software, more problems, help for the reader, additional statistical material, extensions to the software, and much more.

We have used drafts of this textbook for several years as a basis for two different courses in applied regression analysis: One is intended primarily for advanced undergraduate and beginning graduate students in fields other than statistics, while the other is intended for first-year statistics graduate students. Drafts have also been used by others for students in social science, business, and other disciplines. After completing these courses, our students can analyze regression problems with greater ease and more depth than could our students before we began to use this book.

Arc

Integrated into the text is discussion of a computer package called *Arc*. This is a user-friendly program designed specifically for studying this material, as well as for applying the ideas learned to other data sets. The program permits the user to do the analyses discussed in the book easily and quickly. It can be down-loaded for free from the Internet site for this book at the address given in the Appendix; versions are available for Windows, Macintosh, and Unix.

For those readers who prefer to use other statistical packages, we include on our Internet site descriptions of how a few of the major packages can be used for some of the calculations.

ORGANIZATION AND STYLE

When writing the book, we envisioned the reader sitting at a computer and reworking the examples in the text. Indeed, some of the text can be read as if we were next to the reader, suggesting what to do next. To maintain this low-key style, we have tried to avoid heavy algebra. References and technical comments are collected in the complements section at the end of each chapter.

PATHS

The book is divided into four parts:

I: Introduction. The first part consists of five chapters that present many basic ideas of regression to set the stage for later developments, including the one-sample problem, using and interpreting histograms and scatterplots, smoothing including density estimates and scatterplot smoothers, and bivariate distributions like the normal. These chapters weave together standard and non-standard topics that provide a basis for understanding regression. For example, smoothing is presented before simple linear regression.

II: Multiple Linear Regression. The second part of the book weaves standard linear model ideas, starting with simple regression, with the basics of graphics, including 2D and 3D scatterplots and scatterplot matrices. All these graph types are used repeatedly throughout the rest of the book. Emphasis is split between basic results assuming the multiple linear regression model holds, graphical ideas, the use of transformations, and graphs as the basis for diagnostic methods.

III: Graphics. The third part of the book, which is unique to this work, shows how graphs can be used to better understand regression problems in which no model is available, or else an appropriate model is in doubt. These methods allow the analyst to see appropriate answers, and consequently have increased faith that the data sustain any models that are developed.

IV: Other Models. The last part of the book gives the fundamentals of fitting generalized linear models. Most of the graphical methods included here are also unique to this book.

For a one-quarter (30-lecture) course, we recommend a very quick tour of Part I (3–5 lectures), followed by about 18 lectures on Part II, with the remainder of the course spent in Part III. For a semester course, the introduction can be expanded, and presentation of Parts II and III can be slowed down as well. We have used Part IV of the book in the second quarter of a two-quarter course; completing the whole book in one semester would require a fairly rapid pace. Some teachers may prefer substituting Part IV for Part III; we don't recommend this because we believe that the methodologies described in Part III are too useful to be skipped.

Some of the material relating to the construction of graphical displays is suitable for presentation in a laboratory or recitation session by a teaching assistant. For example, we have covered Section 7.1 and most of Chapters 5 and 8 in this way. Further suggestions on teaching from this book are available in a teacher's manual; see the Internet site for more information.

OTHER BOOKS

The first book devoted entirely to regression was probably by the American economist Mordechai Ezekiel (1924, revised 1930 and 1941 and finally as Ezekiel and Fox, 1959). Books on regression have proliferated since the advent of computers in universities. We have contributed to this literature. Weisberg (1985) provided an introduction to applied regression, particularly to multiple linear regression, but at a modestly higher mathematical level. This book includes virtually all the material in Weisberg (1985); even some of the examples, but none of the prose, are common between the two books. Cook and Weisberg (1994b) provided an introduction to graphics and regression. Nearly all of the material in that book, though again little of the prose, has found its way into this book. Chapter 15 contains a low-level introduction to the material in the research monograph on residual and influence analysis by Cook and Weisberg (1982). Finally, Cook (1998b) provides a mathematical and rigorous treatment of regression through graphics that is the core of this book.

On the Internet site for this book, we provide additional references for reading for those who would like a more theoretical approach to this area.

ACKNOWLEDGMENTS

Several friends, colleagues, and students have been generous with their time and ideas in working through this material. Iain Pardoe, Jorge de la Vega, and Bret Musser have helped with many parts of this work. The manuscript was read and used by Bert Steece, Robert Berk, Jeff Witmer, Brant Deppa, and David Olwell. John Fox was particularly helpful in the development of *Arc*. We were stimulated by working with several recent University of Minnesota graduates, including Rodney Croos-Debrara, Francesca Chariomonte, Efstathia Bura, Pawel Stryszak, Giovanni Porzio, Hakbae Lee, Nate Wetzel, Chongsun Park, and Paul Louisell. We'd be glad to have more wonderful students like these!

Throughout this work, we have been supported by the National Science Foundation's Division of Undergraduate Education. Without their support, this book would not exist.

R. DENNIS COOK
SANFORD WEISBERG

St. Paul, Minnesota
May, 1999

Applied Regression
Including Computing and Graphics

PART I

Introduction

In the first part of this book, we set the stage for the kind of problems we will be studying. Chapter 1 contains a brief review of some of the fundamental ideas that are required for the rest of this book. Chapter 2 contains an introduction to the fundamental ideas of regression. This theme is continued in Chapters 3–5 with discussions of smoothing, bivariate distributions, and two-dimensional scatterplots.

CHAPTER 1

Looking Forward and Back

1.1 EXAMPLE: HAYSTACK DATA

Hay farming invites risk. It is complex, requires hard work and also the co-operation of the weather. After cutting and drying, hay is usually stored in small rectangular bales of 25–60 pounds for a "two-string" bale or 80–180 pounds for the large "three-wire" bales favored in California. Modern balers can produce large round bales of up to 2000 pounds. In some areas, hay is stacked in the field today as it was 100 years ago. The enormous haystacks found in the Montana range country are a memorable sight.

The value of a ton of hay can vary considerably, depending on the type of hay, weed content, whether it was cut early to maximize nutrition or late to maximize yield, and whether it was dried properly before baling. In the summer of 1996, baled hay from a first cutting of alfalfa in Scottsbluff County, Nebraska was offered for $120 per ton, while a ton of baled hay from intermediate wheatgrass was offered for $35 in Charles Mix County, South Dakota.

Before balers, hay was gathered loose and was subsequently stacked in the field or stored in a barn mow. While the size of haystacks could vary considerably, hay on the inside of a stack keeps the best, resulting in some preference for large round stacks. This was the prevailing condition in much of the United States around the beginning of the 20th century. Although steam balers were known during the middle part of the 19th century, this technological innovation was slow to be adopted in the West. In particular, farmers in the Great Plains during the 1920s sold hay by the stack, requiring estimation of stack volume to ensure a fair price. Estimating the volume of a haystack was not a trivial task and could require much give-and-take between the buyer and seller to reach a mutually agreeable price.

A study was conducted in Nebraska during 1927 and 1928 to see if a simple method could be developed to estimate the volume of round haystacks. It was reasoned that farmers could easily use a rope to characterize the size

3

of a round haystack with two measurements: The circumference around the base of a haystack and the "over," the distance from the ground on one side of a haystack to the ground on the other side. The haystack study involved measuring the volume, circumference, and "over" on 120 round haystacks. The volume of a haystack was determined by using survey instruments that were not normally available to farmers in the 1920s.

For notational convenience, let *Vol* and *C* represent the volume and circumference of a haystack, and let *Over* represent the corresponding "over" measurement. The issue confronting the investigators in the haystack study was how to use the data, as well as any available prior information, in the development of a simple formula expressing the volume of a haystack as a function of its circumference and "over" to a useful approximation. In other words, can a function f be developed so that

$$Vol \approx f(Over, C)?$$

The symbol "$\approx$" means "is *approximately* equal to." To be most useful, the approximation should hold for all future round haystacks, as well as for the 120 haystacks represented in the data.

The haystack study is an instance of a broad class of experimental problems that can be addressed using *regression*, the topic of this book. There are several types of regression problems. The one that covers the haystack study begins with a population of experimental units, haystacks in our instance. Each experimental unit in the population can be characterized by measuring a *response variable*, represented generically by the letter *y*, and *predictor variables*, represented generically by $x_1, x_2, \ldots, x_p$, where *p* is the total number of predictors used. In broad terms, regression is the study of how the response *y* depends statistically on the *p* predictors: Knowing the values of $x_1, x_2, \ldots, x_p$ for a particular unit, what can we say about the likely value of the response on that unit? In the haystack study the response variable is $y = Vol$ and the $p = 2$ predictors are $x_1 = C$ and $x_2 = Over$. The ordering of the predictors is unimportant, and we could equally have written $x_1 = Over$ and $x_2 = C$. The goal of the haystack study was to use the observed data on 120 haystacks to produce predictions of volume for other haystacks, based on measurements of *C* and *Over*. Prediction is a common goal of regression, but in some problems, description of the dependence of the response on the predictors is all that is desired.

In other literature, the response variable is sometimes called the *dependent variable*, and the predictor variables may be called *independent* variables, *explanatory* variables, *carriers*, or *covariates*. These alternate names may imply slightly different contexts, but any distinctions among them are not important in this book.

As in the haystack problem, regression studies are often based on a random sample of units from the population. There are many ways to use the resulting data to study the dependence of the response on the predictors.

One way begins by postulating a *model*, a physical or mathematical metaphor that can help us tackle the problem. To frame the idea, let's again return to the haystack study, and imagine a haystack as being roughly a hemisphere. The volume of a hemisphere, being half that of the corresponding sphere, can be written as a function of only the circumference C of the sphere,

$$\text{hemisphere volume} = \frac{C^3}{12\pi^2} \tag{1.1}$$

This simple formula gives us one possible way to use C and *Over* to estimate haystack volume: Ignore *Over* and use

$$Vol \approx \frac{C^3}{12\pi^2} \tag{1.2}$$

This representation is an example of a *model* relating *Vol* to C and *Over*. As the book progresses, our models will become more comprehensive.

Approximation (1.2) reflects the fact that if a haystack were really a hemisphere, farmers would not need *Over* because then $C = 2 \times Over$. However, it seems unreasonable to expect haystacks to be exact hemispheres and so we should not expect $C = 2 \times Over$. On the other hand, it may be reasonable to expect $C \approx 2 \times Over$, so that the two predictors are giving essentially the same information about haystack volume. Eventually, we will be using the data to test the adequacy of possibilities like (1.2), to test if haystacks can really be treated as hemispheres, to suggest alternative models, and to quantify uncertainty when predicting the volume of a haystack from its circumference and "over."

We consider in the next section a different example that will allow us to focus on specific ideas in regression, and to review some of the prerequisites for this book. We will return to the haystack data in later chapters.

1.2 EXAMPLE: BLUEGILL DATA

Bluegills aren't the hardest fish in the world to catch, but they won't jump into your boat either. Native to the eastern half of the United States and adjacent portions of Canada and Mexico, bluegills have now been introduced throughout the United States as a sport and forage fish. The scientific name for bluegills is *Lepomis macrochirus*, which in Greek means "scaled gill cover" "large hand." The average weight in a bluegill population varies somewhat with the region. In Minnesota, the typical weight is about 110 grams (0.25 pound); the state record goes to a 1275-gram fish caught in Alice Lake. The state record for Ohio is a whopping 1475 grams (3.25 pounds) for a fish that was more than 30 cm long.

Fishery managers are interested in the way fish, including bluegills, grow. One might wonder, for example, how long a bluegill takes to grow to 25 or 30 cm, and what fraction of a bluegill population would eventually reach such lengths. The maximum life span for a bluegill is 8–10 years. A good understanding of the statistical relationships between bluegill size and age can be a great help in developing answers to such questions.

In 1981 a sample of 78 bluegills was taken from Lake Mary in Minnesota, and the *Length* in mm and *Age* in years of each fish was determined. The method of collection of the data, and the time and location of the data are important because the relationship between *Length* and *Age* can depend on such factors. For example, some lakes will provide better habitat for bluegills than will others. For many species of fish, including bluegills, age can be determined by counting the number of annular rings on a scale, much the same as is done with tree rings. These data can be approached as a regression problem. The response variable is bluegill *Length* and the single predictor is *Age*. We would like to study the statistical dependence of *Length* on *Age*.

We have a long way to go before these issues, or those raised in connection with the haystack data, can be addressed fully. In the rest of this chapter we describe how to perform basic tasks using *Arc*, the regression computer package developed for this book, and we review statistical methodology for a single variable, knowledge of which is the primary prerequisite for this book. We return to the topic of regression in the next chapter.

1.3   LOADING DATA INTO *Arc*

The bluegill data from Lake Mary are available in the data file `lakemary.lsp`. If you are following along on a computer, you may need to read parts of the Appendix. Loading data files is discussed in Section A.5. Data files can be loaded into *Arc* either by selecting "Load" from the Arc menu or by typing the command (`load "lakemary"`) in the text window.

Two menus appear on the menu bar each time a data set is loaded into *Arc*. The first is referred to generically as the *data set menu*. The actual name of this menu is specified in the data file, and it can be set to reflect some aspect of the problem. For the bluegill data we named the data set menu "LakeMary," which is the name of the lake sampled to collect the fish. The items in the data set menu generally allow you to manipulate the data and compute summary statistics. The second menu, which is always named "Graph&Fit," contains items that allow you to plot the data and fit models. The various items within these menus will be discussed at the appropriate time as the book progresses.

All of the data sets discussed in this book come with *Arc*. Instructions on how to prepare your own data to be read by *Arc* are given in Appendix A, Section A.5.

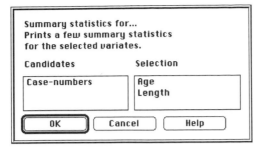

FIGURE 1.1 The "Display summaries" dialog.

TABLE 1.1 Output from "Display summaries" for the Lake Mary Data

Data set = LakeMary, Summary Statistics						
Variable	N	Average	Std. Dev.	Minimum	Median	Maximum
Age	78	3.6282	0.92735	1.	4.	6.
Length	78	143.6	24.137	62.	150.	188.

Data set = LakeMary, Sample Correlations		
Age	1.0000	0.8573
Length	0.8573	1.0000
	Age	Length

1.4 NUMERICAL SUMMARIES

1.4.1 Display Summaries

To compute summary statistics for the bluegill data from Lake Mary, select "Display summaries" from the data set menu, which is called "LakeMary" in this example, to get the dialog shown in Figure 1.1. The items in the "Candidates" list are the available variables in the active data set (see Section A.6), and the items in the "Selection" list are the variables you have selected for which summary statistics will be computed. Variables can be moved from the "Candidates" list to the "Selection" list by double-clicking on the variable name, or by clicking once on the variable name to select it and then clicking once in the "Selection" window. Variables can be moved back to the "Candidates" window in the same way. In Figure 1.1 two variables, *Age* and *Length*, have been moved from the "Candidates" list to the "Selection" list. The variable *Case-numbers* assigns numbers to the fish from 0 up to 77; it is of no interest for the present. Variable selection in other *Arc* dialogs follows the same pattern.

Once your dialog is the same as that in Figure 1.1, click on the "OK" button. The summary statistics shown in Table 1.1 will then be displayed in the text window. Table 1.1 consists of two parts. The first contains univariate summary statistics and the second contains the sample correlations. We discuss the correlation coefficient in Chapter 4. For now we stay with the univariate summaries.

The first column of the summary statistics contains the variable names and the second contains the number of observations. The third and fourth columns contain the average and standard deviation, respectively. As a generic notation, let $y_1, \ldots, y_n$ denote the $n = 78$ values of the response. The *sample mean* $\bar{y}$ is computed as

$$\bar{y} = \frac{\sum_{i=1}^{n} y_i}{n} = 143.6 \text{ mm} \tag{1.3}$$

and the sample standard deviation sd(y) = 24.137 mm is the square root of the sample variance,

$$\text{sd}^2(y) = \frac{\sum_{i=1}^{n}(y_i - \bar{y})^2}{n - 1} = (24.137)^2 \text{ mm}^2 \tag{1.4}$$

In this book we will always use the symbol sd(y) to indicate the sample standard deviation of the samples of y-values; this is often denoted by s_y in other books. We will not have special notation for a sample variance, but rather will always denote it as the square of a standard deviation.

The fifth and seventh columns of Table 1.1 give the minimum and maximum observations in the data. The maximum length is 188 mm, or about 7.4 inches, which seems considerably less than the length of Minnesota's record bluegill, 10.5 inches. The size of the difference $10.5 - 7.4 = 3.1$ inches can be judged in terms of the distribution of *Length* by finding the number of standard deviations that 10.5 is above 7.4. This could be done in millimeters as the original data are recorded, or in inches. To perform the calculation in inches, we need to convert the standard deviation of *Length* from 24.137 mm to inches. In general, if we multiply all observations by a constant c and then add another constant b, the standard deviation sd($b + cy$) in the new scale of $b + cy$ is related to the standard deviation sd(y) in the original scale as follows:

$$\text{sd}(b + cy) = |c| \times \text{sd}(y) \tag{1.5}$$

where $|c|$ denotes the absolute value of c. Now, 1 millimeter equals $c = 0.03937$ inches, so the standard deviation of *Length* in inches is

$$\text{sd}(cy) = |c| \times \text{sd}(y) = 0.03937 \times 24.137 = 0.95$$

The length of the record bluegill is therefore about

$$\frac{10.5 - 7.4}{0.95} = 3.26$$

standard deviations above the maximum length observed in the Lake Mary study. What value would we have obtained if we had performed the computation in millimeters rather than inches?

The sixth column in Table 1.1 gives the sample medians. The median divides a sample into the lower half and the upper half of the data; it is defined to be the middle observation in the ordered data if n is odd, and the average of the two middle observations if n is even.

1.4.2 Command Line

Numerical summaries can also be computed by typing in the text window's command line following the prompt >. For example, the following output shows how to compute the mean, median, and standard deviation of *Length* after loading the data file lakemary.lsp.

```
>(mean length)
143.603
>(median length)
150
>(standard-deviation length)
24.1367
```

The text on each line following a prompt, including the parentheses, is typed by the user. After pressing the "Enter" key, the output is given on the line following the command. Thus, the mean length is 143.603 mm.

Sample quantiles other than the median can be computed by using the quantile command. This command takes two arguments, the name of a variable and a fraction f between 0 and 1 indicating the quantile. Informally, the fth quantile divides a sample so that $f \times 100$ percent of the data fall below the quantile and $(1 - f) \times 100$ percent fall above the quantile. The exact definition needs to be a bit more detailed to account for the fact that a sample is discrete: The fth quantile of a univariate sample is a number q_f that divides the sample into two parts so that

- At most $f \times 100$ percent of the observations are less than q_f
- At most $(1 - f) \times 100$ percent of the observations are greater than q_f

The three sample *quartiles* are the quantiles corresponding to $f = 0.25, 0.5$, and 0.75. They can be computed as follows:

```
>(quantile length .25)
137.5
>(quantile length .5)
150
>(quantile length .75)
160
```

The quartiles divide the sample in four. The first quartile $q_{.25}$ is the median of the lower half of the data, and the third quartile $q_{.75}$ is the median of the upper half of the data. The 0.5 quantile is the same as the median.

Arc has many other commands that can be used to manipulate data and do statistical computations. Many of the statistical commands will be introduced as needed. The syntax for basic numerical calculations is described in Appendix A, Section A.2.

1.4.3 Displaying Data

All the data on selected variables can be displayed in the text window by using the "Display data" item in the "LakeMary" menu. After selecting "Display data" you will be presented with a dialog for choosing the variables to be displayed.

Alternatively, you can display all the data on a variable by simply typing its name without parentheses in the text window:

```
>length
(67 62 109 83 91 88 137 131 122 122 118 115 131
143 142 . . . 130 160 130 170 170 160 180 160 170)
```

1.4.4 Saving Output to a File and Printing

For information on how to save output to a file, perhaps for later printing, see Appendix A, Section A.3.

1.5 GRAPHICAL SUMMARIES

1.5.1 Histograms

To construct a histogram of *Length*, select the item "Plot of" from the Graph&Fit menu. In the resulting dialog shown in Figure 1.2, move *Length* from the "Candidates" list to the "H" box. This can be done by simply double clicking on *Length* or clicking once on *Length* to select it and then clicking once in the "H" box. Generally, "H," "V," and "O" refer to the horizontal, vertical, and out-of-page axes of a plot. Placing variables in the "H" and "V" boxes will result in a 2D plot, while placing variables in all three boxes will result in a 3D plot. If only the "H" box is used, the plot produced will be a histogram. Leave the "V" and "O" boxes empty and place *Length* in the "H" box, as shown in Figure 1.2, and then click "OK." The histogram of *Length* shown in Figure 1.3 will be displayed in its own window on the computer screen. On Macintosh and Windows, the title "Histogram" is added to the menu bar; on Unix a button labeled "Menu" appears on the plot itself. The menu is used to change characteristics of the histogram.

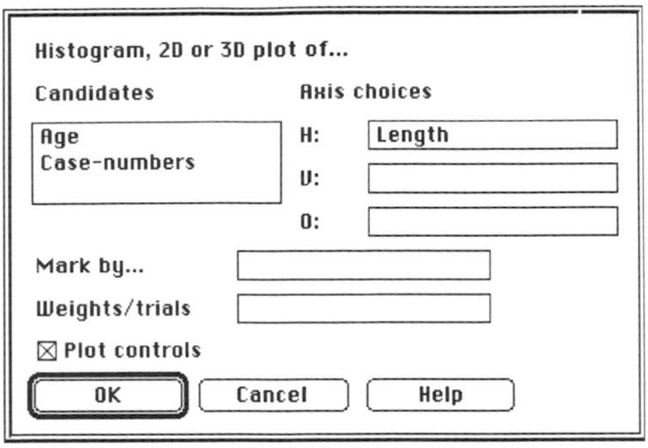

Histogram, 2D or 3D plot of...

Candidates   Axis choices

Age   H: Length
Case-numbers   V:

                       O:

Mark by...

Weights/trials

☒ Plot controls

    OK   Cancel   Help

FIGURE 1.2 The "Plot of" dialog.

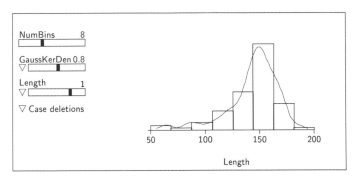

FIGURE 1.3 Histogram of bluegill length.

The purpose of many of the items in this menu can be discovered by trying them.

The intervals on the horizontal axis marked off by the individual rectangles are called *bins*. The area of an individual rectangle gives the fraction of the observations that fall in the corresponding bin. Histograms produced by *Arc* will always have all bins of equal width.

At the left of the histogram are three *slidebars* and three *pop-up menus* marked by triangles. Two of the pop-up menus are located just to the left of the second and third slidebars. The first slidebar, named "NumBins," controls the number of bins in the histogram, as shown at the right of the slidebar. The histogram in Figure 1.3 has 8 bins. The *slider*, the dark area within the slidebar, is used to change the number of bins by changing its position. The position of any slider can be changed by clicking the mouse in the slidebar to the left or right of the slider, or by dragging the slider, holding down the mouse button on the slider and moving it to the right or left.

The second slidebar causes a smooth version of the histogram to be super-imposed on the plot, as shown in Figure 1.3. The slider controls the amount of smoothing. The corresponding pop-up menu allows the smoothing method to be changed. The third slidebar transforms the plotted values, and the pop-up menu "Case deletions" is used to delete and restore observations from a data set. These items will be discussed in more detail when the associated statistical ideas are introduced.

The histogram in Figure 1.3 indicates that the distribution of *Length* is *skewed* to the left, with more small fish than large fish. This asymmetry might have been anticipated from the summary statistics in Table 1.1 because the sample mean of *Length* is less than the sample median of *Length*. Regulations that place a lower limit on the length of fish that can be removed from the lake would contribute to the asymmetry of the distribution.

You can resize a histogram or any other plot by holding down the mouse button on the lower-right corner of the window and dragging the corner. You can move a plot by holding down the mouse button on the top margin of the window and dragging. To remove a plot from the screen, click on the plot's close box. On the Macintosh, the close box is a small square at the upper-left corner of the plot. With Windows, the close box is at the upper-right corner on the plot; it is a small square with a $\times$ in it. On Unix, the close box is a large button labeled "Close."

1.5.2   Boxplots

The boxplot is another option for visualizing the distribution of *Length*. Se-lect the item "Boxplot of" from the Graph&Fit menu. In the resulting dialog, move *Length* to the "Selection" window and then click "OK." A boxplot like the one in Figure 1.4, but without the added commentary, will then be con-structed. A boxplot consists of a box with a horizontal line through it, and two vertical lines extending from the upper and lower boundaries of the box. The line through the box marks the location of the sample median relative to the vertical axis of the plot. The lower edge of the box marks the location of the first quartile $q_{.25}$ and the upper edge marks the third quartile $q_{.75}$. The box contains 50% of the observations. The width of the box contains no relevant information.

The line extending down from the box terminates at the data point which is closest to, but larger than

$$L = q_{.25} - 1.5(q_{.75} - q_{.25})$$

Data points with values smaller than L are called *lower outer values*. Similarly, the line extending up from the box terminates at the data point being the closest to, but smaller than

$$U = q_{.75} + 1.5(q_{.75} - q_{.25})$$

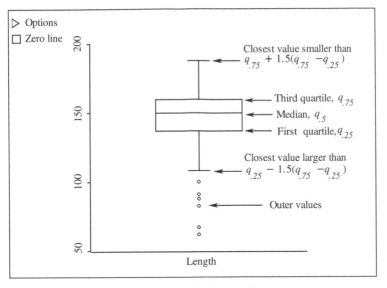

FIGURE 1.4   Boxplot of bluegill length.

Data points with values larger than U are called *upper outer values*. In the boxplot for bluegill length, there are no upper outer values, so the line terminates
at 188 mm, which, from Table 1.1, is the maximum value of *Length*.

The boxplot for bluegill length in Figure 1.4 again leads to the impression
that the data are skewed toward the smaller values of *Length* because there are
no upper outer values while there are several lower outer values.

1.6   BRINGING IN THE POPULATION

We use the notation $E(y)$ and $Var(y)$ to denote the population mean and population variance of a random variable y. The notation $E(y)$ is shorthand for
"the expectation of y." In many texts, the population mean and variance are
denoted using Greek letters, μ (mu) for the mean and σ^2 (sigma squared) for
the variance, but here we will mostly stay with the fuller notation to smooth
the transition to regression in the next chapter. The *population standard deviation* of y is the square root of the population variance. For example, $E(Length)$
and $Var(Length)$ refer to the mean and variance of *Length* in the population of
bluegills in Lake Mary. The population mean and variance are two useful properties indicating how a random variable is distributed across the population.
The distribution of y can be characterized fully by the *cumulative distribution
function* (cdf), $Pr(y \leq a)$, which is read as "the probability that y is less than
or equal to the number a." It is a function of the argument a. If we knew
$Pr(Length \leq \ell)$ for every number ℓ, then the distribution of *Length* would be
known completely.

1.6.1   The Density Function

The distribution of a continuous random variable y is usually represented using a smooth curve with the property that $\Pr(y \leq a)$ is the area under the curve to the left of a. For illustration, think of the smooth curve in Figure 1.3 as representing the distribution of *Length*. Then $\Pr(Length \leq 150)$ is the area under the curve to the left of 150. Similarly, the probability that *Length* exceeds 150, $\Pr(Length \geq 150)$, is the area under the curve to the right of 150, and the probability that *Length* is between 100 and 150, $\Pr(100 \leq Length \leq 150)$, is the area under the curve between 100 and 150. These smooth curves are called *density functions*, or just *densities*, because they show relative density: The curve in Figure 1.3 shows the relative density of bluegills with respect to their length.

Occasionally we will need to construct new random variables by taking a linear function of y, say $w = a + (b \times y)$ where a and b are constants. The population mean and variance of w are related to the population mean and variance of y as follows:

$$E(w) = a + (b \times E(y)) \tag{1.6}$$

and

$$Var(w) = b^2 \times Var(y) \tag{1.7}$$

1.6.2   Normal Distribution

The family of normal distributions has a central place in statistics. It serves as a reference point for many types of investigations and it provides a useful approximation to the random behavior of many types of variables, particularly those associated with measurement error.

Suppose that the random variable y is normally distributed. To characterize fully its normal distribution, we need only the population mean $E(y)$ and the population variance $Var(y)$. Symbolically, we write

$$y \sim N(E(y), Var(y)) \tag{1.8}$$

to indicate that y has a normal distribution with mean $E(y)$ and variance $Var(y)$. If we have a random sample $y_1, y_2, \ldots, y_n$ from the same normal distribution, we will write

$$y_i \sim NID(E(y), Var(y)), \qquad i = 1, \ldots, n \tag{1.9}$$

where the symbol "NID" means normal, independent, and identically distributed.

For the normal distribution $y \sim N(E(y), Var(y))$ with $E(y)$ and $Var(y)$ known, we can compute $\Pr(y \leq a)$ for any value a that we choose. This is one of the distinguishing characteristics of the normal distribution, since for many other

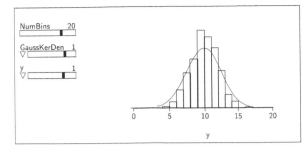

FIGURE 1.5 Histogram of 1000 observations from a normal distribution with mean 10 and variance 4.

distributions knowledge of the mean and variance is not enough to calculate the cumulative distribution function.

Figure 1.5 shows a histogram of 1000 observations from a normal distribution with $E(y) = 10$ and $Var(y) = 4$. The smooth version of the histogram closely matches the normal distribution's bell-shaped density function. As suggested by the figure, the density for the normal distribution is symmetric: If we fold the density at its mean of 10, the two halves will coincide.

The standardized version of y,

$$z = \frac{y - E(y)}{\sqrt{Var(y)}}$$

follows a *standard normal* distribution, $z \sim N(0, 1)$, the normal distribution with mean $E(z) = 0$ and variance $Var(z) = 1$. The mean and variance of z follow from the general rules for determining the mean and variance of a linear function of y: In reference to (1.6) and (1.7), set

$$a = -\frac{E(y)}{\sqrt{Var(y)}}$$

and

$$b = \frac{1}{\sqrt{Var(y)}}$$

We can write any quantile of the normal distribution of y in terms of the quantiles of the standard normal distribution of z in the same way. For example, let $Q(y, 0.5)$ and $Q(z, 0.5)$ denote the second quartiles (medians) for y and z. We have used the uppercase letter Q for quantiles because they are being computed from the population rather than from a sample; we previously used the lowercase letter q for sample quantiles. In this notation,

$$Q(z, 0.5) = \frac{Q(y, 0.5) - E(y)}{\sqrt{Var(y)}}$$

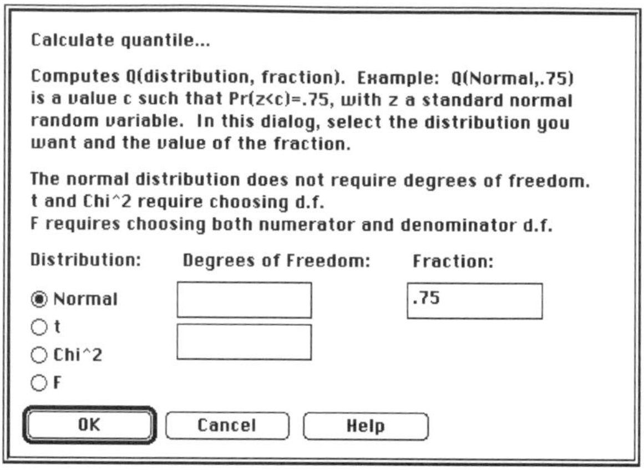

Calculate quantile...

Computes Q(distribution, fraction). Example: Q(Normal,.75) is a value c such that Pr(z<c)=.75, with z a standard normal random variable. In this dialog, select the distribution you want and the value of the fraction.

The normal distribution does not require degrees of freedom. t and Chi^2 require choosing d.f. F requires choosing both numerator and denominator d.f.

Distribution:	Degrees of Freedom:	Fraction:
⦿ Normal		.75
○ t		
○ Chi^2		
○ F		

| OK | Cancel | Help |

FIGURE 1.6   The "Calculate quantile" dialog.

and

$$Q(y, 0.5) = E(y) + Q(z, 0.5)\sqrt{\mathrm{Var}(y)}$$

1.6.3   Computing Normal Quantiles

Arc has a menu item for computing quantiles of standard distributions like the standard normal distribution. For example, to compute $Q(z, 0.75)$, the point with fraction 0.75 of the probability to its left, select the item "Calculate quantile" from the Arc menu. The resulting dialog is shown in Figure 1.6. In this dialog, enter the value 0.75 in the box for the "Fraction," and then select the distribution you want. At this point, we are interested only in the standard normal, but the other distributions will be used later. Click the "OK" button, and the following is displayed in the text window:

```
>Quantile calculation: Q(Normal,0.75) = 0.67449.
```

Thus 75% of the probability under the standard normal density function is to the left of about 0.67. The 0.75 quantile of y, a normally distributed variable with mean $E(y)$ and variance $\mathrm{Var}(y)$, is $E(y) + (0.67 \times (\mathrm{Var}(y))^{1/2})$.

1.6.4   Computing Normal Probabilities

Normal probabilities can be computed in *Arc* using the "Calculate probability" item in the Arc menu. If z is a standard normal random variable, then the dialog can be used to compute any of the lower-tail $\Pr(z \leq c)$, upper-tail $\Pr(z \geq c)$, or two-tail $\Pr(|z| \geq c)$ probabilities. The dialog is shown in Figure 1.7. To use this

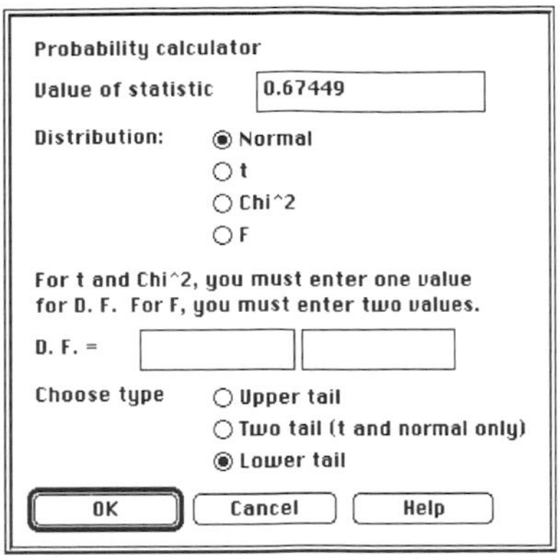

```
┌─────────────────────────────────────────────┐
│ Probability calculator                       │
│                                              │
│ Value of statistic      │ 0.67449        │   │
│                                              │
│ Distribution:   ◉ Normal                     │
│                 ○ t                          │
│                 ○ Chi^2                       │
│                 ○ F                          │
│                                              │
│ For t and Chi^2, you must enter one value    │
│ for D. F. For F, you must enter two values.  │
│                                              │
│ D. F. =    │          │ │          │         │
│                                              │
│ Choose type    ○ Upper tail                  │
│                ○ Two tail (t and normal only)│
│                ◉ Lower tail                  │
│                                              │
│  ┌────────┐   ┌────────┐   ┌────────┐        │
│  │   OK   │   │ Cancel │   │  Help  │        │
│  └────────┘   └────────┘   └────────┘        │
└─────────────────────────────────────────────┘
```

FIGURE 1.7   The "Calculate probability" dialog.

dialog for normal probabilities, select the button for the normal, give the value c, and select the region you want, either the upper-tail, the lower-tail, or two-tails. As shown in the dialog, we have entered the value 0.67449, and selected the lower-tail. After clicking the "OK" button, the following is displayed in the text window:

```
>Normal dist., value = 0.67449, lower-tail probability = 0.75
```

which shows that $\Pr(z \leq 0.67449) = 0.75$. If we had selected the button for a two-tail probability the answer would have been

$$\Pr(z \leq -0.67449 \text{ or } z \geq 0.67449) = 0.50.$$

The answer for the upper-tail probability would have been $\Pr(z \geq 0.67445) = 0.25$.

Normal probabilities can also be computed using typed commands, as will be discussed in Problem 1.3.

1.6.5 Boxplots of Normal Data

Figure 1.8 shows a boxplot of a large ($n = 1000$) sample from a normal distribution with mean 10 and standard deviation 4. Since the normal is symmetric about its mean, the mean and the median coincide, and the center line of the box is close to the actual value of 10. The upper and lower quartiles are at approximately $10 + 4Q(z, 0.75)$ and $10 - 4Q(z, 0.25)$, or at about 12.7 and 7.3, respectively. We can also compute the approximate fraction of outer values in a normal population to be about 0.007, as in Problem 1.3.

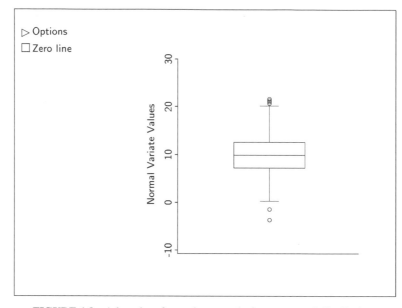

FIGURE 1.8 A boxplot of a random sample from a normal distribution.

In this last calculation we considered the fraction of outer values when constructing a boxplot of an entire normal population. The fraction of outer values in a sample from a normal population is expected to be larger, depending on the value of n. For sample sizes between 10 and 30, the expected fraction of outer values ranges between 0.02 and 0.03. For $n > 30$ the expected fraction is about 0.01, while for $5 \leq n < 10$ the expected fraction of outer values can be as large as 0.09.

1.6.6 The Sampling Distribution of the Mean

Think of a random sample $y_1, \ldots, y_n$ of size n from some population. The sample might be the $n = 120$ observations on haystack volume, for example. The *central limit theorem* tells us that, even if the sampled population is not normal, the *sampling distribution* of the sample mean $\bar{y}$ might be approximated adequately by a normal distribution, the approximation becoming progressively better as the sample size increases.

To illustrate this idea, consider the following process. We took the smoothed version of the histogram in Figure 1.3 as the true density of the bluegill length in the Lake Mary population. This distribution is skewed to the left, so it is not normal. We next randomly selected $n = 2$ bluegill lengths from this distribution and computed the sample mean. Repeating this process 500 times resulted in 500 sample means each based on $n = 2$ observations. The histogram of the 500 sample means is shown in Figure 1.9a. This is essentially the sampling

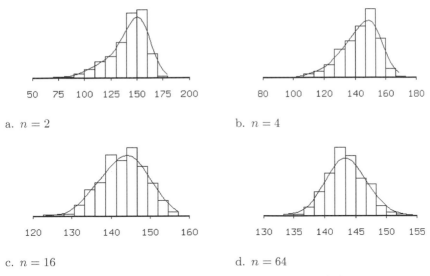

a. $n = 2$ b. $n = 4$

c. $n = 16$ d. $n = 64$

FIGURE 1.9 Four histograms illustrating the central limit theorem.

distribution of a mean of 2 observations from the Lake Mary population. Like the sampled population, it is still skewed to the left. Figures 1.9b–d were constructed in the same way, only the value of n was changed. The sampling distribution for $n = 4$ is still noticeably skewed, but the skewness is no longer evident visually for $n = 16$. The sampling distribution for $n = 64$ is very close to the normal.

The ranges on the horizontal axes of the histograms in Figure 1.9 decrease with n. For example, the range of the data for $n = 2$ is $200 - 50 = 150$, while the range for $n = 64$ is only 25. This reflects the general fact that the variance of a sample mean decreases as the sample size n increases. In particular,

$$\text{Var}(\bar{y}) = \frac{\text{Var}(y)}{n} \tag{1.10}$$

The variance of the sample mean is $1/n$ times the variance in the sampled population, or by taking square roots of both sides of (1.10), the *standard deviation of a sample mean* is $1/\sqrt{n}$ times the standard deviation in the sampled population. Finally,

$$E(\bar{y}) = E(y)$$

which says that the population mean of a sampling distribution is the same as that for the original population.

Putting all this together in a summary statement, we can say that for "large" samples, $\bar{y}$ is approximately normal with mean $E(y)$ and variance $\text{Var}(y)/n$.

1.7   INFERENCE

1.7.1   Sample Mean

The sample mean $\bar{y}$ is an estimate of the population mean $E(y)$, which can be expressed as $\hat{E}(y) = \bar{y}$, where estimation is indicated by the "hat" above the E. Similarly, the sample variance is an estimate of the population variance, $\widehat{\text{Var}}(y) = \text{sd}^2(y)$.

From Section 1.6.6, we have that the standard deviation of the sample mean is equal to $(\text{Var}(y)/n)^{1/2}$. In practice, this standard deviation is estimated by replacing $\text{Var}(y)$ by an estimate $\widehat{\text{Var}}(y)$. The quantity obtained by replacing parameters by estimates in the standard deviation of a statistic is called its *standard error*. The standard error of the sample mean is denoted by $\text{se}(\bar{y})$ and is given by $(\widehat{\text{Var}}(y)/n)^{1/2}$. For the bluegill data, $y = Length$ and

$$\text{se}(\bar{y}) = \frac{\text{sd}(y)}{\sqrt{n}} = \frac{24.137}{\sqrt{78}} = 2.73 \text{ mm}$$

Suppose that h is some hypothesized value for the population mean $E(y)$; for the sake of this discussion, let's assume that $h = 150$ mm. We can form a test of the *null hypothesis*

$$\text{NH:} \qquad E(y) = h$$

against the *alternative hypothesis*

$$\text{AH:} \qquad E(y) \neq h$$

The test is computed using the *t*-statistic,

$$t_{obs} = \frac{\bar{y} - h}{\text{se}(\bar{y})} = \frac{143.6 - 150.0}{2.73} = -2.34$$

To summarize the evidence concerning the null hypothesis in the data, we compute a *p-value*, which is the probability of observing a value of the statistic, here t_{obs}, at least as extreme as the one we actually obtained, given that the null hypothesis in fact holds. If the *p*-value is small, then either we have observed an unlikely event, or our premise that the null hypothesis holds must be false. Consequently, the *p*-value can be viewed as a *weight of evidence*, with small values providing evidence against the null hypothesis.

The statistic t_{obs} has a *t*-distribution with $n - 1$ *degrees of freedom* (df) when the null hypothesis is true. To get a *p*-value, we can use the "Calculate probability" item in the Arc menu. When you select this item, a dialog similar to Figure 1.7 will appear, but now we use it to compute a probability based on a *t*-distribution. Specify the value of the statistic, -2.34, the *t*-distribution with 77 df, where the df is $n - 1 = 78 - 1 = 77$. Select either Lower tail if the alternative hypothesis is AH: $E(y) \leq h$; Upper tail if the alternative is AH: $E(y) \geq h$; or Two tail if the alternative is $E(y) \neq h$. Since we have specified a two-tailed alternative, use this option. After pressing "OK," the following is displayed in the text window:

```
>t dist. with 77 df, value = -2.34,
two-tail probability = 0.0218758.
```

This result indicates that—if the null hypothesis were true—an outcome less favorable to the null hypothesis than the observed outcome would occur about 22 out of 1000 times. This is a fairly rare event, and we might conclude that this provides fairly strong evidence against the null hypothesis. Occasionally, decision rules are used, in which we "reject" the null hypothesis when the p-value is less than some prespecified number, like 0.05. We will rarely use formal accept/reject rules of this type.

1.7.2 Confidence Interval for the Mean

A confidence interval for $E(y)$ can also be computed using a t-distribution. Returning to the bluegill data, a 95% confidence interval for $E(Length)$ is the set of all values for $E(y)$ such that

$$\bar{y} - Q(t_{77}, 0.975)\text{se}(\bar{y}) \leq E(y) \leq \bar{y} + Q(t_{77}, 0.975)\text{se}(\bar{y})$$

$$143.6 - 1.99 \times 2.73 \leq E(y) \leq 143.6 + 1.99 \times 2.73$$

$$143.6 - 5.4 \leq E(y) \leq 143.6 + 5.4$$

$$138.2 \leq E(y) \leq 149.0$$

where $Q(t_{77}, 0.975)$ is the 0.975 quantile of the t-distribution with 77 df. This value can be computed using the "Calculate quantile" dialog, selecting the t-distribution, and setting the df to 77. In repeated data sets, the true population mean $E(y)$ will be included in 95% of all confidence intervals computed this way.

1.7.3 Probability of a Record Bluegill

How likely are we to break the record for bluegill length while fishing in Lake Mary? Phrased somewhat differently, how likely is it that the length of a fish randomly selected from Lake Mary exceeds 10.5 inches or 267 mm, the length of Minnesota's record bluegill? In symbols, what is $\Pr(y > 267)$, where y denotes bluegill length in millimeters?

   For illustration and review, let's pretend that y is normally distributed with population mean $E(y)$ and population variance $Var(y)$. Then the standardized variable

$$z = \frac{y - E(y)}{\sqrt{Var(y)}}$$

is normally distributed with population mean $E(z) = 0$ and population variance $Var(z) = 1$.

We don't know $E(y)$ or $Var(y)$, but they can be estimated by using the corresponding sample statistics from Table 1.1. Thus,

$$Pr(y > 267) = Pr(z > (267 - E(y))/\sqrt{Var(y)})$$

$$\approx Pr(z > (267 - 143.6)/24.137)$$

$$= Pr(z > 5.1)$$

The probability that a standard normal random variable exceeds 5.1, which is quite small, can be computed using the "Calculate probability" dialog. The answer is about 0.00000017. Clearly, we shouldn't expect to catch a record bluegill in Lake Mary.

1.8   COMPLEMENTS

The haystack data used in this chapter were presented by Ezekiel (1930). The bluegill data were presented in Weisberg (1986). Our discussion of the bluegill data is dedicated to the memory of Richard Frie, who collected the data.

Arc is written in a computer language called *Xlisp-Stat*, which was written by Luke Tierney; in fact, *Arc* includes a complete copy of *Xlisp-Stat*. Although not required for using *Arc*, you can learn more about *Xlisp-Stat* from Tierney (1990), by subscribing to the Lisp-Stat listserv on the Internet (send email to lisp-stat@stat.umn.edu containing the one-word message "subscribe"), or in several Web sites, including *StatLib*, http://www.stat.cmu.edu, the UCLA statistics web site, http://www.stat.ucla.edu, and the University of Minnesota statistics web site, http://www.stat.umn.edu.

The results in Section 1.6.5 on the fraction of outer values in normal samples are due to Hoaglin, Iglewicz, and Tukey (1986).

A figure very similar to Figure 1.5 in Section 1.6.2 can be generated with the following statements in *Arc*:

```
>(def x (+ 10 (* 4 (normal-rand 1000)))))    Random data
>(def h (histogram x))                        Draw the histogram
>(send h :plot-controls)                      Add plot controls
```

PROBLEMS

1.1   As in Section 1.4.1, convert all measurements to metric units and show that the record bluegill is 3.26 standard deviations above the largest observed value, the same answer obtained when the measurements were converted to English units.

1.2 The file `haystack.lsp` gives the haystack data described in Section 1.1. Load this file into *Arc*. This will produce two menus, the data set menu which is labeled "Haystacks" in this example, and the Graph&Fit menu.

    1.2.1 Obtain both a histogram and a boxplot of haystack volume, *Vol*. Describe the distribution in these plots. Does *Vol* appear to be approximately normally distributed?

    1.2.2 Give the standard error for the average volume of the 120 haystacks in the study. Give the standard error for the average volume in cubic yards. (A *yard* is a measure of length equal to three feet, or about 0.91 m.)

    1.2.3 Give a 90% confidence interval for E(*Vol*), along with a brief interpretation.

    1.2.4 Construct the *p*-value for the hypothesis that the mean volume is 3500 cubic feet versus the alternative that it does not equal 3500.

    1.2.5 Assume that *Vol* is normally distributed with population mean 3018 cubic feet and population standard deviation 915 cubic feet. Find

        a. The probability that a randomly selected haystack has *Vol* less than 2500 cubic feet.

        b. The fraction of haystacks with *Vol* between 2000 and 4000 cubic feet.

        c. The number v such that

$$\Pr(\mathrm{E}(Vol) - v \le Vol \le \mathrm{E}(Vol) + v) = 0.95$$

        d. The 0.25 quantile of *Vol*.

        e. The median and first quartile of *Vol*.

        f. The distribution (including the mean and variance) of *Vol* when it is measured in cubic yards.

        g. A 95% confidence interval for the volume of a randomly selected haystack.

    1.2.6 If the distribution of *Vol* were skewed to the right, would the median of the distribution be larger or smaller than the mean? Why?

1.3 Normal probabilities and quantiles can be computed using the dialogs described in the text, or they can be computed using typed commands. If z is a standard normal random variable, the command `(normal-cdf c)` will return $\Pr(z \le c)$. The command `normal-cdf` is short for *normal* *cumulative distribution function.* For example,

```
>(normal-cdf 0)
0.5
```

returns the area to the left of 0 for a standard normal random variable, so $\Pr(z < 0) = 0.5$. Similarly, the command (normal-quant f) will return $Q(z,f)$. For example,

```
>(normal-quant 0.5)
0.0
```

The commands normal-cdf and normal-quant are inverses of each other. They both solve the equation

$$\Pr(z \leq c) = f \tag{1.11}$$

with normal-cdf solving for f for given c, and normal-quant solving for c for given f. Taken together, these two commands represent a computerized version of the normal tables found in many textbooks.

1.3.1 Upper-tail probabilities of the form $\Pr(z > c)$ can be computed by writing them in terms of probabilities directly computed by the command normal-cdf,

$$\Pr(z > c) = 1 - \Pr(z \leq c)$$

so the value of $\Pr(z > 1) = 1 - \Pr(z \leq 1)$ can be computed as

```
>(- 1 (normal-cdf 1))
0.158655
```

(How to perform arithmetic with the *Arc* is discussed in Appendix A, Section A.2.) Compute $\Pr(-1 \leq z \leq 1)$, which is approximately equal to 0.68, a reminder that about 68% of a normal distribution falls within one standard deviation of the mean. Also, find the probability that a normal random variable falls within two standard deviations of the mean and then the probability that it is within three standard deviations of the mean.

1.3.2 Consider constructing a boxplot of a standard normal population. About what fraction of the population would we expect to be outer values, either lower outer values or upper outer values? Because the normal is symmetric, the fraction of upper outer values is the same as the fraction of lower outer values. Thus, to find the required fraction we need to compute only the fraction of lower outer values and then multiply by two:

$$2\Pr(z < Q(z, 0.25) - 1.5[Q(z, 0.75) - Q(z, 0.25)]) \qquad (1.12)$$

Show that this value is approximately equal to 0.007, so only about 0.7% of a normal population will appear as outer values in a boxplot.

1.4 The data file demo-clt.lsp is a demonstration program that can be used to illustrate the sampling distribution of the mean and the central limit theorem as in Figure 1.5. When you load this file into *Arc*, you will get a new menu called CLT, which is short for central limit theorem.

From the CLT menu, select the item "Haystack volume." This will create two histograms. The upper histogram is for the raw values of the 120 volumes in the data. (The volumes have all been divided by 1000 to make reading the values on the horizontal axis a bit easier, so, for example, the value 4 means four thousand cubic feet.) The lower histogram was created by plotting 500 means of random samples of size two from these 120 volumes; every time you do the demo, you will get a different set of 500 samples. On both histograms, a normal density has been superimposed with mean and standard deviation that match the sample values (for the upper histogram, the mean and standard deviation of the volumes; for the lower, the mean and standard deviation of the 500 means of size two).

1.4.1 Which appears to be closer to a normal density, the original data or the means of size two? Are the means of size two close to normally distributed? As you push the mouse button to the right of the slider in the slidebar marked "Sample size," the number of observations used to compute each of the sample means will increase. Examine the histograms for each sample size, and give a qualitative description of the agreement between the histogram and the approximating normal. Does the mean stay about the same as sample size increases? Does the standard deviation stay the same? How can you tell?

The slidebar has a pop-up menu obtained by pushing the mouse button on the little triangle under the words "Sample size;" this has one item called "Adjust scale." If you select this item, then every time you change the sample size, the range on the horizontal axis is adjusted so that the histogram fills the plotting area; if you select it a second time, then this automatic adjustment is turned off. You will get better resolution if you select this item, but you may lose visual information on changes in standard deviation. Hence, you may want to look at various sample sizes, sometimes adjusting for scales changes, sometimes not adjusting.

1.4.2   Repeat 1.4.1, but using the item "N(10,4)," which is just a set of normal random numbers with mean 10 and standard deviation 4.

1.4.3   Repeat 1.4.1, but using the item "Highly skewed exponential data," which are random numbers generated from an exponential distribution that looks nothing at all like a normal distribution.

1.4.4   Repeat 1.4.1, but using the item "Cauchy data." If x and y are independently distributed as $N(0,1)$, then the ratio x/y has a Cauchy distribution. This is a favorite among statisticians because it is simple to generate, but both the mean and variance of x/y are undefined. For Cauchy random variables, the central limit theorem does not apply, and so the results should be different from the previous three sections of this problem.

1.5   The data in the file camplake.lsp are similar to the data on Lake Mary, except they were collected on a different lake. The data file also includes the radius of a key fish scale, measured from the center to the edge of the scale.

1.5.1   The length distribution in Lake Mary appeared to be skewed to the left. Is this true also for Camp Lake?

1.5.2   Test the null hypothesis that the mean length in Camp Lake is the same as the mean length in Lake Mary against the alternative hypothesis that the mean lengths are different. This requires a two-sample t-test that is a prerequisite for this book, but was not reviewed in this chapter.

CHAPTER 2

Introduction to Regression

In the previous chapter we saw several different ways of studying the distribution of the length of bluegills, without regard to their age or any other variable. These distributions are often called *marginal distributions* to emphasize that other variables of interest are not simultaneously considered. In contrast, regression is the study of *conditional distributions*. The bluegill population in Lake Mary is made up of several subpopulations, one for each possible value of the predictor *Age*. The distribution of *Length* in any one of these subpopulations is called a conditional distribution to remind us that a condition has been imposed; namely, that all the bluegills are of the same age. The description of a conditional distribution usually includes a name for the subpopulation. For example, the conditional distribution of *Length* given *Age* = 3 is just the distribution of *Length* in the subpopulation of 3-year-old bluegills. The conditional distribution of *Length* given *Age* = 5 is the distribution of *Length* in the subpopulation of bluegills that are 5 years old.

The primary goal in a regression analysis is to understand, as far as possible with the available data, how the conditional distribution of the response y varies across subpopulations determined by the possible values of the predictor or predictors. Since this is the central idea, it will be helpful to have a convenient way of referring to the response variable restricted to a subpopulation in which the predictor does not vary. We will use the notation $y \mid (x = \tilde{x})$ to indicate the response in the subpopulation where the predictor is fixed at the value $\tilde{x}$. The vertical bar in this notation stands for the word *given*. If the particular value of x is unimportant for the discussion at hand, we will use abbreviated notation and write $y \mid x$, understanding that the predictor is held fixed at some value. For example, we may refer to the distribution of $Length \mid (Age = 2)$, which means the distribution of *Length* in the specific subpopulation of 2-year-old bluegills. Or, we may refer to the distribution of $Length \mid Age$, which is the distribution of *Length* in the subpopulation determined by the value of *Age*.

The mean and variance of $y \mid (x = \tilde{x})$, the response variable restricted to the subpopulation in which $x = \tilde{x}$, are represented by $E(y \mid x = \tilde{x})$ and

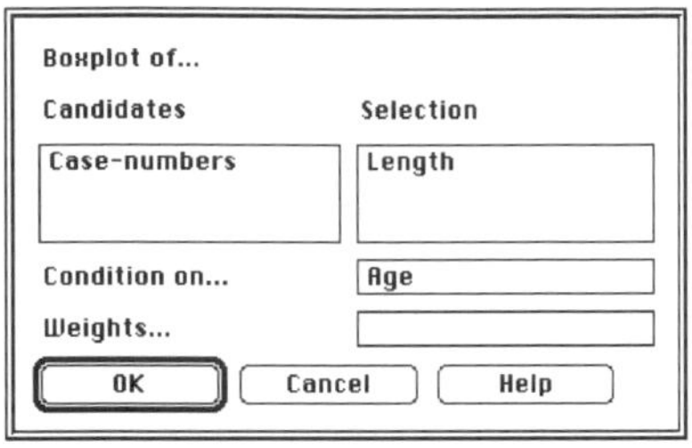

```
┌──────────────────────────────────────────┐
│ ┌──────────────────────────────────────┐ │
│ │  Boxplot of...                       │ │
│ │                                      │ │
│ │  Candidates            Selection     │ │
│ │  ┌────────────────┐  ┌─────────────┐ │ │
│ │  │ Case-numbers   │  │ Length      │ │ │
│ │  │                │  │             │ │ │
│ │  └────────────────┘  └─────────────┘ │ │
│ │                                      │ │
│ │  Condition on...     ┌─────────────┐ │ │
│ │                      │ Age         │ │ │
│ │  Weights...          ┌─────────────┐ │ │
│ │                      │             │ │ │
│ │  ┌──────────┐  ┌────────┐  ┌──────┐│ │
│ │  │    OK    │  │ Cancel │  │ Help ││ │
│ │  └──────────┘  └────────┘  └──────┘│ │
│ └──────────────────────────────────────┘ │
└──────────────────────────────────────────┘
```

FIGURE 2.1   The "Boxplot of" dialog with conditioning.

$\mathrm{Var}(y \mid x = \tilde{x})$, or more simply by $\mathrm{E}(y \mid x)$ and $\mathrm{Var}(y \mid x)$ when the particular value of x is not at issue. As we move from subpopulation to subpopulation by changing the value of x, both $\mathrm{E}(y \mid x)$ and $\mathrm{Var}(y \mid x)$ may change. For example, in the bluegill population it is surely reasonable to expect that 4-year-old fish are, on the average, longer than 1-year-old fish; that is,

$$\mathrm{E}(Length \mid Age = 4) > \mathrm{E}(Length \mid Age = 1)$$

The key point is that both $\mathrm{E}(y \mid x)$ and $\mathrm{Var}(y \mid x)$ are functions of the value of x. To emphasize this fact, we will refer to them as the *mean function* and the *variance function*. Similarly, the *standard deviation function* is simply the square root of the variance function. The mean and variance functions are the most frequently studied properties of the conditional distributions of $y \mid x$, but other properties like quartile functions or the median function may be of interest.

2.1   USING BOXPLOTS TO STUDY *LENGTH | AGE*

After loading the Lake Mary data, select "Boxplot of" from the Graph&Fit menu and move *Length* from the "Candidates" box to the "Selection" box. In addition, place *Age* in the "Condition on" box; this can be done by clicking once on *Age* to select the variable and then clicking once in the "Condition on" box. Click "OK" when your dialog box looks like Figure 2.1.

Conditioning on *Age* in this manner causes *Arc* to produce several boxplots in a single display, one boxplot for each value of the conditioning variable as shown in Figure 2.2.

Each boxplot in Figure 2.2 gives a graphical summary of the observations from the distribution of *Length | Age* for one value of *Age* in the data. Taken together they show how the distribution of *Length | Age* changes with the value of *Age* for the 78 observations from Lake Mary.

In addition to visual representations of conditional distributions provided by the boxplots, numerical summaries can be helpful. These can be obtained

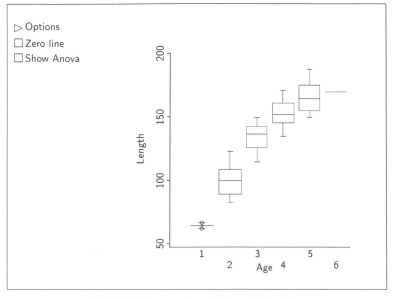

FIGURE 2.2 Boxplots of Length given Age.

in *Arc* by using the "Table data" item in the data set menu, which is labeled "LakeMary." Arrange your dialog to resemble Figure 2.3. To move *Length* to the list of "Variates," click once on the name of the variable, and then click on the list under the heading "Variates." *Age* can be moved to the "Condition on" list in the same way, or you can simply double-click on *Age*. Click on the boxes for the mean and the standard deviation (SD) to select them. Click on the button for "Quantiles" to get the quantiles you type in the text area; in this example we will use the default 0.50 quantile to get the median. Next, click on the button for "Display as list." Finally, click on the "OK" button. This setup will cause the number of observations, mean, median (denoted as $q[.50]$), and standard deviation to be computed and displayed for the variate *Length* at each value of *Age*, as shown in the output of Table 2.1.

Return to the boxplots of Figure 2.2. The first and last of the boxplots look different from the rest. We have only two samples for *Length* | (*Age* = 1), and only one sample for *Length* | (*Age* = 6), as can be seen in the second column of Table 2.1. Otherwise, the boxplots seem remarkably similar, except for their location: Both the sample mean and sample median of *Length* increase with the value of *Age*. Estimates of points along the mean and variance functions can be taken from Table 2.1. For example,

$$\hat{\text{E}}(Length \,|\, Age = 4) = 153.8$$

$$\widehat{\text{Var}}(Length \,|\, Age = 4) = 9.9^2$$

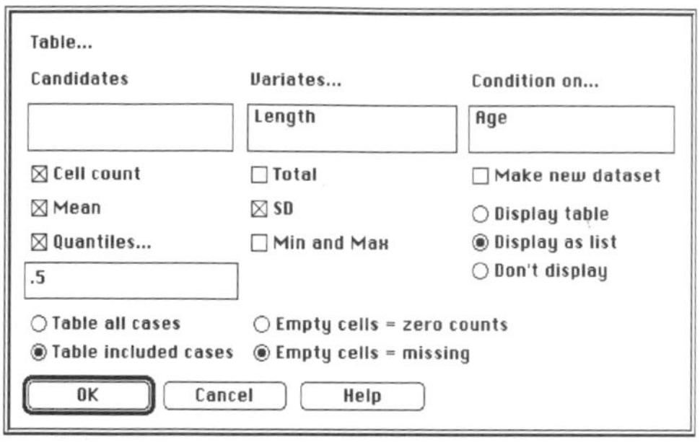

FIGURE 2.3   The "Table data" dialog.

TABLE 2.1   Output from "Table data" for the Setup Shown in Figure 2.3

```
Data set = LakeMary, Table of included cases
Col. 1 = Age
Col. 2 = Count
Col. 3 = Length[Mean]
Col. 4 = Length[SD]
Col. 5 = Length[q(0.5)]
────
1    2     64.5    3.53553    64.5
2    7   100.429   14.164    100.0
3   19   134.895   11.4741   137.0
4   41   153.829    9.90177  152.0
5    8   166.125   13.6532   165.0
6    1   170.          0     170.0
```

and the estimate of the standard deviation at $Age = 4$ is 9.9. Finally, the standard error of $\hat{E}(Length \mid Age = 4)$ is $9.9/\sqrt{41} = 1.6$.

There is no clear indication that other characteristics such as skewness of the conditional distribution of $Length \mid Age$ change with the value of Age. By comparing Figures 1.4 and 2.2, we see that the lower outer values in Figure 1.4 are all measurements on 1- or 2-year-old bluegills. The skewness in the marginal distribution of $Length$ is evidently linked to bluegill age. Going a bit further, after conditioning there is little visual evidence to suggest that the conditional distributions of Figure 2.2 are skewed. This observation is supported by the similarity of the means and medians in Table 2.1.

In summary, regression is the study of how the conditional distribution of $y \mid x$ changes with the value of x. Often special attention is devoted to the mean function and the variance function. For the bluegill data, our visual analysis suggests that the distribution of $Length \mid Age$ changes primarily through its mean function $E(Length \mid Age)$, which increases with the value of Age.

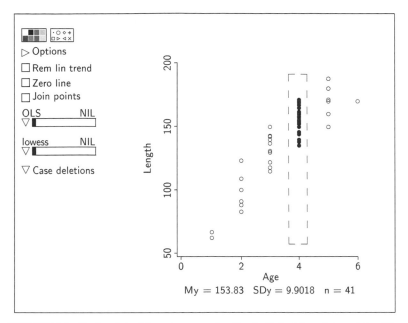

FIGURE 2.4 Scatterplot of *Length* versus *Age* for the Lake Mary data on bluegills.

2.2 USING A SCATTERPLOT TO STUDY *LENGTH | AGE*

A scatterplot is another graphical display that can be quite useful for showing how the conditional distribution of $y \mid x$ changes with the value of x. Two-dimensional scatterplots for a regression with a single predictor are usually constructed with the response assigned to the vertical axis and the predictor assigned to the horizontal axis.

To construct a two-dimensional plot of *Length* versus *Age* for the data on bluegills, return to the Graph&Fit menu and select "Plot of," assigning *Length* to the *V*-axis and *Age* to the *H*-axis. The resulting plot should look like Figure 2.4. Ignore the highlighting of the points in the dashed rectangle and the text just below the plot; these will be discussed shortly.

The scatterplot in Figure 2.4 shows how the distribution of *Length | Age* changes with the value of *Age*. The plot is similar to the boxplots of Figure 2.2 except the data from the conditional distributions have not been graphically summarized with boxplots. Again, we see that the distribution of *Length | Age* changes primarily in the mean, with no notable evidence that the variance function is not constant.

2.3 MOUSE MODES

The controls on plots can be used to convert the visual information in plots into numerical information. Go to the menu called "2Dplot" for the scatter-

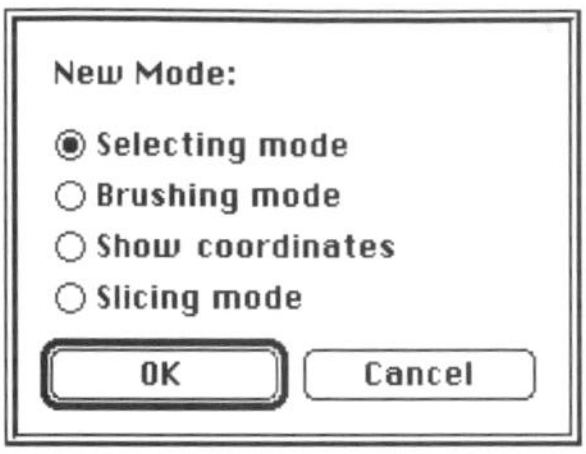

New Mode:

⦿ Selecting mode
○ Brushing mode
○ Show coordinates
○ Slicing mode

[OK   ]   [Cancel   ]

FIGURE 2.5   The "Mouse mode" dialog.

plot of Figure 2.4, and select the item "Mouse mode." You will be presented with a mouse mode dialog like that shown in Figure 2.5. The dialog allows you to choose from four mouse modes by clicking on one of the four radio buttons. The default mouse mode with the cursor appearing as an arrowhead is "Selecting mode."

2.3.1   Show Coordinates Mouse Mode

Click in the radio button for "Show coordinates," and then click the "OK" button. The cursor should now be a hand with a pointing finger. Place the tip of the finger on the single point from the distribution of $Length \mid (Age = 6)$ and press the mouse button. As long as the mouse button is pressed the following information will be displayed in the computer screen:

$$73:(6,170)$$

The first number is the *case name*; unless you specify case names using the item "Set case names" from the data set menu, the program will assign names to cases. If you have text variables, then the program will use the first of these as case names. If you do not have any text variables, then the program will number the cases according to their order in the data file and use these numbers as the case names. Case numbering in *Arc* starts with 0, not with 1. Thus, the 73 above refers to bluegill number 73, which is really the 74th fish in the data file.

The pair of numbers in parentheses following the colon are the coordinates of the point: So this point corresponds to $Age = 6$ on the horizontal axis and $Length = 170$ on the vertical axis. The same operation can be repeated for any point on the plot.

2.3.2   Slicing Mode

Return to the "Mouse mode" menu and select "Slicing mode." The cursor will now change to a paint brush with a long thin rectangle outlined by dashed lines. The rectangle is shown in Figure 2.4 without the cursor. The rectangle itself is called the *brush*. Several things happen as the brush is moved across the plot, a procedure called *brushing*. First, the points within the brush are *highlighted*. Second, the number of points, sample mean and standard deviation are computed and printed at the bottom of the plot for the highlighted points. For example, we see in Figure 2.4 that there are $n = 41$ fish which are 4 years of age in the sample. The mean and standard deviation of the *Length* of those fish are $\bar{y} = My = 153.83$ and $sd(y) = SDy = 9.9018$. Clicking the mouse button while in the slicing mode will display the summary statistics for the slice in the text window. These statistics are also available from Table 2.1.

The slicing mode can be used with boxplots like those in Figure 2.2. In this case the brush serves as a device for selecting a boxplot. The summary statistics printed below the plot are for the data used in the construction of the selected boxplot.

The slicing mouse mode allows you to use both graphical and numerical information when investigating how the conditional distribution of $y \mid x$ changes across a scatterplot, or a series of boxplots. Moving the brush in Figure 2.4 from age to age while observing the summary statistics below the plot can provide information to support visual impressions.

The size of the brush can be changed by selecting the item "Resize brush" from the 2Dplot menu. This item produces a small window with the instructions, "To resize brush click in this window and drag." The idea here should be clear after a little experimentation. The slicing mode works with any brush size and the statistics displayed below the plot are for those points within the brush. A long narrow brush that captures all the points above a short interval on the *x*-axis should be used when studying conditional distributions $y \mid x$ in a scatterplot of y versus x. Smaller rectangular brushes will be used when studying bivariate distributions in Chapter 4.

2.3.3 Brushing Mode

The final mouse mode is the "Brushing mode." This mode works like slicing mode, except no summary statistics are displayed. We return to brushing in Chapter 5.

2.4 CHARACTERIZING *LENGTH | AGE*

Our graphical analysis of the bluegill data supports the notion that the mean function E(*Length | Age*) increases with the value of age, while the variance function Var(*Length | Age*) is constant. The conclusion regarding the mean function agrees with our prior expectation. The conclusion regarding the variance function is less firm for two reasons. First, seeing changes in the variance function is more difficult than seeing changes in the mean function.

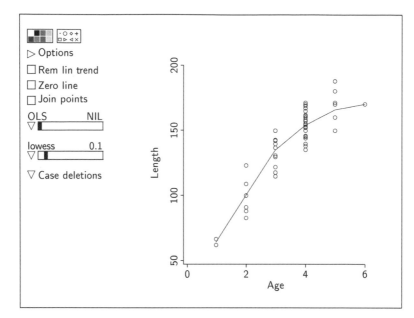

FIGURE 2.6   Scatterplot of *Length* versus *Age* for the bluegill data with a smooth of the mean function.

The slicing mouse mode is often useful for gaining numerical information on how $\text{Var}(y \mid x)$ could be changing across a scatterplot. Second, there are only two observations available to judge $\text{Var}(Length \mid Age = 1)$, and there is only one observation from the distribution of $Length \mid (Age = 6)$. Our tentative conclusion about the constancy of the variance function may well change if we could get more data on the conditional distributions at $Age = 1$ and 6, but for now we have to do the best we can with the information that we have.

At this point it may be useful to characterize how the mean function is increasing. A better visualization of the mean function can be obtained by marking the position of the estimates $\hat{\text{E}}(Length \mid Age)$ on the scatterplot of Figure 2.4 and then connecting the adjacent points with straight lines. This can be done in *Arc* using the following procedure. First, make sure that the mouse is in selecting mode, and then click once in the slidebar titled "lowess." Your plot of *Length* versus *Age* should now look like Figure 2.6. The curve on the plot connects the sample means of the data from the six conditional distributions. Such a curve is called a *smooth* of the data. More will be said about the rationale and construction of smooths in the next chapter. For now, we see that the mean function seems to be a nonlinear function of *Age* because the change in length from one age to the next is larger for some ages than for others. In particular, it seems that older fish do not grow as fast as younger fish.

2.5 MEAN AND VARIANCE FUNCTIONS

The mean and variance functions—$E(y \mid x)$ and $Var(y \mid x)$—are important characteristics of the conditional distribution of $y \mid x$. We study some properties of these functions in this section.

2.5.1 Mean Function

The expected response $E(y)$ in the total population and the mean function $E(y \mid x)$, which gives the expected response in the subpopulation determined by the value of x, are related: The population mean is the expected value of the mean function with respect to the distribution of the predictor. That is,

$$E(y) = E\{E(y \mid x)\}$$

$$= \text{weighted mean of } E(y \mid x)$$

$$= \text{weighted mean of the subpopulation means} \qquad (2.1)$$

This equation says that the mean response $E(y)$ is a mean of means; take the weighted mean of the subpopulation means, where the weights are the relative sizes of the subpopulations. Further, if the mean function is constant, then each subpopulation mean must be the same as the overall mean, $E(y \mid x) = E(y)$.

Let's return to the bluegill data to explore the meaning of this relationship a bit more. We first need to describe the structure of the bluegill population so we can get a handle on the weights. Suppose that nine years is the age of the oldest bluegill in Lake Mary. There are then nine subpopulations of bluegills determined by age. For each subpopulation, let P_k denote the number of k-year-old bluegills divided by the total number of bluegills in the population. P_k is the fraction of the total population of bluegills that are k years old. The fraction P_k is also the probability that a randomly selected bluegill falls in the kth subpopulation.[1]

The mean length $E(Length)$ can be thought of as the expected length of a randomly selected bluegill. A bluegill can be selected randomly in two stages:

- Randomly select a subpopulation according to the probabilities $P_1, \ldots, P_9$.
- Then select a bluegill at random from the selected subpopulation.

The expected length $E(Length)$ is the same as a weighted mean of the mean subpopulation lengths,

$$E(Length) = \sum_{k=1}^{9} (P_k \times E(Length \mid Age = k))$$

[1]As most anglers know, the fish in a lake are not equally likely to be captured, so randomly selecting a bluegill may not be easy in practice.

Comparing this result to the general description in (2.1), we see that interpreting $E(y)$ as the expected value of $E(y \mid x)$ is the same as saying $E(y)$ is a weighted mean of the mean function, where the weights are the relative sizes of the subpopulations.

So far, our discussion has been in terms of the population, but the same ideas work for the data. The average length in the sample is 143.603 mm. This is related to the sample subpopulation means of Table 2.1 in the same way that the population mean is related to the subpopulation means:

$$143.603 = \left(\frac{2}{78} \times 64.5 \right) + \left(\frac{7}{78} \times 100.429 \right) + \cdots + \left(\frac{1}{78} \times 170 \right)$$

2.5.2   Variance Function

The variance function can be written as

$$\text{Var}(y \mid x) = E[(y - E(y \mid x))^2 \mid x] \tag{2.2}$$

This equation may not be as formidable as it might seem at first glance. It describes the following steps to construct a variance function:

- Form the mean function $E(y \mid x)$.
- Form the new variable $e = y - E(y \mid x)$ and square it to get e^2, which is just the squared deviation of an observation from its subpopulation mean.
- Form the mean function $E(e^2 \mid x)$.

The variance function can be viewed as a mean function for the regression of the squared deviations e^2 on x. It describes the mean-squared deviation from the mean function. This characterization of the variance function will be used in subsequent chapters to construct estimates. The first instance of such a construction is in Section 3.6.3.

We saw in the last section that the population mean $E(y)$ is a weighted mean of the subpopulation means. Is the same true for the variance function? Is it true that $\text{Var}(y) = E\{\text{Var}(y \mid x)\}$? The answer is generally no; the relationship between the population variance $\text{Var}(y)$ and the subpopulation variances $\text{Var}(y \mid x)$ is a little more complicated than the relationship between the means that we saw in the last section. But there is a very useful relationship nevertheless:

$$
\begin{aligned}
\text{Var}(y) &= E\{\text{Var}(y \mid x)\} + \text{Var}\{E(y \mid x)\} \\
&= \text{weighted mean of } \text{Var}(y \mid x) + \text{variance of } E(y \mid x) \\
&= \text{weighted mean of } \text{Var}(y \mid x) \\
&\quad + \text{weighted mean of } \{E(y \mid x) - E(y)\}^2
\end{aligned} \tag{2.3}
$$

This can be written more explicitly for the bluegill data,

$$\text{Var}(Length) = \sum_{k=1}^{9} (P_k \times \text{Var}(Length \mid Age = k))$$

$$+ \sum_{k=1}^{9} P_k \times \{\text{E}(Length \mid Age = k) - \text{E}(Length)\}^2 \qquad (2.4)$$

These various expressions tell us that the variance of y is equal to the expected variance function plus the variance of the mean function. If the mean function is constant, $\text{E}(y \mid x) = \text{E}(y)$, then the variance of the mean function is zero and in this special case

$$\text{Var}(y) = \text{E}\{\text{Var}(y \mid x)\}$$

But otherwise the variance of y will be greater than the expected variance function.

The relationships described in this section also work for the data, as discussed in Problem 2.3

2.6 HIGHLIGHTS

Regression is the study of how the *conditional distribution* of $y \mid x$ changes with the value of x. Special emphasis is often placed on the *mean function* $\text{E}(y \mid x)$ and on the *variance function* $\text{Var}(y \mid x)$. When there are only a few possible values of x, as in the bluegill data, a *boxplot* display of the data from the various conditional distributions can be useful for visualizing how the distribution of $y \mid x$ changes with the value of x. A *scatterplot* of y versus x can be used in the same way.

When there are *many* values of the predictor in the data, the fundamental ideas of this chapter do not change, but regression is a bit more complicated. We use the haystack data to address associated issues in the next chapter.

2.7 COMPLEMENTS

Boxplots are a relatively recent invention due to John Tukey. An early description of them is given in Tukey (1977).

PROBLEMS

2.1 The data file `camplake.lsp` contains data on a sample of fish similar to the Lake Mary data described in the text, except from a different lake in Minnesota.

2.1.1 Reproduce the equivalents of Table 1.1, Figure 1.3, Figure 2.2, Table 2.1 and Figure 2.6.

2.1.2 Give a brief description of the *major* differences between the Lake Mary and Camp Lake populations with respect to the distribution of *Length* | *Age*.

2.1.3 Suppose that the distributions of *Length* and of *Length* | (*Age* = 5) are normal, with means and variances given by the sample values. Without doing any calculations, which will be larger, Pr(*Length* > 167.5) or Pr(*Length* > 167.5 | *Age* = 5)? Why? Compute estimates of these two quantities and verify your conjecture.

2.2 Construct the *p*-value for the hypothesis that E(*Length* | *Age* = 5) is the same for the Lake Mary and Camp Lake populations, versus the alternative that these subpopulation means differ. (*Hint*: This requires use of a test not described in the text.)

2.3 Equation 2.4 describes what happens in the population of bluegills. Verify that this relationship holds for the Lake Mary sample of bluegills by following these steps.

a. Compute the sample mean and variance of *Length*,

$$\hat{E}(Length) = 143.603 \quad \text{and} \quad \widehat{Var}(Length) = (24.137)^2$$

b. Compute the weighted mean of the sample variances from the subpopulations,

$$\text{weighted mean of } \widehat{Var}(Length \mid Age) = \left(\frac{2}{78} \times (3.53553)^2\right)$$
$$+ \left(\frac{7}{78} \times (14.164)^2\right)$$
$$+ \cdots + \left(\frac{1}{78} \times (0)^2\right)$$

c. Compute the weighted mean of the squared deviations of the sample mean function from $\hat{E}(Length)$,

$$\text{sample variance of } \hat{E}(Length \mid Age) = \left(\frac{2}{78} \times (64.5 - 143.603)^2\right)$$
$$+ \left(\frac{7}{78} \times (100.429 - 143.603)^2\right)$$
$$+ \cdots + \left(\frac{1}{78} \times (170 - 143.603)^2\right)$$

d. Apart from rounding error, the sample variance

$$\widehat{\text{Var}}(Length) = (24.137)^2 = 582.595$$

should equal the sum of the quantities computed in steps (b) and (c).

2.4 Reproduce Figure 2.2. After printing the figure, construct estimates of the median function and the third quartile function by connecting the corresponding estimates for the individual subpopulations as illustrated in Figure 2.6.

CHAPTER 3

Introduction to Smoothing

The haystack data introduced in Section 1.1 consist of the response *Vol* and two predictors, *C* and *Over*. In this chapter we ignore *Over* and consider only the regression of *Vol* on *C*, because we are not yet ready to tackle regressions with $p = 2$ predictors. The data are available in the file haystack.lsp.

3.1 SLICING A SCATTERPLOT

Shown in Figure 3.1 is a scatterplot of *Vol* versus *C*. Unlike the bluegill data, there are many different values of the predictor in the haystack data. This makes seeing how the conditional distributions change a bit more difficult, but the fundamental regression problem is the same: We want to study how the distribution of *Vol* | *C* changes across the subpopulations determined by the value of *C*.

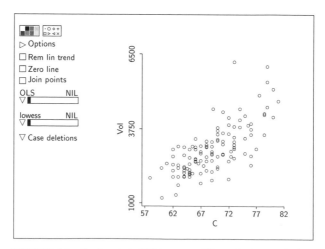

FIGURE 3.1 Scatterplot of *Vol* versus *C* from the haystack data.

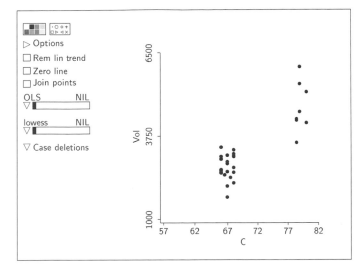

FIGURE 3.2 Scatterplot of *Vol* versus *C* from the haystack data with slices at *C* = 67 and 80.

Suppose that we want to use Figure 3.1 to compare the distribution of *Vol* | (*C* = 67) to the distribution of *Vol* | (*C* = 80). There are few, if any, values in the data at which *C* is exactly 67 or 80, but there are many values of *C* close to these values, and we will treat nearby values as if they have the same value of *C*. This procedure is called *slicing*.

To compare the distributions of *Vol* | (*C* = 67) and *Vol* | (*C* = 80), we sliced the plot in Figure 3.1 about the values 67 and 80 on the horizontal axis, and we deleted the other points for visual clarity. The resulting plot is shown in Figure 3.2. The portion of the horizontal axis covered by a slice is called the *slice window*, and the width of a slice window is called the *window width*. The two slice windows in Figure 3.2 are centered at 67 and 80, and the common window width is about 4 feet. The slices in Figure 3.2 are visual aids that enable us to focus on the data in question. The average of *Vol* for the slice at *C* = 67 is clearly less than the average for the slice at *C* = 80.

Figure 3.2 was constructed by holding down the mouse button while dragging the cursor over the points that correspond roughly to the slice at *C* = 67, and then, with the shift key depressed, selecting points in a similar fashion that correspond to the slice at *C* = 80. Holding down the shift key extends the selection to include both sets of points. From the plot's menu, choose the item "Focus on Selection" to remove all points that are not selected. To return to the plot in Figure 3.1, select the "Show all" item from the plot's menu.

The slices in Figure 3.2 serve as graphical enhancements to aid in the discussion of statistical interpretations of a scatterplot. They are usually not needed in practice because our eyes provide a smooth transition between the distributions of *Vol* | *C* for adjacent values of *C*.

3.2   ESTIMATING $E(Y \mid X)$ BY SLICING

Estimating the mean function for the bluegill data was fairly straightforward because there were only six conditional distributions, one for each value of *Age* in the data. The situation is a bit more complicated for the haystack regression because there are many more conditional distributions represented in the data.

Consider the slice about $C = 67$ in Figure 3.2 with a window width of about 4 feet. We use this slice to approximate observations from the distribution of $Vol \mid (C = 67)$ with observations from the distribution of $Vol \mid (65 \le C \le 69)$, where $69 - 65 = 4$ is the window width. In effect, we are approximating the subpopulation in which $C = 67$ with observations from the somewhat larger subpopulation in which $65 \le C \le 69$.

It appears from Figure 3.2 that $E(Vol \mid C)$ is fairly constant within the slice, $65 \le C \le 69$, and that any within-slice change in $E(Vol \mid C)$ is surely small relative to the within-slice standard deviation in *Vol*, or in symbols

$$\frac{E[Vol \mid C = 69] - E[Vol \mid C = 65]}{\text{within-slice SD of } Vol}$$

is small. Since the change in the regression function within the slice is relatively small, we summarize the data in this slice by using the average of the responses, which is about 2,716 cubic feet, and the midpoint of the slice window, which is 67. Thus 2,716 cubic feet should be a useful estimate of $E(Vol \mid C = 67)$, the mean volume of haystacks with a circumference of 67 feet. The slicing mode introduced in Section 2.3.2 allows this general operation to be performed rapidly as the brush is moved across the plot.

3.3   ESTIMATING $E(Y \mid X)$ BY SMOOTHING

We can use slicing to construct a crude estimate of the mean function by partitioning the range of the predictor variable into several nonoverlapping slices. Within any slice, estimate $E(y \mid x)$ to be constant and equal to the average of the response for all points within the slice.

Return to Figure 3.1 and, with the mouse in selecting mode, place the cursor in the small triangle indicating a pop-up menu just to the left of the *smoother slidebar*, the slidebar that is initially marked "lowess." Press and hold the mouse button. While the mouse button is pressed, a pop-up window will be displayed. Move the mouse to select the item "SliceSmooth" and then release the mouse button. The name above the slidebar should now be "SliceSmooth." Finally, click once in the slidebar to the right of the slider. Two things should have happened: The number 32 should now be above the slidebar, and a series of connected line segments should have appeared on the plot, as shown in Figure 3.3.

The number of slices is equal to 32, the number shown. The slice smoother is defined to put as close to an equal number of observations in each slice

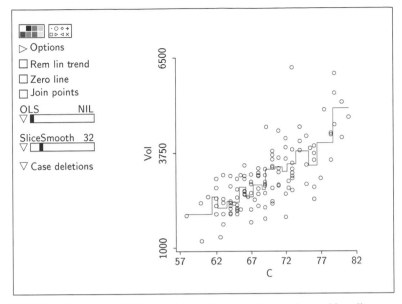

FIGURE 3.3   Scatterplot of *Vol* versus *C* for the haystack data along with a slice smooth.

as possible by varying the slice widths. There are 120 observations in the
haystack data, so with 32 slices each slice should contain about four observations, subject to the limitations imposed by ties among the values of the
predictor on the horizontal axis. The horizontal line segments on the plot
mark the width of the slices at the slice means. The vertical line segments that
join the horizontal segments aid visualization by giving your eye something
to follow. The result is called (somewhat confusingly) a "curve." This curve
gives a rough idea about how the mean function changes with the circumference.

The procedure leading to the curve on Figure 3.3 is an instance of *smoothing*. The curve traced on the plot is called a *smooth* of the data. The name
comes from the process of representing each slice by a smooth curve, in this
case a horizontal line segment. The final smooth need not actually be smooth,
as should be clear from Figure 3.3. The curve on Figure 2.6 is also a smooth
constructed in a related but somewhat different way. There are many different
smoothing methods. Most come with a *smoothing parameter* that allows the
user to control the amount of smoothing.

The smoothing parameter for the slice smooth shown in Figure 3.3 is the
number of slices. Moving the slider changes the number of slices and the
amount of smoothing. A large number of slices as in Figure 3.3 provides
minimal smoothing, leading to a very rough curve that is too variable and is
called *undersmooth*. Too few slices will *oversmooth*, giving very smooth curves
that may miss important features of the mean function, resulting in bias in estimation of the mean function. In the extreme case of a single slice, the smooth

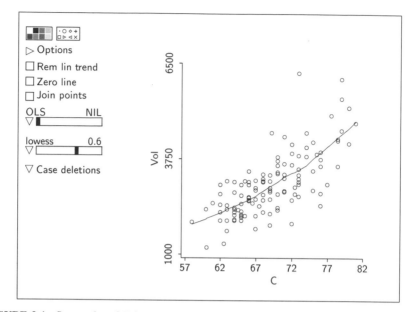

FIGURE 3.4   Scatterplot of *Vol* versus *C* for the haystack data along with a *lowess* smooth.

is simply a horizontal line at the average of *Vol* for all 120 observations, which is a very poor estimate of E(*Vol* | *C*). A reasonable value for the smoothing parameter that balances between undersmoothing and oversmoothing, or between variability and bias, can usually be obtained by interactively changing the smoothing parameter and visually judging the smooth against the data. The references in the complements to this chapter discuss numeric techniques for selecting a smoothing parameter.

The slice smoother is pretty crude compared to others available, but it serves to illustrate basic ideas underlying smoothing. *Arc* has a more sophisticated smoother, called the *lowess* smoother. The construction of this smoother relies on techniques presented later in the book, but the underlying ideas are like those for the basic slice smoother. We will discuss construction in a later chapter, after developing necessary background material. For now we will use smoothers as visual aids for understanding mean and variance functions.

The *lowess* smoother can be selected from the pop-up menu next to the smoother slidebar. The *lowess* smoothing parameter is always between 0 and 1. Values close to 0 usually produce undersmoothed jagged curves, while values close to 1 usually produce oversmoothed curves. Values of the smoothing parameter between 0.4 and 0.7 usually produce smooths that match the data well, as shown in Figure 3.4. The *lowess* smooth gives a much cleaner impression of the mean function than the slice smooth in Figure 3.3. Nevertheless, both smooths suggest that the mean volume increases nonlinearly with circumference, as might be expected.

Throughout this book, we will call the slidebar that is used to add a smoother to a plot the *smoother slidebar*.

3.4 CHECKING A THEORY

During our initial discussion of the haystack data in Section 1.1, we conjectured that the simple approximation

$$Vol \approx \frac{C^3}{12\pi^2}$$

might give a useful prediction of the volume of a haystack based on its circumference. We can now make this possibility a bit more precise by imagining that if haystacks are hemispheres on the average, then

$$E(Vol \mid C) = \frac{C^3}{12\pi^2} \tag{3.1}$$

One way to check this conjecture is to add a plot of this mean function on a scatterplot of *Vol* versus *C*.

A curve can be added to a scatterplot in *Arc* using the "Options" dialog. Remove any smooths by moving the smoother slidebar to the left, and then click on the triangle by the word "Options." The resulting dialog is shown in Figure 3.5. This dialog, which is described more fully in Appendix A, Section A.8.2, allows changing many features of a plot, including the range and labels on the axes. To add the curve specified by (3.1), type the equation as shown in Figure 3.5 in the long narrow text window in the dialog and click "OK." A plot of the mean function (3.1) will be added to the scatterplot. For purposes of comparison, a *lowess* smooth with smoothing parameter 0.6 is also shown on the plot in Figure 3.6.

The plot of the mean function given by (3.1) is the lower curve in Figure 3.6. It passes though the bulk of the data and suggests that treating round haystacks as hemispheres gives predictions that might be useful. However, since the *lowess* smooth passes uniformly *above* the conjectured mean function, treating haystacks as hemispheres will uniformly *underestimate* the mean function E(*Vol* | *C*), leading to underpayment to farmers. There is room for improvement.

3.5 BOXPLOTS

In the previous section we relied on a scatterplot of *Vol* versus *C* to study the conditional distribution of *Vol* | *C*. Boxplots can also be used to study this regression problem.

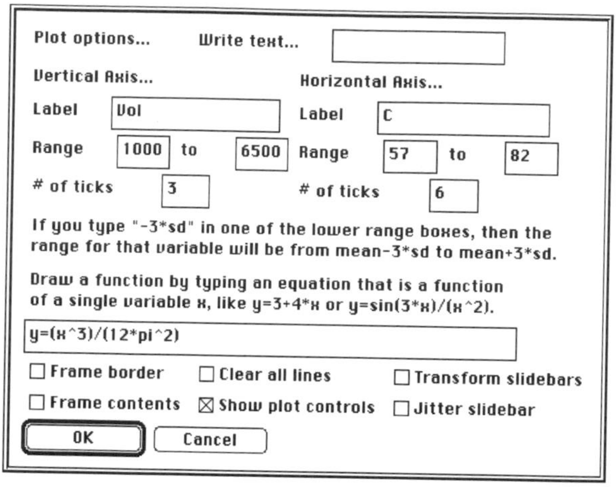

Plot options...   **Write text...**

Vertical Axis...   **Horizontal Axis...**

Label   Vol   Label   C

Range   1000 to   6500   Range   57   to   82

# of ticks   3   # of ticks   6

If you type "-3*sd" in one of the lower range boxes, then the range for that variable will be from mean-3*sd to mean+3*sd.

Draw a function by typing an equation that is a function of a single variable x, like y=3+4*x or y=sin(3*x)/(x^2).

y=(x^3)/(12*pi^2)

☐ Frame border   ☐ Clear all lines   ☐ Transform slidebars
☐ Frame contents   ☒ Show plot controls   ☐ Jitter slidebar

[OK]   (Cancel)

FIGURE 3.5   Options dialog for constructing the curve $y = C^3/(12\pi^2)$ on the scatterplot of *Vol* versus C from the haystack data.

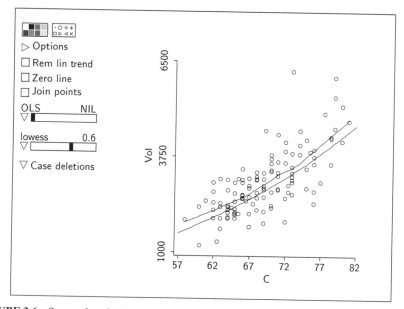

FIGURE 3.6   Scatterplot of *Vol* versus C for the haystack data. The lower curve is the conjectured mean function (3.1). The upper curve is a *lowess* smooth.

As discussed in Section 1.5.2, boxplots for the regression of *Vol* on C can be constructed by using the "Boxplot of" item in the Graph&Fit menu for the haystack data by moving *Vol* to the "Selection" list, and moving C to the "Condition on" box. The dialog shown in Figure 3.7 will then appear. The top

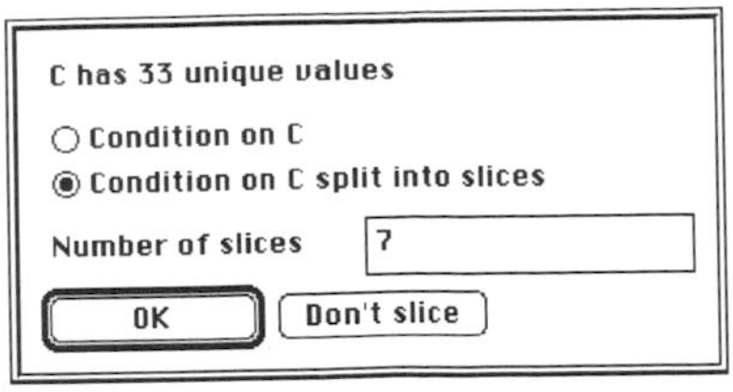

┌───┐
│ C has 33 unique values   │
│   │
│ ○ Condition on C   │
│ ◉ Condition on C split into slices   │
│   │
│ Number of slices   │ 7 │   │
│   │
│ ┌──────────────┐ ┌─────────────┐   │
│ │   OK   │ │ Don't slice │   │
│ └──────────────┘ └─────────────┘   │
└───┘

FIGURE 3.7   The dialog for slicing prior to the construction of boxplots.

line in this dialog tells us that there are 33 unique values of C in the data set. Conditioning on C will result in 33 boxplots in a single display. The dialog gives us an opportunity to view the 33 boxplots by clicking the top button in the dialog. By selecting the second button, we can slice C into the number of slices we specify. One boxplot will be drawn for each slice. Conditioning on 33 values of C will not be very effective because each boxplot will be based on very few observations; slicing will be more effective. As shown in the dialog, we have selected seven slices.

After clicking the "OK" button, the program will create a new variable called $C[7\_slc]$ that divides C into seven groups, each containing about the same number of observations. Ties may cause these numbers to be uneven. The mean value of C for all the observations in the first slice is 62.57, and this is the value of $C[7\_slc]$ for all these observations. Similarly, in the last slice the mean is 77.85, and the value of $C[7\_slc]$ for the corresponding observations is 77.85. You can view all the values of $C[7\_slc]$ using the "Display data" item in the data set menu.

Figure 3.8 is the set of boxplots produced by this process. This display consists of seven boxplots, as expected. The horizontal axis is labeled according to the within-slice circumference mean. The number of observations per slice, as well as the sample mean and variance of Vol within each slice, can be found by using the slicing mouse mode. The slice and *lowess* smooths provide information on the mean function $E(Vol \mid C)$. In contrast, the boxplots of Figure 3.8 show medians and consequently allow us to judge the behavior of the *median function* of $Vol \mid C$. The median function would be a linear function of C if the rate of increase in Vol was the same at all values of C. The plots suggest that the median function is a nonlinear, increasing function of the value of C. The nonlinearity may be more apparent if you imagine a straight line from the median of the first slice to the median of the last slice. The medians of the other slices all fall below the line.

For symmetric distributions, the mean and the median are the same. Since the seven boxplots of Figure 3.8 all seem fairly symmetric, the median function is about the same as the mean function estimated by the *lowess* smoother for the haystack regression in Figure 3.4.

Finally, since the boxes at the right of Figure 3.8 are longer than the boxes at the left, the variance function $Var(Vol \mid C)$ increases with C. This conclusion

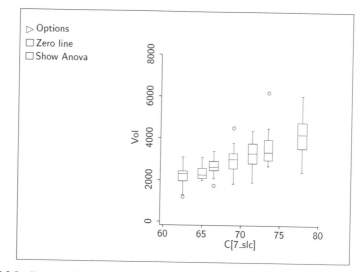

FIGURE 3.8   Boxplots for visualizing the conditional distribution of *Vol* | *C* in the haystack data.

might be reached also by visual inspection of the scatterplot in Figure 3.4, but the visual evidence doesn't seem nearly as strong. Additional enhancements to scatterplots are usually needed for visual inference about the variance function. This is discussed in the next example.

3.6   SNOW GEESE

Snow geese are large, impressive birds that can have wing spans of three feet and weigh six pounds. Traveling a mile high at 50 miles per hour, their yearly migration cycle can cover 6000 miles. Snow geese arrive at their breeding grounds in Arctic regions of North America and extreme Siberia during late May or early June, and they begin their journey south in late August or early September. The snow geese of North America winter in the south-central United States near the Gulf of Mexico, particularly in Louisiana and Texas. Some venture into Mexico.

In 1985 nearly half a million snow geese nested in the McConnell River area on the west coast of Hudson Bay near Arviat (formerly Eskimo Point) in the Northwest Territories. The coastal plain of Hudson Bay's west coast is covered with countless shallow lakes interspersed with meadows of wet and dry tundra. The snow geese of the McConnell River area are likely the most intensively studied in the Canadian Arctic. Biological studies began in the 1950s, but it wasn't until the late 1970s that aerial survey methodology was developed to track the size of the McConnell River population. Aerial photo surveys are currently conducted every five or six years during July and August when the birds are in moult and flightless.

Several different methods were tried during the development of the aerial survey methodology. Visual aerial surveys, in which observers visually estimated flock size, were relatively fast and cheap, and the results were immediately available. But flocks of flightless snow geese can range in size from a few to several hundred and so the accuracy of visual estimates was at issue. Aerial photo surveys were known to be accurate, but also more time consuming and costly.

An experiment was conducted in 1978 to study the ability of trained observers to estimate flock size visually. Two trained aerial observers estimated the sizes of $n = 45$ flocks of snow geese. The actual size of each flock was determined from photographs taken during the survey. The data from this experiment are available in the *Arc* file geese.lsp.

3.6.1 Snow Goose Regression

The results from the snow goose survey consist of three variables, the actual flock size *Photo* determined from the aerial photographs, and the estimated flock sizes, *Obs1* and *Obs2*, from each of the observers. The issue in this experiment can be approached as a regression with response $y = Photo$ and $p = 2$ predictors $x_1 = Obs1$ and $x_2 = Obs2$. However, we will stay with a single predictor and consider only the regression of *Photo* on *Obs2*.

The regression problem is to study the distribution of *Photo* | *Obs2*: If the observer estimates that a particular flock has 100 geese, what can we say about the actual flock size? As in our previous examples, the mean and variance functions E(*Photo* | *Obs2*) and Var(*Photo* | *Obs2*) may be of particular interest, but the goals of this regression study are somewhat different from those for the haystack or bluegill data. Here the scientists were interested only in gaining information to aid in choosing between photo and visual surveys. Once their decision was made, the results of the experiment would cease to be of much interest, except as might be needed in support of that decision.

A plot of *Photo* versus *Obs2* is shown in Figure 3.9. The points with the four largest values of *Obs2* account for a substantial portion of the available plotting region. Because of this it is difficult to see the data from the conditional distributions having small values of *Obs2*. The visual resolution in the plot can be improved by removing the four points in question. This can be done by first selecting the four points so that they become highlighted, as shown in Figure 3.9. Next, from the plot's menu select the item "Remove Selection." This causes the four selected points to be removed from the plot. The item "Focus on Selection" would cause all points *except* those selected to be removed from the plot. After removing the selected points, the plot will look just like that in Figure 3.9 but the points in question will no longer be present. Finally, return to the plot's menu once again and select the item "Rescale plot." This final action rescales the plot so the space is filled by the remaining points. Your plot should now look like that in Figure 3.10 but without the two superimposed lines.

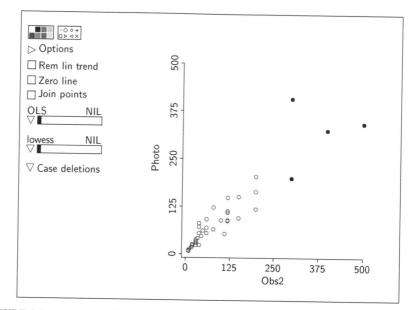

FIGURE 3.9   Scatterplot of the photo count *Photo* versus the visual estimate *Obs2* by observer 2 from the snow geese data. The points with the four largest values of *Obs2* are selected.

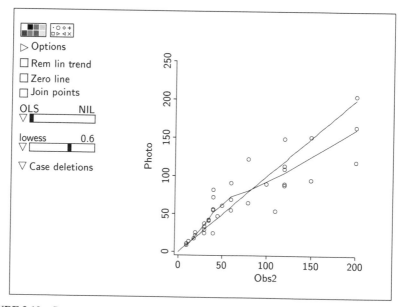

FIGURE 3.10   Scatterplot of the photo count *Photo* versus the visual estimate by observer 2 *Obs2* from the snow geese data. The four points with the largest values of *Obs2* have been removed for visual clarity.

3.6.2   Mean Function

If the second observer's estimates were very accurate, we could expect the mean function to be close to a straight line with intercept zero and slope one. We begin the analysis by thinking of the possibility that

$$E_U(Photo \mid Obs2) = Obs2 \qquad (3.2)$$

Given this mean function, the observer would be correct on the average. For example, if the observer were to estimate each of 500 flocks to be of size 40, we would expect the average photo count for these flocks to be close to 40, although the photo counts for the individual flocks would surely vary about this average. Any systematic deviation of the actual mean function from the possibility given in (3.2) would represent a *bias* in the observer's estimates. Consistent over- or underestimation of flock size could certainly have a serious impact on estimates of the size of the snow goose population produced from a full survey by the second observer. We can therefore think of the mean function in (3.2) as the *unbiased* mean function; the subscript U on the mean function is intended to remind us of this characterization.

The scatterplot of Figure 3.10 contains two lines. The curved line is the *lowess* estimate $\hat{E}(Photo \mid Obs2)$ with smoothing parameter 0.6. The straight line is simply $y = x$; it corresponds to the unbiased mean function (3.2). This line was placed on the plot by typing y = x into the text area of the plot's "Options" dialog, as in Section 3.4. We see from the plot that for small values of *Obs2*

$$E_U(Photo \mid Obs2) \approx \hat{E}(Photo \mid Obs2)$$

where $\hat{E}(Photo \mid Obs2)$ represents the *lowess* estimate of the mean function $E(Photo \mid Obs2)$. This makes sense because the observer could actually count the number of geese in very small flocks. However, the observer seems to underestimate the size of flocks containing 15 to 80 geese, and to overestimate the size of larger flocks. There may be good reason to question the accuracy of visual estimates by the second observer.

3.6.3   Variance Function

In addition to providing information on the regression function, Figure 3.10 also suggests that the variance function increases with the second observer's count. The telling characteristic of the plot is the increasing variation about the *lowess* smooth. Again this seems reasonable if the observer could essentially count the number of geese in very small flocks.

Remove the lines on your copy of Figure 3.10. The *lowess* smooth can be removed by moving its slider to the extreme left position. The straight line can be removed by checking the box "Clear all lines" in the plot's "Options" dialog. Next, go to the pop-up menu on the smoothing slidebar and select the

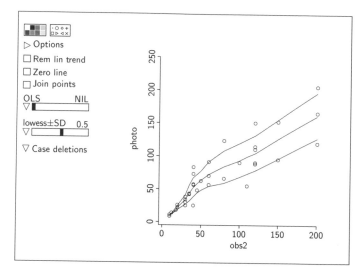

FIGURE 3.11   Scatterplot of the photo count *Photo* versus the visual estimate by observer 2 *Obs2* from the snow geese data. The lines superimposed on the plot are *lowess* smooths for the mean and variance function. The four points with the largest values of *Obs2* have been removed for visual clarity.

item "lowess+−SD." Finally, move the slider to select a smoothing parameter of 0.5. Your plot should now look like that in Figure 3.11.

Figure 3.11 contains three curves. The middle curve is just the *lowess* estimate of the mean function, which is currently represented by $\hat{E}(Photo \mid Obs2)$. The upper and lower curves are of the form

$$\hat{E}(Photo \mid Obs2) + \sqrt{\widehat{\text{Var}}(Photo \mid Obs2)}$$

and

$$\hat{E}(Photo \mid Obs2) - \sqrt{\widehat{\text{Var}}(Photo \mid Obs2)}$$

where $(\widehat{\text{Var}}(Photo \mid Obs2))^{1/2}$ is a *lowess* estimate of the standard deviation function. The two curves together are called *variance smooths*.

Estimates of the standard deviation function are constructed by *Arc* according to the discussion of (2.2) in Section 2.5.2:

1. A smooth estimate of the mean function $\hat{E}(y \mid x)$ is constructed using the smoothing parameter selected with the slider.
2. The new variable

$$\hat{e}^2 = (y - \hat{E}(y \mid x))^2$$

is formed.

3. A smoothed estimate of $E(\hat{e}^2 \mid x)$ is formed using the same smoothing parameter selected for the mean function in the first step. This step is just like smoothing a scatterplot of $\hat{e}^2$ versus x.

4. Finally, $(\widehat{\text{Var}}(y \mid x))^{1/2}$ is set equal to the square root of the smooth estimate of $E(\hat{e}^2 \mid x)$ from the third step.

The variance smooths on Figure 3.11 support our previous visual impression that the variance function is increasing. Additionally, it seems to increase rapidly for values of *Obs2* between 25 and 80 geese and then become relatively stable.

3.7 COMPLEMENTS

Regression smoothing has become a major focus of statistical research, with a recent change in emphasis from theoretical to practical results. Altman (1992) provides an introduction to the area. Book-length treatments with an applied orientation include Härdle (1990) and Simonoff (1996). Bowman and Azzalini (1997) provide interesting discussions of the use of smoothing in inference, and they also provide excellent computer code that can be used with the S-plus computer package.

The slice-smoother described in the early part of this chapter is not used very often in practice, but it does provide an introduction to the ideas of smoothing. The *lowess* smoother was suggested by Cleveland (1979), and if the mean function is smooth, *lowess* will give similar results to other popular smoothing techniques. The variance smoothers described in Section 3.6.3 can be improved upon at the cost of more calculations; see Ruppert, Wand, Holst and Hössjer (1997).

The ozone data used in Problem 3.2 were discussed by Breiman and Friedman (1985). The snow geese experiment was designed and supervised by one of the authors (RDC).

PROBLEMS

3.1 The following problems relate to the snow goose study.

 3.1.1 Construct a plot of *Photo* versus *Obs1* and remove the four points with the largest values of *Obs1*. Next, construct a *lowess* estimate of $E(Photo \mid Obs1)$ with smoothing parameter 0.6. Compare this estimate of the first observer's mean function to the unbiased mean function. Does observer 1 seem to be a "better" estimator of flock size than observer 2? Why?

 3.1.2 Describe the *lowess* estimate of $\text{Var}(Photo \mid Obs1)$. Do the predictions by observer 1 seem more or less variable than those by observer 2? Why?

3.1.3 Construct a plot of *Photo* versus *Obs1*, remove the 4 points with the largest values of *Obs1*, and construct a *lowess* estimate $h(Obs1)$ of E(*Photo* | *Obs1*) with smoothing parameter 0.6. Next, from the pop-up menu on the smoother slidebar, select the item "Extract mean." This item allows you to save the values $h(Obs1_i)$, for $i = 1, \ldots, 45$, as the variable h, or the name you type in the text area of the dialog that appears on the screen. Next, draw a plot with *Obs1* on the vertical axis and h on the horizontal axis. What does this plot show? Finally, draw a plot of *Photo* versus h. What information about the regression is available in this last plot?

3.2 The file `ozone.lsp` provides daily measurements on ozone level and several other quantities for 330 days in 1976 near Upland, in southern California. Load this data file, and then draw the plot of the ozone level *Ozone* versus the temperature in degrees Fahrenheit *Temp*. Use smoothers to describe the dependence of *Ozone* on *Temperature*.

3.3 A sample of $n = 52$ faculty members was selected from a small midwestern college. For each faculty member in the sample the investigator determined their sex (*Sex* = 0 for males and *Sex* = 1 for females), their salary, and their years of service. The data are available in file `salary.lsp`. The data were collected in the early 1980s, so the salaries are much lower than those today.

Suppose we select another faculty member at random from the college and determine only her/his salary. Using the data collected, can we say anything about whether the faculty member is more likely to be male or female? We can approach this as a regression with *Sex* as the response and *Salary* as the predictor. To gain information to help us answer the question, we need to study the conditional distribution of *Sex* | *Salary*. So far in this book we have considered response variables that are continuous or many-valued. The regression in this problem is a bit different because the response is *binary* taking two values zero or one. This does not change the general issues, but special interpretations are possible and useful.

Because the response is binary, *Sex* | *Salary* has a Bernoulli distribution with a probability of female ("success"), Pr(*Sex* = 1 | *Salary*), that can be a function of the salary. (Bernoulli distributions, which are the same as binomial distributions with one trial, are assumed as part of the prerequisites for this book, but they are reviewed in Section 21.2.) This probability is the same as the regression's mean function.

3.3.1 Show that

$$E(Sex \mid Salary) = Pr(Sex = 1 \mid Salary)$$

Hint: Calculate the expectation of a Bernoulli random variable.

3.3.2 Construct a scatterplot of *Sex* on the vertical axis versus *Salary* on the horizontal axis. Does the highest salary belong to a female or male faculty member. What is the lowest salary of a male faculty member? (*Hint*: Use the "Show coordinates" mouse mode.)

3.3.3 Place a *lowess* smooth with tuning parameter 0.6 on your plot. Describe what the smooth is estimating. What does the smooth say about the mean function? What does this imply about the salary structure at the college? Do the results necessarily imply anything about discrimination at the college?

3.3.4 Continuing with the same plot, delete the point corresponding to the highest female salary. How did the smooth change and why did it do so?

3.3.5 Estimate the mean function for the regression of *Sex* on *Year* using a slice smooth with 6 slices. Combining this information with that from the regression of *Sex* on *Salary*, what can you infer about the equity of the salary structure at the college?

CHAPTER 4

Bivariate Distributions

In Section 1.6 we reviewed ways of thinking about the distribution of a single random variable and then applied some of those ideas to the normal distribution. Our goal in this chapter is to study a pair of random variables, with emphasis on the normal. This study can help gain insights into the nature of regression.

Regressions often begin with a population of experimental units, such as haystacks in Nebraska, bluegills in Lake Mary, or some well-defined population of people. The experimental units might be characterized by several measurements, but in this chapter we will be concerned with only two at a time. We denote these two measurements by x and y. Selecting an experimental unit at random from the population gives an observation on the random pair (x, y). Both x and y are random variables and they vary together from experimental unit to experimental unit. The *bivariate distribution* of (x, y) is relevant when it is important to understand how x and y vary together. For example, as part of the haystack study we might wish to investigate the joint distribution of circumference and "over."

Not all regressions are of this type. Consider an experiment to investigate the effect on grain yield (Y) of adding 5, 10, 15 or 20 pounds of nitrogen (N) per acre. The experiment is conducted by applying each rate of nitrogen to 10 fields, for a total of 40 fields. In this experiment it is valid to speak of the distribution of $Y \mid N$. But the experiment does not allow a bivariate distribution for (Y, N) because N is purposefully selected by the experimenter and therefore is not random.

In this chapter we restrict attention to studies in which both x and y are random variables.

4.1 GENERAL BIVARIATE DISTRIBUTIONS

Using the haystack data for illustration, a basic way of understanding the joint distribution of (x, y) is through probability statements of the form

$$\Pr(c_l \leq C \leq c_u \quad \text{and} \quad v_l \leq Over \leq v_u) \tag{4.1}$$

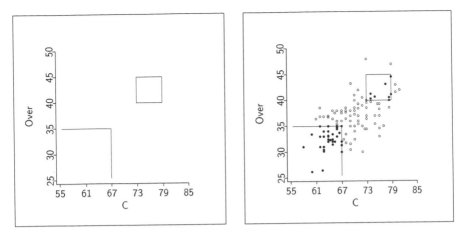

a. Regions.   b. Regions plus data.

FIGURE 4.1   Representative regions for illustrating joint distributions.

where the c's and v's are constants to be specified. For example, the statement

$$\Pr(C < 67 \quad \text{and} \quad Over < 35) \tag{4.2}$$

represents the probability that a randomly selected haystack has circumference less than 67 feet and "over" less than 35 feet. It can be visualized as the probability that a randomly selected haystack falls within the rectangular region in the lower left of the plot shown in Figure 4.1a. Similarly, the statement

$$\Pr(73 \le C \le 79 \quad \text{and} \quad 40 \le Over \le 45)$$

represents the probability that a randomly chosen haystack falls within the rectangular region in the upper right of Figure 4.1a. The regions in Figure 4.1a are shown again in Figure 4.1b superimposed on a scatterplot of the circumference and "over" for the 120 haystacks in the data set. The point cloud gives information on how C and $Over$ vary together. Haystacks with relatively large values of C tend to have relatively large values of $Over$, for example. In addition, while the probability in (4.2) is unknown, it can be estimated by using the fraction of sample haystacks that fall within the region. Since there are 35 haystacks in the region, the estimated probability is $35/120 = 0.29$. The number of haystacks in rectangular regions can be determined quickly by using the "slicing mode" and resizing the brush to the desired shape, as described in Section 2.3.2.

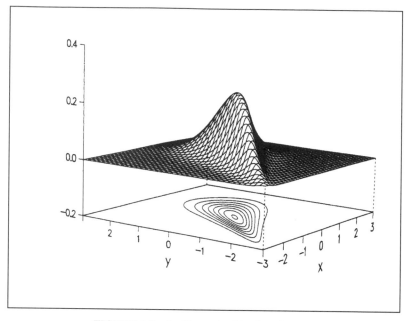

FIGURE 4.2   An illustrative bivariate density.

4.1.1   Bivariate Densities

If we could calculate probability statements like in (4.1) for any values of the constants, we would know the joint distribution of y and x, although it could still be difficult to gain a useful mental image of how y and x vary together. In some cases the nature of the random variables themselves allows more intuitive representations of their joint distributions. In particular, the joint distribution of continuous random variables (x, y) is usually expressed in terms of their *bivariate density function*. The bivariate density for C and *Over* from the haystack data gives the relative density of haystacks with respect to their circumference and "over." Generally, bivariate densities are interpreted much like the univariate densities reviewed in Section 1.6, except they characterize the joint variation of two random variables rather than the variation of a single random variable.

An illustrative bivariate density for two generic random variables is shown in Figure 4.2, which consists of a 3D surface and a series of contours drawn in the horizontal xy-plane. Ignore the contours for the moment and focus on the surface. The height of the surface gives the relative density in the population; the most likely values are in the region of the xy-plane where the density is the highest. In the case of a single random variable, probabilities are associated with *area* under its density. Similarly, for joint distributions, probabilities are associated with *volume* under the density. The total volume under the density of Figure 4.2 is 1, and the probability of an observation falling in any region in the xy-plane is the volume under the density and above that region.

Bivariate densities can also be represented as a series of concentric curves that trace contours of constant height of the density function. Each of the contour curves in the horizontal plane of Figure 4.2 traces a path of constant height on the density function. The smallest contour in the center marks the region of highest density. That contour encloses the most likely values of (x, y). The probability that (x, y) falls within the region enclosed by the smallest contour is the volume under the density and above that region.

4.1.2 Connecting with Regression

Recall that regression is the study of conditional distributions. When x and y are both random, so that (x, y) follows a bivariate distribution, there are two possible regressions. We could study either the regression of y on x or the regression of x on y, although usually the nature of the study dictates which variable will be the response and which will be the predictor. For example, in the haystack data the pair (Vol, C) follows a bivariate distribution, and we could study either the regression of Vol on C or the regression of C on Vol, although the goal of predicting Vol from C suggests that Vol should be the response.

Suppose now that y is the response variable. Creating a regression setting from a bivariate distribution requires creating conditional densities for $y \mid x$ from the bivariate density for (x, y). This can be done by slicing the bivariate density at the desired value of x. For example, to get the conditional density of $y \mid (x = 0)$ from Figure 4.2, mentally slice the density perpendicular to the xy-plane with another plane containing the line $x = 0$. The intersection of the slicing plane with the bivariate density forms a curve that, when normalized to have area one, gives the conditional univariate density of $y \mid (x = 0)$. Figure 4.3 shows a stylized representation of five conditional densities formed in this way. The means of these five conditional distributions fall on a common line which corresponds to the mean function $E(y \mid x)$; hence, for this regression the mean function is linear. The mean function could well be nonlinear in other regressions, in which case the means would fall on a curve rather than a straight line. The variances of the five conditional distributions are not the same, and thus the variance function $Var(y \mid x)$ is nonconstant.

4.1.3 Independence

The notion of independence is an important special case for bivariate random variables: The random variables x and y are *independent* if the conditional distribution of $y \mid x$ does not depend on the value of x. Equivalently, x and y are independent if the conditional distribution of $y \mid x$ is the same as the marginal distribution of y regardless of the value of x. The roles of x and y

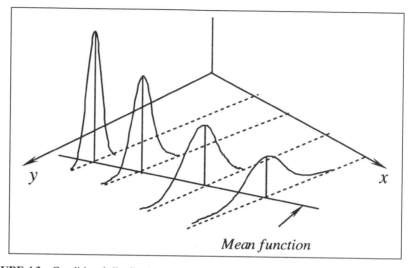

FIGURE 4.3   Conditional distributions whose means lie on a line with variance that changes as the mean changes.

can be changed in this definition without changing its meaning; saying that the conditional distribution of $y \mid x$ is the same as the marginal distribution of y is the same as saying that the conditional distribution of $x \mid y$ is the same as the marginal distribution of x.

Independence means just what the name implies: Knowing the value of x gives absolutely no information about y. If x furnishes any information at all about y, then x and y are said to be *dependent*.

In some regressions it may be obvious that two random variables are dependent. It seems clear that *Vol* and *Over* are dependent in the haystack data, and that *Length* and *Age* are dependent in the Lake Mary data. However, this will not be so clear in other regressions that we encounter in this book.

Two random variables can be dependent in many different and complicated ways. Consequently, there are no general measures of the strength of dependence between two random variables. But there are measures for special types of dependence. We discuss in the next section a way of measuring the strength of *linear dependence* between two random variables.

4.1.4   Covariance

In this section we introduce the *population covariance*, denoted $\text{Cov}(x, y)$, which is a numerical measure of how two random variables x and y vary together. Specifically, $\text{Cov}(x, y)$ measures the degree to which x and y are *linearly related*: How tightly would the points in a scatterplot of y versus x cluster about a common straight line with nonzero slope?

The population covariance is defined as

$$\mathrm{Cov}(x,y) = \mathrm{E}\{[x - \mathrm{E}(x)][y - \mathrm{E}(y)]\} \tag{4.3}$$

$$= \text{mean of } \{[x - \mathrm{E}(x)][y - \mathrm{E}(y)]\}$$

$$= \mathrm{E}(xy) - \mathrm{E}(x)\mathrm{E}(y) \tag{4.4}$$

$$= \text{mean of the product} - \text{the product of the means}$$

Equations (4.3) and (4.4) are alternate expressions for the same quantity. We can see from (4.3) that $\mathrm{Cov}(x,y)$ is positive if x above its mean tends to occur with y above its mean, and x below its mean tends to occur with y below its mean. If x tends to be below its mean when y is above its mean, or vice versa, then $\mathrm{Cov}(x,y)$ is negative. In both cases, the absolute value of $\mathrm{Cov}(x,y)$ indicates the strength of the linear relationship between x and y.

The absolute value of a covariance can never be larger than the product of the standard deviations,

$$-\sqrt{\mathrm{Var}(x)\mathrm{Var}(y)} \le \mathrm{Cov}(x,y) \le \sqrt{\mathrm{Var}(x)\mathrm{Var}(y)} \tag{4.5}$$

If $\mathrm{Cov}(x,y) = (\mathrm{Var}(x)\mathrm{Var}(y))^{1/2}$, then there must be an exact *linear* relationship between x and y, and all pairs of observations must fall on a straight line with *positive* slope. If $\mathrm{Cov}(x,y) = -(\mathrm{Var}(x)\mathrm{Var}(y))^{1/2}$, then again there must be an exact linear relationship between x and y, but now all pairs of observations must fall on a straight line with *negative* slope. If $\mathrm{Cov}(x,y) = 0$, then there is no *linear* relationship between x and y, although generally there could be some more complicated type of dependence.

The population covariance can be estimated using either (4.3) or (4.4) by replacing the population means with sample averages,

$$\widehat{\mathrm{Cov}}(x,y) = \frac{1}{n-1} \sum_{i=1}^{n} (x_i - \bar{x})(y_i - \bar{y})$$

$$= \frac{1}{n-1} \left(\sum_{i=1}^{n} (x_i y_i) - n\bar{x}\bar{y} \right) \tag{4.6}$$

The divisor of $n-1$, instead of n, is conventional and brings about certain desirable statistical properties. In any event, the effect of using $n-1$ instead of n is negligible, unless n is quite small.

Here is a list of additional properties of the population covariance. We will not be using all of them immediately, but it should be helpful to have them collected in one place for future reference.

- The covariance is a symmetric measure of linear dependence: The covariance between x and y is the same as the covariance between y and x. In symbols, $\mathrm{Cov}(x,y) = \mathrm{Cov}(y,x)$.

- The covariance of a random variable x with itself is the same as its variance,

$$\text{Cov}(x,x) = \text{Var}(x)$$

- Define the new random variable $w = a + (b \times x)$, where a and b are constants. Then

$$\text{Cov}(w,y) = \text{Cov}(a + (b \times x), y) = b \times \text{Cov}(x,y)$$

Adding a constant to either x or y does not change their covariance, but multiplying either x or y by a constant results in the covariance being multiplied by the same constant. This property, which is similar to that for standard deviations discussed in Section 1.6, reflects the fact that the magnitude of the covariance depends on the units in which x and y are measured. Changing the units of either x or y will change the value of the covariance, unless $\text{Cov}(x,y) = 0$.

- Let x, y, and z be three random variables. Then

$$\text{Cov}(x, y + z) = \text{Cov}(x,y) + \text{Cov}(x,z)$$

- The variance of the sum of two random variables is equal to the sum of their variances plus twice their covariance:

$$\text{Var}(x + y) = \text{Var}(x) + \text{Var}(y) + 2\text{Cov}(x,y) \tag{4.7}$$

- The last equation generalizes: If $y_1, y_2, \ldots, y_k$ are k random variables and $b_1, b_2, \ldots, b_k$ are any k constants, then

$$\text{Var}\left(\sum_{i=1}^{k} b_i y_i\right) = \sum_{i=1}^{k} b_i^2 \text{Var}(y_i) + 2\sum_{i=1}^{k-1} \sum_{j=i+1}^{k} b_i b_j \text{Cov}(y_i, y_j) \tag{4.8}$$

4.1.5   Correlation Coefficient

Although the covariance measures the linear relationship between two variables, it is hard to use because interpretation of its value depends on the product of the standard deviations: We can't tell if $\text{Cov}(x,y) = 1$ is large or small without knowing the values of $\text{Var}(x)$ and $\text{Var}(y)$. If $(\text{Var}(x)\text{Var}(y))^{1/2} = 1$, then $\text{Cov}(x,y) = 1$ has attained its upper limit, and we know that y is an exact linear function of x, but if $(\text{Var}(x)\text{Var}(y))^{1/2} = 1000$ there is only a weak linear association between the random variables. It will be easier to interpret the population covariance if we standardize it by dividing by the product of the standard deviations. The standardized population covariance is called the *population correlation coefficient* and is denoted by using the Greek letter ρ

(rho):

$$\rho(x,y) = \frac{\text{Cov}(x,y)}{\sqrt{\text{Var}(x)\text{Var}(y)}} \tag{4.9}$$

When the random variables x and y are clear from context, we will simply write ρ for $\rho(x,y)$.

The population correlation coefficient can also be expressed in terms of the *standardized variables*:

$$Z_x = \frac{x - \text{E}(x)}{\sqrt{\text{Var}(x)}}$$

and

$$Z_y = \frac{y - \text{E}(y)}{\sqrt{\text{Var}(y)}}$$

The correlation between x and y is the same as the covariance between Z_x and Z_y:

$$\rho(x,y) = \text{Cov}(Z_x, Z_y)$$

The sample correlation coefficient is just the sample covariance divided by the product of the sample standard deviations:

$$\hat{\rho}(x,y) = \frac{\widehat{\text{Cov}}(x,y)}{\text{sd}(x)\text{sd}(y)} \tag{4.10}$$

The sample correlation coefficient can be obtained in *Arc* as described in Section 1.4 and illustrated in Table 1.1.

Because we have divided by the product of the standard deviations, the correlation coefficient is a unitless number that always falls between -1 and 1. If $\rho(x,y) = 0$, the random variables are said to be *uncorrelated*. Otherwise, the magnitude of $\rho(x,y)$ indicates the degree to which x and y are linearly related, just as the population covariance does. If $\rho(x,y) = 1$, then (x,y) always falls on a line with positive slope. If $\rho(x,y) = -1$, then (x,y) always falls on a line with negative slope. Beyond that, careful interpretation of the correlation coefficient depends on characteristics of the bivariate distribution. In the next section we study the meaning of the correlation coefficient in the family of normal distributions.

4.2   BIVARIATE NORMAL DISTRIBUTION

The family of bivariate normal distributions is frequently encountered in statistics. Five numbers are required to specify a particular bivariate normal distribution for (x, y):

- The marginal population mean $E(y)$ and variance $\text{Var}(y)$ of y
- The marginal population mean $E(x)$ and variance $\text{Var}(x)$ of x
- The correlation coefficient $\rho(x, y)$

These five numbers are the only population information that is required to compute the joint probability statements discussed in Section 4.1.

Of these five numbers the first four are used to characterize the marginal distributions of x and y. In particular, if (x, y) has a bivariate normal distribution, then the marginal distributions of x and y are each univariate normal. This means that only one additional number, the correlation coefficient $\rho(x, y)$, is needed to specify the joint variation for a bivariate normal. This is one of the distinguishing properties of the normal distribution. For many other distributions, the correlation coefficient is not enough to characterize joint variation fully. We will return to this idea in Section 4.2.2.

A *standard bivariate normal distribution* is any bivariate normal in which x and y have each been standardized to have mean 0 and standard deviation 1. There are many standard bivariate normal distributions, one for each possible value of $\rho(x, y)$.

4.2.1   Correlation Coefficient in Normal Populations

The following cases serve as first benchmarks for understanding what the correlation coefficient measures in bivariate normal populations:

Independence, $\rho = 0$. If x and y are uncorrelated so that $\rho(x, y) = 0$, then they are *independent*. This is another special property of the bivariate normal distribution. Uncorrelated random variables need not be independent unless they follow a bivariate normal distribution.

The density for a standard bivariate normal distribution with $\rho = 0$ is shown in Figure 4.4. In this case the contours of constant density are concentric circles, although the perspective view shown in the figure makes them appear as ellipses. Figure 4.5a shows contours along with a sample of 100 points from the standard bivariate normal with $\rho = 0$. Reading from the inside out, the contours enclose 10%, 30%, 50%, 70%, 90%, and 98% of the population. The probability that an observation on the random pair (x, y) falls within the smallest contour is 0.10, and the probability that it falls outside the largest contour is 0.02. Thus for our sample of 100 points, we should expect about 10 to fall within the smallest contour and about 2 to fall outside the largest contour.

The Sign of ρ. If $\rho \neq 0$ then x and y are *dependent*. The sign of ρ indicates the qualitative nature of the dependence. Positive correlations indicate that larger values of x tend to occur with larger values of y. Similarly, negative

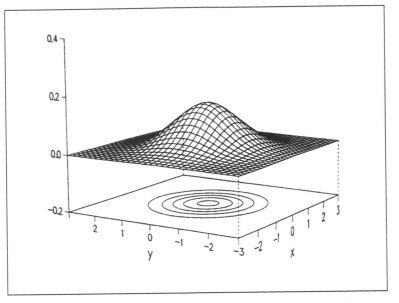

FIGURE 4.4 Standard bivariate normal density function for $\rho = 0$.

correlations indicate that larger values of x tend to occur with smaller values of y. This is the same interpretation that we discussed for the population covariance in Section 4.1.4.

The Magnitude of ρ. When the correlation coefficient is positive the contours of the bivariate normal distribution change from circles to ellipses that become increasingly narrow as ρ approaches 1. Figures 4.5b–4.5d, which were constructed in the same way as Figure 4.5a, show how standard bivariate normal distributions change as ρ increases; when $\rho = 1$ all contours collapse to the line with slope 1 that passes through the origin, and all sample points must fall on this line.

The same comments apply when $\rho < 0$ except the contours are oriented from upper left to lower right, as illustrated in Figure 4.6.

Units of ρ. The correlation coefficient is a unitless number between -1 and 1. Figure 4.6a shows contours of the standard bivariate normal distribution with $\rho(x,y) = -0.75$, along with a sample of 100 observations. Figure 4.6b was constructed in the same way, except the variance of y was set to $\text{Var}(y) = 100^2$. Except for the numbers on the vertical axes, the visual impressions of these two figures are nearly identical. The plot in Figure 4.6c was constructed from a bivariate normal distribution with $\rho(x,y) = -0.75$. The numbers on the axes have been removed so we have no information on the mean or variance of x or y. However, the plot still allows an impression of the relationship between x and y, and of the correlation coefficient. The particular

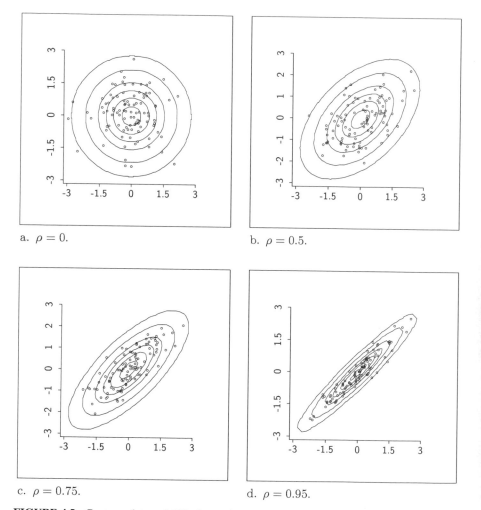

a. $\rho = 0$.

b. $\rho = 0.5$.

c. $\rho = 0.75$.

d. $\rho = 0.95$.

FIGURE 4.5   Contour plots and 100 observations from four standard bivariate normal distributions. In each case the contours enclose 10%, 30%, 50%, 70%, 90%, and 98% of the population.

units attached to x and y do not matter for interpretation of the correlation coefficient.

Plotting Guidelines. The plot in Figure 4.6d is a little different. Like Figure 4.6a, the plot is of a sample from a standard bivariate normal distribution with $\rho = -0.75$. However, the range on the y-axis of Figure 4.6a is -3 to 3, while the range on the y-axis of Figure 4.6d is -6 to 6. Such differences in plot construction can change the visual impressions of the plot, the strength of relationship, and of the magnitude of the correlation coefficient. For ease and consistency of interpretation, scatterplots should be constructed so that (1) the physical length of the vertical axis is the same as that of the horizontal axis,

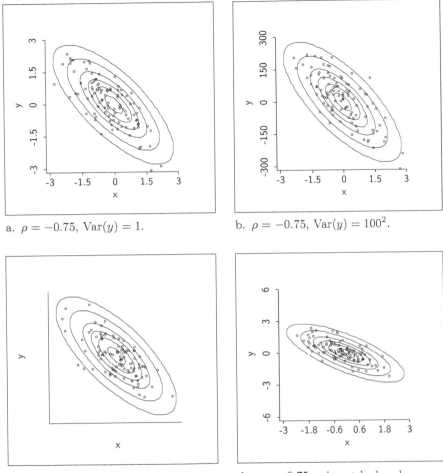

a. $\rho = -0.75$, $\mathrm{Var}(y) = 1$.

b. $\rho = -0.75$, $\mathrm{Var}(y) = 100^2$.

c. $\rho = -0.75$, missing scales.

d. $\rho = -0.75$, mismatched scales.

FIGURE 4.6 Contour plots and 100 observations from four standard bivariate normal distributions. In each case the contours enclose 10%, 30%, 50%, 70%, 90%, and 98% of the population.

and (2)

$$\frac{\text{Vertical axis range}}{\mathrm{sd}(y)} \approx \frac{\text{Horizontal axis range}}{\mathrm{sd}(x)}$$

so that the ranges on the vertical and horizontal axes cover about the same number of standard deviations.

The plots in Figures 4.6a–4.6c were constructed so that

$$\frac{\text{Vertical axis range}}{\sqrt{\mathrm{Var}(y)}} = \frac{\text{Horizontal axis range}}{\sqrt{\mathrm{Var}(x)}} = 6$$

For this reason the contours look identical from plot to plot. However, the range for the vertical axis of the plot in Figure 4.6d covers 12 standard deviations. This difference in plot construction accounts for the different appearance of Figures 4.6a and 4.6d.

Arc uses rules for determining axis ranges that attempt to balance several competing objectives, including consistent visualization of strength of relationship, but changing these ranges is sometimes required. The range of either axis of a 2D plot can be changed from the "Options" dialog as shown in Figure 3.5, and typing the upper and lower limits of the range in the "Range" boxes. In particular, typing 3sd or -3sd in the first of the two range boxes for an axis will give that axis a range of six standard deviations. Similarly, for any number k, typing $k$sd will result in a range of $2k$ standard deviations for the axis.

4.2.2   Correlation Coefficient in Non-normal Populations

When the population has a bivariate normal distribution, the correlation coefficient is all that is needed to characterize fully the joint variation. The mental image is of an elliptical cloud of points that becomes narrower as the magnitude of the correlation coefficient increases. When the population is non-normal, the correlation coefficient still gives a measure of the strength of any linear relationship, but it may miss important additional information about joint variation.

Figures 4.7a–4.7c show scatterplots of 100 observations from each of three non-normal populations. The structure of these plots is quite different, and yet the sample correlation coefficient for each plot is the same, 0.75. All the points in Figure 4.7a fall on a curve, but the correlation coefficient cannot recognize this. The data in Figure 4.7b fall in two distinct clusters. The 50 observations in each cluster are from bivariate normal populations with $\rho = 0$, but each cluster has a different mean and standard deviation. Again, the correlation coefficient cannot tell us of this structure. In Figure 4.7c all points except one, the point at $x = 8$, fall on a line. If this point were removed the correlation coefficient would be 1, but with the point the correlation coefficient is again 0.75. Unlike the normal case, the correlation coefficient provides an incomplete description of the joint variation.

Figure 4.8 gives plots of 100 observations from each of two non-normal populations. The population correlation coefficient is zero for each plot. If the populations were normal, this would imply that x and y are independent. But the populations are not normal, and x and y are dependent in each plot. The dependence in Figure 4.8a comes about because there is an exact quadratic relationship between x and y; given x you know the exact value of y because $y = x^2$. In Figure 4.8b, $\text{Var}(y \mid x = 0) < \text{Var}(y \mid x = 10)$ and thus x and y are again dependent. In summary, a correlation of zero does not imply that x and y are independent, unless they follow a bivariate normal distribution.

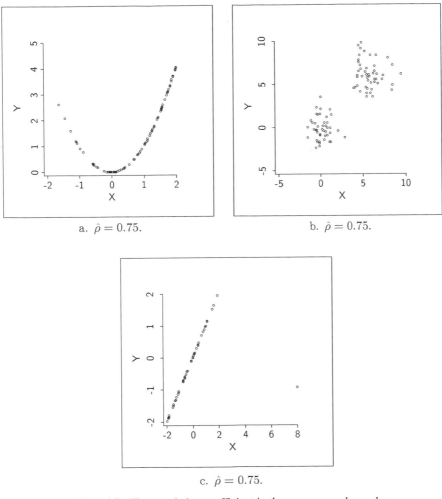

a. $\hat{\rho} = 0.75$.   b. $\hat{\rho} = 0.75$.

c. $\hat{\rho} = 0.75$.

FIGURE 4.7 The correlation coefficient in three non-normal samples.

4.3 REGRESSION IN BIVARIATE NORMAL POPULATIONS

When sampling from a bivariate population there are two possible regressions, the regression of y on x and the regression of x on y. In this section we focus mostly on the regression of y on x, understanding that the roles of x and y can be interchanged at any point in the discussion.

If (x, y) follows a bivariate normal distribution, then the conditional distribution of $y \mid (x = \tilde{x})$ is univariate normal for each value $\tilde{x}$ of x. A univariate normal distribution is completely characterized by its mean and variance. Thus, when (x, y) follows a bivariate normal distribution, the mean function $E(y \mid x = \tilde{x})$ and the variance function $\text{Var}(y \mid x = \tilde{x})$ completely characterize

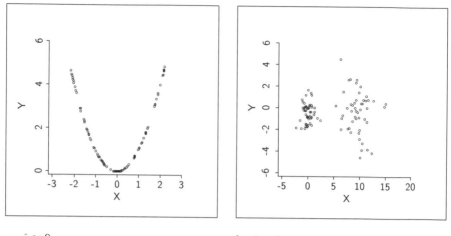

a. $\hat{\rho} \approx 0$.   b. $\hat{\rho} \approx 0$.

FIGURE 4.8   Two non-normal samples, each with a sample correlation of 0.

the distribution of $y \mid (x = \tilde{x})$; equivalently, they completely characterize the regression of y on x.

4.3.1   Mean Function

Beginning with a bivariate normal random variable (x, y), the mean function for the regression of y on x is

$$
E(y \mid x = \tilde{x}) = E(y) + \rho(x, y) \sqrt{\frac{\text{Var}(y)}{\text{Var}(x)}} (\tilde{x} - E(x)) \tag{4.11}
$$

The mean function depends on all five parameters of the bivariate normal distribution. If $\rho = 0$, then $E(y \mid x = \tilde{x}) = E(y)$ and x and y are independent. If $\tilde{x} = E(x)$, then at this point $E(y \mid x = \tilde{x}) = E(y)$. Beyond these first properties, there are two different ways to understand the mean function.

4.3.2   Mean Function in Standardized Variables

Rearranging terms in (4.11) gives a form for the mean function in terms of standardized variables:

$$
\frac{E(y \mid x = \tilde{x}) - E(y)}{\sqrt{\text{Var}(y)}} = \rho(x, y) \left(\frac{\tilde{x} - E(x)}{\sqrt{\text{Var}(x)}} \right) \tag{4.12}
$$

Imagine that we wish to predict y when $x = \tilde{x}$. A reasonable prediction for y is just $E(y \mid x = \tilde{x})$, the mean of the conditional distribution of $y \mid (x = \tilde{x})$.

TABLE 4.1 Output from "Display summaries ... " for the Berkeley Guidance Study

Data set = BGS-girls, Summary Statistics						
Variable	N	Average	Std. Dev.	Minimum	Median	Maximum
Ht2	70	87.253	3.3305	80.9	87.1	97.3
Ht18	70	166.54	6.0749	153.6	166.75	183.2
Data set = BGS-girls, Sample Correlations						
Ht2	1.0000	0.6633				
Ht18	0.6633	1.0000				
	Ht2	Ht18				

Equation (4.12) tells us two basic results:

- If $\tilde{x}$ is one standard deviation above the mean of x, then $E(y \mid x = \tilde{x})$ will be ρ standard deviations above the mean of y.
- More generally, if $\tilde{x}$ is $z_x = (\tilde{x} - E(x))/(\text{Var}(x))^{1/2}$ standard deviations from the mean of x, then $E(y \mid x = \tilde{x})$ will be $\rho \times z_x$ standard deviations from the mean of y.

A brief example may illustrate why the interpretation of the standardized form is important in regression. Table 4.1 gives the *Arc* summary of the height at age 2, *Ht2*, and the height at age 18, *Ht18*, both in cm, of 70 girls born in Berkeley, California in 1928–1929 and monitored on a number of growth measurements periodically until age 18 in the Berkeley Guidance Study. These data, along with additional description, are available in *Arc* data file BGSgirls.lsp. The scatterplot of *Ht18* versus *Ht2* shown in Figure 4.9 has an elliptical shape, indicating that the joint distribution could be bivariate normal. Approximate bivariate normality is typical for this type of data.

The five population parameters for the height data are unknown, but they can be estimated by using the sample versions from Table 4.1. The estimated version of the mean function (4.11) is obtained by simply replacing the unknown parameters by their estimates.

Suppose we wish to predict the height at age 18 of a girl whose height was one standard deviation above the average height at age two, or $87.253 + 3.331 = 90.584$ cm. According to (4.12), the estimate of $E(Ht18 \mid Ht2 = 90.584)$ is $\hat{\rho}(Ht2, Ht18) = 0.6633$ standard deviations above the mean of *Ht18*, or $166.54 + 0.6633 \times 6.075 = 170.57$ cm.

Suppose we turn the prediction problem around, and start with a girl whose height is 170.57 cm at age 18. What do we expect was her height at age two? Interchanging the roles of variables, we can again use (4.12): At age 18 the girl is 0.6633 standard deviations above the average. We should expect that she was $\hat{\rho}(Ht2, Ht18) \times 0.6633 = 0.4400$ standard deviations above the average at age two, or $87.253 + 0.4400 \times 3.331 = 88.719$ cm. There is an inherent asymmetry

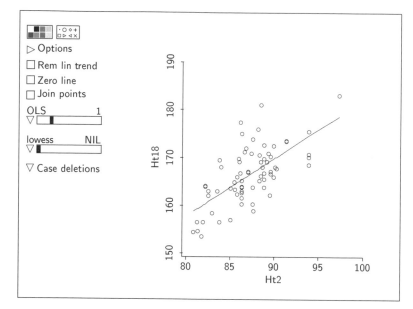

FIGURE 4.9   Scatter plot of *Ht18* versus *Ht2* for the Berkeley Guidance study.

in our predictions:

$$\hat{E}(Ht18 \mid Ht2 = 90.584) = 170.57 \qquad \text{but}$$

$$\hat{E}(Ht2 \mid Ht18 = 170.57) = 88.719$$

This arises because there are really two different regressions. We will return to this issue after discussing an alternate form for the mean function.

4.3.3   Mean Function as a Straight Line

The mean function shown in (4.11) is a linear function of $\tilde{x}$. We can emphasize this fact by rearranging the terms in (4.11), rewriting the mean function as

$$E(y \mid x = \tilde{x}) = E(y) + \rho\sqrt{\frac{\text{Var}(y)}{\text{Var}(x)}}(\tilde{x} - E(x))$$

$$= E(y) - \rho\sqrt{\frac{\text{Var}(y)}{\text{Var}(x)}}E(x) + \rho\sqrt{\frac{\text{Var}(y)}{\text{Var}(x)}}\tilde{x}$$

$$= E(y) - \beta_{y|x}E(x) + \beta_{y|x}\tilde{x}$$

$$= \alpha_{y|x} + \beta_{y|x}\tilde{x} \qquad\qquad (4.13)$$

We have defined two new symbols, using the Greek letters alpha (α) and beta (β):

$$\alpha_{y|x} = E(y) - \beta_{y|x}E(x) \tag{4.14}$$

and

$$\beta_{y|x} = \rho\sqrt{\frac{\text{Var}(y)}{\text{Var}(x)}} \tag{4.15}$$

The subscript "$y \mid x$" is meant as a reminder that the intercept and slope are for the regression of y on x.

The final equation (4.13) shows that the mean function is a straight line with slope $\beta_{y|x}$ and intercept $\alpha_{y|x}$. The slope can be interpreted as the increase in $E(y \mid x = \tilde{x})$ per unit increase in $\tilde{x}$. The sign of the correlation coefficient determines the sign of the slope of the line. We can obtain estimates of the intercept and slope by substituting estimates for the quantities that define them. For the Berkeley Guidance Study the estimated slope is

$$\hat{\beta}_{Ht18|Ht2} = 0.6633 \times \frac{6.075}{3.331} = 1.21$$

and the estimated intercept is

$$\hat{\alpha}_{Ht18|Ht2} = 166.54 - 1.21 \times 87.25 = 60.97$$

The line drawn on Figure 4.9 is the estimated mean function

$$\hat{E}(Ht18 \mid Ht2) = \hat{\alpha}_{Ht18|Ht2} + \hat{\beta}_{Ht18|Ht2}Ht2$$
$$= 60.97 + 1.21Ht2$$

Suppose that we now consider the regression of x on y. The linear form of the regression equation can be obtained by simply interchanging the roles of the variables in (4.13):

$$E(x \mid y = \tilde{y}) = \alpha_{x|y} + \beta_{x|y}\tilde{y} \tag{4.16}$$

where

$$\alpha_{x|y} = E(x) - \beta_{x|y}E(y)$$

and

$$\beta_{x|y} = \rho\sqrt{\frac{\text{Var}(x)}{\text{Var}(y)}}$$

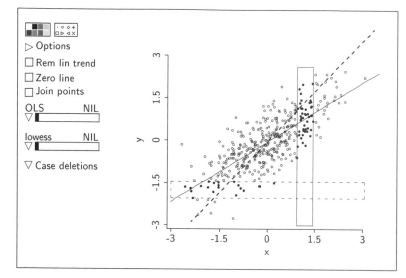

FIGURE 4.10   Scatterplot illustrating two regressions. Solid line is the mean function for the regression of y on x. Dashed line is the mean function for the regression of x on y.

   The crucial point here is that the two mean functions $E(x \mid y = \tilde{y})$ and $E(y \mid x = \tilde{x})$ can be quite different, and one cannot be obtained directly from the other. This difference can be traced to two causes: The basic definition of a mean function and the elliptical nature of the bivariate normal density. To illustrate how the difference arises, consider Figure 4.10, which consists of a sample of 400 observations from a standard bivariate normal population with $\rho = 0.65$. The mean function $E(y \mid x) = 0.65x$ is shown as the solid line on the plot, and the other mean function $E(x \mid y) = 0.65y$ is shown as the dashed line on the plot. The mean of the y values over the highlighted points in the vertical slice on the plot is an estimate of $E(y \mid x = 1.25)$. As the slice is moved across the plot, the slice means will fall close to the line $E(y \mid x) = 0.65x$. The horizontal slice on the plot is interpreted similarly, except the slice means will fall close to the dashed line, the mean function for the regression of x on y.

4.3.4   Variance Function

If (x, y) follows a bivariate normal distribution, then the variance function for the regression of y on x is

$$\text{Var}(y \mid x = \tilde{x}) = [1 - \rho^2(x, y)]\text{Var}(y) \qquad (4.17)$$

Three properties of this variance function are particularly important.

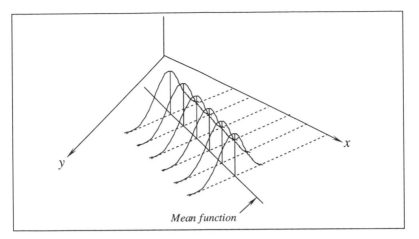

FIGURE 4.11   Conditional distributions from a bivariate normal.

- The variance is constant since it does not depend on $\tilde{x}$. Regressions with nonconstant variance functions are generally more difficult to analyze than regressions with a constant variance function.
- $\text{Var}(y \mid x)$ decreases as the magnitude of the correlation coefficient increases. We again see that if $\rho = 1$ or -1, y will be completely determined by x because in those cases $\text{Var}(y \mid x) = 0$.
- Because $0 \leq \rho^2 \leq 1$, the constant variance function is never greater than the marginal variance of the response,

$$\text{Var}(y \mid x) \leq \text{Var}(y)$$

Shown in Figure 4.11 is a stylized representation of the regression of y on x when (x, y) follows a bivariate normal distribution. The figure shows five densities of $y \mid x$ corresponding to five values of x. Each conditional distribution is normal and each has the same variance. The means of the five representative conditional densities lie on a line, reflecting the mean function.

As with the mean function, the variance function can be estimated by substituting estimates for the unknown parameters. For the regression of *Ht18* on *Ht2* from the Berkeley Guidance Study, the estimated variance function is

$$\widehat{\text{Var}}(y \mid x) = (1 - \hat{\rho}^2)\widehat{\text{Var}}(y)$$
$$= (1 - 0.6633^2)6.075^2$$
$$= 20.67$$

which is considerably less than the estimated marginal variance of y, $\widehat{\text{Var}}(y) = 36.90$.

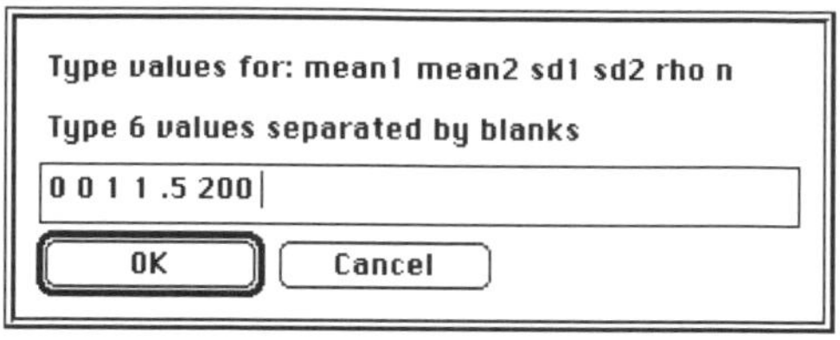

```
┌─────────────────────────────────────────────────┐
│ ┌───────────────────────────────────────────┐   │
│ │                                           │   │
│ │  Type values for: mean1 mean2 sd1 sd2 rho n │ │
│ │                                           │   │
│ │  Type 6 values separated by blanks        │   │
│ │  ┌──────────────────────────────────────┐ │   │
│ │  │ 0 0 1 1 .5 200│                       │ │   │
│ │  └──────────────────────────────────────┘ │   │
│ │  ┌─────────────┐    ┌─────────────┐       │   │
│ │  │     OK      │    │   Cancel    │       │   │
│ │  └─────────────┘    └─────────────┘       │   │
│ └───────────────────────────────────────────┘   │
└─────────────────────────────────────────────────┘
```

FIGURE 4.12   Dialog for the demonstration program `demo-bn.lsp` that generates bivariate normal samples.

To summarize the Berkeley Guidance Study regression, the distribution of *Ht18* | *Ht2* is approximately normal with estimated mean function

$$\hat{\text{E}}(Ht18 \mid Ht2) = 60.98 + 1.21 Ht2$$

and estimated variance function $\widehat{\text{Var}}(y \mid x) = 20.67 \approx 4.5^2$. The predicted height at age 18 of a girl who was 100 cm tall when she was two years old is 181.98 cm. Further, as a first approximation, there is about a 68% chance that the girl's height at age 18 will be within one standard deviation of the estimate. The predicted height is 181.98 cm $\pm$ 4.5 cm or so. Refined standard errors will be discussed in later chapters.

4.4   SMOOTHING BIVARIATE NORMAL DATA

The bivariate normal distribution is special because the associated regressions have linear mean functions and constant variance functions. These two properties can be used to gain insights about the behavior of the smooths discussed in Chapter 3 by using a demonstration program that comes with *Arc*. This demonstration will generate samples from bivariate normal distributions, and display them for you to examine.

After loading the file `demo-bn.lsp`, the dialog of Figure 4.12 will appear on the computer screen. This dialog is used to specify six items, the five parameters of the bivariate normal and the sample size; these must be entered in the order indicated in the dialog's instructions. The six numbers that have been typed into the text window of the dialog in Figure 4.12 instruct *Arc* to generate 200 samples from a standard bivariate normal distribution with correlation coefficient $\rho = 0.5$. After typing the desired values and sample size, click the "OK" button, and a plot similar to Figures 4.5 and 4.6 will appear on your screen. The plot shows the *n* points and six contours of the specified bivariate density function. The contours will always be the same as those shown in Figures 4.5 and 4.6: Reading from the inside out, the contours enclose 10%, 30%, 50%, 70%, 90%, and 98% of the population. If desired, these contours, along with any other lines that may have been subsequently

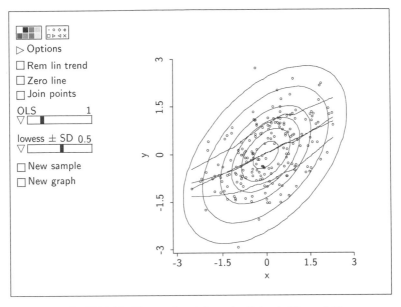

FIGURE 4.13 Plot produced from the dialog in Figure 4.12.

drawn on the plot, can be removed by checking the "Clear all lines" box in the plot's "Options" dialog.

Figure 4.13 contains the plot generated from the dialog of Figure 4.12, along with 4 lines that were added subsequently. The straight line passing through the middle of the point cloud is the true mean function for the regression of *y* on *x*. It was added by typing y = .5*x into the curve-generating text area of the plot's "Options" dialog, as demonstrated previously in Figure 3.5. The other three lines are mean and standard-deviation smooths from *Arc* smoothing options discussed in Section 3.6.3. Would you have thought that a linear mean function and a constant variance function were reasonable possibilities in view of the mean and standard-deviation smooths?

The plots produced from the demonstration module contain two special buttons. Clicking in the button for "New sample" will result in a new sample of *n* observations from the same bivariate normal distribution. Any smooths present will be updated automatically, but any curves drawn on the plot with the "Options" dialog will not be updated. In this way, smooths of many different samples can be viewed in a short period of time. This should provide some intuition about the behavior of smooths when the underlying regression has a linear mean function and a constant variance function.

The second special button "New graph" allows the generating bivariate normal distribution to be changed. Samples from different bivariate normal distributions can be viewed at the same time in separate plots by loading the demonstration module as necessary.

4.5   COMPLEMENTS

4.5.1   Confidence Interval for a Correlation

Given a bivariate normal population, a confidence interval for the population correlation ρ can be obtained by (1) transforming $\hat{\rho}$ to a scale in which it is approximately normal, (2) computing a confidence interval for the transformed correlation, and (3) back-transforming the end-points of the interval to the original scale. Fisher (1921) showed that the quantity, usually called *Fisher's z-transform*,

$$z_r = \frac{1}{2} \log \left(\frac{1 + \hat{\rho}}{1 - \hat{\rho}} \right)$$

is approximately normally distributed with variance $1/(n-3)$. For example, suppose that the sample size is $n = 9$, and $\hat{\rho} = -0.889$. Then $z_r = 1.417$, and $(1/(n-3))^{1/2} = 0.408$. For a 95% confidence interval based on the normal, the appropriate multiplier is $Q(z, 0.975) = 1.96$, and the confidence interval in the transformed scale has lower end-point $z_L = 1.417 - 1.96(0.408) = 0.617$ and upper end-point $z_U = 1.417 + 1.96(0.408) = 2.217$. The end-points of the interval in the correlation scale are obtained by using the inverse of the z_r-transformation. For the lower end-point, we get $(\exp(2z_L) - 1)/(\exp(2z_L) + 1)$ $= 0.549$ and $(\exp(2z_U) - 1)/(\exp(2z_U) + 1) = 0.976$. Restoring the sign, the confidence interval for ρ is then from -0.976 to -0.549.

4.5.2   References

Figures 4.2 and 4.4 are reproduced from Johnson (1987) with permission ©1987 John Wiley & Sons, Inc. The Berkeley Guidance Study data are taken from Tuddenham and Snyder (1954).

PROBLEMS

4.1   State whether the conclusion in each of the following situations is true or false, and provide a brief one or two sentence justification for your answer. Each statement concerns a bivariate random sample (x_i, y_i), $i = 1, \ldots, n$, from the joint distribution of (x, y).

   a. A scatterplot of y_i versus x_i clearly shows that $\text{Var}(y \mid x)$ is not constant. Therefore, (x, y) cannot have a bivariate normal distribution.

   b. It is known that (x, y) follows a bivariate normal distribution, and a scatterplot of y_i versus x_i clearly shows that $\text{E}(y \mid x)$ is constant. Therefore, x and y are independent.

   c. It is known that (x, y) follows a standard bivariate normal distribution with $\rho = 0.5$. Therefore, to predict x from y we should use $x = y/2$, and to predict y from x we should use $y = 2x$.

d. It is known that $\rho(x,y) > 0$. Therefore, $\text{Var}(x + y) > \text{Var}(x) + \text{Var}(y)$. *Hint*: Recall (4.7).

e. It is known that $\rho(x,y) < 0$. Therefore, $\text{Var}(x + y) > \text{Var}(x) + \text{Var}(y)$.

f. It is known that x and y are uncorrelated. Therefore, $\text{Var}(x + y) = \text{Var}(x) + \text{Var}(y)$.

4.2 Suppose (x,y) follows a bivariate normal distribution with $E(x) = E(y) = 0$, $\text{Var}(x) = 4$, $\text{Var}(y) = 9$ and $\rho = 0.6$. Find the variance of the sum, $\text{Var}(x + y)$.

4.3 Suppose (x,y) follows a standard bivariate normal distribution with $\rho = 0$. Find $\text{Pr}(y < 0.5)$. Give the mean and variance functions $E(y \mid x)$ and $\text{Var}(y \mid x)$.

4.4 In the haystack data given in file `haystack.lsp`, assume that the joint distribution of $(C, Over)$ is bivariate normal.

    4.4.1 Estimate the mean functions $E(C \mid Over)$ and $E(Over \mid C)$.

    4.4.2 If haystacks were hemispheres on the average, we would expect $E(C \mid Over) = 2 \times Over$. Judging from the data, does this seem a reasonable possibility? Why?

    4.4.3 A particular haystack has a circumference that is 1.5 standard deviations above the average circumference. How many standard deviations is $E(Over \mid C)$ above the mean of $Over$?

4.5 In the haystack data, obtain the p-value for a test of NH: $E(C) = 2 \times E(Over)$ against the alternative $E(C) \neq 2 \times E(Over)$.

4.6 Suppose we approximate a haystack with a hemisphere having circumference $C_1 = (C + 2Over)/2$.

    4.6.1 Construct a scatterplot of Vol versus C_1 and describe the *lowess* mean and variance functions.

    4.6.2 Estimate the correlation coefficient $\rho(Vol, C_1)$. Is the correlation coefficient sufficient to describe the joint variation of Vol and C_1?

    4.6.3 Estimate the variance of C_1 directly by using the sample values of C_1 and indirectly by substituting estimates of the variances and covariance on the right side of (4.7). In (4.7) take $x = C$ and $y = 2 \times Over$. Should the two estimates agree?

    4.6.4 If haystacks can be well approximated by hemispheres with circumference C_1 then we should have

$$E(Vol \mid C, Over) \approx \frac{C_1^3}{12\pi^2} \qquad (4.18)$$

Plot Vol versus $V = C_1^3/(12\pi^2)$. Do the data provide clear visual information to contradict the possibility that (Vol, V) follows a bivariate normal distribution? Is the correlation coefficient a good way to characterize the joint variation of these random variables?

4.6.5   Does the scatterplot of *Vol* versus *V* support or contradict (4.18)? *Hint*: Compare (4.18) to a *lowess* estimate of the mean function.

4.6.6   Assuming that (Vol, C_1) follows a bivariate normal distribution, estimate the standard deviation $(\mathrm{Var}(Vol \mid C_1))^{1/2}$. Next, use the slicing mode in a scatterplot of *Vol* versus C_1 to estimate $(\mathrm{Var}(Vol \mid C_1))^{1/2}$ for various values of C_1. Do the estimates obtained while slicing seem to agree with the estimate obtained under the assumption of bivariate normality, remembering that there will be variation from slice to slice? What does this say about the assumption of bivariate normality?

4.7   As illustrated in Figure 4.10, the two population mean functions (4.13) and (4.16) intersect when drawn on the same graph. What is their point of intersection? When will they be perpendicular? Is it possible for them to be the same?

4.8   Using the haystack data, construct a scatterplot of *C* versus *Over* including *lowess* estimates with smoothing parameter 0.6 of the mean and variance functions. Next, using the bivariate normal data generator in file `demo-bn.lsp`, generate several different plots of 120 bivariate normal observations having means, variances, and correlation the same as the sample values from the plot of *C* versus *Over*. Based on visual comparison of the normal plots with the haystack data, does it seem reasonable that the joint distribution of $(C, Over)$ is close to bivariate normal?

CHAPTER 5

Two-Dimensional Plots

Not all 2D scatterplots are equivalent for the purposes of gaining an understanding of the conditional distribution of $y \mid x$ or of the mean or variance functions. Two scatterplots with the same statistical information can appear different because our ability to process and recognize patterns depends on how the data are displayed. We encountered an example of this in Section 4.2 when considering how to visualize strength of linear dependence in scatterplots. At times the default display produced by a computer package may not be the most useful.

In this chapter we discuss various graphical problems that may be encountered when using a 2D plot to study a regression with a single predictor.

5.1 ASPECT RATIO AND FOCUSING

An important parameter of a scatterplot that can greatly influence our ability to recognize patterns is the *aspect ratio*, the physical length of the vertical axis divided by that of the horizontal axis. Most computer packages produce plots with an aspect ratio near one, but this is not always the best. The ability to change the aspect ratio interactively can be important.

As an example, consider Figure 5.1, which is a plot of the monthly U.S. births per thousand population for the years 1940 to 1948. The horizontal axis is labeled according to the year. The plot indicates that the U.S. birth rate was increasing between 1940 and 1943, decreasing between 1943 and 1946, rapidly increasing during 1946, and then decreasing again during 1947–1948. These trends seem to deliver an interesting history lesson since the U.S. involvement in World War II started in 1942 and troops began returning home during the first part of 1945, about nine months before the rapid increase in the birth rate. A copy of Figure 5.1 can be obtained from *Arc* by loading the file birthrt1.lsp and then using the item "Plot of" in the Graph&Fit menu to plot *Birth Rate* on the vertical axis versus *Year* on the horizontal axis. This plot has plot controls removed by selecting the item "Hide plot

81

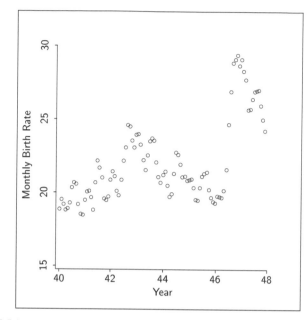

FIGURE 5.1   Monthly U.S. birth rate per 1000 population for the years 1940–1948.

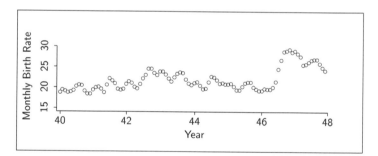

FIGURE 5.2   Monthly U.S. birth rate with a small aspect ratio.

controls" from the plot's menu. Selecting this item again will restore the plot controls.

Let's now see what happens to Figure 5.1 when the aspect ratio is changed. Hold down the mouse button in the lower right corner and drag up and to the right. One reshaped plot is shown in Figure 5.2 which has an aspect ratio of about 1:4. The visual impact of the plot in Figure 5.2 is quite different from that in Figure 5.1. The global trends apparent in Figure 5.1 no longer dominate our visual impression, as Figure 5.2 reveals many peaks and valleys. Is it possible that there are relatively minor within year trends in addition to the global trends described in connection with Figure 5.1?

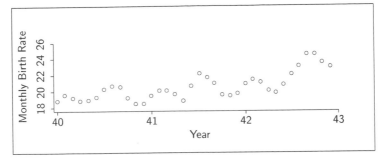

FIGURE 5.3   Monthly U.S. birth rate for 1940–1943.

To answer this question we can *focus* on part of the data. Select the points corresponding to the years 1940–1943. Now from the plot's menu select the item "Focus on selection." This will remove the points in the plot that are not currently selected. The menu item "Remove selection" would, as the name implies, remove the selected points, and leave the rest. Return to the plot's menu and select the item "Rescale plot," which will recompute the values on the axes so the remaining data fill the plotting area. The result is shown in Figure 5.3. A within-year cycle is clearly apparent, with the lowest within-year birth rate at the beginning of summer and the highest occurring some time in the fall. This pattern can be enhanced with plot controls showing by pushing the "Join points" button. This will draw lines between adjacent points. This gives the eye a path to follow when traversing the plot and can visually enhance the peaks and valleys.

The aspect ratio for the plot in Figure 5.3 is again about 1 : 4. To obtain this degree of resolution in a plot of all the data would require an aspect ratio around 1 : 8. To return the plot of Figure 5.3 to its original state, choose the "Show all" item from the plot's menu and reshape it so that the aspect ratio is again about 1 : 1.

Changing the aspect ratio and focusing are useful methods for changing the visual impact of a plot, but they will not always work. Figure 5.4 contains a plot of body weight *BodyWt* in kilograms and brain weight *BrainWt* in grams for sixty-two species of mammal. This plot can be obtained by loading the file brains.lsp, and then drawing the plot of *BrainWt* versus *BodyWt*. The plot consists of three separated points and a large cluster of points at the lower left of the plot, and because of the uneven distribution of points it contains little useful visual information about the conditional distribution of brain weight given body weight. Removing the three separated points and rescaling the plot helps a bit, but a large cluster near the origin remains. Repeating the procedure does not seem to help. The problem in this example is that the measurements range over several orders of magnitude. Body weight ranges from 0.01 kg to 6654 kg, for example. Transformations are needed to bring the data into a usable form.

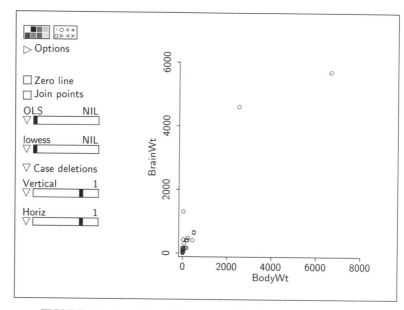

FIGURE 5.4   Plot of *BrainWt* versus *BodyWt* for 62 mammal species.

5.2   POWER TRANSFORMATIONS

The most common transformations are *power transformations*. A power transformation of a variable v is simply v^λ, where λ is the *transformation parameter*. For example, the notation $(BodyWt)^{0.5}$ refers to a variable with values equal to the square root of *BodyWt*. We call this a *basic power transformation*. The transformation parameter can take any value, but the most useful values are frequently found in the interval $-1 \le \lambda \le 2$. The variable v must be positive, or we will end up with complex numbers when λ is not an integer.

Another type of power transformation is called a *scaled power transformation*, for which we use the notation $v^{(\lambda)}$. For a given λ, we define the transformation to be

$$v^{(\lambda)} = \begin{cases} (v^\lambda - 1)/\lambda & \text{if} \quad \lambda \ne 0 \\ \log(v) & \text{if} \quad \lambda = 0 \end{cases} \tag{5.1}$$

When $\lambda \ne 0$, the scaled power transformation differs from the basic power transformation only by subtracting one and dividing by λ. These changes have no important effect on the analysis, so when $\lambda \ne 0$ the scaled power transformations and the basic power transformations are practically equivalent. When $\lambda = 0$, the basic power transformation has the value $v^0 = 1$; but for the scaled power transformation $v^{(0)} = \log(v)$, using natural logarithms. This form therefore adds logarithms to the family of power transformations. Also, unlike

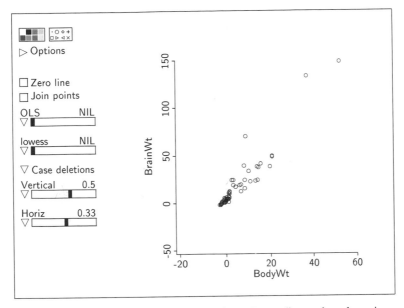

FIGURE 5.5 Both *BodyWt* and *BrainWt* are transformed according to the values given on the transformation slidebars.

v^λ, $v^{(\lambda)}$ is a continuous function of λ, so varying λ will not produce jumps when using the transformation sliders described next.

Return to the brain weight data displayed in Figure 5.4. Select the item "Options" from the plot controls, and in the resulting dialog check the item "Transform slidebars" and then push the "OK" button. This will add two slidebars to the plot, as shown in Figure 5.4. The slidebars on the plot are used to choose a scaled power transformation. They are labeled according to the axis and the current value of the transformation parameter λ. Initially $\lambda = 1$, which means that no transformation has been applied. Each time you move the slider the value of λ is changed and the plot is updated to display $v^{(\lambda)}$ in place of v, unless the new value of λ equals 1. For this case, the original data are displayed. The labels on the axes are not changed after a transformation; to see the power you must look at the numbers above the slidebars.

Here and elsewhere in *Arc*, λ can take the values ± 1, ± 0.67, ± 0.5, ± 0.33, 0, 1.25, 1.5, 1.75, and 2. These values of λ should be sufficient for most applications, but additional values of λ can be added to a slidebar by selecting "Add power choice to slidebar" from the slidebar's pop-up menu.

One possible transformed plot is shown in Figure 5.5, where the scaled power transformation with $\lambda = 0.33$ has been applied to *BodyWt*, while the scaled power transformation with $\lambda = 0.5$ has been applied to *BrainWt*. Figure 5.5 is thus a plot of *BrainWt*$^{(0.5)}$ versus *BodyWt*$^{(0.33)}$. This figure doesn't give a very good representation of the data, although it does seem a little better than the plot of the untransformed data. Manipulate the sliders a bit

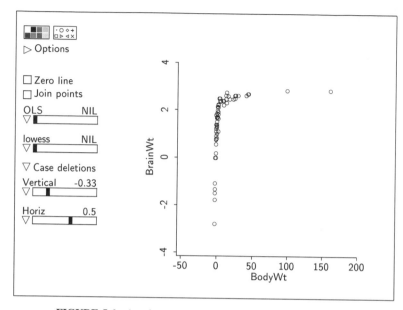

FIGURE 5.6   Another transformation of the body weight data.

and see if you can find a better pair of transformations for the brain weight data.

5.3   THINKING ABOUT POWER TRANSFORMATIONS

There are two simple rules that can make manipulating the power choice sliders easier:

- To spread the *small* values of a variable, make the power λ *smaller*.
- To spread the *large* values of a variable, make the power λ *larger*.

Most of the values of $BrainWt^{(0.5)}$ in Figure 5.5 are clustered close to zero, with a few larger values. To improve resolution, we need to spread the smaller values of $BrainWt$, so λ should be smaller. The values of $BodyWt^{(0.33)}$ are also mostly small, so we should spread the small values of body weight as well, again requiring a smaller power.

Decrease the transformation parameter for $BrainWt$ in Figure 5.5 to $\lambda = 0.33$. The resulting plot is an improvement, but we still need to spread the small values of both $BodyWt$ and $BrainWt$, and this indicates that we need to make both transformation parameters smaller still.

Set the transformation parameter for $BodyWt$ to 0.5 and for $BrainWt$ to -0.33, as shown in Figure 5.6. We now need to spread the larger values of

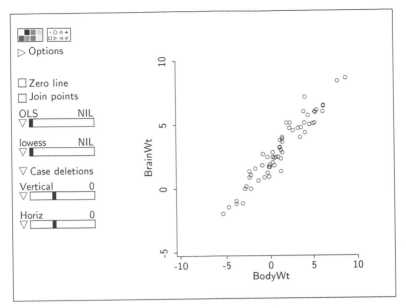

FIGURE 5.7   Log transformed brain weight data.

BrainWt, and thus we should make λ larger. At the same time we need to spread the small values of *BodyWt*, and this is accomplished by decreasing λ.

Figure 5.7 shows a plot of the brain data after applying the log transformation to both variables, $BodyWt^{(0)} = \log(BodyWt)$ and $BrainWt^{(0)} = \log(BrainWt)$. This plot gives a good depiction of the data and strongly suggests that there is a linear relationship in the log-log scale; that is,

$$\text{E}(\log(BrainWt) \mid \log(BodyWt))$$

is a linear function of $\log(BodyWt)$. In addition, the elliptical shape of the point cloud supports the possibility that $(\log(BrainWt), \log(BodyWt))$ is a bivariate normal random variable. The sample correlation coefficient 0.96 therefore provides a useful characterization of the dependence between the transformed variables. In contrast, the sample correlation between the untransformed variables falls considerably short of providing a useful numerical characterization of the dependence shown in the scatterplot of Figure 5.4.

5.4   LOG TRANSFORMATIONS

Transformation of variables to log scale will frequently make regression problems easier. Here is a general rule: Positive predictors that have the ratio between their largest and smallest values equal to 10 and preferably 100 or more should very likely be transformed to logarithms. This rule is satisfied for both

BodyWt, with range 0.005 kg to 6654 kg, and for *BrainWt*, with range 0.14 g to 5712 g, so log transformations are clearly indicated as a starting point.

5.5   SHOWING LABELS AND COORDINATES

Every case in a data file has a *case name*. If your data file has one or more text variables, then the first text variable read is used as case names; if no text variables are included, then the case names are the numbers from 0 to $n - 1$. You can change the case names to another variable using the item "Set case names" from the data set menu.

Selecting the "Show labels" item from the plot's menu will cause the case names to be displayed as point labels for any points selected in the plot. When the "Show labels" option is on, a check will appear beside the "Show labels" menu item. To turn the option off, select "Show labels" again.

To find the coordinates of a point, select the item "Mouse mode" from the plot's menu. A dialog box will appear with four choices. Choose "Show coordinates" and then push "OK." The cursor will change from an arrow to a hand with a pointing finger. Pointing at any data point and clicking the mouse button will show the point label and the coordinates as long as the mouse button is depressed; holding the shift key while depressing the mouse button will make the coordinates remain on the plot. To remove the coordinates, shift-click again on the data point. To return to the usual selecting mode, choose "Selecting" from the "Mouse mode" dialog box.

5.6   LINKING PLOTS

Linking refers to connecting two or more plots, so that actions such as focusing, selecting, and removing points in one plot are automatically applied in the others. The applicability of linking plots in simple regression is somewhat limited, but another set of birth rate data provides a convenient opportunity to introduce the idea.

Load the data file `birthrt2.lsp`. We begin by constructing a plot of *BirthRate* versus *Year* for the U.S. birth data between 1956 and 1975. The plot is shown in Figure 5.8a, which is constructed by plotting each data point against the year in which it was obtained, ignoring the month. This plot provides information on the distribution of monthly birth rates given the year but not the month. There is still one point for each month in the plot but the correspondence between points and months is lost. Additional information might be obtained from Figure 5.8a, if we could tell which points correspond to each month. There are a variety of ways to do this. One is to construct a linked plot of *Month* versus *Year* as shown in Figure 5.8b. Most plots in *Arc* for a single data set are linked automatically; to unlink a plot, select the "Unlink view" item from the plot's menu. This is an uninteresting plot by

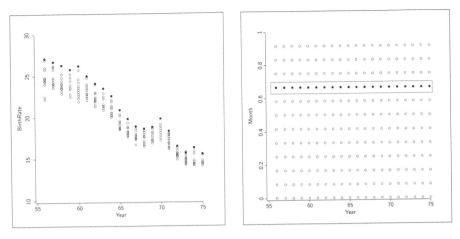

a. Birth Rate versus Year   b. Month versus Year

FIGURE 5.8   U.S. birth rate data.

itself, but it is useful because it is linked to the plot in Figure 5.8a. Select-ing all the points for any month in Figure 5.8b will cause the corresponding points in Figure 5.8a to be highlighted, as demonstrated for the month of September.

5.7 POINT SYMBOLS AND COLORS

The point symbols on a scatterplot can be changed using the *symbol palette*, and, if your computer has a color monitor, the color can be changed using the *color palette*. First select the point or points you want to change, and then click the mouse button on a color or symbol in the palettes at the top of the plot controls. Colors and symbols are *inherited* by all plots linked to the plot you changed. Later in the book, we will learn about methods that set colors and symbols of points automatically using *point marking*.

5.8 BRUSHING

An alternative method of selecting points is called *brushing*. In this method, the mouse pointer is changed into a selection rectangle. As the rectangle is moved across the screen, all points in the selection rectangle are highlighted, as are the corresponding points in all plots linked to it.

To use brushing, select "Mouse mode" from the plot's menu, and select "Brushing" in the resulting dialog box. After pushing the "OK" button, the mouse pointer is changed into a paintbrush, with an attached selection rect-angle. The size and shape of the rectangle can be changed by selecting the

"Resize brush" item from the plot's menu and following the instructions given to resize the brush. Usually, long, narrow brushes are the most useful.

In the plot shown in Figure 5.8b, change the mouse mode to brushing, and make the brush a narrow horizontal rectangle. Then, brush across the plot from bottom to top. As you do so, examine the plot shown in Figure 5.8a. The within-year trends are easy to spot as you move the brush. In particular, it appears that September always had the highest birth rate.

5.9   NAME LISTS

A *name list* is used to keep track of case labels. It is displayed as a separate window by selecting the "Display case names" item from the data set menu. Since this window is linked with all graphics windows, point selection, focusing, and coloring will be visible in this window.

5.10   PROBABILITY PLOTS

The primary use of the 2D plots discussed so far in this book has been to study the conditional distribution of the variable on the vertical axis given a fixed value of the variable on the horizontal axis—in short, regression. Two-dimensional plots can have other uses as well, where conditional distributions are not the primary focus. One plot of this type is the *probability plot*.

A probability plot is used to study the distribution of a random variable. Suppose we have n observations $y_1, \ldots, y_n$, and we want to examine the hypothesis that these are a sample from a specific distribution, such as a normal distribution. Let $y_{[1]} \leq \cdots \leq y_{[n]}$ be the y_i's reordered from smallest to largest. Let $q_{[1]} \leq \cdots \leq q_{[n]}$ be the expected values of an ordered sample of size n from the hypothesized distribution. That is, $q_{[i]} = E(z_{[i]})$ where $z_1, \ldots, z_n$ is a sample from the hypothesized distribution. For a given standard hypothesized distribution, the q's can be easily computed, or at least closely approximated, so we can assume that the q's are known.

If the true distribution of the y's is in fact the same as the hypothesized distribution, we would expect the ordered y's to be linearly related to the q's. A probability plot is a plot of $y_{[i]}$ versus $q_{[i]}$. If the sampling distribution of the y's is the same as the hypothesized distribution, then the plot should be approximately straight; if the plot is significantly curved, then we have evidence that the y's are not from the hypothesized distribution.

Judging if a probability plot is straight requires practice. To help gain the necessary experience, load the demonstration file demo-prb.lsp, and select the item "Probability plots" from the "ProbPlots" menu. You will get the dialog shown in Figure 5.9. You can specify the sample size, the true sampling distribution, and the hypothesized distribution.

For the first try, set both distributions to be normal, and leave the sample size at 50. A probability plot similar to Figure 5.10 will result. The points in this

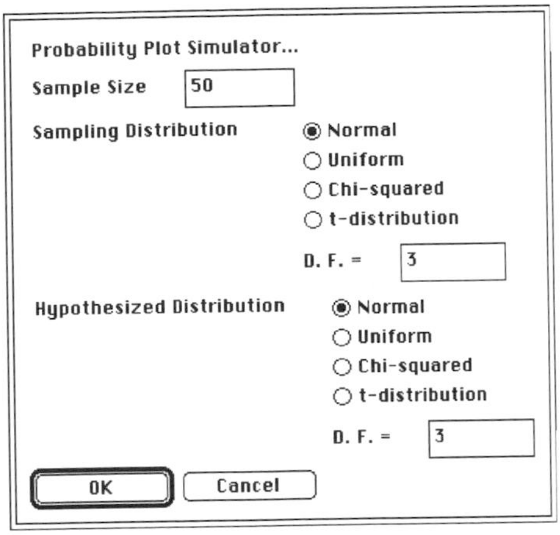

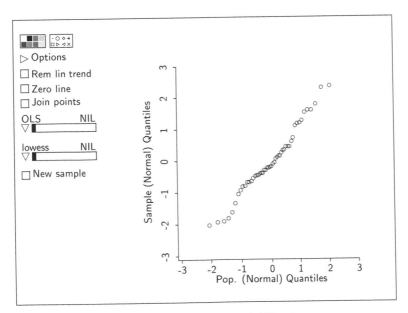

FIGURE 5.9 Dialog for the probability plot demonstration.

FIGURE 5.10 A sample probability plot.

plot should lie close to a straight line because the sampling and hypothesized distributions are the same. A new sample from the sampling distribution can be obtained by using the "New sample" button on the plot; this can be repeated many times.

After looking at several plots with sample size 50, start the demonstration over by again selecting "Probability plots" from the menu, but this time set the sample size to 10. Can you judge normality as easily for a sample of size 10 as you could for a sample of size 50?

Non-null shapes of probability plots can be studied by setting the sampling distribution and the hypothesized distribution to be different in the dialog; this will require selecting the "Probability plots" item again. Relative to the normal, the uniform distribution has short tails, that is, few extreme values. The t-distributions, for which you must specify the degrees of freedom, have long tails for small df. The χ^2 distributions are skewed to the right; the smaller the degrees of freedom, the greater the skewness.

Probability plots are obtained in *Arc* by selecting the item "Probability plot of" from the Graph&Fit menu. Choose one or more variables to be plotted, and the target distribution to get the $q_{[i]}$. If the distribution chosen is χ^2 or t, then the number of degrees of freedom must be specified as well. Only one probability plot is shown in the window at a time; use the slidebar on the plot to cycle between the plots.

5.11   COMPLEMENTS

Becker and Cleveland (1987) discuss scatterplot brushing and Stuetzle (1987) discusses plot linking. The birth rate data are taken from Velleman (1982). The brain weight data are from Allison and Cicchetti (1976). The soil temperature data in Problem 5.1 was provided by Orvin Burnside.

The probability plot described here is often called a *QQ-plot*, as it is a plot of the quantiles of the hypothesized distribution against the observed quantiles of the sample. Gnanadesikan (1977) is a standard reference for these plots.

PROBLEMS

5.1   Load the file `aspect.lsp`. This file will produce one plot. Can you see any pattern in this plot? Now, using the mouse, change the aspect ratio so the plot becomes increasingly long and narrow. At what point do you see the pattern, and how would you describe that pattern? Are there any other graphical tools that might have helped you find the pattern? This plot provides a demonstration that interaction between the user and the plot may be required for the user to extract information.

The figure you just examined is based on simulated data, but the same phenomenon can occur with real data. Load the file `mitchell.lsp`, which gives average monthly soil temperature for Mitchell, NE over 17 years during a weed longevity experiment. Draw the plot of *Temp* versus *Month*. After attempting to summarize the information in the original plot, change the aspect ratio to get the plot into a more useful form, and summarize the (not very surprising) information that it contains.

5.2 Take a closer look at the data of Figure 5.8 by using the graphical tools discussed in this chapter. Is it really true that the highest birth rate always occurred in September? Which month tended to have the lowest birth rates?

5.3 Redraw Figure 5.7. Which species has the highest brain weight and the highest body weight? Which species has the lowest brain weight and body weight? Find the point for humans on the plot and give the coordinates for the point.

5.4 The discussion of the brain weight data suggests that $\log(BrainWt)$ and $\log(BodyWt)$ appear to be approximately bivariate normally distributed. If they are approximately bivariate normal, then each of their marginal distributions must be approximately univariate normal. Use probability plots to explore normality of each of these marginal distributions.

 While it is true that (x, y) bivariate normal implies that both x and y have univariate normal distributions, the converse is not true: x and y both univariate normal does not guarantee that (x, y) is bivariate normal.

PART II

Tools

This part of the book presents tools for studying regression. Many of the tools make use of the linear regression model, as linear models continue to have a central role in the analysis of regression. In the next few chapters, we present many of the basic ideas behind this approach.

CHAPTER 6

Simple Linear Regression

In Part I of this book we introduced regression, which is the study of how the conditional distribution of a response variable y changes as a predictor x is varied. We have seen in Chapter 3 how the conditional distribution of y given x can be summarized graphically using boxplots and scatterplots, and how the mean function can be estimated by using a smoother. This method of summarization requires only minimal assumptions and is primarily graphical. At another extreme, we have seen in Chapter 4 that if the pair (x, y) follows a bivariate normal distribution, then the conditional distribution of $y \mid x$ is completely specified as another normal distribution, with mean function $E(y \mid x)$ given by (4.11) and with constant variance function $\text{Var}(y \mid x)$ given by (4.17). Consequently, when (x, y) has a bivariate normal distribution, we end up with a simple and elegant summary of the statistical dependence of y on x. In this case, the use of a smooth to represent the mean function remains valid, but the resulting summary of a regression by a plot with a smooth is more complicated than is needed.

The assumption of bivariate normal data is an example of a *statistical model*. In this chapter, we will study the use of models for regressions with one predictor that do not require bivariate normal data.

Models consist of assumptions about how the data behave, and one or more equations that describe their behavior. Models can be based on theoretical considerations like physical laws or economic theories, or they can be empirical, found while examining a scatterplot. For example, in the discussion of the haystack data in Section 1.1, we were led to Equation (3.1), which relates haystack volume *Vol* to the circumference C of the haystack:

$$E(Vol \mid C) = \frac{C^3}{12\pi^2}$$

This is a model for the mean value of *Vol* for a given value of C. It is an example of a model based on theoretical considerations since the actual data played no role in its formulation. It is somewhat unusual for a statistical model because it has no free *parameters*, unknown quantities that must be estimated to tailor a model to fit the data at hand. In this chapter we begin by considering an important class of models called *simple linear regression* models.

6.1   SIMPLE LINEAR REGRESSION

In the 1840s and 1850s a Scottish physicist, James D. Forbes, wanted to estimate altitude above sea level from measurement of the boiling point of water. He knew that altitude could be determined accurately with a barometer, with lower atmospheric pressures corresponding to higher altitudes. More precisely, he stated in a paper published in 1857 that the logarithm of barometric pressure is proportional to altitude; that is,

$$\log(\text{barometric pressure}) = c \times \text{altitude}$$

for some constant *c*. Forbes didn't give this constant at the time, but instead referred the reader to some of his earlier work. In any event, he considered the logarithm of barometric pressure as an effective substitute for altitude. With this substitution, Forbes modified his goal to that of estimating the logarithm of barometric pressure from the boiling point of water. His interest in using boiling point to measure altitude via the logarithm of barometric pressure was in part motivated by the expense and difficulty of transporting the barometers of the time, as well as by the relative ease of boiling water and measuring temperature.

The data file `forbes.lsp` contains measurements that Forbes took at 17 locations in Scotland and in the Alps, from his 1857 paper. Forbes drew a scatterplot equivalent to the one shown in Figure 6.1 of $\log_{10}(Pressure)$ versus temperature in degrees Fahrenheit (*Temp*). Forbes' paper is one of the earliest published uses of a scatterplot to study a regression, and indeed it predates the coining of the term regression. Following Forbes, we will also use base 10 logarithms.

Examine Figure 6.1. As *Temp* increases, so does $\log_{10}(Pressure)$. A smooth curve fit by eye for the mean function looks like it might be a straight line. Consequently, we proceed by assuming that the mean function is a linear function of *Temp*, understanding that we will need to check this assumption eventually. A straight line is determined by two *parameters*, an *intercept* and a *slope*. When a straight line is appropriate, the essential information about the mean function is contained in estimates of these two parameters.

We now have the first ingredients for building a more comprehensive model. Here is the general idea. We want to study $E(y \mid x)$ using data that consist of *n* pairs of observations on (x, y). In Forbes' data, $x = Temp$ and $y = \log_{10}(Pressure)$. We then write the identity

$$y \mid x = E(y \mid x) + [y \mid x - E(y \mid x)]$$

$$= E(y \mid x) + e \mid x \tag{6.1}$$

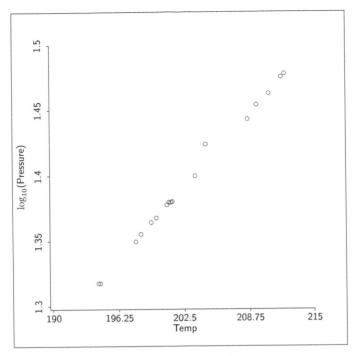

FIGURE 6.1   Forbes' data.

so each observation is viewed as a sum of two terms. The first term is $E(y \mid x)$, which is the same for all observations taken at the same value of x. The function $E(y \mid x)$ is unknown, and estimating it from data is often of primary interest. The quantity $e \mid x$, which is the difference between the response $y \mid x$ and the mean function $E(y \mid x)$ at the same value of x, is usually called an *error*. Don't confuse an *error* with a *mistake*; statistical errors are devices that account for variation about a central value, usually the mean function. The errors are random variables with mean $E(e \mid x) = 0$ and variance $\text{Var}(e \mid x) = \text{Var}(y \mid x)$; these results can be verified by rearranging terms and using (1.6) and (1.7). Relationship (6.1) is theoretical because neither $E(y \mid x)$ nor $e \mid x$ are observable directly; we only get to see x and y. The derivation (6.1) does not require any assumptions; all that we have done so far is to define notation.

To make further progress, we need to say something about $E(y \mid x)$. For Forbes' data, Figure 6.1 suggests that $E(y \mid x)$ may be replaced by a straight line, so we write

$$E(y \mid x) = \eta_0 + \eta_1 x \qquad\qquad (6.2)$$

Equivalently, using (6.1), we may write

$$y \mid x = \eta_0 + \eta_1 x + e \mid x \qquad\qquad (6.3)$$

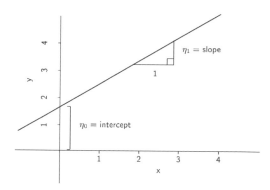

FIGURE 6.2   The equation of a straight line.

In equations (6.2) and (6.3), η_0 is the *intercept*, the value of $E(y \mid x)$ when $x = 0$. The symbol η is a Greek letter, read *eta*; it is roughly equivalent to the Roman letter h. The intercept η_0 and the *slope* η_1 are called *regression coefficients*. The slope can be interpreted as the change in $E(y \mid x)$ per unit change in x: the difference between population means $E(y \mid x = x_0 + 1)$ and $E(y \mid x = x_0)$ is

$$E(y \mid x = x_0 + 1) - E(y \mid x = x_0) = \eta_0 + \eta_1(x_0 + 1) - [\eta_0 + \eta_1 x_0] = \eta_1$$

and so when the predictor increases by one unit the expected response increases by η_1 units, as illustrated in Figure 6.2. This change will be the same for any initial value x_0. As the intercept and slope, η_0 and η_1, range over all possible values, they give all possible straight lines. The statistical estimation problem is to pick the best values for the slope and intercept using the data.

Using (6.2) for the mean function is our first real assumption, and like any assumption it represents an approximation that may or may not be adequate. Our specification for the mean function comes from examining the scatterplot of Figure 6.1, not from any theory. Forbes apparently did have a theoretical reason for choosing this mean function; Forbes' theory has been improved considerably over the last 140 years, as will be explained later in Section 6.5.

At this point, our regression model (6.3) can be visualized using Figure 4.3, page 60, as a prototype. The line in that plot is a representation of the mean function (6.2). The four different distributions about the mean function represent distributions of the error $e \mid x$ at four different values of x. As shown in the figure, those distributions have different variances, so the distribution of $e \mid x$ depends on the value of x. For example, the variation about the mean function is much smaller at the smallest value of x represented in the figure than it is at the largest value of x.

It is often assumed that the distribution of $e \mid x$ does *not* depend on x at all, so that the error is independent of x. Under this assumption, we can write e in place of $e \mid x$ without confusion. The condition that $e \mid x$ is independent of

x implies that the variance function is constant:

$$\text{Var}(y \mid x) = \sigma^2 \tag{6.4}$$

As is conventional, we use the symbol σ^2, where σ is the Greek letter *sigma*, to denote the common value of the variance. Equation (6.4) says that for each value of x, the variance function $\text{Var}(y \mid x)$ has the same value, σ^2.

Assuming that the mean function is linear and that $e \mid x$ is independent of x, the simple linear regression model can be written as

$$y \mid x = \eta_0 + \eta_1 x + e \tag{6.5}$$

with $\text{Var}(e) = \sigma^2$. Figure 4.11, page 75, provides a schematic representation of the simple linear regression model. The main difference between Figures 4.11 and 4.3 has to do with the error structure. The error distributions are all the same in Figure 4.11 because $e \mid x$ is independent of x. The error distributions differ in Figure 4.3, reflecting that the distribution of $e \mid x$ depends on the value of x.

The simple linear regression model (6.5) is appropriate when (x, y) follows a bivariate normal distribution. In that case, the parameters of (6.5)—η_0, η_1, and σ—can be interpreted in terms of the parameters of the underlying bivariate normal distribution, as in Sections 4.3.3 and 4.3.4. But bivariate normality is not a necessary condition for using (6.5), and the simple linear regression model may hold when (x, y) is not bivariate normal. For example, consider the experiment mentioned briefly at the beginning of Chapter 4 to investigate the effect on grain yield (Y) of adding 5, 10, 15, or 20 pounds of nitrogen (N) per acre. The simple linear regression model (6.5) could be appropriate for $Y \mid N$, but (Y, N) cannot be bivariate normal because N is selected by the experimenter and therefore is not random.

We are now ready to think about estimating η_0, η_1, and σ^2, which are the three parameters that characterize the simple linear regression model (6.5).

6.2 LEAST SQUARES ESTIMATION

Suppose that the simple linear regression mean function (6.2) is in fact appropriate for our data. We need to estimate the unknown parameters η_0 and η_1. The general idea is to choose estimates to make the estimated mean function match the data as closely as possible. The most common method of doing this uses *least squares* estimation. Later in the book we will use other estimation methods that take advantage of additional information about a regression.

6.2.1 Notation

Recall that our data consist of n independent observations on (x, y). We call a pair of observations on (x, y) a *case*. When necessary, we use the subscript i

to label the cases in the data, so (x_i, y_i) are the values of the predictor and the response for case i, $i = 1, \ldots, n$. In terms of our full notation, x_i is the ith value of x, and y_i means the same thing as $y \mid (x = x_i)$. When we leave subscripts off, y and x are not restricted to the data, but they are restricted to the data when the subscripts are present.

To maintain a clear distinction between parameters and their estimates, we will continue the practice of using Greek letters like η_0, η_1, and σ to represent unobservable parameters. Estimates will be indicated with a *hat*, so $\hat{\eta}_1$, which is read *eta-one hat*, denotes an estimate of η_1. Upon occasion, we will need a value to replace an unknown parameter in an equation, and for this we will use a Roman letter like h to replace a Greek letter like η.

With simple linear regression, the mean function is $\mathrm{E}(y \mid x) = \eta_0 + \eta_1 x$. Given estimates $\hat{\eta}_0$ and $\hat{\eta}_1$, the *estimated mean function* is $\hat{\mathrm{E}}(y \mid x) = \hat{\eta}_0 + \hat{\eta}_1 x$. This equation for the estimated mean function can be evaluated at any value of x. When it is evaluated at the observed values x_i of x, we get the *fitted values*,

$$\hat{y}_i = \hat{\mathrm{E}}(y \mid x = x_i) = \hat{\eta}_0 + \hat{\eta}_1 x_i \tag{6.6}$$

The *residuals*, $\hat{e}_i$, are defined as the differences between the observed values of the response and the fitted values,

$$\hat{e}_i = y_i - \hat{y}_i \tag{6.7}$$

There is a parallel between equation (6.1) that defines the errors and the definition of the fitted values and residuals. We can write the two equations:

$$y \mid x = \mathrm{E}(y \mid x) + e \tag{6.8}$$

$$y_i = \hat{y}_i + \hat{e}_i \tag{6.9}$$

Equation (6.8), which is the same as (6.1), divides the *observable* random variable $y \mid x$ into two unknown parts, a mean and an error. Equation (6.9) divides the *observed* values of y_i into two parts, but now the parts are numbers that can be computed with estimates of the parameters.

6.2.2 The Least Squares Criterion

Figure 6.3 shows two candidates for an estimate of the simple regression mean function for a hypothetical regression. The fit in Figure 6.3a is considerably better than the fit in Figure 6.3b because the line in Figure 6.3a more closely matches the data. The least squares line corresponds to choosing values for the intercept and slope that minimize a particular function of the data, called an *objective function*.

Suppose now that (h_0, h_1) is a pair of candidate values for the parameters (η_0, η_1). The least squares summary of the fit of the candidate straight-line

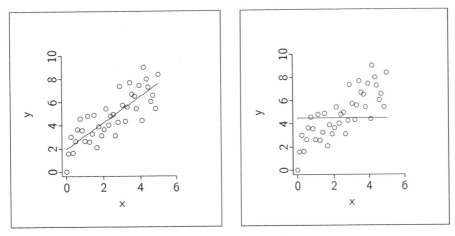

a. Good fit.

b. Poor fit.

FIGURE 6.3   Two candidate fitted lines.

mean function $h_0 + h_1 x$ to the data is

$$RSS(h_0, h_1) = \sum_{i=1}^{n} (y_i - [h_0 + h_1 x_i])^2 \tag{6.10}$$

The symbol $RSS(h_0, h_1)$ stands for "the *residual sum of squares function* evaluated at h_0 and h_1." Better choices for (h_0, h_1) will have smaller values of $RSS(h_0, h_1)$. The values $(\hat{\eta}_0, \hat{\eta}_1)$ that *minimize* $RSS(h_0, h_1)$ will be called the *ordinary least squares estimates*, or the OLS estimates of the parameters η_0 and η_1. Thinking in terms of a scatterplot with y on the vertical axis and x on the horizontal axis, the least squares criterion chooses estimates to minimize the sum of the squared vertical distances between the observed values y_i, and corresponding point $h_0 + h_1 x_i$ on the estimated line, as represented in Figure 6.4.

Finding the OLS estimates is equivalent to solving the calculus problem of minimizing the residual sum of squares function $RSS(h_0, h_1)$. For the simple linear regression model, one pair $(\hat{\eta}_0, \hat{\eta}_1)$ minimizes RSS, and there is a simple formula for the estimates. Other mean functions may not have simple formulas for the estimates, and a solution to the calculus problem can be achieved only numerically, usually on a computer. Even for simple regression, where the estimates have simple formulas, using a computer is nearly universal. Least squares is a purely mathematical formulation that does not depend on any assumptions concerning the errors, or on whether or not the simple linear regression mean function is reasonable.

The least squares estimates for simple linear regression depend on a few summary statistics defined in Table 6.1. The summaries include the sample averages $\bar{x}$ and $\bar{y}$, and sample variances $sd^2(x)$ and $sd^2(y)$, which are just the

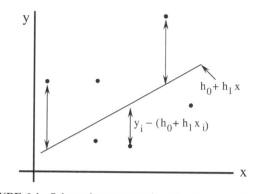

FIGURE 6.4   Schematic representation of ordinary least squares.

TABLE 6.1   Definition of Symbols[a]

Quantity	Definition	Description
$\bar{x}$	$\sum x_i/n$	Sample average of x
$\bar{y}$	$\sum y_i/n$	Sample average of y
SXX	$\sum(x_i - \bar{x})^2 = \sum(x_i - \bar{x})x_i$	Sum of squares for the x's
$sd^2(x)$	$SXX/(n-1)$	Sample variance of the x's
$sd(x)$	$\sqrt{SXX/(n-1)}$	Sample standard deviation of the x's
SYY	$\sum(y_i - \bar{y})^2 = \sum(y_i - \bar{y})y_i$	Sum of squares for the y's
$sd^2(y)$	$SYY/(n-1)$	Sample variance of the y's
$sd(y)$	$\sqrt{SYY/(n-1)}$	Sample standard deviation of the y's
SXY	$\sum(x_i - \bar{x})(y_i - \bar{y}) = \sum(x_i - \bar{x})y_i$	Sum of cross products
$r(x,y)$	$SXY/\{(n-1)sd(x)sd(y)\}$	Sample correlation

[a]In each equation, the symbol $\sum$ means to add over all the n values or pairs of values in the data.

squares of the sample standard deviations, and the sample correlation, $r(x,y)$. The hat rule described earlier would suggest that different symbols should be used for these quantities. For example, $\hat{\rho}(x,y)$ might be more appropriate notation for the sample correlation if the population correlation is $\rho(x,y)$. This apparent inconsistency is deliberate, because the hat notation makes sense only if $\rho(x,y)$ is meaningful in the first place. For example, in Forbes' experiments, data were collected at 17 selected locations. As a result, the sample variance of boiling points, $sd^2(x) = 33.17$, does not estimate any meaningful population variance because no clear population has been defined and sampled. Similarly, the value of $r(x,y)$ can be made larger by choosing experimental units whose values of x are as different as possible, or smaller by choosing experimental units whose values of x are as alike as possible. The value of the sample correlation may reflect the method of sampling as much as it does the population value $\rho(x,y)$, should such a population value make sense. In short, a sample correlation can always be computed, but it may not be a sensible estimate of a population correlation.

6.2.3 Ordinary Least Squares Estimators

The formulas for the OLS estimators, which are derived in Section 6.7.1, are

$$\hat{\eta}_1 = \frac{SXY}{SXX} = r(x,y)\frac{sd(y)}{sd(x)} = \sum_{i=1}^{n}\left(\frac{x_i - \bar{x}}{SXX}\right)y_i \qquad (6.11)$$

$$\hat{\eta}_0 = \bar{y} - \hat{\eta}_1\bar{x} \qquad (6.12)$$

According to (6.11), the estimated slope is the ratio SXY/SXX. The second form in (6.11) shows the close relationship between the slope and the sample correlation $r(x,y)$: The slope equals the sample correlation multiplied by the ratio of sample standard deviations $sd(y)/sd(x)$. This second form was used in Section 4.3.1 during the earlier discussion of bivariate normal populations. If the data were rescaled so both x and y had sample standard deviations equal to 1, then the estimated slope would equal the sample correlation for the rescaled data (see Problem 6.2).

The third form for the estimated slope in (6.11), which can be derived by substituting $\sum(x_i - \bar{x})y_i$ for SXY, shows that the estimated slope is a *linear combination* of the y_i with coefficients

$$c_i = (x_i - \bar{x})/SXX$$

That is, $\hat{\eta}_1 = \sum c_i y_i$. The coefficients always sum to 0, $\sum c_i = 0$. These facts are useful when working out the properties of these estimates.

To interpret the intercept parameter, we compute the estimated population mean at $x = \bar{x}$,

$$\hat{\eta}_0 + \hat{\eta}_1\bar{x} = \bar{y} - \hat{\eta}_1\bar{x} + \hat{\eta}_1\bar{x} = \bar{y}$$

Thus the estimated intercept ensures that the point $(\bar{x}, \bar{y})$ falls exactly on the estimated regression line, and the line passes through the center of the data.

We next find the smallest possible value of the residual sum of squares function, $RSS(\hat{\eta}_0, \hat{\eta}_1)$, and we will call this number simply RSS, the residual sum of squares. Algebra given in Section 6.7.1 gives the formula

$$RSS = \sum_{i=1}^{n}\hat{e}_i^2 = SYY(1 - r^2(x,y)) \qquad (6.13)$$

The residual sum of squares is the sum of the squared residuals. It can be computed by multiplying SYY by $(1 - r^2)$. The relationship between RSS and

SYY is similar to the relationship between Var($y \mid x$) and Var(y) given in (4.17). Equation (6.13) can be solved for r^2 to give

$$r^2(x,y) = 1 - \frac{RSS}{SYY} = \frac{SYY - RSS}{SYY} \tag{6.14}$$

If we think of *SYY* as the total variability in the sample, and *RSS* as the variability remaining after conditioning on x, then we see from (6.14) that r^2 can be interpreted as the *proportion of variability explained by conditioning on x*. If r^2 has its maximum value of one, then the data must fall exactly on a straight line with no variation; if r^2 has its minimum value of zero, then conditioning on x explains none of the variability in y.

6.2.4 More on Sample Correlation

Every sample correlation coefficient $r(x,y)$ is a one-number summary of the 2D scatterplot of y versus x. This correlation summarizes the plot through the proportion of variability explained by the simple linear regression of y on x. As with any one-number summary of such a complex object, the correlation can be misleading. As described in Section 4.1.5, the correlation coefficient measures only how well the scatterplot is approximated by a straight line. This was illustrated in Figure 4.7 (page 69) and Figure 4.8 (page 70), where plots with very different appearance have the same correlation. Without examining the corresponding scatterplot, the correlation may be of little value.

6.2.5 Some Properties of Least Squares Estimates

If the simple linear regression mean function is appropriate, then the OLS estimates are *unbiased*, meaning that on the average the estimates will equal the true values of the parameters, $E(\hat{\eta}_1) = \eta_1$ and $E(\hat{\eta}_0) = \eta_0$. The proofs of this and the other results of this section and the next are given in Section 6.7.2.

We next turn to the variances of the estimates. Remembering that we have assumed Var($y \mid x$) = σ^2 to be constant, we can get formulas for the variances and covariance of the OLS estimates:

$$\text{Var}(\hat{\eta}_1) = \sigma^2 \frac{1}{SXX} \tag{6.15}$$

$$\text{Var}(\hat{\eta}_0) = \sigma^2 \left(\frac{1}{n} + \frac{\bar{x}^2}{SXX} \right) \tag{6.16}$$

$$\text{Cov}(\hat{\eta}_0, \hat{\eta}_1) = -\sigma^2 \frac{\bar{x}}{SXX} \tag{6.17}$$

The covariance between the estimates is negative if $\bar{x}$ is positive, and it is positive if $\bar{x}$ is negative.

6.2.6 Estimating the Common Variance, σ^2

Under the constant variance assumption and the simple linear mean function (6.2), the expectation of the squared difference $e^2 = (y - \eta_0 - \eta_1 x)^2$ is equal to σ^2. The usual estimate of σ^2 is based on the squared residuals $\hat{e}_i^2 = (y_i - \hat{\eta}_0 - \hat{\eta}_1 x_i)^2$, since these estimate the squared errors. The residual sum of squares then provides a basis for estimating the variance,

$$\hat{\sigma}^2 = \sum \hat{e}_i^2 / (n - 2) = RSS/(n - 2) \tag{6.18}$$

Using $(n - 2)$ as a divisor gives an unbiased estimate, $E(\hat{\sigma}^2) = \sigma^2$. Estimated variances of the OLS estimates are obtained by substituting $\hat{\sigma}^2$ for σ^2 in equations (6.15) and (6.16). Taking square roots, we get the *standard errors* of the estimates,

$$se(\hat{\eta}_1) = \hat{\sigma} \left(\frac{1}{SXX} \right)^{1/2} \tag{6.19}$$

$$se(\hat{\eta}_0) = \hat{\sigma} \left(\frac{1}{n} + \frac{\bar{x}^2}{SXX} \right)^{1/2} \tag{6.20}$$

We use the term *standard error* of an estimate to refer to the square root of the variance of an estimate with all parameters like σ replaced by estimates like $\hat{\sigma}$.

6.2.7 Summary

The simple linear regression model is usually viewed as consisting of both the linear mean function and the constant variance function,

$$E(y \mid x) = \eta_0 + \eta_1 x \quad \text{and} \quad \text{Var}(y \mid x) = \sigma^2 \tag{6.21}$$

The parameters in the mean function can be estimated by using least squares, and the variance can be estimated from the minimized value of $RSS(h_0, h_1)$. Checks on the adequacy of any model are usually necessary, and they will be discussed in later chapters.

6.3 USING *Arc*

Arc can be used to obtain the statistics discussed in the last few sections. Load the file forbes.lsp to create the Forbes data set. First, we compute the logarithm of pressure. Select the item "Transform" from the data set menu, which for the example is called Forbes. This gives the dialog shown in Figure 6.5. The dialog can be used to transform variables via power transformations or

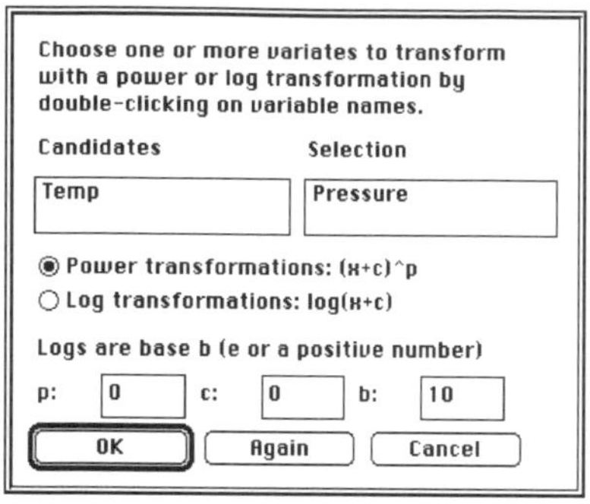

```
Choose one or more variates to transform
with a power or log transformation by
double-clicking on variable names.

Candidates                  Selection

┌─────────────────┐  ┌─────────────────┐
│ Temp            │  │ Pressure        │
└─────────────────┘  └─────────────────┘

● Power transformations: (H+c)^p
○ Log transformations: log(H+c)

Logs are base b (e or a positive number)

p:  [0]     c:  [0]     b:  [10]

┌──────────┐  ┌──────────┐  ┌──────────┐
│    OK    │  │   Again  │  │  Cancel  │
└──────────┘  └──────────┘  └──────────┘
```

FIGURE 6.5   The "Transform" dialog.

TABLE 6.2   Output from "Display summaries" for Forbes' Data

```
Dataset = Forbes, Summary Statistics
Variable          N    Average  Std. Dev  Minimum  Median   Maximum
Temp              17   202.95    5.7597   194.30   201.30   212.20
log10[Pressure]   17   1.3960   0.051715  1.3179   1.3804   1.4780

Data set = Forbes, Sample Correlations
Temp              1.0000   0.9975
log10[Pressure]   0.9975   1.0000
                  Temp     log10[
```

by taking logarithms. More general transformations require using the "Add a variate" item in the data set menu.

Move the variables you want to transform from the "Candidates" list to the "Selection" list. All selected variables will be transformed in the same way, as specified with the buttons and text areas at the bottom of the dialog. If you are transforming a variate z, the transformation will be a *shifted power transformation* $(z + c)^p$ if $p \neq 0$, or it will be $\log(z + c)$ if either $p = 0$ or the "Log transformations" button is selected. The base of the logarithms is chosen in the text area for b; the default is to use natural logarithms with base $e = 2.718...$, but any positive number will work. In this book, we will use natural logarithms, and logs to base 2 and base 10. To be consistent with Forbes, we will use base 10 logarithms in this regression and set $b = 10$.

Next, use the "Display summaries" item from the data set menu to obtain the summary statistics shown in Table 6.2. Among other values, this output gives the sample means and standard deviations. The sample correlation $r(Temp, \log_{10}(Pressure))$ is given in the correlation matrix in Table 6.2. Since $r(Temp, \log_{10}(Pressure)) = r(\log_{10}(Pressure), Temp)$, the same value of 0.9975 is given for both off-diagonal entries. This output does not give the sample covariance, but it can be determined from the output by using its definition (4.6).

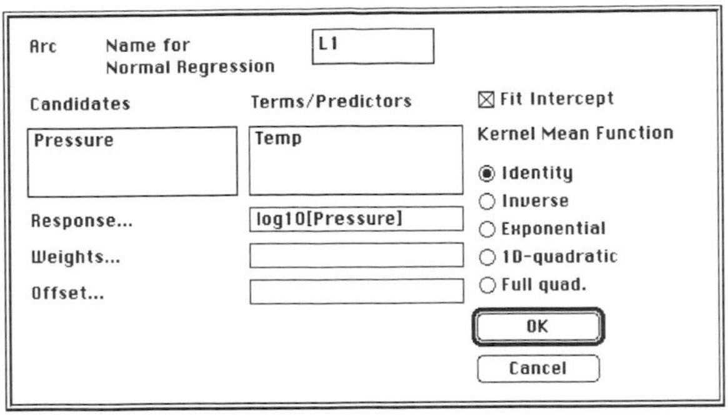

Arc	Name for Normal Regression	L1	
Candidates	Terms/Predictors	☒ Fit Intercept	
Pressure	Temp	Kernel Mean Function	
		⦿ Identity	
		○ Inverse	
Response...	log10[Pressure]	○ Exponential	
Weights...		○ 1D-quadratic	
Offset...		○ Full quad.	
		OK	
		Cancel	

FIGURE 6.6   The OLS regression dialog.

To get *Arc* to compute the OLS estimates, you need to fit a model. Select the item "Fit linear LS" from the "Graph&Fit" menu to get a dialog like the one shown in Figure 6.6. This dialog is used to select the predictors and the response. As shown in the figure, $\log_{10}(Pressure)$ has been specified as the response, and *Temp* as the predictor using the usual mouse-click commands. At the top of the dialog you can specify a name for the model; the default name is L1, indicating that it is the first model of type "L" (Linear) for this data set. The other items in the dialog—setting weights and an offset, selecting an alternative form for the mean function, and toggling the intercept—will be used later and are not needed for this example.

When you click on the "OK" button, the simple linear regression model for $\log_{10}(Pressure)$ on *Temp* will be calculated and summary statistics will be displayed in the text window, as shown in Table 6.3. Besides giving the output, a *model menu* will be created with the name L1 or whatever you typed for the name of the regression model; the items in this model menu will be of interest in later chapters.

In the output, the column marked "Estimate" in Table 6.3 gives the estimated parameter values, -0.421642 for the intercept and 0.00895618 for the slope. Be careful when working with such small numbers: It is very easy to drop one of the leading zeroes. The column marked "Std. Error" gives the standard errors of the estimated intercept and slope. These are both very small relative to the values of the estimates, as can be seen from the numbers in the column labeled "t-value," which are the estimates divided by their standard errors. The standard error for $\hat{\eta}_1$ is given as 0.000164575. Many computer programs would write this in scientific notation as 1.64575E-4. *Arc* generally avoids scientific notation except for very small or very large numbers.

The next number in the output is labeled "R Squared," which is 0.994961. For simple linear regression, this is just $r^2(x,y)$, and so almost all (99.5%) of the variability in $\log_{10}(Pressure)$ is explained by the value of *Temp*. "Sigma hat" is $\hat{\sigma}$, the square root of the estimated variance assuming that the variance function is constant. $\hat{\sigma}$ has the same units as the response. The "Degrees of freedom" is equal to the number of observations used in the calculations, which

TABLE 6.3 Simple Linear Regression Output for Forbes' Data

```
Data set = Forbes, Name of Fit = L1
Normal Regression
Kernel mean function = Identity
Response    = log10[Pressure]
Terms       = (Temp)
Coefficient Estimates
Label         Estimate       Std. Error      t-value
Constant      -0.421642      0.0334136       -12.619
Temp          0.00895618     0.000164575      54.420

R Squared:              0.994961
Sigma hat:              0.00379159
Number of cases:            17
Degrees of freedom:         15

Summary Analysis of Variance Table
Source        df        SS            MS             F        p-value
Regression    1      0.0425757     0.0425757      2961.55    0.0000
Residual      15     0.000215643   0.0000143762
```

is 17, minus the number of regression coefficients estimated, which for simple linear regression is two, namely the intercept and the slope. The "Summary Analysis of Variance Table" in Table 6.3 is described in Section 6.6.

We can get a visual impression of how well the OLS line matches the data. From the Graph&Fit menu, select "Plot of," and then draw the plot with *Temp* on the horizontal axis and $\log_{10}(Pressure)$ on the vertical axis. We want to add the OLS line to this plot. The slidebar on the plot that is initially marked "OLS" can be used for this; we call this the *parametric smoother slidebar*. If you click the mouse button on this slidebar, the OLS simple linear regression with the quantity on the vertical axis as response and the quantity on the horizontal axis as predictor will be computed, and the OLS line will be displayed on the figure. More general use of this control will be discussed later. The resulting plot is shown in Figure 6.7. Apart from one point, the OLS line matches the data closely.

6.3.1 Interpreting the Intercept

The equation for the OLS line shown in Figure 6.7 is

$$\hat{E}(y \mid x) = -0.42 + 0.009x$$

where the estimates of the slope and intercept have been rounded for this discussion. From the slope estimate we expect an increase of 0.009 in log pressure per degree increase in the boiling point of water. At first glance, the

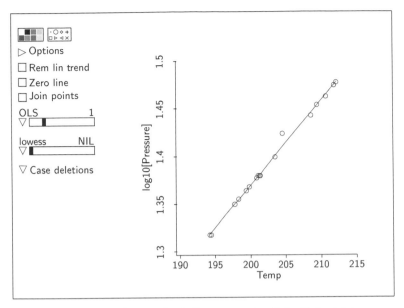

FIGURE 6.7 The OLS simple linear regression fit for Forbes' data.

intercept indicates that the expected log pressure is −0.42 when the boiling point of water is 0 degrees F. How do we know that this interpretation of the intercept is scientifically reasonable? The answer is that we don't, at least not from statistics alone. We do know from Figure 6.7 that the fitted line is a close approximation of the expected response over the range of temperatures $194.3 < Temp < 212.2$ observed by Forbes. But we have no data, and thus no statistical information, on the response for temperatures outside this range. $\hat{E}(y \mid x = 0) = -0.42$ might be a good estimate, or it could be way off because the linear relation that is so clear in Figure 6.7 breaks down for small temperatures. A key point is that *extrapolation*, using a fitted model as the basis for inferences outside the range of the data, should be avoided unless there is external subject-matter information to support it. A second point is that in some regressions the intercept may be only a tuning parameter with no intrinsic scientific meaning.

Interpreting the intercept can often be made easier by reparameterizing the model by *centering* by subtracting the average $\bar{x}$ from x_i. Beginning with the simple linear regression model,

$$y \mid x = \eta_0 + \eta_1 x + e$$
$$= \eta_0 + \eta_1 (x - \bar{x} + \bar{x}) + e$$
$$= \eta_0 + \eta_1 \bar{x} + \eta_1 (x - \bar{x}) + e$$
$$= \alpha + \eta_1 \tilde{x} + e \tag{6.22}$$

Here we have defined the new intercept using the Greek letter alpha ($\alpha = \eta_0 + \eta_1\bar{x}$) and the centered predictor $\tilde{x} = (x - \bar{x})$. α is the expected response when the centered predictor equals 0, $\tilde{x} = 0$. Equivalently, α is the expected response when the original predictor equals its sample average, $x = \bar{x}$. Because $\bar{x}$ is always within the observed range of x, the new intercept α should always have a reasonable interpretation.

6.4   INFERENCE

Several types of inference might be of interest in simple linear regression, including hypothesis tests and confidence statements concerning regression coefficients, or prediction of future values of y for new cases. To develop inference statements, we use an additional assumption beyond those concerning the mean function and variance function summarized by (6.21).

The assumption we now add is that $y \mid x$ is normally distributed (see Section 1.6.2), which we write as

$$y \mid x \sim N(\eta_0 + \eta_1 x, \sigma^2) \tag{6.23}$$

This is called the *normal simple linear regression model*.

As a result of the normality assumption, the coefficient estimates $(\hat{\eta}_0, \hat{\eta}_1)$ follow a bivariate normal distribution with the means, variances, and covariance as given in Section 6.2.5. The correlation between $\hat{\eta}_0$ and $\hat{\eta}_1$ can be found in the usual way, by dividing the covariance by the product of the standard deviations.

The estimated variance $\hat{\sigma}^2$ depends on the squares of the y_i, so it will not have a normal distribution. However, its distribution is related to a *Chi-squared distribution*, written using a Greek letter as χ^2. A Chi-squared distribution has a single parameter called the *degrees of freedom*, or df. We denote the Chi-squared distribution with d degrees of freedom as χ_d^2.

We can now write

$$(n-2)\frac{\hat{\sigma}^2}{\sigma^2} \sim \chi_{n-2}^2 \tag{6.24}$$

which means that $(n-2)\hat{\sigma}^2/\sigma^2$ is distributed as a Chi-squared random variable with $n-2$ df. In addition, $\hat{\sigma}^2$, the estimate of σ^2, is statistically independent of the estimates of the regression coefficients $\hat{\eta}_0$ and $\hat{\eta}_1$.

More discussion on the role of normality is available in Section 6.7.6.

6.4.1   Inferences about Parameters

Hypothesis tests and confidence statements concerning η_0, η_1, and linear combinations of them are based on a t-distribution. As in Section 1.7.2, page 21, we will use the notation t_d to denote the t-distribution with d df, and the notation $Q(t_d, f)$ for the fth quantile of the t_d distribution. This means that the following equation is satisfied: If a random variable V has a t_d distribution, then

$$\Pr(V \leq Q(t_d, f)) = f$$

The most common inference method for regression coefficients and linear combinations of them is based on a general paradigm used throughout this book. Suppose we have constructed a statistic W that is (approximately) normally distributed with mean $E(W)$ and variance $\sigma^2 c$, where c is a known number that can depend on the sample size and any predictors, but not on the response. For example, W could be the average $\bar{y}$ of a simple random sample $y_1, \ldots, y_n$ from a population with mean $E(y_i) = E(\bar{y})$ and variance $\sigma^2 = \text{Var}(y_i)$. We know from the discussion of Section 1.6.6 that $\bar{y}$ will be approximately normally distributed with mean $E(y_i)$ and variance σ^2/n. Thus, $W = \bar{y}$ is approximately normally distributed with $\text{Var}(W) = \sigma^2 c$ where $c = 1/n$.

The standard error of W is estimated by $\text{se}(W) = \hat{\sigma} c^{1/2}$. Suppose that the estimate $\hat{\sigma}^2$ has d df and is independent of W. This holds for the statistics $\hat{\eta}_0$ and $\hat{\eta}_1$ based on data from the normal simple linear regression model (6.23). Then a $(1 - \alpha) \times 100\%$ confidence interval for $E(W)$ is

$$W - Q(t_d, 1 - \alpha/2)\text{se}(W) \leq E(W) \leq W + Q(t_d, 1 - \alpha/2)\text{se}(W) \quad (6.25)$$

We interpret this confidence statement to mean that, if the same experiment was repeated many times, then $(1 - \alpha) \times 100\%$ of the confidence intervals computed from those experiments would include the population mean $E(W)$.

We can apply this result to find a confidence interval for the slope in Forbes' data. Our statistic W in this case is $\hat{\eta}_1$, and $E(W) = \eta_1$. From Table 6.3, $\hat{\eta}_1 = 0.00896$ and $\text{se}(\hat{\eta}_1) = 0.000165$, rounding the results to three significant digits. If we take $\alpha = 0.05$, we seek a 95% confidence interval. The estimate $\hat{\sigma}^2$ has $d = n - 2 = 17 - 2 = 15$ df. We can find the value of $Q(t_{15}, 0.975)$ using the "Calculate quantile" item in the Arc menu. The result is $Q(t_{15}, 0.975) = 2.13$, rounded to two decimal digits. Then the 95% confidence interval is

$$0.00896 - 2.13(0.000165) < \eta_1 < 0.00896 + 2.13(0.000165)$$

$$0.00861 < \eta_1 < 0.00931$$

The t_d distribution can also be used to assess competing hypotheses concerning regression coefficients, using a strategy paralleling that for the mean of a normal distribution reviewed in Section 1.7.1. The usual procedure is based on using the data to compare a null hypothesis we label NH to an alternative hypothesis AH. The null hypothesis specifies a restriction on a model, while the alternative gives the unrestricted model. For example, hypotheses concerning the slope are

$$NH : \eta_1 = h_1, \quad \eta_0 \text{ arbitrary}$$
$$AH : \eta_1 \neq h_1, \quad \eta_0 \text{ arbitrary}$$

(6.26)

where h_1 is some specified value of interest.

Null hypotheses of the form given in (6.26) can be judged against an alternative hypothesis by using the general *Wald test statistic*

$$t = \frac{\text{estimate} - \text{hypothesized value}}{\text{standard error of the estimate}}$$
$$= \frac{W - h}{se(W)}$$

(6.27)

where W is still the general estimate defined at the beginning of this section and h is some specified value. In normal linear regression, this statistic is conventionally labeled as t, as we do here. Recalling the estimate $\hat{\sigma}^2$ has d df, a p-value is obtained by comparing t to the t_d distribution.

For example, consider again the slope η_1 in Forbes' data and the hypothesis given by (6.26). The general test statistic t applied to the slope becomes

$$t = \frac{\hat{\eta}_1 - h_1}{se(\hat{\eta}_1)}$$

(6.28)

with $d = n - 2$ df. The context for Forbes' data does not furnish a particular value of h_1, so for the purpose of this example we will choose $h_1 = 0.0095$ to illustrate the computations. The t-statistic for the null hypothesis $\eta_1 = 0.0095$ is

$$t = \frac{0.00896 - 0.0095}{0.000165} \approx -3.3$$

We can use the "Calculate probability" item in the Arc menu to compute the p-value. After selecting this item, you will get a dialog like the one in Figure 1.7 on page 17. Enter the value -3.30 for the value of the statistic, choose the t-distribution, and set the df to 15. The alternative hypothesis we have specified is two-tailed, so click the "Two tail" radio button. Use the lower tail for AH: $\eta_1 < 0.0095$, and use the upper tail for AH: $\eta_1 > 0.0095$. After pressing "OK," the following is displayed in the text window:

```
t dist. with 15 df, value = -3.3,
two-tail probability = 0.0048589
```

Against the general two-tailed alternative, the finding of $p \approx 0.005$ provides reasonably strong evidence against the null hypothesis that $\eta_1 = 0.0095$, since a value of t this extreme would be observed only about five times in a thousand if the null hypothesis were true.

The regression output for Forbes' data shown in Table 6.3 gives a "t-value" for each estimated regression coefficient. These are the values of t for the default hypotheses that the regression coefficients equal 0. For example, the *Arc* t-value for the slope is

$$Arc \text{ t-value} = \frac{\text{estimate} - 0}{\text{standard error of the estimate}}$$

$$= \frac{\hat{\eta}_1}{\text{se}(\hat{\eta}_1)}$$

$$= \frac{0.00896}{0.00165} = 54.42$$

This result shows that the estimate of the slope is about 54 standard errors above 0, giving a very strong indication that $\eta_1 \neq 0$, which should agree with the visual impression of the fit shown in Figure 6.7.

6.4.2 Estimating Population Means

From time to time, we may be interested in obtaining an estimate of $E(y \mid x)$, the population mean of y at a given value of x. The estimated mean at x is $\hat{E}(y \mid x) = \hat{\eta}_0 + \hat{\eta}_1 x$, with standard error $\text{se}(\hat{E}(y \mid x))$ given by the standard error of the linear combination $\hat{\eta}_0 + \hat{\eta}_1 x$. The general inference paradigm discussed in Section 6.4 applies to this regression. In particular, with $E(W) = E(y \mid x)$ equations (6.25) and (6.27) can be used to construct confidence intervals and test hypotheses about $E(y \mid x)$. The only missing ingredient is the computation of

$$\text{se}(\hat{E}(y \mid x)) = \text{se}(\hat{\eta}_0 + \hat{\eta}_1 x)$$

We can compute this standard error from the basics by first writing

$$\text{Var}(\hat{\eta}_0 + \hat{\eta}_1 x) = \text{Var}(\hat{\eta}_0) + \text{Var}(\hat{\eta}_1 x) + 2\text{Cov}(\hat{\eta}_0, \hat{\eta}_1 x)$$

$$= \text{Var}(\hat{\eta}_0) + x^2 \text{Var}(\hat{\eta}_1) + 2x\text{Cov}(\hat{\eta}_0, \hat{\eta}_1)$$

These expressions are based on the properties of covariance discussed in Section 4.1.4. Estimates of the two variances and covariance in the second expression can be computed from (6.15), (6.16), and (6.17) after substituting $\hat{\sigma}^2$ for σ^2. We then get $\text{se}(\hat{E}(y \mid x))$ by taking the square root of the estimate of $\text{Var}(\hat{\eta}_0 + \hat{\eta}_1 x)$ obtained by adding the estimates of the individual terms in its sum.

Arc automates this procedure. Suppose that the name of the model menu for the Forbes data is L1. Select the item "Prediction" from the model menu L1.

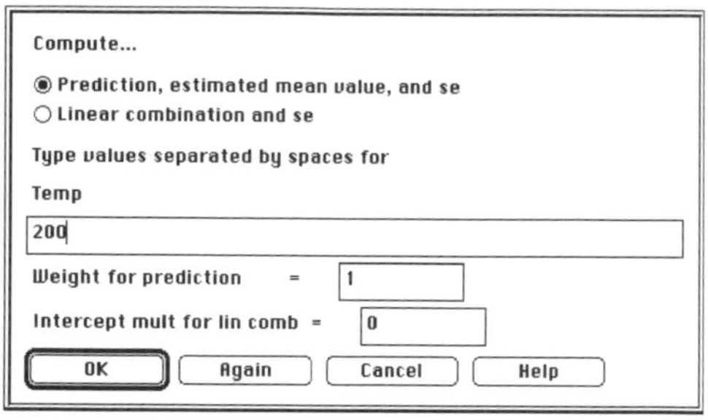

FIGURE 6.8   The "Prediction" dialog.

**TABLE 6.4   Output from the Prediction Dialog Shown in Figure 6.8 for Forbes'
Data**

```
Data set = Forbes, Name of Fit = L1
Normal Regression
Kernel mean function = Identity
Response        = log10[Pressure]
Terms           = (Temp)
Term values     = (200)
Prediction = 1.36959, with se(pred) = 0.00393167
Leverage = 0.0752518, Weight = 1
Est. population mean value = 1.36959, se = 0.00104011
```

The resulting dialog shown in Figure 6.8 can be used to calculate an estimated mean and its standard error. For example, to get the estimated population mean of $\log_{10}(Pressure)$ at *Temp* = 200, type 200 into the text area of the dialog and then click "OK." Table 6.4 gives the results that will be displayed in the text window. The estimated population mean is $\hat{E}(y \mid x = 200) = 1.370$ and its standard error $\text{se}(\hat{E}(y \mid x = 200)) = 0.001$. A $(1 - \alpha) \times 100\%$ confidence interval for $E(y \mid x = 200)$ is

$$\hat{E}(y \mid x = 200) - Q(t_d, 1 - \alpha/2) \times \text{se}(\hat{E}(y \mid x = 200))$$

$$\leq E(y \mid x = 200)$$

$$\leq \hat{E}(y \mid x = 200) + Q(t_d, 1 - \alpha/2) \times \text{se}(\hat{E}(y \mid x = 200))$$

$$1.37 - Q(t_{15}, 1 - \alpha/2) \times 0.001$$

$$\leq E(y \mid x = 200) \leq 1.37 + Q(t_{15}, 1 - \alpha/2) \times 0.001$$

The remaining information in the table is discussed in the next section.

6.4.3 Prediction

An estimated model is often used to predict $y \mid x$, a future value of y for a new case with predictor x. We assume that the data used to estimate $E(y \mid x)$ is relevant to prediction. In the Forbes example, we might not expect reasonable predictions of pressure at temperatures very different from those in Forbes' experiment. However, we can expect the model to be valid for temperatures in experiments conducted at similar altitudes; we will make this notion of extrapolation more precise in a later chapter. With this understanding, a point prediction y_{pred} at x is equal to the estimated population mean at x:

$$y_{pred} \mid x = \hat{E}(y \mid x) = \hat{\eta}_0 + \hat{\eta}_1 x \tag{6.29}$$

A prediction differs from an estimated population mean in that prediction refers to a single as yet unobserved response $y \mid x$, while an estimated population mean corresponds to $E(y \mid x)$. This difference affects the standard errors.

Suppose first that we know the variance function and the mean function exactly; in simple linear regression this means knowing η_0, η_1 and σ. Taking a single future observation is the same as sampling from a population with mean $E(y \mid x)$ and standard deviation $(\text{Var}(y \mid x))^{1/2} = \sigma$. Our prediction, the population mean, will not agree exactly with the future value of $y \mid x$, because of the inherent population variation represented by σ. In this simple case, the standard error of a prediction is just the known population standard deviation. Since we don't actually know the regression coefficients, there will be an *additional* error due to estimating them. This additional error is the same as the standard error of the estimated mean, $se(\hat{E}(y \mid x))$; this quantity was discussed earlier in Section 6.4.2. These two components are added in the variance scale:

$$\text{Var}(y_{pred} \mid x) = \text{population variance} + \text{estimation variance}$$
$$= \text{Var}(y \mid x) + \text{Var}(\hat{\eta}_0 + \hat{\eta}_1 x)$$
$$= \sigma^2 + \text{Var}(\hat{\eta}_0 + \hat{\eta}_1 x) \tag{6.30}$$

The standard error of a prediction is now found by taking the square root of the prediction variance, with an estimate $\hat{\sigma}^2$ used in place of the unknown value of σ^2. This standard error is given by

$$se(y_{pred} \mid x) = (\hat{\sigma}^2 + [se(\hat{\eta}_0 + \hat{\eta}_1 x)]^2)^{1/2} \tag{6.31}$$

This expression shows that the standard error of a prediction will always be larger than the standard error of the corresponding population mean estimate because of the inherent variability in the population.

The item "Prediction" in a model menu gives results for population mean values and predictions at the same time. Table 6.4 contains the prediction and its standard error for the example with Forbes' data. For this example, the standard error of the prediction is nearly four times the standard error of the estimated population mean value.

A confidence interval for a prediction can be obtained using a t-distribution because the prediction is a linear combination of normally distributed estimates, and the variance estimate $\hat{\sigma}^2$ is independent of the estimates. To compute of a 90% prediction interval, for example, we need $Q(t_{15}, 0.95)$, which, from the "Calculate quantile" item in the Arc menu, is about 1.75. The confidence interval is then

$$y_{pred} \mid (x = 200) - Q(t_{15}, 0.95) \times se(y_{pred} \mid (x = 200))$$
$$\leq y \mid (x = 200)$$
$$\leq y_{pred} \mid (x = 200) + Q(t_{15}, 0.95) \times se(y_{pred} \mid (x = 200))$$

where $y_{pred} \mid (x = 200)$ is calculated using (6.29), and $se(y_{pred} \mid (x = 200))$ is calculated using (6.31). Then

$$1.370 - 1.75 \times 0.0039 \leq y \mid (x = 200) \leq 1.370 + 1.75 \times 0.0039$$
$$1.363 \leq y \mid (x = 200) \leq 1.377$$

A 90% confidence interval for pressure $= 10^y$ can be obtained by exponentiating the end-points of the interval:

$$10^{1.363} \leq 10^{y \mid (x = 200)} \leq 10^{1.377}$$
$$23.05 \leq Pressure \mid (Temp = 200) \leq 23.82$$

This last interval is in inches of mercury, the units that would be of interest to most investigators. Although the regression model appeared to be very precise, a fair amount of variability remains as the prediction interval has a width of about 0.75 inches of mercury.

6.5   FORBES' EXPERIMENTS, REVISITED

We chose to use simple linear regression for Forbes' data because the plot for the 17 data points appears to be linear. Forbes apparently selected this model from theoretical considerations not explicitly given in his paper. Advances in physics since Forbes' time provide us with additional theory in the form of the Clausius–Clapeyron formula of classical thermodynamics. According to this formula, we should find that

$$E_C(\log_{10}(Pressure) \mid Temp) = \eta_0 + \eta_1 \frac{1}{Ktemp} \tag{6.32}$$

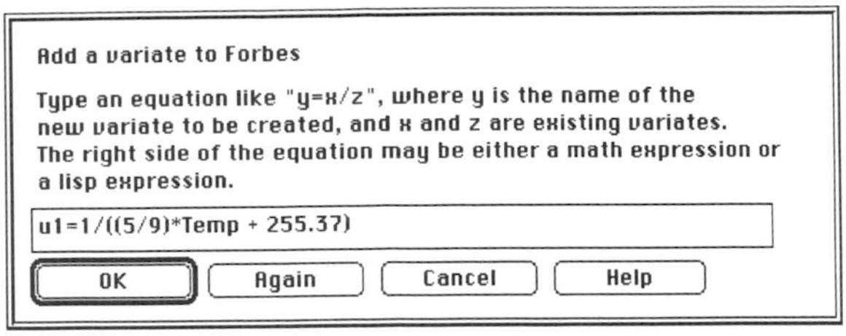

```
┌──────────────────────────────────────────────────────────┐
│ Add a variate to Forbes                                  │
│                                                          │
│ Type an equation like "y=x/z", where y is the name of the│
│ new variate to be created, and x and z are existing      │
│ variates.                                                │
│ The right side of the equation may be either a math      │
│ expression or                                            │
│ a lisp expression.                                       │
│ ┌──────────────────────────────────────────────────────┐ │
│ │ u1=1/((5/9)*Temp + 255.37)                           │ │
│ └──────────────────────────────────────────────────────┘ │
│ ┌─────────┐  ┌─────────┐  ┌─────────┐  ┌─────────┐       │
│ │   OK    │  │  Again  │  │ Cancel  │  │  Help   │       │
│ └─────────┘  └─────────┘  └─────────┘  └─────────┘       │
└──────────────────────────────────────────────────────────┘
```

FIGURE 6.9   The "Add a variate" dialog.

where *Ktemp* is temperature in degrees Kelvin, which are degrees Celsius above absolute zero. The subscript C on E_C is intended as a reminder that this mean function is implied by the Clausius–Clapeyron formula. If we were to graph this mean function on a plot of $\log_{10}(Pressure)$ versus *Ktemp*, we would get a curve, not a straight line.

Even though this mean function is curved, we can estimate the parameters η_0 and η_1 using simple linear regression methods. Define the new variable u_1 to be the inverse of temperature in degrees Kelvin,

$$u_1 = \frac{1}{Ktemp} = \frac{1}{(5/9)Temp + 255.37} \tag{6.33}$$

Then the mean function (6.32) can be rewritten as

$$E_C(\log_{10}(Pressure) \mid Temp) = \eta_0 + \eta_1 u_1 \tag{6.34}$$

for which simple linear regression is again suitable. This technique of replacing the predictor by a transformation of it allows us to adapt simple linear regression models to a variety of situations. Notice also the notation we have used in (6.34): The left side of the equation says we are conditioning on *Temp*, but the variable *Temp* does not appear explicitly on the right side of the equation. The conditioning comes about because the predictor u_1 on the right side is a function of *Temp* as given by (6.33).

Computing u_1 is too complicated for the "Transform" menu item, but general transformations like this can be computed using the "Add a variate" item. After selecting this item from the data set menu, you will get a dialog like Figure 6.9. In the text area of this dialog, first type a name for the new variate, then an equal sign, and then an expression that defines the new variate. The expression in Figure 6.9 transforms to the inverse of degrees Kelvin. The expression can be in regular mathematical format, or a statement in a computer language called *lisp*; this latter type of expression is illustrated in Sections A.2 and A.7.2. Following computation of u_1, the simple linear regression of $\log_{10}(Pressure)$ on u_1 can be carried out as described previously in this chapter.

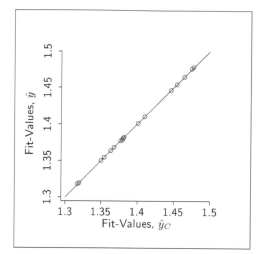

FIGURE 6.10 Scatterplot of fitted values $\hat{y}$ from the fit shown in Table 6.3 versus the fitted values $\hat{y}_C$ from the fit of the Clausius–Clapeyron formula. The reference line $y = x$ was added to the plot using the plot's options menu.

We now have two possible models for the same data, the model

$$E(\log_{10}(\textit{Pressure}) \mid \textit{Temp}) = \eta_0 + \eta_1 \textit{Temp} \tag{6.35}$$

used by Forbes, and model (6.34) based on the Clausius–Clapeyron formula. Which is better? Forbes approach is supported by empirical evidence, but (6.34) is supported by theory.

Shown in Figure 6.10 is a plot of the fitted values $\hat{y}$ from the fit of model (6.35) shown in Table 6.3 versus the fitted values $\hat{y}_C$ from model (6.34). The diagonal reference line on the plot is $y = x$. The plot shows that the two sets of fitted values are nearly identical, falling very close to the diagonal reference line. Although model (6.35) may not be supported by present-day theory, it seems to provide a very good approximation over the range of temperatures covered by Forbes' data. Empirical regression models can provide very good approximations even if they are not fully supported by substantive theory.

6.6 MODEL COMPARISON

6.6.1 Models

The discussion so far has been mostly in the context of the simple linear regression model $y \mid x = \eta_0 + \eta_1 x + e$, which can be stated equivalently as the pair of conditions $E(y \mid x) = \eta_0 + \eta_1 x$ and $\text{Var}(y \mid x) = \sigma^2$. Sometimes, even simpler models might be useful. For example, for the haystack data we might study

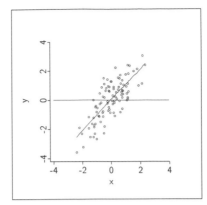

FIGURE 6.11   Representation of model (6.38) with a constant regression function and the simple linear regression model (6.41).

the model

$$E(Vol \mid C) = C^3/12\pi^2 \qquad \text{and} \qquad \text{Var}(Vol \mid C) = \sigma^2 \qquad (6.36)$$

For the snow geese data (Section 3.6.2), the model

$$E(Photo \mid Obs2) = Obs2 \qquad \text{and} \qquad \text{Var}(Photo \mid Obs2) = \sigma^2 \quad (6.37)$$

might be considered. These two models contain no unknown regression coefficients, but σ^2 is still unknown. The general ideas discussed in this chapter apply to these simpler models.

Returning to simple linear regression, we might want to consider any of the four models

$$E(y \mid x) = 0 \qquad \text{and} \qquad \text{Var}(y \mid x) = \sigma^2 \qquad\qquad (6.38)$$

$$E(y \mid x) = \eta_0 \qquad \text{and} \qquad \text{Var}(y \mid x) = \sigma^2 \qquad\qquad (6.39)$$

$$E(y \mid x) = \eta_1 x \qquad \text{and} \qquad \text{Var}(y \mid x) = \sigma^2 \qquad\qquad (6.40)$$

$$E(y \mid x) = \eta_0 + \eta_1 x \qquad \text{and} \qquad \text{Var}(y \mid x) = \sigma^2 \qquad (6.41)$$

The mean function is constant in (6.38) and in (6.39) and does not depend on the predictor. In model (6.38) the constant value of the mean function is specified as 0. In model (6.39) the constant value of the mean function is an unknown value represented by the regression coefficient η_0. The mean function for model (6.40) is linear in the predictor with a single regression coefficient η_1. This model is often referred to as *regression through the origin* since the intercept is zero. Model (6.41) is the simple linear regression model. Figure 6.11 illustrates models (6.38) and (6.41) for a particular data set.

We can use OLS to fit any of these models by choosing the regression coefficients to minimize the residual sum of squares function

$$RSS(h_0, h_1) = \sum_{i=1}^{n} (y_i - E(y \mid x_i))^2$$

where $E(y \mid x_i)$ can be set to any of the four mean functions described previously, and (h_0, h_1) is a pair of candidate values for the parameters (η_0, η_1). In the case of models (6.36), (6.37), and (6.38), there are no regression coefficients and thus no minimization is necessary because the residual sum of squares function is constant.

In the case of model (6.39), the residual sum of squares depends only on h_0 and is minimized at $\hat{\eta}_0 = \bar{y}$. Thus, $\hat{E}(y \mid x) = \hat{\eta}_0 = \bar{y}$ and the residual sum of squares is just

$$RSS = \sum_{i=1}^{n} (y_i - \hat{E}(y \mid x_i))^2$$

$$= \sum_{i=1}^{n} (y_i - \bar{y})^2$$

$$= SYY \tag{6.42}$$

Arc can be used to fit model (6.40) by un-checking the box "Fit Intercept" in the regression dialog shown in Figure 6.6.

The estimate of σ^2 follows the same form for any of the models we have discussed:

$$\hat{\sigma}^2 = \frac{RSS}{df}$$

where the degrees of freedom

$$df = n - \text{number of } \eta\text{-terms estimated}$$

For models (6.36), (6.37), and (6.38), $df = n$. For models (6.39) and (6.40), $df = n - 1$; and for the simple linear regression model (6.41), $df = n - 2$.

Finally, tests of hypotheses and confidence intervals can be constructed following the general steps of Section 6.4.

6.6.2   Analysis of Variance

Analysis of variance is used to summarize the calculations necessary for comparing competing models on the same data. For example, consider the following two models stated as hypotheses:

$$NH : E(y \mid x) = \eta_0 \quad \text{with} \quad Var(y \mid x) = \sigma^2$$
$$AH : E(y \mid x) = \eta_0 + \eta_1 x \quad \text{with} \quad Var(y \mid x) = \sigma^2$$

(6.43)

The mean function of the null hypothesis is a constant. Under the simple linear regression model of the alternative hypothesis, there is no restriction on the estimated mean function as long as it is a straight line. The NH is a special case of the AH, obtained by setting the parameter $\eta_1 = 0$. Because of this we say that the model of the NH is *nested* within the model of the AH. Analysis of variance is appropriate only for such nested hypotheses.

Before turning to the analysis of variance for (6.43), let's consider some other hypothesis that might be of interest:

$$NH : E(y \mid x) = 0 \quad \text{with} \quad Var(y \mid x) = \sigma^2$$
$$AH : E(y \mid x) = \eta_0 + \eta_1 x \quad \text{with} \quad Var(y \mid x) = \sigma^2$$

(6.44)

Here the null hypothesis contains no unknown parameters, but it is still nested within the model AH, and analysis of variance is appropriate.

$$NH : E(y \mid x) = \eta_1 x \quad \text{with} \quad Var(y \mid x) = \sigma^2$$
$$AH : E(y \mid x) = \eta_0 + \eta_1 x \quad \text{with} \quad Var(y \mid x) = \sigma^2$$

(6.45)

Now the null hypothesis specifies a straight line with intercept equal to zero and slope unrestricted. Under the AH the slope and intercept are both unrestricted. Again, the model of the NH is nested within the model of the AH, and analysis of variance is again appropriate. Finally, consider the hypotheses

$$NH : E(y \mid x) = \eta_0 \quad \text{with} \quad Var(y \mid x) = \sigma^2$$
$$AH : E(y \mid x) = \eta_1 x \quad \text{with} \quad Var(y \mid x) = \sigma^2$$

(6.46)

These hypotheses differ from those above because the model of the NH is not nested within the model of the AH. Analysis of variance cannot be used for this test.

Consider any of the hypotheses described above except (6.46). If the AH model provides a much better fit than does the NH model, we would expect that the residual sum of squares under the null hypothesis RSS_{NH} will be considerably larger than the residual sum of squares under the alternative hypothesis RSS_{AH}. This comparison provides the basis for a test of the competing models. The test statistic is

$$F = \frac{(RSS_{NH} - RSS_{AH})/(\text{df}_{NH} - \text{df}_{AH})}{\hat{\sigma}^2_{AH}}$$

(6.47)

TABLE 6.5   **Generic Analysis of Variance Table**

Source	df	SS	MS	F
Regression	$\mathrm{df}_{NH} - \mathrm{df}_{AH}$	$RSS_{NH} - RSS_{AH}$	SS/df	See (6.47)
Residual	df_{AH}	RSS_{AH}	$SS/\mathrm{df} = \hat{\sigma}^2$	

TABLE 6.6   **Analysis of Variance for Forbes' Data, Repeated from Table 6.3**

```
Summary Analysis of Variance Table
Source        df      SS          MS           F         p-value
Regression     1   0.0425757   0.0425757    2961.55     0.0000
Residual      15   0.000215643 0.0000143762
```

where $\hat{\sigma}^2_{AH}$ is the estimate of σ^2 under the AH model. Large values of F are evidence against the NH and in favor of the AH. When the null hypothesis is true, this statistic has an F_{d_1,d_2} distribution with $d_1 = \mathrm{df}_{NH} - \mathrm{df}_{AH}$ and $d_2 = \mathrm{df}_{AH}$. We have used df_{AH} to denote the df for the residual sum of squares under the AH. Similarly, df_{NH} is the df for the residual sum of squares under the NH. The p-value of the test is computed as the probability that an F_{d_1,d_2} random variable is as large or larger than the observed value of F. The calculations for this F-test are traditionally summarized in an analysis of variance as shown in Table 6.5.

Now, consider the hypotheses in (6.43). We can get least squares estimates under the NH by minimizing the residual sum of squares $\sum (y_i - h_0)^2$ over all possible values of h_0. This leads to $RSS_{NH} = SYY$ with $\mathrm{df}_{NH} = n - 1$ df, as shown in (6.42). For the simple linear regression model of the AH, we have previously seen that the residual sum of squares is given by (6.13) with $\mathrm{df}_{AH} = n - 2$.

The elements needed for the test of (6.43) on Forbes' data are summarized in the Analysis of Variance table shown in Table 6.6. The column marked "Source" refers to descriptive labels given to the sums of squares. The df column gives the number of degrees of freedom associated with each named source. The next column gives the values of the sums of squares. A *mean square* is defined to be a sum of squares divided by its df. The mean square on the residual line is $\hat{\sigma}^2$.

The sum of squares in this table on the line marked "Regression" gives the difference $RSS_{NH} - RSS_{AH}$ while the associated degrees of freedom is $\mathrm{df}_{NH} - \mathrm{df}_{AH}$. The NH and AH here are those in (6.43). Consequently, the F-test for comparing the two models is just the ratio of the two mean squares shown in the Analysis of Variance table. This ratio, shown in the column marked "F," is 2961.55, and its p-value rounds to zero in the four significant digits shown. The evidence against the NH is very strong, as might have been expected from inspection of Figure 6.7.

The F-test for the hypotheses in (6.44) and (6.45) are *not* automatically summarized in a single Analysis of Variance table by *Arc*. Instead, two models need to be fit in each case, and the test statistics calculated by hand for each using (6.47). Results can then be summarized, again by hand, using an Analysis of Variance table.

For the hypothesis tests specified by (6.26) with $h_1 = 0$, we used the statistic labeled t, and compared it to the t_d distribution to get a p-value. For (6.43) we used an F-test. These two approaches appear to be identical: Both allow the intercept to be arbitrary, but under the NH the slope is zero. For the simple linear regression model, these two tests give the same information and $t^2 = 54.42^2 = 2961.55 = F$.

6.7 COMPLEMENTS

6.7.1 Derivation of Estimates

In this section we derive OLS estimates. Understanding regression models is quite possible without learning any of these details, but some readers may find that the algebra deepens their understanding.

Deriving the OLS estimates for the simple linear regression model is simplified by centering the data as in (6.22). Given the simple linear regression mean function and data (x_i, y_i), $i = 1, \ldots, n$, we can write the model as

$$y_i = \alpha + \eta_1(x_i - \bar{x}) + e_i \tag{6.48}$$

where

$$\alpha = \eta_0 + \eta_1 \bar{x} \tag{6.49}$$

Once we have determined the OLS estimates of α and η_1, we can get the OLS estimate of η_0 by substituting these estimates into (6.49).

The least squares estimates minimize the residual sum of squares function,

$$RSS(a, h_1) = \sum_{i=1}^{n} [y_i - a - h_1(x_i - \bar{x})]^2$$

$$= \sum_{i=1}^{n} [(y_i - \bar{y}) - (a - \bar{y}) - h_1(x_i - \bar{x})]^2$$

where the second equation is obtained by adding and subtracting $\bar{y}$, and we have replaced the parameters α and η_1 by place holders a and h_1 because the residual sum of squares function is defined for any values of a and h_1, not just for the true values of the parameters. Squaring the term in the square brackets and then adding, this can be rewritten after a modest amount of algebra as

$$RSS(a, h_1) = SYY + n(a - \bar{y})^2 - 2h_1 SXY + h_1^2 SXX \tag{6.50}$$

The notation SYY, SXY, and SXX is defined in Table 6.1, page 104. The least squares estimates will make (6.50) as small as possible. Since a enters (6.50) only through $(a - \bar{y})^2$, setting $a = \bar{y}$ will make (6.50) as small as possible. To find the least squares estimate of the slope parameter, we must find the value of h_1 that minimizes the quadratic equation $SYY - 2h_1 SXY + h_1^2 SXX$. Recalling that the minimum of a quadratic equation $c_0 + c_1 z + c_2 z^2$ occurs at $-c_1/2c_2$, provided that c_2 is positive, the minimum residual sum of squares occurs at

$$\hat{\eta}_1 = -\frac{-2SXY}{2SXX} = \frac{SXY}{SXX} \tag{6.51}$$

The estimate of the intercept is obtained by substituting the estimates obtained so far into (6.49):

$$\hat{\eta}_0 = \bar{y} - \hat{\eta}_1 \bar{x} \tag{6.52}$$

To find the residual sum of squares RSS, substitute $a = \bar{y}$ and $h_1 = \hat{\eta}_1 = SXY/SXX$ into (6.50), and simplify:

$$RSS = SYY - 2\left(\frac{SXY}{SXX}\right) SXY + \left(\frac{SXY}{SXX}\right)^2 SXX$$

$$= SYY - \left[\frac{SXY^2}{(SXX)(SYY)}\right] SYY$$

$$= SYY(1 - r^2(x, y))$$

as given by (6.13).

6.7.2 Means and Variances of Estimates

The least squares estimates are linear combinations of the y_i, so results given previously can be used to find moments of the estimates. Assume the simple linear regression model

$$E(y \mid x) = \eta_0 + \eta_1 x \quad \text{and} \quad \text{Var}(y \mid x) = \sigma^2$$

We recall from (6.51) that the OLS estimate of η_1 is $\hat{\eta}_1 = SXY/SXX$. Writing out

$$SXY = \sum_{i=1}^{n} (x_i - \bar{x})(y_i - \bar{y}) = \sum_{i=1}^{n} (x_i - \bar{x})y_i$$

we can write $\hat{\eta}_1$ as given by (6.11):

$$\hat{\eta}_1 = \sum_{i=1}^{n} \left(\frac{x_i - \bar{x}}{SXX} \right) y_i \qquad (6.53)$$

Viewing the x_i as fixed numbers, we recognize $\hat{\eta}_1$ as a linear combination of $y_1, \dots, y_n$. We now turn to finding the mean and variance of $\hat{\eta}_1$. As notation we use only in complements sections, let X stand for all the observed values $x_1, \dots, x_n$. Using (6.53) and applying the formula $E(\sum_{i=1}^{n} c_i y_i) = \sum_{i=1}^{n} c_i E(y_i)$, we obtain

$$E(\hat{\eta}_1 \mid X) = E\left[\sum_{i=1}^{n} \left(\frac{x_i - \bar{x}}{SXX} \right) y_i \mid X \right]$$

$$= \sum_{i=1}^{n} \left(\frac{x_i - \bar{x}}{SXX} \right) E(y_i \mid X)$$

Now $E(y_i \mid X) = E(y_i \mid x_i)$ because y_i depends on X only through x_i. Substituting, we get

$$E(\hat{\eta}_1 \mid X) = \sum_{i=1}^{n} \left(\frac{x_i - \bar{x}}{SXX} \right) E(y_i \mid x_i)$$

$$= \sum_{i=1}^{n} \left(\frac{x_i - \bar{x}}{SXX} \right) (\eta_0 + \eta_1 x_i)$$

$$= \eta_0 \sum_{i=1}^{n} \left(\frac{x_i - \bar{x}}{SXX} \right) + \eta_1 \sum_{i=1}^{n} \left(\frac{(x_i - \bar{x})x_i}{SXX} \right)$$

To complete the computation we need two further results that can be proved by direct summation; $\sum(x_i - \bar{x}) = 0$, and $\sum(x_i - \bar{x})x_i = \sum(x_i - \bar{x})^2 = SXX$. Substituting these results into the last equation gives

$$E(\hat{\eta}_1 \mid X) = \eta_1$$

showing that $\hat{\eta}_1$ is an unbiased estimate of η_1. As notation for the main part of this book, we will not give the conditioning on X explicitly, and we write this result simply as $E(\hat{\eta}_1) = \eta_1$. A similar proof will show that the intercept is also unbiased.

We next turn to finding the variances of the estimates. Assuming that observations are independent, using (4.8), page 62, we can write

$$Var(\hat{\eta}_1 \mid X) = Var\left[\sum_{i=1}^{n} \left(\frac{x_i - \bar{x}}{SXX} \right) y_i \mid X \right]$$

$$= \sum_{i=1}^{n} \left(\frac{x_i - \bar{x}}{SXX} \right)^2 Var(y \mid x_i)$$

where once again we used the fact that the distribution of y_i depends on X only through x_i. We next substitute $\mathrm{Var}(y_i \mid x_i) = \sigma^2$:

$$\mathrm{Var}(\hat{\eta}_1 \mid X) = \sum_{i=1}^{n} \left(\frac{x - \bar{x}}{SXX} \right)^2 \sigma^2$$

$$= \sigma^2 \frac{SXX}{SXX^2}$$

$$= \sigma^2 \frac{1}{SXX}$$

as was given by (6.15). A similar derivation under the same assumptions gives both the variance of the intercept (6.16) and the covariance of the estimates (6.17).

Finally, we apply equation (4.8) to find the variance of an arbitrary linear combination of the OLS estimates, $b_0 \hat{\eta}_0 + b_1 \hat{\eta}_1$, where b_0 and b_1 are any real numbers:

$$\mathrm{Var}(b_0 \hat{\eta}_0 + b_1 \hat{\eta}_1 \mid X) = b_0^2 \mathrm{Var}(\hat{\eta}_0 \mid X) + b_1^2 \mathrm{Var}(\hat{\eta}_1 \mid X)$$

$$+ 2b_0 b_1 \mathrm{Cov}(\hat{\eta}_0, \hat{\eta}_1 \mid X)$$

$$= \sigma^2 \left(b_0^2 \left[\frac{1}{n} + \frac{\bar{x}^2}{SXX} \right] + b_1^2 \frac{1}{SXX} - 2b_0 b_1 \frac{\bar{x}}{SXX} \right)$$

This formidable-looking equation simplifies greatly in special cases. In the case of a fitted value, $b_0 = 1$ and $b_1 = x$, and $\hat{y} = \hat{\eta}_0 + \hat{\eta}_1 x$. We get the simple formula

$$\mathrm{Var}(\hat{y} \mid X) = \sigma^2 \left(\frac{1}{n} + \frac{(x - \bar{x})^2}{SXX} \right)$$

This variance depends on the unknown value of σ^2 and on the known value of x. To get an estimated variance that can be used in practice, $\hat{\sigma}^2$ is substituted for σ^2. The standard error of $\hat{y}$ is

$$\mathrm{se}(\hat{y} \mid X) = \hat{\sigma} \sqrt{\frac{1}{n} + \frac{(x - \bar{x})^2}{SXX}}$$

The item "Prediction" from a regression model's menu can be used to compute the estimate and standard error for any linear combination, and also to compute estimates and standard errors for fitted values and for predictions, as described in Sections 6.4.2–6.4.3.

6.7.3 Why Least Squares?

The least squares criterion (6.10) chooses estimates that minimize the squares of the *vertical* differences between the observed values of y and the values that fall on the fitted line. This is but one of many possible ways of defining a objective function to estimate parameters. For example, we could minimize the sum of the absolute values of the vertical differences, or we could use the differences in some other way. Why choose this particular criterion function?

Most of the methodology outlined in this book doesn't depend very strongly on using least squares. However, the estimates we use must have a few properties. In the "must have" category is *consistency*: Roughly speaking, this means that if the simple linear regression model is true, then as the sample size increases, the difference between the estimates and the true values of the parameters should get smaller. If we could collect enough data and if the model we use were true, we could eventually know the values of the parameters. OLS shares consistency with many other estimation methods. See Problem 6.3 for more on this topic.

A second "must have" criterion is called *invariance*: The results of an analysis should be unchanged, or changed in a deterministic way, when the data are scaled, for example by changing from Imperial to metric units. Again, least squares shares invariance with many other estimation methods. See Problem 6.2 for more on this topic.

One of the principal theoretical justifications for the use of OLS estimates is the *Gauss–Markov theorem*, which states that if the model (6.21) holds, then the OLS estimates have the smallest variance among estimates that are unbiased and that can be written as a linear combination of the elements of y.

Also, there is the connection between OLS and the bivariate normal distribution described in Section 4.3.3. The estimates of the slope and the intercept in that section are identical to the OLS estimates.

In later chapters of this book we will consider other estimators besides least squares. We will do this because we want to take advantage of additional information, often about the relationship between the mean function and the variance function. Least squares is based only on the mean function, essentially assuming that the variance function is constant; later we will generalize to weighted least squares to allow the variance function to vary independently, but not as a function of the mean. When the mean and the variance are related, then other estimation methods should be used.

The method of least squares was apparently first proposed by C. F. Gauss in a series of papers and letters presented to the Royal Society of Göttingen in 1821, 1823, and 1826.

6.7.4 Alternatives to Least Squares

There are many estimation methods other than the methods described in this book. A good introduction to alternative methods of regression is available in the book by Birkes and Dodge (1993).

6.7.5   Accuracy of Estimates

In this book, we often present tables of numbers just as they are presented by the computer. This is done both to allow users to reproduce our results and to allow the output shown to be used in intermediate calculations. Results presented with too many digits are hard to read. For this reason we often round computer output when illustrating calculations. Ehrenberg (1981) presents some simple rules for presenting tables of numbers that make them easy to read and to facilitate comparisons.

6.7.6   Role of Normality

We started our discussion of inference in Section 6.4 by assuming that the conditional distribution of $y \mid x$ is normal (see equation (6.23)). When this assumption holds, tests and confidence intervals have the properties we described. When normality does not hold, the inference procedures are approximate, which means that the level of a test or confidence interval may not be as the normal theory says. How accurate are inference procedures for regression coefficients when normality does not hold? When the sample size is reasonably large relative to the number of regression coefficients, they are generally quite accurate and the assumption of normality is not really crucial. This conclusion holds for all of the regression coefficients studied in this chapter, as well as for the coefficients in more complicated models discussed in later chapters.

The reason for this conclusion begins with the form of an estimated regression coefficient. For OLS, estimated regression coefficients are always linear combinations of the responses, $\hat{\eta} = \sum_{i=1}^{n} c_i y_i$, where the c_i's are constants that can depend on the predictor and the sample size, but not on the response. Linear combinations of independent normal random variables are themselves normally distributed. Thus, if the y_i are independent and normal, then $\hat{\eta}$ is normally distributed. The inference procedures discussed in this chapter are based on this fact. If the y_i are not normally distributed, $\hat{\eta}$ will still be approximately normal, the accuracy of approximation increasing with sample size n. This is the same phenomenon described during our discussion of the central limit theorem in Section 1.6.6. Thus, as long as the sample size is reasonably large, the inference can proceed as if the responses were normally distributed.

There are exceptions to this general conclusion. One is confidence intervals for a prediction, because here we do not gain the benefits of the central limit theorem. Generally, confidence limits for prediction require that we know the distribution of the response so we can adequately characterize the uncertainty in a single future observation.

6.7.7   Measurement Error

Measurement error refers to the idea that we can't measure with sufficient precision what theory says we need to measure. To illustrate the complications that can be caused by measurement error, consider a response y and a predictor x that follow a bivariate normal distribution. We know from the results of Section 4.3 that the mean function is linear in x,

$$E(y \mid x) = \eta_0 + \eta_1 x$$

where

$$\eta_1 = \frac{\text{Cov}(y,x)}{\text{Var}(x)}$$

In addition, the variance function is constant:

$$\text{Var}(y \mid x) = \sigma^2 = [1 - \rho^2(x,y)]\text{Var}(y)$$

Suppose we can't determine y precisely, so instead of observing y we observe $\tilde{y} = y + \delta$, where δ represents an independent normal measurement error with mean 0 and variance σ_δ^2. For example, y might represent the weight of a light object. If our scale isn't sufficiently precise, we could obtain different independent readings each time the same object is weighted. The variance σ_δ^2 represents the variation from weighing to weighing of a single object. The condition $E(\delta) = 0$ means that our scale is unbiased; the average of many different weighings of the same object would give its actual weight.

Using $\tilde{y}$ in place of y, the simple linear regression model becomes

$$E(\tilde{y} \mid x) = \eta_0 + \eta_1 x \tag{6.54}$$

again with constant variance

$$\text{Var}(\tilde{y} \mid x) = \sigma^2 + \sigma_\delta^2 \tag{6.55}$$

The mean function for $\tilde{y} \mid x$ is the same as the mean function for $y \mid x$, but the variance has increased by σ_δ^2. The result of measurement error in y is to increase the variation about the mean function.

Measurement error in the predictor has different consequences. Suppose we observe $\tilde{x} = x + \varepsilon$ instead of x. The independent measurement error in the predictor is represented by ε with mean $E(\varepsilon) = 0$ and variance σ_ε^2. Assuming that ε is normally distributed, $(y,\tilde{x})$ has a bivariate normal distribution that is the same as the bivariate normal distribution of (y,x), except that $\text{Var}(\tilde{x}) = \sigma_x^2 + \sigma_\varepsilon^2$, where $\sigma_x^2 = \text{Var}(x)$. Thus, the mean function for the regression of y on $\tilde{x}$ is

$$E(y \mid \tilde{x}) = \tilde{\eta}_0 + \tilde{\eta}_1 \tilde{x} \tag{6.56}$$

where

$$\tilde{\eta}_1 = \eta_1 \frac{\sigma_x^2}{\sigma_x^2 + \sigma_\varepsilon^2} \tag{6.57}$$

As a consequence, the slope from the OLS fit of y on $\tilde{x}$ is not an estimate of η_1 but instead is an estimate of $\tilde{\eta}_1 \leq \eta_1$. The implication of this calculation is that measurement error in the predictor can result in a biased estimate of the regression coefficient. The estimated regression coefficient from the regression of y on $\tilde{x}$ will be too small on the average, the magnitude of the bias depending on the relative sizes of σ_x^2 and σ_ε^2. If σ_ε^2 is small relative to σ_x^2 the measurement error could be unimportant, but otherwise the bias may be substantial. In the case of simple linear regression, measurement error in the predictor causes the estimated slope to be too small. In regressions with multiple predictors, estimated regression coefficients can be too small or too large on the average.

There is a substantial literature on measurement error in linear regression. Fuller (1987) provided a comprehensive reference for the area. Measurement error in the predictors is not considered further in this book.

6.7.8   References

Forbes' data are from Forbes (1857), with some of the discussion in Section 6.5 adapted from Brown (1993). The Clausius–Clapeyron formula goes back to Clausius (1850). The Old Faithful Geyser data used in Problem 6.8 were provided by Roderick Hutchinson, the Yellowstone Park Geologist. The data in Problem 6.9 are from Sacher and Staffeldt (1974). The paddlefish data in Problem 6.10 were provided by Ann Runstrom.

PROBLEMS

6.1 Suppose we fit a simple linear regression model with data (x_i, y_i), $i = 1, \ldots, n$. Write out a table like Table 6.3, using the formulas based on the summaries given in Table 6.1 in place of numbers.

6.2 **Invariance.** Suppose we have data (x_i, y_i), $i = 1, \ldots, n$, and we use OLS to fit the simple linear regression model

$$E(y_i \mid x_i) = \eta_0 + \eta_1 x_i \qquad \text{with} \qquad \text{Var}(y_i \mid x_i) = \sigma^2 \qquad (6.58)$$

Let $y_i^* = a + by_i$ and $x_i^* = c + dx_i$ for known numbers a, b, c, d with $b \neq 0$ and $d \neq 0$. Suppose we fit the simple linear regression model

$$E(y_i^* \mid x_i^*) = \gamma_0 + \gamma_1 x_i^* \qquad \text{with} \qquad \text{Var}(y_i^* \mid x_i^*) = \tau^2 \qquad (6.59)$$

What is the relationship between $\hat{\eta}_1$ estimated from (6.58) and $\hat{\gamma}_1$ estimated from (6.59)? Between $\hat{\eta}_0$ and $\hat{\gamma}_0$? Is r^2 the same in the two regressions? What about the relationship between $\hat{\sigma}^2$ and $\hat{\tau}^2$? Are tests (both t and F-tests) the same or different (consider the tests for $\eta_1 = 0$ for example)?

This problem can be done two ways. First, you can use the formulas for the various quantities given in Section 6.2. Alternatively, you can choose any data set, fit the simple linear regression model for y on x, then fit the simple linear regression model for y^* on x^* for some a, b, c, d, and see what happens. This also provides a way of checking the formulas obtained using the first method.

6.3 The variance of the OLS estimate $\hat{\eta}_1$ is σ^2/SXX. Under what condition *on the values of* x will this variance approach zero as the sample size n grows toward infinity? What happens to $\hat{\eta}_1$ as σ^2/SXX approaches zero?

Can you construct an example (that is, a sequence of values $x_1, x_2, \ldots$) such that as the sample size goes to infinity the variance of $\hat{\eta}_1$ will not approach zero? If you can find such an example (and this is possible), then you will have shown that for some configurations of x, the OLS estimates will not converge to the true value.

6.4 An experiment was conducted to estimate the slope η_1 in the simple linear regression of y on x. The predictor had five possible values, -1, -0.5, 0, 0.5, 1, and one observation was sampled from each of the corresponding five conditional distributions of $y \mid x$. The resulting data were of the form (x_i, y_i), $i = 1, \ldots, 5$.

After the experiment, the investigator decided to collect additional data. One assistant suggested that two additional observations be taken, one at $x = -1$ and one at $x = 1$. Another assistant suggested that 20 additional observations be taken, all at $x = 0$. Which design is better? Why? This problem illustrates a general result on how to select predictor values in simple linear regression. Can you guess the result?

6.5 The file `forbes1.lsp` repeats Forbes' data plus additional observations taken by Joseph Hooker at about the same time, but at generally higher altitudes in India.

    6.5.1 In principal, any one case could be *influential*, and its deletion could cause large changes in parameter estimates. Using the model summarized in Table 6.3 fit to Forbes' data only, how does deletion of case 11 change the estimates with these data? You can answer this by comparing (a) the fitted model obtained with all the data and (b) the fitted model obtained with the suspect case not used.

    6.5.2 The Clausius–Clapeyron equation of thermodynamics suggests that the coefficient η_1 in the fit of (6.34) should be equal to -2111, which is the gas constant, the heat of evaporation of water around 373 K. Test the hypothesis that η_1 has this value, against the alternative that η_1 has some other value. Also, obtain a confidence interval for η_1. Do your computations based on (1) the combined data of Forbes and Hooker in data file `forbes1.lsp`; (2) the combined data, but without case 11; (3) Forbes' data alone; and (4) Hooker's data alone. Summarize your results.

6.6 **Regression through the origin.** Occasionally, it may be useful to consider a model in which the intercept is zero: $E(y \mid x) = \eta_1 x$, and $Var(y \mid x) = \sigma^2$. The residual sum of squares function for such *regression through the origin* is $RSS(h) = \sum (y_i - hx_i)^2$.

6.6.1 Show that the least squares estimate of η_1 is given by

$$\hat{\eta}_1 = \sum x_i y_i / \sum x_i^2$$

Show that $\hat{\eta}_1$ is unbiased and that $Var(\hat{\eta}_1) = \sigma^2 / \sum x_i^2$. Find an expression for $\hat{\sigma}^2$. How many df does it have?

6.6.2 Derive the analysis of variance table for the null hypothesis that $E(y \mid x) = \eta_1 x$ versus the alternative $E(y \mid x) = \eta_0 + \eta_1 x$. Show that the F-statistic derived from this table is equal to the square of $t = \hat{\eta}_0 / se(\hat{\eta}_0)$ computed under the alternative model.

6.6.3 In the file `rivlevel.lsp`, the predictor x is equal to the water content of snow on April 1 and the response y is equal to water yield in inches from April to July in Wyoming's Snake River Watershed for the $n = 17$ years from 1919 to 1935.

Draw the plot of y versus x. On the basis of the plot, is linear regression through the origin plausible? Why or why not? The plot's options menu can be used to change the ranges on the axes so 0 is included.

Test the hypothesis that the intercept is zero against the alternative that it is not zero. To compute regression through the origin with *Arc*, in the regression dialog un-check the item "Fit Intercept."

6.7

6.7.1 For the haystack data, let $u = C^3$, and consider the regression of *Vol* on u. Draw the scatterplot of *Vol* versus u. Does a simple linear regression model seem plausible? Why or why not? One of the haystacks seems to be well-separated from the others, with a value of *Vol* that is too large for its circumference. Find its case number.

6.7.2 Based on the simple linear regression model for *Vol* on u, provide 90% prediction intervals for haystacks with values of C equal to 60, 70, and 80 feet. Do you think the prediction interval equation used here would be accurate for a very small haystack with $C = 5$, or a very large one with $C = 150$? Why or why not?

6.7.3 Using the analysis of variance discussed in Section 6.6, perform the following hypothesis test:

$$NH: E(Vol \mid C) = \eta_1 u$$

$$AH: E(Vol \mid C) = \eta_0 + \eta_1 u$$

where under both NH and AH we assume that $Var(Vol \mid C)$ is constant. Summarize the result of the test by giving the p-value, and a one-sentence summary of the result of the test. Also, obtain a t-test of the same hypothesis, and thus show numerically that $t^2 = F$.

6.7.4 Consider the hypothesis test:

$$NH: E(Vol \mid C) = (1/12\pi^2)u$$

$$AH: E(Vol \mid C) = \eta_0 + \eta_1 u$$

This hypothesis test compares the mean function suggested by assuming that a haystack is a hemisphere, as in Section 3.4, to a linear mean function.

Perform the F-test of this hypothesis. (*Hint*: To get the F-statistic, equation (6.47) can be used. You must figure out how to compute RSS_{NH} and df_{NH}.)

6.7.5 Give the estimated slope for the simple linear regression of *Vol* on C^3 when C is measured in yards rather than feet. Will the estimate of σ^2 change when the units if C are changed? (The answer to this question is related to Problem 6.2.)

6.8 The data in the data file `oldfaith.lsp` gives information about eruptions of Old Faithful Geyser during October 1980. Variables are the *duration* in seconds of the current eruption and the *interval*, the time in minutes to the next eruption. The data were collected by volunteers. Apart from missing data for the period from midnight and 6 AM, this is a complete record of eruptions for that month.

Old Faithful Geyser is an important tourist attraction, with up to several thousand people watching it erupt on pleasant summer days. The park service uses data like these to obtain a prediction equation for the time to the next eruption.

6.8.1 Use simple linear regression methodology to obtain a prediction equation for *interval* from *duration*. Summarize your results in a way that might be useful for the nontechnical personnel who staff the Old Faithful Visitor's Center. The remaining problems on the Old Faithful data use these results.

6.8.2 Construct a 95% confidence interval for

$$E(interval \mid duration = 250)$$

6.8.3 An individual has just arrived at the end of an eruption that lasted 250 seconds. Give a 95% confidence interval for the time the individual will have to wait for the next eruption.

6.8.4 Estimate the 0.90 quantile of the conditional distribution of

$$interval \mid (duration = 250)$$

assuming that the population is normally distributed.

6.9 The data in the data file `allomet.lsp` contains information on brain weight, body weight, gestation period, and litter size for 96 placental mammal species. These are not the same data used in Chapter 5, and so some of the values for brain weight and body weight may differ from the values given there. In this problem, consider only the relationship between brain weight and gestation period.

Draw the plot of brain weight versus gestation period. Does the simple linear regression model appear to be satisfactory for these data? If not, perhaps transforming the response, the predictor, or both will help. Add transformation slidebars to the plot using the "Options" plot control, and the methodology of Section 5.2 can be used to try to achieve linearity. When you are satisfied that a simple linear regression mean function might be appropriate in the transformed scale, fit this model, and provide a summary of your fit that you think includes the relevant summary features and plots.

6.10 The paddlefish, *Polydon spathula*, is a large North American freshwater fish. The fossil record shows that it has existed for at least 300 million years. It has a limited range, primarily in the central United States, and exists in very small numbers.

Even though they are protected by law in some states, the paddlefish may become endangered, a victim of past overfishing, pollution, and poaching. Their meat and eggs are considered a delicacy by many, and the value of a single large paddlefish can be several thousand dollars.

The data in the file `paddle.lsp` represents one of the largest collections of data on paddlefish, collected in 1970 on the Mississippi River along the Iowa–Illinois border. For this problem, consider the regression of weight on length. Draw the scatterplot of weight versus length, and find a transformation of one or both of the variables that linearizes the mean function and, if possible, gives a constant variance function. Present both numerical and graphical summaries.

6.11 The data in the file `wine.lsp` gives the average per capita consumption of wine in liters and the mortality rate from heart disease per thousand for 18 countries. Analyze the data with the goal of summarizing the dependence of mortality on wine consumption. If you find a dependence, do you think you can infer *causation*; that is, does drinking wine change (either decrease or increase) the risk of heart disease?

6.12

6.12.1 In the simple linear regression model, show that

$$\rho(\hat{\eta}_0, \hat{\eta}_1) = -\bar{x}\frac{\sqrt{\text{Var}(\hat{\eta}_1)}}{\sqrt{\text{Var}(\hat{\eta}_0)}}$$

where $\rho(a,b)$ is the population correlation. How can the values of x be chosen (1) to make $\rho(\hat{\eta}_0, \hat{\eta}_1)$ arbitrarily close to zero and (2) to make $\rho(\hat{\eta}_0, \hat{\eta}_1)$ close to $+1$ or -1?

6.12.2 *In your own words*, describe what $\rho(\hat{\eta}_0, \hat{\eta}_1)$ means.

6.12.3 Let $z_i = x_i - \bar{x}$ be the predictor centered to have average value zero. An alternative version of the simple regression model is

$$y \mid z = \alpha + \eta_1 z + e$$

where $\alpha = \eta_0 + \eta_1\bar{x}$. Write $\text{Var}(\hat{\alpha})$ as a function of n and σ^2, and find the value of $\rho(\hat{\alpha}, \hat{\eta}_1)$ for this model.

6.13 Suppose you fit a simple regression, $y \mid x = \eta_0 + \eta_1 x + e$. Obtain a 95% confidence interval for a fitted value at $x = 25$ given the following values: $n = 42$; $\bar{x} = 20$; $\text{se}(\hat{\eta}_0) = 6.5$; $\text{se}(\hat{\eta}_1) = 0.3$, and $\hat{\eta}_0 + \hat{\eta}_1 \times 25 = 60$.

6.14 Consider the geese data discussed in Section 6.6. In the OLS fit of the simple linear regression model for *Photo* on *Obs1*, the estimate of σ^2 is 1971.86, and the correlation $r(Photo, Obs1) = 0.8662$. Also, $r(Photo, Obs2) = 0.9245$. Using only these three numbers and without using *Arc*, give the estimate of σ^2 from the OLS fit of the simple linear regression model for *Photo* on *Obs2*.

6.15

6.15.1 Prove that in the OLS fit of a simple linear regression model with an intercept, we have

$$\sum_{i=1}^{n}\hat{e}_i = 0 \qquad \text{and} \qquad \sum_{i=1}^{n}\hat{y}_i = \sum_{i=1}^{n}y_i$$

6.15.2 Show that the sample covariance between $\hat{e}$ and $\hat{y}$ equals zero, and hence that the residuals and the fitted values are uncorrelated. (*Hint*: This problem is easy in the matrix formulation of linear regression, but it is more challenging using the notation of this chapter. It is helpful to write $\hat{y}_i = \bar{y} + \hat{\eta}_1(x_i - \bar{x})$. You will also need to use the result that $\sum(a_i - \bar{a})(b_i - \bar{b}) = \sum(a_i - \bar{a})b_i.$)

6.16   Verify equation (6.57) using the results of Section 4.3 and the fact that $(y, \tilde{x})$ has a bivariate normal distribution. You may need to refer to Section 4.1.4 to find how to evaluate expressions like $\text{Cov}(y, \tilde{x})$.

CHAPTER 7

Introduction to Multiple Linear Regression

We now begin the study of problems with more than one predictor, presenting here the *multiple linear regression model*. This model generalizes simple linear regression in two ways: It allows the mean function to depend on more than one predictor, and to have shapes other than straight lines, although it does not allow for arbitrary shapes.

We first introduce a graphical device called a *scatterplot matrix* that can be used to view many variables at once. Then, we describe how mean functions can be constructed for multiple linear regression, making a distinction between (a) *predictors*, which are measured variables, and (b) *terms*, which are functions of the predictors. We next show how to fit multiple linear regression models and how to do standard operations like estimating parameters, determining their standard errors, and obtaining tests and confidence intervals. Deciding if a particular linear regression model is useful in any given problem and deciding what it means are more challenging questions that we will address in the next few chapters.

The multiple *linear* regression model is not the only approach to studying the dependence of a response on a set of predictors, and later in this book we provide other approaches to studying this type of dependence. However, the tools of linear regression are fundamental to the methodology, and they deserve careful study in their own right.

In a general regression, we have a response y, and p predictors $x_1, \ldots, x_p$. For compact notation, we collect the predictors into a $p \times 1$ vector given by the boldface symbol

$$\mathbf{x} = \begin{pmatrix} x_1 \\ x_2 \\ \vdots \\ x_p \end{pmatrix}$$

TABLE 7.1   The Big Mac Data

BigMac	Minutes of labor required by an average worker to buy a BigMac and French fries.
Bread	Minutes of labor required to buy one kilogram of bread.
BusFare	The lowest cost of a ten-kilometer bus, tram, or subway ticket, in U.S. dollars.
EngSal	The average annual salary of an electrical engineer, in thousands of U.S. dollars.
EngTax	The average tax rate paid by engineers.
Service	Annual cost of 19 services, primarily relevant to Europe and North America.
TeachSal	The average annual salary of a primary school teacher, in thousands of U.S. dollars.
TeachTax	The average tax rate paid by primary teachers.
VacDays	Average days of vacation per year.
WorkHrs	Average hours worked per year.
City	Name of city.

With p predictors, the general goal in regression is to study how the conditional distribution of $y \mid \mathbf{x}$ changes as the value of $\mathbf{x}$ changes, often concentrating on the mean function

$$E(y \mid \mathbf{x}) = E(y \mid x_1, \ldots, x_p)$$

and less frequently on the variance function $\text{Var}(y \mid \mathbf{x})$. These functions now depend on p arguments, the values of the individual predictors. Regression with one predictor is the special case when $p = 1$.

7.1   THE SCATTERPLOT MATRIX

A scatterplot matrix is a 2D array of 2D plots. To introduce the idea, load the file `big-mac.lsp`. This file includes economic data on 45 world cities from the period 1990–1991. The Big Mac hamburger is a simple commodity that is virtually identical throughout the world. One might expect that the price of a Big Mac should therefore be the same everywhere, but of course it is not the same. *The Economist* magazine has published a Big Mac parity index, which compares the costs of a Big Mac in various places, as a measure of inefficiency in currency exchange. We will use these data to study how the cost of a Big Mac varies with economic indicators that describe each city. The variables in the data file are described in Table 7.1; in this discussion we will use *BigMac*, *Bread*, *TeachSal*, *TeachTax*, and *BusFare*.

Select "Scatterplot matrix of" from the Graph&Fit menu. In the resulting dialog, move variable names from the left list to the right in the order *BigMac*, *Bread*, *BusFare*, *TeachSal*, and *TeachTax*, and then push the "OK" button. The

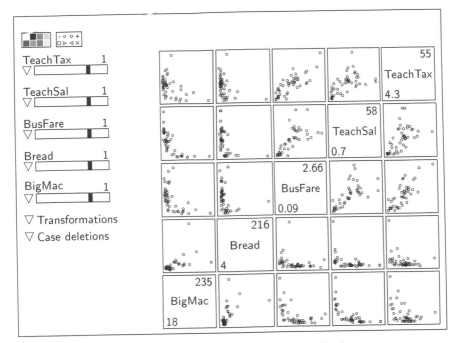

FIGURE 7.1 Scatterplot matrix of the Big Mac data.

plot on your computer screen should look like Figure 7.1. You may wish to make the plot larger by resizing. You can't change the aspect ratio in scatterplot matrices.

Except for the diagonal, each frame of the matrix in Figure 7.1 contains a scatterplot. The variable names on the diagonal label the axes; the variables appear in the order selected beginning in the lower left and proceeding up the diagonal. The numbers in the diagonal cells are the minimum and maximum of the corresponding variable. *TeachTax*, for example, ranges between 4.3% and 55%, while *BusFare* varies from $0.09 to $2.66. The plots above the diagonal are "inverses" of the plots below the diagonal. For example, the bottom right plot is *BigMac* versus *TeachTax*, while the top left plot is *TeachTax* versus *BigMac*. Variable names may be truncated to fit along the diagonal.

7.1.1 Pairs of Variables

Scatterplot matrices tell us about *marginal* relationships between each pair of variables without reference to the other variables, particularly the mean and variance functions. For example, the plot of *TeachTax* versus *BusFare* appears to have an approximately linear mean function with constant variance, which means that, ignoring all other variables, the regression of *TeachTax* on *BusFare* can be described by the simple linear regression model. The plot for *TeachSal* versus *TeachTax* may also be approximately linear, but the variability appears to increase as *TeachTax* increases. The 2D plots that include *Bread* appear to be nonlinear, although the impression of these plots is strongly influenced by at least one isolated point. The characterizations of these plots provide an interesting descriptive summary of the 2D relationships, but how this is related to the larger regression issue of understanding how *BigMac* depends on all four predictors simultaneously is not obvious. For example, does the curvature in the plots including *BigMac* imply that the relationship between *BigMac* and all four predictors is curved? We will return to this important point later.

7.1.2 Separated Points

Scatterplot matrices can be helpful in finding isolated points. In several frames in the scatterplot matrix for the Big Mac data, one point appears to be separated from the others. Is it always the same point? To answer this, simply select the point and it will be highlighted in all frames of the scatterplot matrix. Assuming that you highlighted the point with the extremely high cost of bread, you will see that this point is very low on *TeachSal* and *TeachTax*, moderate on *BusFare*, and high, but not extreme, on *BigMac*. Which city is this? Select "Show labels" from the plot's menu, and then select the point again: This city is Lagos, Nigeria. One might wonder if our impression of these plots would be altered if Lagos were removed from the plot. As with other plots, a point can be removed by selecting the point, and then selecting the menu item "Remove selection" from the plot's menu. After selecting "Rescale plot," the plot is shown in Figure 7.2. Removing Lagos does not change most of the qualitative judgments concerning the bivariate relationships between these variables. Also, we see that the next largest value of *Bread*, for Manila, is 86 minutes.

We are now faced with a common problem in regression analysis: an apparently unusual point. A value of 216 for *Bread* means that the average worker must work for more than 3.5 hours to buy one loaf of bread. At this price, bread must be a luxury item in the Nigerian diet, or the value of 216 is an error. In any case, we choose to continue analysis without Lagos included in the data.

Using the "Remove selection" item in the plot's menu makes a point invisible in that plot and all other plots that are linked to it, but it does not delete the point from the data set. To remove Lagos from the data set, first restore it to the plot by selecting the "Show all" item from the plot's menu. Then, select the point for Lagos again, and finally hold down the mouse button over the plot control called "Case deletions" and select the item "Delete selection from data set." Lagos will now be ignored in all future calculations. You can restore Lagos to the data by selecting the "Restore all" item from the "Case deletions" plot control.

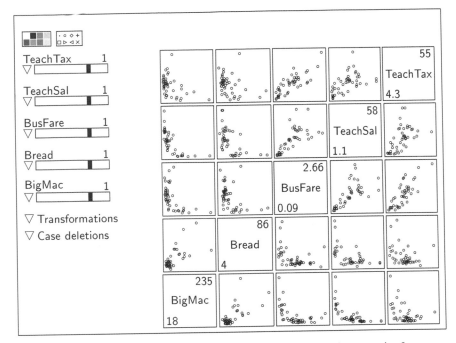

FIGURE 7.2   Rescaled scatterplot matrix of the Big Mac data after removing Lagos.

7.1.3 Marginal Response Plots

Figure 7.3 gives the scatterplot matrix for the Big Mac data, still excluding Lagos, with all the predictors transformed to a log scale. The 2D plots of predictors appear to be more informative in Figure 7.3 than were the corresponding plots in Figure 7.2, because the points in each frame of the plot are more evenly spread out, and relationships between predictors appear to be more nearly linear than were the relationships between the untransformed predictors. We will see later that this is a very desirable situation for understanding a regression.

The last row of Figure 7.3 gives the 2D plots of the response versus each of the transformed predictors. We call these *marginal response plots* because they display the dependence of the response y on each transformed predictor $\log(x_j)$ without regard for any of the other predictors. Each of these four plots shows curvature to some degree. For example, the marginal response plot *Big-Mac* versus $\log(TeachSal)$ suggests that *BigMac* decreases with $\log(TeachSal)$, but the decrease is not necessarily linear in $\log(TeachSal)$. Similar statements can be made concerning the other three marginal response plots, and each of these can give useful information if the plot behaves in an unexpected way. For example, the cost of a Big Mac is lower in cities with higher taxes, a conclusion that might not have been anticipated before looking at the plot.

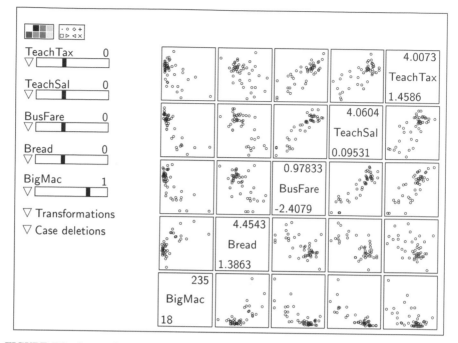

FIGURE 7.3   Scatterplot matrix for the Big Mac data with the predictors in log scale. The transformations are indicated by the numbers above the transformation slidebars. The value 0 corresponds to log transformations.

The marginal response plots display information about the marginal regression of the response on each of the predictors. What can we learn from this row of plots about the full p-dimensional mean function

$$E(y \mid \mathbf{x}) = E(y \mid x_1, x_2, \ldots, x_p)?$$

The marginal response plots always provide a visual lower bound for the goodness-of-fit that can be achieved with the full regression. If x_j does a good job explaining the response y, then a set of predictors that includes x_j shouldn't do worse than x_j alone. Without further knowledge of the regression relationships among the predictors, the marginal response plots tell only about bivariate relationships. For example, the marginal response plots in Figure 7.3 are all curved, indicating that $E(y \mid x_j)$ is nonlinear for each predictor. Can we take this as evidence that the full regression function $E(y \mid \mathbf{x})$ is curved? Similarly, if each of marginal response plots had been linear, could we have concluded that the regression function $E(y \mid \mathbf{x})$ is linear? Without further qualifications that will come in later chapters, the answer to these questions is no.

When inspecting a scatterplot matrix of a response and p predictors, it is important to remember the general purpose of the regression, to study the conditional distribution of $y \mid \mathbf{x}$. Try not to get sidetracked by studying marginal regressions shown in a scatterplot matrix unless there is good reason to do so.

7.1.4 Extracting Plots

Individual plots in a scatterplot matrix can be enlarged by moving the mouse over a plot and holding down both the option and shift keys on the Macintosh, or the control and shift keys on Windows or Unix, and pushing the mouse button. A window will open containing the plot you selected. For example, a larger version of the plot of log(*TeachSal*) versus log(*TeachTax*) can be obtained in this way to examine the relationship between these two variables more closely and to take advantage of the plot controls available on 2D plots.

If you click the mouse on a variable name with the appropriate keys held down, a histogram of that variable or of its transformation will be created in a window.

7.2 TERMS AND PREDICTORS

A general regression has a response variable y and p predictors collected into the vector $\mathbf{x}$, and the goal is to study how the conditional distribution of $y \mid \mathbf{x}$ changes as the value of $\mathbf{x}$ changes, usually concentrating on the mean function $E(y \mid \mathbf{x})$. For example, in the Big Mac data, the response is the minutes of labor required to buy a Big Mac hamburger and French fries, and the four predictors describe economic variables—either prices, taxes, or salaries—for each of the cities in the study.

In a study of the effects of environmental factors on fish growth in several lakes, the response might be the average size or growth rate of a particular species of fish in the lake. The predictors might consist of characteristics of the lakes, including size, location, maximum depth, concentrations of pollutants, and pH. Additional predictors might describe the management practices for the lake, such as whether or not fishing is permitted.

The definition of the predictors may not be unique; for example, a concentration could be replaced by its logarithm, or several predictors might be combined in some way into an index. For now we will take the predictors as given.

Predictors can be numerical measurements or they can be categorical. Measured predictors include age, weight, length, temperature, and so on. A categorical predictor need not be numerical and could be a variable like sex, which has the two categories male and female, or eye color, which may have many categories like blue, gray, and green. Some categorical predictors have ordered categories; for example, judgment of health made by a physician may be chosen from poor, good, very good, and excellent. While very good is better than good, the categories may not be assigned numerical scores easily because the difference between poor and good may be quite different from the difference between good and very good.

Terms are built from the predictors. In its most general form, the multiple linear regression mean function is

$$E(y \mid \mathbf{x}) = \eta_0 u_0 + \eta_1 u_1 + \cdots + \eta_{k-1} u_{k-1} \qquad (7.1)$$

There are k *terms* in this mean function, $u_0, \ldots, u_{k-1}$. Each term u_j is a function of the predictors. A more complete representation for the jth term would be $u_j(\mathbf{x})$, to show the dependence on $\mathbf{x}$ explicitly. We will generally use the simpler notation, but keep in mind that terms always depend on the predictors. Conditioning on $\mathbf{x}$ on the left side of (7.1) may seem a bit odd, since $\mathbf{x}$ does not appear explicitly on the right side of this equation, but if we condition on $\mathbf{x}$, then the u-terms are determined. This mean function is said to be *linear* because the *regression coefficients* $\eta_0, \eta_1, \ldots, \eta_{k-1}$ enter the equation linearly. Examples of linear and nonlinear regression models are give in Problem 7.4.

Here are definitions of terms that can appear in multiple linear regression models.

The Intercept. The term u_0 for the intercept is a constant, always equal to one, $u_0 = 1$. Most of the models we study will include an intercept. As in simple regression, η_0 is the the value of $E(y \mid \mathbf{x})$ when all the other terms equal zero. The multiple linear regression model with an intercept can be written as

$$E(y \mid \mathbf{x}) = \eta_0 u_0 + \eta_1 u_1 + \cdots + \eta_{k-1} u_{k-1} \qquad (7.2)$$

$$= \eta_0 + \eta_1 u_1 + \cdots + \eta_{k-1} u_{k-1} \qquad (7.3)$$

Since $u_0 = 1$, we usually write just η_0 instead of $\eta_0 u_0$, as in (7.3). Models without an intercept can also be expressed in the form (7.2) setting $u_0 = 0$.

Predictors. The simplest type of term is equal to one of the predictors, x_j, $u_1 = u_1(\mathbf{x}) = x_j$. Many multiple linear regressions will consist solely of terms of this type.

Powers of Predictors. Terms can be constructed from powers of a predictor. For example, $u_2 = x_j^2$ has values that are the squares of the values of x_j; other powers can be used as well. Mean functions that can be expressed as polynomials in a predictor are therefore included in the multiple linear regression family.

Transformations of Predictors. A term might consist of a transformation of a predictor, like $u_3 = \log(x_j)$ if x_j is strictly positive. The transformation can be as simple as the logarithm or more complex. For example, the two predictors height and weight might be used to create a body mass index, given by height/weight^2.

Binary Predictors. A binary predictor consists of two categories, such as success or failure, living or dead, treated or not. A binary predictor can be included by defining a term that is coded with the values 0, perhaps for failure, or 1, perhaps for success.

Factors. Categorical predictors are called *factors*. A factor with only two categories is equivalent to a binary predictor, but with more than two categories several binary predictors are required. A factor with ℓ distinct levels usually requires the addition of $\ell - 1$ terms to a mean function. We defer the details of using factors to Chapter 12.

Interactions. An interaction term is equal to the product of two or more other terms. The simplest interaction is the two-term interaction, consisting of the product of two predictors, $u_4 = x_1 x_2$, but more complex interactions like $u_5 = \log(x_1) \times x_2^2$ are possible.

A regression with p predictors can have any number of terms in the mean function, but fitting problems result if the number of terms is greater than the number of cases n. The mean function can have fewer than p terms if some of the predictors are ignored or appear in the mean function only through combinations of the predictors. There can be more than p terms if the mean function includes polynomials, factors, and interactions. The distinction between predictors and terms plays an important role in understanding regression through plots. For example, a regression with one predictor can always be studied using the 2D scatterplot of the response versus the predictor, regardless of the number of terms required in the mean function.

7.3 EXAMPLES

To illustrate the use of terms and predictors, we briefly describe several regressions that are discussed at greater length elsewhere in this book.

7.3.1 Simple Linear Regression

Let's begin by reviewing the mean functions used for Forbes' data in Chapter 6, and recast them using the notation of this chapter. The response is $y = \log_{10}(\text{pressure})$, and the single predictor is $x = \text{temperature}$ in degrees Fahrenheit. The first mean function considered for Forbes' data was, repeating (6.2),

$$E(y \mid x) = \eta_0 u_0 + \eta_1 u_1 = \eta_0 + \eta_1 x \qquad (7.4)$$

This mean function has two terms. The term u_0 for the intercept is always equal to 1. The second term is $u_1 = x$, so this term is equal to the predictor. A plot of this mean function against x will trace a straight line.

An alternative mean function was used for Forbes' data in Section 6.5. That mean function is given by

$$E(y \mid x) = \eta_0 + \eta_1 u_2 = \eta_0 + \eta_1 \left(\frac{1}{255.37 + \frac{5}{9}x} \right) \qquad (7.5)$$

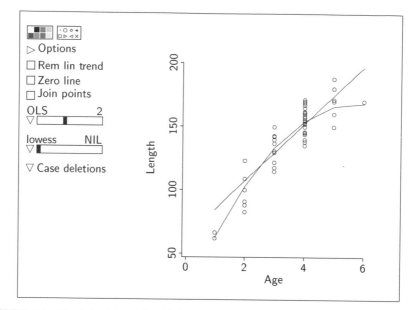

FIGURE 7.4   The Lake Mary bluegill data. The straight line is obtained by using OLS to fit simple linear regression. The curved line is obtained by using OLS to fit the quadratic polynomial (7.8).

This is also a linear regression mean function, again with two terms. As before, the first term is the constant $u_0 = 1$. The second term is $u_2 = 1/(255.37 + (5/9)x)$, the inverse of the temperature measured in degrees Kelvin. The same symbols have been used for the parameters η_0 and η_1 in (7.4) and (7.5), but in each the parameters have different meanings and values. In (7.4), η_1 gives the change in $E(y \mid x)$ for a unit change in temperature in degrees Fahrenheit, while in (7.5), η_1 gives the change in $E(y \mid x)$ for a unit change in the inverse of the temperature in degrees Kelvin. η_0 is an intercept parameter in each of the two mean functions, but the values of the intercept are likely to differ. Perhaps some clarity could have been obtained if we had used different symbols for the parameters in these two mean functions, but this can become cumbersome when we consider several mean functions at once. Using the same symbols in many mean functions, even though their meanings may change, is a common practice that we will follow in this book.

7.3.2   Polynomial Mean Functions with One Predictor

Figure 7.4 is a plot of y = length against x = age for the Lake Mary bluegill data. Superimposed on this plot is the OLS fit of

$$E(y \mid x) = \eta_0 + \eta_1 x \tag{7.6}$$

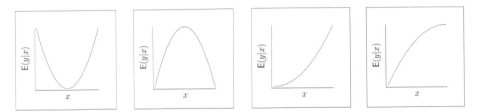

FIGURE 7.5 Four polynomial mean functions. In (a) and (b) the minimum or maximum is in the range of x, while in (c) and (d) the minimum or maximum is outside the range of x.

which is the simple linear regression mean function with the two terms $u_0 = 1$ and $u_1 = x$. This function is shown as a straight line on the plot. The mean function does not appear to match the data very well, as the line is above all the observed values at ages one and six, and does not appear to match the subpopulation averages for ages two and five. A straight-line mean function specifies that the rate of growth is the same for all ages of fish, but judging from the plot, growth rates appear to be slower for older ages.

Lacking a specific functional form for the mean function, we might try to match the data by using a polynomial. *Polynomial regression* provides a rich class of mean functions to fit as alternatives to the straight line used in simple linear regression. A polynomial is often used to approximate a mean function for which we don't have a specific form, and they work well because a function can often be approximated adequately by a polynomial. When using a polynomial of degree d for the mean function we have

$$E(y \mid x) = \eta_0 + \eta_1 x + \eta_2 x^2 + \cdots + \eta_d x^d \tag{7.7}$$

The polynomial of degree d with one predictor has $k = d + 1$ terms, one for the intercept and one for each of the d powers of x.

The shape of the polynomial depends on both the degree d and on the η's. When $d = 2$, the polynomial is a quadratic in x and, depending on the η's, the mean function can have any of the shapes shown in Figure 7.5. In the first two frames of Figure 7.5, we see the full quadratic curve, which is symmetric about a minimum or maximum. Quadratic polynomials are often used in experiments where the goal is to determine the value of x where the mean function is minimized or maximized. The quadratic curve is one of the simplest functions that has a single minimum or maximum.

The remaining two curves in Figure 7.5 are obtained by using only a part of the quadratic curves; in these figures, the minimum or maximum occurs outside the range of the data. A quadratic can therefore be used to provide an empirical mean function that increases at an increasing rate as x increases or a mean function that has a decreasing rate of change with x. Polynomials with degree $d > 2$ are more flexible than quadratics and include (a) mean functions that are not symmetric about a minimum or maximum and (b) functions with local minima or maxima.

In *Arc*, polynomial mean functions with one predictor are fit on a scatterplot of the response against the predictor using the parametric smoother slidebar. The degree of the polynomial, as given by the number above the slidebar, increases as the slidebar is moved to the right. For the Lake Mary data in Figure 7.4, the second-degree polynomial fit is similar in shape to the quadratic in Figure 7.5d. There appears to be an improvement over the straight-line mean function of (7.6) by fitting this quadratic instead, as given by

$$E(y \mid Age) = \eta_0 + \eta_1 Age + \eta_2 Age^2 \qquad (7.8)$$

7.3.3 Two Predictors

An important component in making a business more efficient is to understand its costs. In some manufacturing businesses, determining costs can be fairly easy, but in the service industry where the main component of cost is in time spent serving customers, determining costs can be harder. Imagine a large enterprise like a bank whose employees perform two types of transactions with customers. In this example, data were collected on $n = 261$ branch banks; on each we have the response *Time* = total minutes of labor spent on transactions, along with two predictors T_1 and T_2 that give the number of transactions of each type for the 1985–1986 financial year. Since cost is determined largely by transaction time, our goal is to study *Time* as a function of the counts of the two types of transactions. All transactions in these branches are either of type one or type two.

Suppose that, on average, a transaction of type one takes η_1 minutes, and a transaction of type two takes η_2 minutes. If all transactions are independent, the average time spent on transaction type one in a given branch is expected to be $\eta_1 T_1$, but any specific transaction may take more or less time. Similarly, the time spent on type two transactions is expected to be $\eta_2 T_2$ minutes; thus the total time on transactions is expected to be $\eta_1 T_1 + \eta_2 T_2$, giving the mean function

$$E(Time \mid T_1, T_2) = \eta_0 + \eta_1 T_1 + \eta_2 T_2 \qquad (7.9)$$

The additional parameter η_0 represents a fixed cost of doing business in a branch, and it is the number of minutes of labor even when no transactions occur; possibly $\eta_0 = 0$. This example has two predictors T_1 and T_2 and three terms, $u_0 = 1$, $u_1 = T_1$, and $u_2 = T_2$. The mean function is constructed from considerations about the predictors. Even here, however, the mean function could be wrong if transactions were not independent, if the branches performed some other type of transaction not included in the data, or if the complexity of transactions was different in each branch.

In (7.9), each of the parameters has a clear meaning: η_0 is the minutes of labor required in the absence of transactions; η_1 is the average minutes of labor per transaction of type one, so it has the units of minutes per transaction, and similarly η_2 is in minutes per transaction. The coefficients η_1 and η_2 convert the units of T_1 and T_2, which are the number of transactions, to the units of the response which are minutes.

7.3.4 Polynomial Mean Functions with Two Predictors

The terms in a polynomial mean function with two predictors, x_1 and x_2, are generally powers and interactions. The *full quadratic polynomial mean function* has six terms: an intercept, two linear terms x_1 and x_2, two squared terms x_1^2 and x_2^2, and one interaction term $x_1 x_2$:

$$E(y \mid \mathbf{x}) = \eta_0 + \eta_1 x_1 + \eta_2 x_2 + \eta_{11} x_1^2 + \eta_{22} x_2^2 + \eta_{12} x_1 x_2 \qquad (7.10)$$

The double subscripts on the η's are intended as reminders about the corresponding terms. For example, the term for η_{11} is $x_1 x_1 = x_1^2$, while the term for η_{12} is $x_1 x_2$. The squared terms allow for nonlinearity in the corresponding predictors, as shown in Figure 7.5, while the interaction term allows for twisting of the surface in a 3D plot of $E(y \mid \mathbf{x})$ versus the two predictors. 3D plots will be discussed in Chapter 8.

To gain a little understanding of the full quadratic mean function, imagine holding x_2 fixed and allowing x_1 to vary. With this in mind, rewrite the mean function as

$$E(y \mid \mathbf{x}) = \{\eta_0 + \eta_2 x_2 + \eta_{22} x_2^2\} + \{\eta_1 + \eta_{12} x_2\} x_1 + \{\eta_{11}\} x_1^2$$

Holding x_2 fixed, this is a quadratic polynomial in x_1. The intercept is the first quantity in curly brackets, the first regression coefficient is the second quantity $\eta_1 + \eta_{12} x_2$ in curly brackets, and the coefficient of the quadratic term is η_{11}. As we change the value of x_2, this intercept changes as a quadratic function of x_2, while the first regression coefficient changes linearly. The coefficient of x_1^2 is constant. If $\eta_{12} = 0$ so the interaction term is not needed, then the coefficient of x_1 is constant as well. We return to the interpretation of the full quadratic mean function in Chapter 8.

Polynomial mean functions in two predictors can contain higher powers and interactions as well. For example, they might contain the terms x_1^3, $x_1 x_2^2$ or $x_1^3 x_2^3$. Mean functions containing such higher-order terms can be quite difficult to interpret, and they tend to be the exception rather than the rule in practice.

7.3.5 Many Predictors

Return to the Big Mac data presented in Section 7.1. Lacking a theory that connects *BigMac* to the predictors, we are apparently free to contemplate a wide variety of mean functions; in particular, we have no reason to believe that the mean function with each term equal to a predictor,

$$E(BigMac \mid \mathbf{x}) = \eta_0 + \eta_1 Bread + \eta_2 TeachSal + \eta_3 TeachTax + \eta_4 BusFare$$

will match the data or provide any useful information. For example, the alternative mean function

$$E(BigMac \mid \mathbf{x}) = u_0 + \eta_1 u_1 + \eta_2 u_2 + \eta_3 u_3 + \eta_4 u_4$$

with the five terms

$$u_0 = 1, \qquad u_1 = \log(Bread), \qquad u_2 = \log(TeachSal),$$

$$u_3 = \log(TeachTax), \qquad \text{and} \qquad u_4 = \log(BusFare)$$

may provide a better description. This latter mean function uses terms corresponding to the plotted quantities in the scatterplot matrix in Figure 7.3. Part III of this book presents methods for building mean functions when a general theory is lacking.

7.4 MULTIPLE LINEAR REGRESSION

Before proceeding to estimation in the next section, we are in a position to give a more comprehensive statement of the family of multiple linear regression models. All the above examples can be approached using the same general framework. We begin with p predictors, $x_1, x_2, \ldots, x_p$, which will be collectively referred to by using the vector $\mathbf{x}$. The response is called y. All the variables have been measured on each of n independent cases. From the predictors, we create the terms $u_0, u_1, u_2, \ldots, u_{k-1}$; for this discussion we will assume that the term u_0 is included for the intercept, so there are $k - 1$ regression coefficients in addition to the intercept. Most regression fitting procedures require specifying at least the mean function and the variance function. Recall that the mean function for multiple linear regression with an intercept is

$$E(y \mid \mathbf{x}) = \eta_0 + \eta_1 u_1 + \cdots + \eta_{k-1} u_{k-1} \qquad (7.11)$$

The variance function is for now assumed to be constant,

$$Var(y \mid \mathbf{x}) = \sigma^2 \qquad (7.12)$$

Equations (7.11) and (7.12) define a multiple linear regression model.

For notational simplicity, we will rewrite (7.11) with the predictors, terms, and parameters expressed as *vectors*. In addition to $\mathbf{x}$, $\mathbf{u}$ is the $k \times 1$ vector of terms and η is a $k \times 1$ vector whose elements are $\eta_0, \ldots, \eta_{k-1}$. Using the convention that a vector is always a column, we can write these vectors in full as

$$\mathbf{x} = \begin{pmatrix} x_1 \\ x_2 \\ \vdots \\ x_p \end{pmatrix}, \qquad \mathbf{u} = \begin{pmatrix} u_0 \\ u_1 \\ \vdots \\ u_{k-1} \end{pmatrix} = \begin{pmatrix} 1 \\ u_1 \\ \vdots \\ u_{k-1} \end{pmatrix}, \qquad \text{and} \qquad \eta = \begin{pmatrix} \eta_0 \\ \eta_1 \\ \vdots \\ \eta_{k-1} \end{pmatrix}$$

Also, we will write, for example, $\mathbf{x} = (x_1, \ldots, x_p)^T$, since the transpose T changes a column vector into a row vector. A brief introduction to matrices and vectors is given in Section 7.9.1.

We can now rewrite (7.11) as

$$E(y \mid \mathbf{x}) = \eta^T \mathbf{u}$$
$$= \eta_0 + \eta_1 u_1 + \cdots + \eta_{k-1} u_{k-1} \qquad (7.13)$$

once again with the reminder that $\mathbf{u}$ depends on $\mathbf{x}$. Apart from the complements sections, about the only matrix algebra used is the definition of $\eta^T \mathbf{u}$, which is shown in Section 7.9.1 to equal the final equation of (7.13).

When referring to the data from a particular regression, we may add the subscript "i" to indicate the ith case, just as we did in simple linear regression. For example, (7.13) may be rewritten as

$$E(y_i \mid \mathbf{x}_i) = \eta^T \mathbf{u}_i$$
$$= \eta_0 + \eta_1 u_{i1} + \cdots + \eta_{k-1} u_{i,k-1}$$

where $\mathbf{x}_i = (x_{i1}, \ldots, x_{ip})^T$ is the vector of predictor values for the ith case, and x_{ij} is the ith value of the jth predictor, $i = 1, \ldots, n$, $j = 1, \ldots, p$. Similarly, $\mathbf{u}_i = (u_{i0}, u_{i1}, \ldots, u_{i,k-1})^T$ is the vector of term values for the ith case in the data, and u_{ij} is the ith value of the jth term, where again $i = 1, \ldots, n$, $j = 0, \ldots, k - 1$. These subscripts will be used when needed to clarify exposition, or to describe algebraic operations with the data.

Even allowing for a wide variety of definitions of terms from the predictors, there is no intrinsic reason why this family of models should be appropriate for every regression. For now, however, its usefulness is unquestioned, and we seek estimates assuming that this mean function is correct.

7.5 ESTIMATION OF PARAMETERS

We now turn to estimating the unknown parameters in a mean function for multiple linear regression. As with simple regression, we use least squares estimates that minimize the residual sum of squares function,

$$RSS(\mathbf{h}) = \sum_{i=1}^{n}(y_i - \mathbf{h}^T\mathbf{u}_i)^2 \tag{7.14}$$

The symbol $\mathbf{h}$ is used rather than η because η is a fixed though unknown parameter vector. We will write $\hat{\eta}$ for the value of $\mathbf{h}$ that minimizes (7.14), and

$$\hat{\eta} = \begin{pmatrix} \hat{\eta}_0 \\ \hat{\eta}_1 \\ \vdots \\ \hat{\eta}_{k-1} \end{pmatrix}$$

As long as some of the terms are not exact linear combinations of each other, and $k < n$, this minimization problem has a unique solution $\hat{\eta}$, and there is an equation that can be used to compute the estimates. We will call these the ordinary least squares (OLS) estimates as before. Deriving the estimates and giving their formulas is greatly simplified using matrix and vector notation, and we reserve this to Section 7.9.5.

Once $\hat{\eta}$ is determined, we have

$$i\text{th fitted value} = \hat{y}_i = \hat{\eta}^T\mathbf{u}_i$$

$$i\text{th residual} = \hat{e}_i = y_i - \hat{y}_i$$

$$RSS = \sum_{i=1}^{n}(y_i - \hat{y}_i)^2 = \sum_{i=1}^{n}\hat{e}_i^2 \tag{7.15}$$

$$\hat{\sigma}^2 = \frac{RSS}{n-k}$$

To illustrate the methodology, we return to the transactions example presented in Section 7.3.3. For the remainder of this chapter, we will assume that the variance $Var(Time \mid T_1, T_2)$ is constant, although we might expect that variability will be larger in branches with many transactions. If the assumption of constant variance is not satisfactory, then the estimates, and particularly standard errors, can be improved upon by using information about the variances in computing estimates.

The OLS estimates and related statistics depend on just a few basic summary statistics. After loading the file `transact.lsp`, those summary statistics can be obtained by selecting the item "Display summaries" from the Transactions menu. This gives a dialog to choose the variables of interest; select $Time$, T_1 and T_2. The summary statistics are shown in Table 7.2.

TABLE 7.2 Summary Statistics for the Transactions Data

Data set = Transactions, Summary Statistics					
Variable	N Average	Std. Dev	Minimum	Median	Maximum
Time	261 6607.4	3774.	487.	5583.	20741.
T1	261 281.21	257.08	0.	214.	1450.
T2	261 2421.7	1180.7	148.	2192.	5791.
Data set = Transactions, Sample Correlations					
Time	1.0000	0.8632	0.9236		
T1	0.8632	1.0000	0.7716		
T2	0.9236	0.7716	1.0000		
	Time	T1	T2		

From Table 7.2 we see that the range of each of the variables is quite large. The values of *Time* range from about 487 to over 20,000 minutes. The standard deviation of the *Time* measurements is about 3800 minutes. The number of type one transactions T_1 ranges from no transactions in some branches to 1450 in other branches. T_2 also has a wide range of values. As in Section 5.4, these large ranges are an indication that the logarithm may be a useful transformation of the predictors. We don't pursue transformations here because our emphasis is on basic fitting and because the mean function we use has a theoretical basis.

Also included in Table 7.2 is a matrix of sample correlations. For example, the sample correlation between T_1 and *Time* is 0.86, and the correlation between T_1 and T_2 is 0.77. This matrix of correlations is symmetric because the correlation between two variables v_1 and v_2 is the same as the correlation between v_2 and v_1. The correlations are fairly large in absolute value. From Section 6.2.4, we know that each correlation is a summary statistic for the 2D scatterplot of the variables used in the correlation, and that these statistics cannot be easily interpreted without examining the scatterplot itself. All the 2D scatterplots are given in the scatterplot matrix in Figure 7.6. One interesting feature of these plots is the large number of cases for which $T_1 = 0$: many branches do not perform type one transactions. While this does not invalidate the fitting of multiple linear regression, it does make the use of a correlation as a summary statistic questionable because the correlation is a measure of the linear relationship between variables.

Computing least squares estimates for multiple linear regression using *Arc* is similar to computing estimates for simple linear regression. As shown in Section 7.9.5, the least squares estimates depend only on the summary statistics shown in Table 7.2; any two data sets for which these values are identical will give the same least squares estimates. After loading the file transact.lsp, select the item "Fit linear LS" from the Graph&Fit menu. In the resulting dialog, move *Time* to be the response and move T_1 and T_2 to make them predictors. The output shown is in Table 7.3.

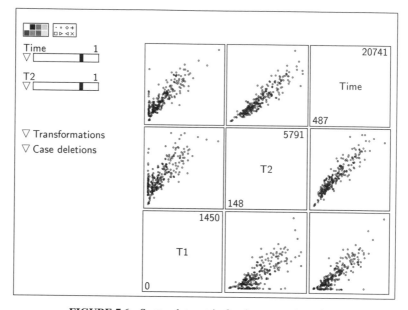

FIGURE 7.6   Scatterplot matrix for the transactions data.

TABLE 7.3   The OLS **Regression of** *Time* **on** T_1 **and** T_2

```
Data set = Transactions, Name of Model = L1
Normal Regression Model
Kernel Mean function = Identity
Response     = Time
Terms        = (T1 T2)
Coefficient Estimates
Label         Estimate     Std. Error    t-value
Constant      144.369      170.544       0.847
T1            5.46206      0.433268      12.607
T2            2.03455      0.0943368     21.567

R Squared:              0.909053
Sigma hat:              1142.56
Number of cases:          261
Degrees of freedom:       258

Summary Analysis of Variance Table
Source     df        SS            MS             F         p-value
Regression 2    3366491409.  1683245705.    1289.42    0.0000
Residual   258  336801747.   1305433.
```

The first part of Table 7.3 summarizes the regression, giving the type of mean function, the response, and the terms in the model. The information given about the mean function is not yet relevant, but will be so in later chapters. The next part of the table, headed "Coefficient Estimates," gives information about the $\hat{\eta}$'s. In particular,

$$\hat{\eta} = \begin{pmatrix} \hat{\eta}_0 \\ \hat{\eta}_1 \\ \hat{\eta}_2 \end{pmatrix} = \begin{pmatrix} 144.369 \\ 5.46206 \\ 2.03455 \end{pmatrix}$$

The estimated transaction time for type one is about 5.5 minutes per transaction, and for type two it is about 2.0 minutes per transaction. The intercept is about 144 minutes. After rounding the coefficient estimates, the mean function estimated using OLS is thus

$$\hat{E}(Time \mid T_1, T_2) = \hat{\eta}_0 + \hat{\eta}_1 T_1 + \hat{\eta}_2 T_2$$
$$= 144.4 + 5.5 T_1 + 2.0 T_2$$

The estimate of σ^2 depends on the residual sum of squares and is given by the formula

$$\hat{\sigma}^2 = \frac{\sum_{i=1}^n (y_i - \hat{y}_i)^2}{n-k} = \frac{\sum_{i=1}^n \hat{e}_i^2}{n-k} = \frac{RSS(\hat{\eta})}{n-k} = \frac{RSS}{n-k} \qquad (7.16)$$

where the fitted values and residuals were defined at (7.15). Additional short-cut formulas for computing RSS are given in Section 7.9.5. The denominator $n-k$ is the degrees of freedom for $\hat{\sigma}^2$. There are $n-k = 261 - 3 = 258$ df for the transactions data. The value of $\hat{\sigma}$ is about 1143 minutes from Table 7.3. The value of $\hat{\sigma} = 1143$ is much smaller than the standard deviation of $Time$, 3774 minutes, given in Table 7.2. Given T_1 and T_2, the standard deviation of $Time$ is 1143; without knowledge of T_1 and T_2, the standard deviation of $Time$ is 3774.

Recall that the standard error of a statistic is the estimated standard deviation of that statistic. The computer output in Table 7.3 has a column marked "Std. Error," and this column gives the standard errors. Both $\hat{\eta}_1$ and $\hat{\eta}_2$ are large relative to their standard errors. The column labeled "t-value" gives the ratio of each estimate to its standard error. The intercept $\hat{\eta}_0$ is somewhat smaller than its standard error.

A complete description of the variability of the coefficient estimates requires determining both the estimated variances of the estimates, which are just the squares of the standard errors, and the estimated covariances between the estimates. The variances and covariances are usually collected into a $k \times k$ matrix. Using equation (7.42) in Section 7.9.5, the variance–covariance matrix can be written as

$$Var(\hat{\eta}) = \sigma^2 \mathbf{M} \qquad (7.17)$$

TABLE 7.4　**The Variance–Covariance Matrix, and Correlation Matrix of the Coefficient Estimates, for the Transactions Data**

```
Variance–covariance matrix of the coefficient estimates
Constant       29085.          23.582        -12.683
T1             23.582           0.18772       -0.031536
T2            -12.683          -0.031536       0.0088994
               Constant        T1             T2
Correlation matrix of the coefficient estimates
Constant      1.0000           0.3191         -0.7883
T1            0.3191           1.0000          -0.7716
T2           -0.7883          -0.7716          1.0000
              Consta           T1              T2
```

where $\mathbf{M}$ is a known $k \times k$ matrix that can be computed from the observed values of the terms. The estimated variance-covariance matrix is

$$\widehat{\mathrm{Var}}(\hat{\boldsymbol{\eta}}) = \hat{\sigma}^2 \mathbf{M} \tag{7.18}$$

The elements of the matrix $\hat{\sigma}^2\mathbf{M}$ can be viewed by selecting the item "Display variances" from the regression menu. The output is shown for the transaction data in Table 7.4. The diagonal elements of the estimated variance–covariance matrix $\hat{\sigma}^2\mathbf{M}$ are the squares of the standard errors of the regression coefficient estimates; for example, $\mathrm{se}(\hat{\eta}_1) = 0.1877^{1/2} = 0.4332$. The off-diagonal elements of $\hat{\sigma}^2\mathbf{M}$ are the estimated covariances between estimated coefficients. The matrix $\hat{\sigma}^2\mathbf{M}$ is symmetric because the covariance between $\hat{\eta}_j$ and $\hat{\eta}_k$ is the same as the covariance between $\hat{\eta}_k$ and $\hat{\eta}_j$.

The second matrix shown in Table 7.4 takes $\hat{\sigma}^2\mathbf{M}$ and converts it to a correlation matrix, using the formula $\rho(\hat{\eta}_i,\hat{\eta}_j) = v_{ij}/(v_{ii}v_{jj})^{1/2}$ as described in Section 7.9.3. While the elements of $\hat{\sigma}^2\mathbf{M}$ are occasionally needed in computations, the rescaling of them into a correlation matrix can make interpretation of the elements easier. We see that $\hat{\eta}_2$, the coefficient estimate for T_2, is negatively correlated with both the intercept and the coefficient estimate for T_1, while $\rho(\hat{\eta}_0,\hat{\eta}_1)$ is considerably smaller.

7.6　INFERENCE

Making inferences requires an assumption about the distribution of the response given the predictors. The assumption we need, in addition to fitting the correct mean function and the correct variance function, is that *Time* given T_1 and T_2 is normally distributed:

$$Time \mid (T_1, T_2) \sim \mathrm{N}(\boldsymbol{\eta}^T\mathbf{u}, \sigma^2) \tag{7.19}$$

This is a direct generalization of the assumption used for simple regression in Section 6.4, and the results here generalize the simple regression results. In Section 6.7.6 the discussion of the importance, or more precisely the lack of importance, of this assumption applies to multiple regression as well.

As a result of the normality assumption, the coefficient estimates are normally distributed:

$$\hat{\eta} \sim N(\eta, \sigma^2 \mathbf{M}) \tag{7.20}$$

Assuming that the mean function is appropriate, that the variance function is constant, and that $y \mid \mathbf{x}$ is normally distributed, the estimate $\hat{\eta}$ is normally distributed, with mean η and variance–covariance matrix $\sigma^2 \mathbf{M}$. Since σ^2 is unknown, we estimate it using $\hat{\sigma}^2$. Additionally, in parallel to (6.24), we have

$$(n - k)\frac{\hat{\sigma}^2}{\sigma^2} \sim \chi^2_{n-k} \tag{7.21}$$

7.6.1 Tests and Confidence Statements about Parameters

Given the results of the last section, all the results for simple linear regression of Section 6.4.1 apply without modification. For example, a $(1 - \alpha) \times 100\%$ confidence interval for η_j, $j = 0, 1, \ldots, k - 1$, is

$$\hat{\eta}_j - Q(t_d, 1 - \alpha/2)\text{se}(\hat{\eta}_j) \leq \eta_j \leq \hat{\eta}_j + Q(t_d, 1 - \alpha/2)\text{se}(\hat{\eta}_j) \tag{7.22}$$

where the df $d = n - k$, and as before $Q(t_d, f)$ is the fth quantile of the t_d distribution. This is the same as (6.25), with a slight change of notation.

For the transactions data, $d = 258$, so to get 90% confidence intervals we obtain $Q(t_{258}, 0.95) = 1.65$ using the "Calculate quantile" item. The confidence intervals for the two transaction time parameters are

$$5.46 - (1.65 \times 0.43) \leq \eta_1 \leq 5.46 + (1.65 \times 0.43)$$

$$4.75 \leq \eta_1 \leq 6.17$$

and

$$2.03 - (1.65 \times 0.094) \leq \eta_2 \leq 2.03 + (1.65 \times 0.094)$$

$$1.88 \leq \eta_2 \leq 2.19$$

respectively.

Tests of hypotheses concerning individual coefficients parallel those for simple linear regression. For example, to test

$$\text{NH} : \eta_0 = 0, \quad \eta_1, \eta_2 \text{ arbitrary}$$

$$\text{AH} : \eta_0 \neq 0, \quad \eta_1, \eta_2 \text{ arbitrary}$$

we can use a Wald test based on the test statistic

$$t = \frac{\text{estimate} - \text{hypothesized value}}{\text{se(estimate)}}$$

The column marked "t-value" in Table 7.3 gives the Wald test statistic that is appropriate for testing each of the individual coefficients to be zero. For the intercept, the value is $t = 0.847$. To get a p-value, use the "Calculate probability" item. Since this is a two-tailed alternative, the p-value is about 0.4. There is little reason to think the intercept is different from zero.

7.6.2   Prediction

Moving from simple regression to multiple regression requires only minor modification of the results of Section 6.4.3. Suppose we have observed values of the predictors $\mathbf{x}$ for a new case for which the response is as yet unobserved. To predict the response, first compute the corresponding terms $\mathbf{u}$ from $\mathbf{x}$. The point prediction is just

$$y_{pred} \,|\, \mathbf{x} = \hat{\eta}^T \mathbf{u}$$

The standard error of prediction has two components: a part due to uncertainty in estimating η and a part due to the variability in the future observation. Combining these, the standard error of prediction is

$$\text{se}(y_{pred} \,|\, \mathbf{x}) = \sqrt{\hat{\sigma}^2 + [\text{se}(\hat{\eta}^T \mathbf{u})]^2}$$
$$= \sqrt{\hat{\sigma}^2 + \hat{\sigma}^2 h}$$
$$= \hat{\sigma}\sqrt{1 + h} \qquad (7.23)$$

The quantity h in (7.23) depends on the values of the predictors for all the cases in the data, but it does not depend on the response. It is called a *leverage* and is discussed in the next section.

The item "Prediction" in the regression menu can be used to obtain predictions, estimated mean values, and their standard errors, as described in Section 6.4.2. The only change is that now a value for each term in the model except the intercept must be entered in the text area of the dialog of Figure 6.8, page 116. For example, the results for a new branch with $T_1 = 500$ and $T_2 = 1000$ are

```
Response = Time
Terms = (T1 T2)
Term values = (500 1000)
Prediction = 4909.95, with se(pred) = 1164.91,
Leverage = 0.0395223, Weight = 1
Estimated population mean value = 4909.95, se = 227.143
```

The predicted number of minutes of labor is about 4910 with a standard error of about 1165 minutes. Prediction intervals can be computed as discussed in Section 6.4.3, page 117.

7.6.3 Leverage and Extrapolation

Equation (7.23) shows that the standard error of prediction depends on the leverage h. The variance of the ith residual $\hat{e}_i$ also depends on the leverage:

$$\text{Var}(\hat{e}_i \mid \mathbf{u}_i) = \sigma^2(1 - h_i) \tag{7.24}$$

An understanding of how the leverages behave is useful for understanding when predictions are likely to be reliable, as well as for understanding residuals and other diagnostic statistics to be introduced in Chapter 15. A fuller notation for the leverage would be $h_i = h(\mathbf{u}_i)$, since the leverage may be different for each value of $\mathbf{u}$.

When the leverages are computed at one of the $\mathbf{u}_i$, the resulting h_i is always between zero and one, $0 \leq h_i \leq 1$. The leverages depend only on the terms in the model, not on the response, and are the diagonal elements of the *Hat matrix* defined at (7.44) in the Complements to this chapter. For a mean function with k terms, $\sum h_i = k$, so the average value of h_i is k/n. If n is large compared to k, then the average leverage k/n is small and the individual leverages will often deviate little from the average.

As h_i approaches one, the residual variance approaches zero and the fitted regression surface will approach y_i. In this sense, a point with h_i large has more "leverage" on the fitted surface than a point with h_i small. In the extreme, if $h_i = 1$, then the fitted regression surface must pass through y_i; that is, $\hat{y}_i = y_i$ and thus $\hat{e}_i = 0$.

The leverage is largest for cases with $\mathbf{u}_i$ farthest from $\bar{\mathbf{u}}$ and is smallest for cases with $\mathbf{u}_i$ closest to $\bar{\mathbf{u}}$. The measure of closeness depends on the sample correlations between the terms in the mean function. For example, shown in Figure 7.7a is a scatterplot of u_1 versus u_2 for a mean function with an intercept and $k = 3$. The contours are determined from the values of the terms. They characterize the density of points in the plot just as the contours discussed in Section 4.2 characterize the density in a bivariate normal population. These contours are used to measure the distance between $\mathbf{u}_i$ and $\bar{\mathbf{u}}$ in terms of the density of the sample. *All points on the same contour are the same distance from $\bar{\mathbf{u}}$ and all have the same leverage.* Leverage increases as we move from the inner contours to the outer contours. The highlighted points in Figure 7.7a that lie near the outer contour have about the same leverage, near the maximum leverage for these data. The point in the lower left corner of the plot has the largest leverage. The points within the inner contour have the smallest leverage.

The contours in Figure 7.7b are circles because u_1 and u_2 are uncorrelated. As suggested in Figure 7.7, contours of equal leverage will always be ellipsoids

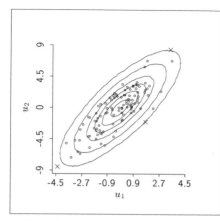

a. Correlated u-terms.

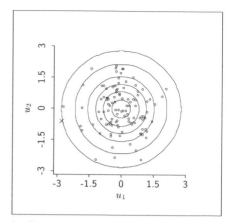

b. Uncorrelated u-terms.

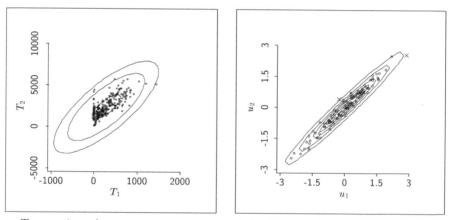

c. Transactions data.

d. Highly correlated u-terms.

FIGURE 7.7 Scatterplots of u-terms with contours of constant leverage. In plots a, b, and d, points marked with a $\times$ have the highest leverage.

centered at $\bar{\mathbf{u}}$ for mean functions with an intercept. Figure 7.7c gives contours of constant h for the transactions data. Even though the data are not elliptically distributed, contours of constant h remain elliptical.

The u-terms in Figure 7.7d are highly correlated, and the density of points is highest along a line. The point with the highest leverage is the highlighted point that is outside the last contour and that lies just above and to the left of the point cloud. This point lies close to the center in terms of Euclidean distance, but lies quite far from the center when distance is measured in terms of the density contours. The leverage for the point in the upper right of the plot is slightly less than that for the highlighted point.

Plots of leverage versus case number are useful for identifying high-leverage cases. After fitting a model, leverages are available in *Arc* and can be plotted using the "Plot of" item in the Graph&Fit menu.

The standard error of a prediction corresponding to any future value **x** is

$$se(y_{pred} \mid \mathbf{x}) = \hat{\sigma}\sqrt{1 + h}$$

where h is the leverage corresponding to $\mathbf{u} = \mathbf{u(x)}$. When $\mathbf{u}$ is not equal to one of the $\mathbf{u}_i$, the value of h can possibly exceed one; the larger the value of h, the farther from $\mathbf{u}$ to $\bar{\mathbf{u}}$, according to the contours determined by the data. Generally, predictions will be reliable only if the corresponding value of h is sufficiently small. If h is larger than the maximum value of h_i observed in the data, the corresponding prediction should be termed an *extrapolation* and may be unreliable.

7.6.4 General Linear Combinations

In some regressions we may be interested in estimating a linear combination of the coefficients. In the transactions data, for example, the difference $\eta_1 - \eta_2$ is the expected number of additional minutes for a type one transaction. Linear combinations can be estimated by replacing parameters by estimates, so

$$\widehat{\eta_1 - \eta_2} = \hat{\eta}_1 - \hat{\eta}_2 = 5.46 - 2.03 = 3.43 \text{ minutes per transaction}$$

On the average, a type one transaction takes about 3.4 minutes longer than a type two transaction.

In general notation, a linear combination is given by

$$L = \sum_{j=0}^{k-1} a_j \hat{\eta}_j \tag{7.25}$$

which is a weighted sum of the random variables, $\hat{\eta}_0, \ldots, \hat{\eta}_{k-1}$. The expected value of L is obtained by replacing each of the random variables in the sum by its expectation, so

$$E(L) = \sum_{j=0}^{k-1} a_j E(\hat{\eta}_j) = \sum_{j=0}^{k-1} a_j \eta_j \tag{7.26}$$

The variance of a general linear combination is computed as an application of (4.8) to be

$$\text{Var}(L) = \sum_{j=0}^{k-1} a_j^2 \text{Var}(\hat{\eta}_j) + 2 \sum_{j=0}^{k-2} \sum_{l=j+1}^{k-1} a_j a_l \text{Cov}(\hat{\eta}_j, \hat{\eta}_l) \tag{7.27}$$

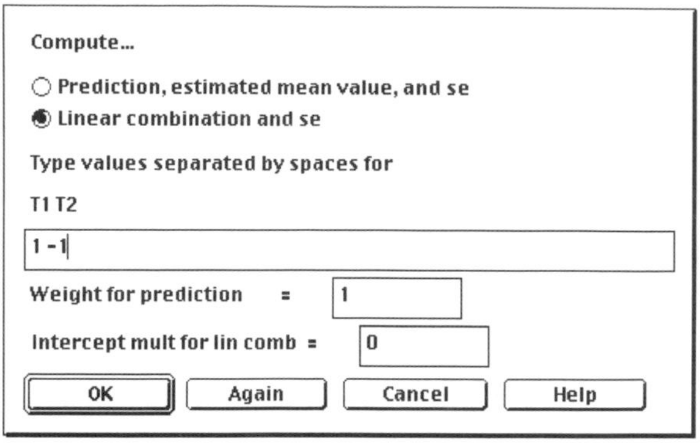

```
Compute...

○ Prediction, estimated mean value, and se
● Linear combination and se

Type values separated by spaces for

T1 T2

┌─────────────────────────────────────────────┐
│ 1 -1│                                        │
└─────────────────────────────────────────────┘

Weight for prediction      =   ┌───────────────┐
                               │ 1             │
                               └───────────────┘

Intercept mult for lin comb =  ┌───────────────┐
                               │ 0             │
                               └───────────────┘

┌──────────┐ ┌──────────┐ ┌──────────┐ ┌──────────┐
│    OK    │ │  Again   │ │  Cancel  │ │  Help    │
└──────────┘ └──────────┘ └──────────┘ └──────────┘
```

FIGURE 7.8   Prediction dialog used to find an estimated linear combination and its standard error.

The standard error of L can be found by first substituting estimates of $\mathrm{Var}(\hat{\eta}_j)$ and $\mathrm{Cov}(\hat{\eta}_j, \hat{\eta}_l)$, which can be obtained from the elements of the matrix $\hat{\sigma}^2 \mathbf{M}$, and then taking the square root of the resulting estimate of $\mathrm{Var}(L)$.

The item "Prediction" in an LS regression menu can be used to obtain L and its standard error. The resulting dialog is illustrated in Figure 7.8. As shown, the estimate and standard error of $\eta_1 - \eta_2$ will be computed and displayed. To get the computer to work with linear combinations instead of fitted values, you must select the appropriate button in the dialog and provide multipliers for each of the η's. The multiplier for the intercept, if present, is entered into a separate text area, and in the example this multiplier is zero. The output is in part

```
Response = Time
Terms = (T1 T2)
Term values = (1 -1)
Multiplier for intercept = 0
Lin Comb = 3.42751, with se = 0.509601
```

so the estimated difference is about 3.4 minutes per transaction, and its standard error is 0.5 minutes per transaction. The t-distribution with df equal to the df for $\hat{\sigma}^2$ could now be used to construct a confidence interval.

7.6.5   Overall Analysis of Variance

The overall analysis of variance is used to summarize information needed to compare the mean functions,

$$
\begin{aligned}
&\mathrm{NH}: \mathrm{E}(y \mid \mathbf{x}) = \eta_0 \quad \text{with} \quad \mathrm{Var}(y \mid \mathbf{x}) = \sigma^2 \\
&\mathrm{AH}: \mathrm{E}(y \mid \mathbf{x}) = \boldsymbol{\eta}^T \mathbf{u} \quad \text{with} \quad \mathrm{Var}(y \mid \mathbf{x}) = \sigma^2
\end{aligned}
\tag{7.28}
$$

TABLE 7.5 Overall Analysis of Variance for the Transaction Data

```
Summary Analysis of Variance Table
Source       df        SS              MS            F         p-value
Regression    2    3366491409.     1683245705.    1289.42     0.0000
Residual    258     336801747.     1305433.
```

The mean function of the null hypothesis does not depend on **x**, and we have seen in Section 6.6 that for this mean function the residual sum of squares is $RSS_{NH} = SYY$ with $df_{NH} = n - 1$. The residual sum of squares from the fit of the alternative hypothesis model is just $RSS_{AH} = RSS$ with $df_{AH} = n - k$ df. The F-test and analysis of variance table are now constructed just as described in Section 6.6.

The analysis of variance table for the transactions data is shown in Table 7.5. The value $F = 1289$ should be compared to the $F_{k-1,n-k}$ distribution, or in this case the $F_{2,258}$ distribution. The p-value is computed as an upper-tail test. This p-value, which is 0 to four digits, is given in the analysis of variance table and suggests that the mean function ignoring the predictors provides a much worse fit than the mean function including the predictors, as would certainly be expected by the nature of the regression.

7.6.6 The Coefficient of Determination

As with simple regression, the ratio

$$R^2 = \frac{SYY - RSS}{SYY} = \frac{SSreg}{SYY} \tag{7.29}$$

gives the fraction of variability in the response explained by adding the terms $u_1, \ldots, u_{k-1}$ to the constant mean function. To distinguish multiple linear regression from simple linear regression, we now use R^2 instead of r^2. Since the residual sum of squares for the constant mean function $E(y \mid \mathbf{x}) = \eta_0$ is just SYY, R^2 can also be viewed as a summary statistic comparing the two models of (7.28).

The value of R^2, also called the *squared multiple correlation*, is between zero and one. If $R^2 = 1$, then $RSS = 0$, and the mean function matches the data perfectly. If $R^2 = 0$, then $SYY = RSS$, and knowledge of the terms tells us nothing about the response. The value of R^2 in the standard regression output shown in Table 7.3 equals 0.909 for the transactions data, indicating that 90.9% of the marginal variation in *Time* is explained by T_1 and T_2.

Suppose we draw a scatterplot with the fitted values $\hat{y}_i = \hat{\eta}^T \mathbf{u}_i$ on the horizontal axis, and the responses y_i on the vertical axis, as shown in Figure 7.9. As long as the mean function includes an intercept, R^2 is the square of the sample correlation between y_i and $\hat{y}_i$ associated with this scatterplot. The correlation is a meaningful summary of the scatterplot only if the points fall into

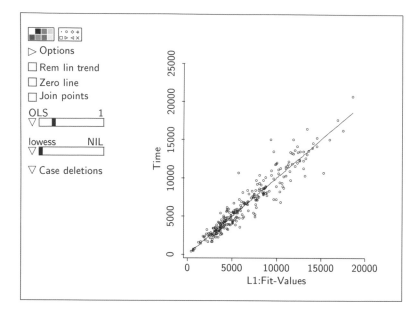

FIGURE 7.9   A summary plot for the regression of *Time* on T_1 and T_2 in the transactions data.

an elliptical cluster about the fitted line. By careful examination of the figure, it appears that points at the upper right of the plot are farther from the fitted line than are points in the lower left, suggesting nonconstant variance. If this were so, the assumption that $\mathrm{Var}(Time \mid T_1, T_2)$ was constant would be called into question. Many of the results in this chapter that depend on the constant variance assumption would no longer hold. First, R^2 would not be an appropriate measure of variability explained. Second, the computations made in the last few sections concerning predictions and variances of estimates need modifications that account for the nonconstant variance. We return to this issue in later chapters.

7.7   THE LAKE MARY DATA

In this section we return to the Lake Mary data in file `lakemary.lsp`. The fitted quadratic polynomial is shown in Figure 7.4 on page 148. This fit was obtained using the parametric smoother slidebar which can be used to fit polynomials up to degree 5; the degree of the polynomial is shown above the right end of the slidebar. The degree can be changed by moving the slider to the left or right. The quadratic fit shown in Figure 7.4 is nearly identical to a lowess fit (not shown), and the quadratic mean function matches the data quite well.

   To get the parameter estimates for a quadratic polynomial, it is necessary to create a new term Age^2 using the "Transform" menu item. The quadratic

TABLE 7.6   Quadratic Fit for the Lake Mary Data

```
Data set = LakeMary, Name of Fit = L1
Normal Regression
Kernel mean function = Identity
Response      = Length
Terms         = (Age Age^2)
Coefficient Estimates
Label         Estimate     Std. Error    t-value
Constant      13.6224      11.0164       1.237
Age           54.0493      6.48884       8.330
Age^2         -4.71866     0.943959      -4.999

R Squared:               0.801138
Sigma hat:               10.9061
Number of cases:         78
Degrees of freedom:      75
```

Summary Analysis of Variance Table

Source	df	SS	MS	F	p-value
Regression	2	35938.0	17969.0	151.07	0.0000
Residual	75	8920.70	118.943		
Lack of fit	3	108.012	36.0040	0.29	0.8295
Pure Error	72	8812.68	122.398		

mean function is

$$E(Length \mid Age) = \eta_0 + \eta_1 Age + \eta_2 Age^2$$

so the mean function has $k = 3$ terms. OLS estimates can be obtained using the "Fit linear LS" item in the Graph&Fit menu. The output is shown in Table 7.6. The estimated mean function is

$$\hat{E}(Length \mid Age) = 13.6 + 54.0 Age - 4.7 Age^2$$

This model specifies slower growth rates for older fish than for younger fish, as one might expect for biological organisms. The model is empirical, however, and could give misleading answers for fish outside the range of the data. For example, because of the general behavior of a quadratic curve, the model must predict that very old fish have shorter lengths. The estimated length of a five-year-old fish is 165.9 mm, while the estimated length of a seven-year-old fish, according to the quadratic model, is only 160.8 mm.

7.8   REGRESSION THROUGH THE ORIGIN

Occasionally, we may like to fit a mean function that does not include the term u_0 for the intercept. This causes no particular problem in either the theory or the computations. We interpret the vector **u** for terms having the $k - 1$ components, $\mathbf{u} = (u_1, \ldots, u_{k-1})^T$, and similarly $\boldsymbol{\eta} = (\eta_1, \ldots, \eta_{k-1})^T$. With these minor modifications, all the results of this chapter apply. In *Arc*, mean functions without an intercept can be obtained by un-checking the box for "Fit Intercept" in the regression dialog.

7.9   COMPLEMENTS

7.9.1   An Introduction to Matrices

Many of the results in multiple linear regression use matrix and vector notation for compact representations. However, nearly all the results in the main part of the text can be understood without any familiarity with manipulating matrices. We provide only a brief introduction to matrices and vectors, so everything we do is defined. More complete references include Graybill (1992), Searle (1982), Schott (1996), or any good linear algebra book.

Boldface type is used to indicate matrices and vectors. We will say that **X** is an $r \times c$ matrix if it is an array of numbers with r rows and c columns. A specific 4×2 matrix **X** is

$$\mathbf{X} = \begin{pmatrix} 2 & 1 \\ 1 & 5 \\ 3 & 4 \\ 8 & 6 \end{pmatrix} = \begin{pmatrix} x_{11} & x_{12} \\ x_{21} & x_{22} \\ x_{31} & x_{32} \\ x_{41} & x_{42} \end{pmatrix} = (x_{ij})$$

An element of a matrix **X** is given by x_{ij}, which is the number in the ith row and the jth column of **X**. For example, in the preceding matrix, $x_{32} = 4$.

A *vector* is a special matrix with just one column. A specific 4×1 matrix **y**, which is a vector of length four, is given by

$$\mathbf{y} = \begin{pmatrix} 2 \\ 3 \\ -2 \\ 0 \end{pmatrix} = \begin{pmatrix} y_1 \\ y_2 \\ y_3 \\ y_4 \end{pmatrix}$$

The elements of a vector are generally singly subscripted; thus, $y_3 = -2$. A special vector written as $\mathbf{1}_r$ is the $r \times 1$ vector with all elements equal to one. Similarly, $\mathbf{0}_r$ is the $r \times 1$ vector with all elements equal to zero.

A *row vector* is a matrix with one row. We don't use row vectors in this book. If a vector is needed to represent a row, a transpose of a column vector will be used (see below).

A *square matrix* has the number of rows r equal to the number of columns c. A square matrix **Z** is *symmetric* if $z_{ij} = z_{ji}$ for all i and j. A square matrix is *diagonal* if all elements off the main diagonal are zero, $z_{ij} = 0$, unless $i = j$. The matrices **C** and **D** below are symmetric and diagonal, respectively:

$$\mathbf{C} = \begin{pmatrix} 7 & 3 & 2 & 1 \\ 3 & 4 & 1 & -1 \\ 2 & 1 & 6 & 3 \\ 1 & -1 & 3 & 8 \end{pmatrix} \qquad \mathbf{D} = \begin{pmatrix} 7 & 0 & 0 & 0 \\ 0 & 4 & 0 & 0 \\ 0 & 0 & 6 & 0 \\ 0 & 0 & 0 & 8 \end{pmatrix}$$

The diagonal matrix with all elements on the diagonal equal to one is called the *identity matrix*, for which the symbol **I** is used. The 4×4 identity matrix is

$$\mathbf{I} = \begin{pmatrix} 1 & 0 & 0 & 0 \\ 0 & 1 & 0 & 0 \\ 0 & 0 & 1 & 0 \\ 0 & 0 & 0 & 1 \end{pmatrix}$$

A *scalar* is a 1×1 matrix, an ordinary number.

Addition and Subtraction. Two matrices can be added or subtracted only if they have the same number of rows and columns. The sum $\mathbf{C} = \mathbf{A} + \mathbf{B}$ of $r \times c$ matrices is also $r \times c$. Addition is done elementwise:

$$\mathbf{C} = \mathbf{A} + \mathbf{B} = \begin{pmatrix} a_{11} & a_{12} \\ a_{21} & a_{22} \\ a_{31} & a_{32} \end{pmatrix} + \begin{pmatrix} b_{11} & b_{12} \\ b_{21} & b_{22} \\ b_{31} & b_{32} \end{pmatrix} = \begin{pmatrix} a_{11}+b_{11} & a_{12}+b_{12} \\ a_{21}+b_{21} & a_{22}+b_{22} \\ a_{31}+b_{31} & a_{32}+b_{32} \end{pmatrix}$$

Subtraction works the same way, with the $+$ signs changed to $-$ signs. The usual rules for addition of numbers apply to addition of matrices, namely commutativity, $\mathbf{A} + \mathbf{B} = \mathbf{B} + \mathbf{A}$, and associativity, $(\mathbf{A} + \mathbf{B}) + \mathbf{C} = \mathbf{A} + (\mathbf{B} + \mathbf{C})$.

Multiplication by a Scalar. If k is a number and **A** is an $r \times c$ matrix with elements (a_{ij}), then $k\mathbf{A}$ is an $r \times c$ matrix with elements (ka_{ij}). For example, the matrix $\sigma^2 \mathbf{I}$ has all diagonal elements equal to σ^2 and all off-diagonal elements equal to zero.

Matrix Multiplication. Multiplication of matrices follows rules that are more complicated than are the rules for addition and subtraction. For two matrices to be multiplied together in the order **AB**, the number of columns of **A** must equal the number of rows of **B**. For example, is **A** is $r \times c$, and **B** is $c \times q$, then we can form the product $\mathbf{C} = \mathbf{AB}$, which will be an $r \times q$ matrix. If the elements of **A** are (a_{ij}) and the elements of **B** are (b_{ij}), then the elements of $\mathbf{C} = (c_{ij})$ are given by the formula

$$c_{ij} = \sum_{k=1}^{c} a_{ik} b_{kj}$$

This formula says that c_{ij} is formed by taking the ith row of $\mathbf{A}$ and the jth column of $\mathbf{B}$, multiplying the first element of the specified row in $\mathbf{A}$ by the first element in the specified column in $\mathbf{B}$, multiplying second elements, and so on, and then adding the products together.

Possibly the simplest case of multiplying two matrices $\mathbf{A}$ and $\mathbf{B}$ together occurs when $\mathbf{A}$ is $1 \times c$ and $\mathbf{B}$ is $c \times 1$. For example, if $\mathbf{A}$ and $\mathbf{B}$ are

$$\mathbf{A} = (1 \quad 3 \quad 2 \quad -1) \qquad \mathbf{B} = \begin{pmatrix} 2 \\ 1 \\ -2 \\ 4 \end{pmatrix}$$

then the product $\mathbf{AB}$ is

$$\mathbf{AB} = (1 \times 2) + (3 \times 1) + (2 \times -2) + (-1 \times 4) = -3$$

$\mathbf{AB}$ is not the same as $\mathbf{BA}$. For the preceding matrices, the product $\mathbf{BA}$ will be a 4×4 matrix:

$$\mathbf{BA} = \begin{pmatrix} 2 & 6 & 4 & -2 \\ 1 & 3 & 2 & -1 \\ -2 & -6 & -4 & 2 \\ 4 & 12 & 8 & -4 \end{pmatrix}$$

Consider a small example that can illustrate what happens when all the dimensions are bigger than one. Symbolically, a 3×2 matrix $\mathbf{A}$ times a 2×2 matrix $\mathbf{B}$ is given as

$$\begin{pmatrix} a_{11} & a_{12} \\ a_{21} & a_{22} \\ a_{31} & a_{32} \end{pmatrix} \begin{pmatrix} b_{11} & b_{12} \\ b_{21} & b_{22} \end{pmatrix} = \begin{pmatrix} a_{11}b_{11} + a_{12}b_{21} & a_{11}b_{12} + a_{12}b_{22} \\ a_{21}b_{11} + a_{22}b_{21} & a_{21}b_{12} + a_{22}b_{22} \\ a_{31}b_{11} + a_{32}b_{21} & a_{31}b_{12} + a_{32}b_{22} \end{pmatrix}$$

Using numbers, an example of multiplication of two matrices is

$$\begin{pmatrix} 3 & 1 \\ -1 & 0 \\ 2 & 2 \end{pmatrix} \begin{pmatrix} 5 & 1 \\ 0 & 4 \end{pmatrix} = \begin{pmatrix} 15 + 0 & 3 + 4 \\ -5 + 0 & -1 + 0 \\ 10 + 0 & 2 + 8 \end{pmatrix} = \begin{pmatrix} 15 & 4 \\ -5 & -1 \\ 10 & 10 \end{pmatrix}$$

In this example, $\mathbf{BA}$ is not defined because the number of columns of $\mathbf{B}$ is not equal to the number of rows of $\mathbf{A}$. However, the associative law holds: If $\mathbf{A}$ is $r \times c$, $\mathbf{B}$ is $c \times q$, and $\mathbf{C}$ is $q \times p$, then $\mathbf{A}(\mathbf{BC}) = (\mathbf{AB})\mathbf{C}$, and the result is an $r \times p$ matrix.

Transpose of a Matrix. The transpose of an $r \times c$ matrix $\mathbf{X}$ is a $c \times r$ matrix called $\mathbf{X}^T$ such that if the elements of $\mathbf{X}$ are (x_{ij}), then the elements of $\mathbf{X}^T$ are (x_{ji}). For the matrix $\mathbf{X}$ given earlier,

$$\mathbf{X}^T = \begin{pmatrix} 2 & 1 & 3 & 8 \\ 1 & 5 & 4 & 6 \end{pmatrix}$$

The transpose of a column vector is a row vector. The transpose of a product $(\mathbf{AB})^T$ is the product of the transposes, in *opposite order*, so $(\mathbf{AB})^T = \mathbf{B}^T\mathbf{A}^T$.

Suppose that $\mathbf{a}$ is an $r \times 1$ vector with elements $a_1, \ldots, a_r$. Then the product $\mathbf{a}^T\mathbf{a}$ will be a 1×1 matrix or scalar, given by

$$\mathbf{a}^T\mathbf{a} = a_1^2 + a_2^2 + \cdots + a_r^2 = \sum_{i=1}^{r} a_i^2 \qquad (7.30)$$

Thus $\mathbf{a}^T\mathbf{a}$ provides a compact notation for the sum of the squares of the elements of a vector $\mathbf{a}$. The square root of this quantity $(\mathbf{a}^T\mathbf{a})^{1/2}$ is called the *norm* or *length* of the vector $\mathbf{a}$. Similarly, if $\mathbf{a}$ and $\mathbf{b}$ are both $r \times 1$ vectors, then we obtain

$$\mathbf{a}^T\mathbf{b} = a_1 b_1 + a_2 b_2 + \cdots + a_n b_n = \sum_{i=1}^{r} a_i b_i = \sum_{i=1}^{r} b_i a_i = \mathbf{b}^T\mathbf{a}$$

The fact that $\mathbf{a}^T\mathbf{b} = \mathbf{b}^T\mathbf{a}$ is often quite useful in manipulating the vectors used in regression calculations.

Another useful formula for regression calculation is obtained by applying the distributive law:

$$(\mathbf{a} - \mathbf{b})^T(\mathbf{a} - \mathbf{b}) = \mathbf{a}^T\mathbf{a} + \mathbf{b}^T\mathbf{b} - 2\mathbf{a}^T\mathbf{b} \qquad (7.31)$$

Inverse of a Matrix. For any scalar $c \neq 0$ there is another number called the *inverse* of c, say d, such that the product $cd = 1$. For example, if $c = 3$, then $d = 1/c = 1/3$, and the inverse of 3 is $1/3$. Similarly, the inverse of $1/3$ is 3.

A square matrix can also have an inverse. We will say that the inverse of a matrix $\mathbf{C}$ is another matrix $\mathbf{D}$, such that $\mathbf{CD} = \mathbf{I}$, and we write $\mathbf{D} = \mathbf{C}^{-1}$. Just as not all scalars have an inverse—zero is excluded—not all square matrices have an inverse. The collection of matrices that have an inverse are called *full rank*, *invertible*, or *nonsingular*. A square matrix that is not invertible is of less than full rank, or *singular*. If a matrix has an inverse, it has a unique inverse.

The inverse is easy to compute only in special cases, and its computation in general requires a computer. The easiest case is the identity matrix $\mathbf{I}$, which is its own inverse. If $\mathbf{C}$ is a diagonal matrix, say

$$\mathbf{C} = \begin{pmatrix} 3 & 0 & 0 & 0 \\ 0 & -1 & 0 & 0 \\ 0 & 0 & 4 & 0 \\ 0 & 0 & 0 & 1 \end{pmatrix}$$

then $\mathbf{C}^{-1}$ is the diagonal matrix

$$\mathbf{C} = \begin{pmatrix} \frac{1}{3} & 0 & 0 & 0 \\ 0 & -1 & 0 & 0 \\ 0 & 0 & \frac{1}{4} & 0 \\ 0 & 0 & 0 & 1 \end{pmatrix}$$

as can be verified by direct multiplication. For any diagonal matrix with nonzero diagonal elements, the inverse is obtained by inverting the diagonal elements. If any of the diagonal elements are zero, then no inverse exists.

7.9.2 Random Vectors

An $n \times 1$ vector $\mathbf{y}$ is a *random vector* if each of its elements is a random variable. The mean of an $n \times 1$ random vector $\mathbf{y}$ is also an $n \times 1$ vector whose elements are the means of the elements of $\mathbf{y}$, so symbolically

$$E(\mathbf{y}) = \begin{pmatrix} E(y_1) \\ E(y_2) \\ \vdots \\ E(y_n) \end{pmatrix}$$

The variance of an $n \times 1$ vector $\mathbf{y}$ is an $n \times n$ square symmetric matrix, often called a *covariance matrix*, written $\mathrm{Var}(\mathbf{y})$. The diagonal elements of $\mathrm{Var}(\mathbf{y})$ are the variances of the elements of $\mathbf{y}$; if we write $\mathbf{V} = \mathrm{Var}(\mathbf{y})$ and if the elements of $\mathbf{V}$ are called v_{ij}, then $v_{ii} = \mathrm{Var}(y_i)$ for $i = 1, 2, \ldots, n$. The off-diagonal elements of $\mathbf{V}$ are the covariances between the elements of $\mathbf{y}$, so $v_{ij} = \mathrm{Cov}(y_i, y_j) = \mathrm{Cov}(y_j, y_i) = v_{ji}$. Since the covariance function is symmetric, so is the covariance matrix.

The rules for working with means and variances of random vectors are matrix versions of the equations given previously in scalar form. If $\mathbf{y}$ is an $n \times 1$ random vector with mean $E(\mathbf{y})$ and variance $\mathrm{Var}(\mathbf{y})$, and $\mathbf{b}_0$ is any $n \times 1$

vector of constants and $\mathbf{B}$ is any $n \times n$ matrix of constants, then

$$E(\mathbf{b}_0 + \mathbf{B}\mathbf{y}) = \mathbf{b}_0 + \mathbf{B}E(\mathbf{y}) \tag{7.32}$$

$$\text{Var}(\mathbf{b}_0 + \mathbf{B}\mathbf{y}) = \mathbf{B}\text{Var}(\mathbf{y})\mathbf{B}^T \tag{7.33}$$

Many of the random variables in this book are conditional, given the fixed values of predictors. The same rules apply to conditional random variables as to unconditional ones. For example, suppose $\mathbf{x}$ is a vector of predictors. Then $E(\mathbf{y} \mid \mathbf{x})$ is a vector the same length as $\mathbf{y}$ with ith element $E(y_i \mid \mathbf{x})$.

7.9.3 Correlation Matrix

A *correlation matrix* is simply a rescaling of a covariance matrix. Given a covariance matrix $\mathbf{V}$ with elements v_{ij}, a correlation matrix $\mathbf{R}$ with elements r_{ij} is obtained by the rule $r_{ij} = v_{ij}/(v_{ii}v_{jj})^{1/2}$. The diagonal elements of $\mathbf{R}$ are all equal to one, while the off-diagonal elements are correlations.

7.9.4 Applications to Multiple Linear Regression

We begin with data $(\mathbf{x}_i, y_i)$ observed on each of n cases. Each y_i is a scalar, and each $\mathbf{x}_i$ is a $p \times 1$ vector of values for the predictors for the ith case:

$$\mathbf{x}_i = \begin{pmatrix} x_{i1} \\ x_{i2} \\ \vdots \\ x_{ip} \end{pmatrix}$$

To save space we would usually write $\mathbf{x}_i$ as $\mathbf{x}_i^T = (x_{i1}, x_{i2}, \ldots, x_{ip})$. From each $\mathbf{x}_i$, we compute a vector of k terms $\mathbf{u}(\mathbf{x}_i)$, as defined in Section 7.2, where each element of $\mathbf{u}(\mathbf{x}_i)$ is a function of $\mathbf{x}_i$:

$$\mathbf{u}(\mathbf{x}_i) = \begin{pmatrix} u_0 \\ u_1(\mathbf{x}_i) \\ \vdots \\ u_{k-1}(\mathbf{x}_i) \end{pmatrix} = \begin{pmatrix} u_0 \\ \mathbf{u}_1(\mathbf{x}_i) \end{pmatrix}$$

The first element of $\mathbf{u}(\mathbf{x}_i)$ is $u_0 = 1$ for fitting an intercept, and the remaining $k - 1$ terms are sometimes collected into a $(k - 1) \times 1$ vector $\mathbf{u}_1(\mathbf{x}_i)$. To make notation less cluttered, we generally will just write $\mathbf{u}_i$ rather than $\mathbf{u}(\mathbf{x}_i)$ and $\mathbf{u}_{i1}$ rather than $\mathbf{u}_1(\mathbf{x}_i)$, but the terms are always derived from the underlying predictors.

Let η be a $k \times 1$ vector of parameters. We can partition $\eta^T = (\eta_0, \eta_1^T)$ to conform to the partition of $\mathbf{u}$, so η_0 is the parameter for the intercept and

η_1 is the vector of the remaining regression coefficients. The multiple linear regression mean function is given by

$$E(y_i \mid \mathbf{u}_i) = \eta_0 + \eta_1 u_{i1} + \cdots + \eta_{k-1} u_{i,k-1}$$
$$= \eta^T \mathbf{u}_i$$
$$= \eta_0 + \eta_1^T \mathbf{u}_{i1}$$

In this equation we have conditioned on $\mathbf{u}_i$, but we could have conditioned on $\mathbf{x}_i$.

To write the mean function in matrix form, we first collect the response into an $n \times 1$ vector $\mathbf{y}^T = (y_1, y_2, \ldots, y_n)$. Next, we will collect the terms $\mathbf{u}_i$ into an $n \times k$ *model matrix* $\mathbf{U}$:

$$\mathbf{U} = \begin{pmatrix} 1 & u_{11} & u_{12} & \cdots & u_{1,k-1} \\ 1 & u_{21} & u_{22} & \cdots & u_{2,k-1} \\ \vdots & \vdots & \vdots & \vdots & \vdots \\ 1 & u_{n1} & u_{n2} & \cdots & u_{n,k-1} \end{pmatrix} \tag{7.34}$$

The ith row of $\mathbf{U}$ consists of a one for the intercept, followed by $\mathbf{u}_{i1}^T$, the values of the terms for the ith case. We can write the mean function in matrix form as

$$E(\mathbf{y} \mid \mathbf{U}) = \mathbf{U}\eta \tag{7.35}$$

The multiple linear regression model is sometimes written as

$$\mathbf{y} \mid \mathbf{U} = E(\mathbf{y} \mid \mathbf{U}) + \mathbf{e} = \mathbf{U}\eta + \mathbf{e} \tag{7.36}$$

where $\mathbf{e}$ is an $n \times 1$ random vector with elements e_i. Equation (7.36) specifies that the random vector $\mathbf{y} \mid \mathbf{U}$ can be expressed as a fixed part $E(\mathbf{y} \mid \mathbf{U}) = \mathbf{U}\eta$, plus a random vector $\mathbf{e}$, the error term, similar to the derivation of the simple linear regression model at (6.1). This additive form for $\mathbf{y} \mid \mathbf{U}$ equal to mean plus error is a special feature of linear models that does not carry over to many nonlinear models for data.

Since $\mathbf{e}$ is a random vector, it has a mean vector and a covariance matrix. We assume it has mean zero, $E(\mathbf{e} \mid \mathbf{U}) = \mathbf{0}_n$, and that all elements of $\mathbf{y} \mid \mathbf{U}$ have variance σ^2 and are uncorrelated, so $Var(\mathbf{e} \mid \mathbf{U}) = \sigma^2 \mathbf{I}$.

7.9.5 Ordinary Least Squares Estimates

The OLS estimate $\hat{\eta}$ of η is given by the arguments that minimize the residual sum of squares function,

$$RSS(\mathbf{h}) = (\mathbf{y} - \mathbf{U}\mathbf{h})^T (\mathbf{y} - \mathbf{U}\mathbf{h})$$

Using (7.31), we obtain

$$RSS(\mathbf{h}) = \mathbf{y}^T\mathbf{y} + \mathbf{h}^T(\mathbf{U}^T\mathbf{U})\mathbf{h} - 2\mathbf{y}^T\mathbf{U}\mathbf{h} \qquad (7.37)$$

Before finding the OLS estimates, we examine (7.37). $RSS(\mathbf{h})$ depends on the data only through three functions of the data: $\mathbf{y}^T\mathbf{y}$, $\mathbf{U}^T\mathbf{U}$, and $\mathbf{y}^T\mathbf{U}$. Any two data sets that have the same values of these three quantities will have the same least squares estimates. Consulting the formulas in Table 6.1, we can write

$$\mathbf{y}^T\mathbf{y} = \sum_{i=1}^{n} y_i^2 = SYY + n\bar{y}^2$$

so apart from a centering constant $n\bar{y}^2$, this is just the total sum of squares SYY. The $k \times k$ matrix $\mathbf{U}^T\mathbf{U}$ has entries given by the sum of squares and cross-products of the columns of $\mathbf{U}$, and the $1 \times k$ matrix $\mathbf{y}^T\mathbf{U}$ contains cross-products of the columns of $\mathbf{U}$ and the response. The information in these quantities is equivalent to the information contained in the sample means of the predictors plus the sample covariances of the predictors and the response. In *Arc*, the averages, standard deviations and correlation matrix can be obtained using the "Display summaries" item in a data set menu. Any two regressions that have identical values for sample averages, standard deviations, and sample correlation matrix will have identical OLS estimates of coefficients.

Using calculus, minimization of (7.37) is straightforward. Differentiating (7.37) with respect to the elements of h and setting the result equal to zero leads to the *normal equations*,

$$\mathbf{U}^T\mathbf{U}\mathbf{h} = \mathbf{U}^T\mathbf{y} \qquad (7.38)$$

The OLS estimates are any solution to these equations. If the inverse of $(\mathbf{U}^T\mathbf{U})$ exists, the OLS estimates are unique and are given by

$$\hat{\eta} = (\mathbf{U}^T\mathbf{U})^{-1}\mathbf{U}^T\mathbf{y} = \mathbf{M}\mathbf{U}^T\mathbf{y} \qquad (7.39)$$

We have defined

$$\mathbf{M} = (\mathbf{U}^T\mathbf{U})^{-1} \qquad (7.40)$$

because this matrix will be used elsewhere. If the inverse does not exist, then the matrix $(\mathbf{U}^T\mathbf{U})$ is of less than full rank, and the OLS estimate is not unique. In this case, *Arc* and most other computer programs will use a subset of the columns of $\mathbf{U}$ in fitting the model, so that the reduced model matrix does have full rank. This is discussed in Section 10.1.3.

Properties of Estimates. Using the rules for means and variances of random vectors, (7.32) and (7.33), we find

$$
\begin{aligned}
\mathrm{E}(\hat{\eta} \mid \mathbf{U}) &= \mathrm{E}((\mathbf{U}^T\mathbf{U})^{-1}\mathbf{U}^T\mathbf{y} \mid \mathbf{U}) \\
&= (\mathbf{U}^T\mathbf{U})^{-1}\mathbf{U}^T\mathrm{E}(\mathbf{y} \mid \mathbf{U}) \\
&= (\mathbf{U}^T\mathbf{U})^{-1}\mathbf{U}^T\mathbf{U}\eta \\
&= \eta
\end{aligned}
\tag{7.41}
$$

so $\hat{\eta}$ is unbiased for η, as long as the mean function that was fit is the true mean function. The variance of $\hat{\eta}$ is

$$
\begin{aligned}
\mathrm{Var}(\hat{\eta} \mid \mathbf{U}) &= \mathrm{Var}((\mathbf{U}^T\mathbf{U})^{-1}\mathbf{U}^T\mathbf{y} \mid \mathbf{U}) \\
&= (\mathbf{U}^T\mathbf{U})^{-1}\mathbf{U}^T\mathrm{Var}(\mathbf{y} \mid \mathbf{U})\mathbf{U}(\mathbf{U}^T\mathbf{U})^{-1} \\
&= (\mathbf{U}^T\mathbf{U})^{-1}\mathbf{U}^T\sigma^2\mathbf{I}\mathbf{U}(\mathbf{U}^T\mathbf{U})^{-1} \\
&= \sigma^2(\mathbf{U}^T\mathbf{U})^{-1}\mathbf{U}^T\mathbf{U}(\mathbf{U}^T\mathbf{U})^{-1} \\
&= \sigma^2(\mathbf{U}^T\mathbf{U})^{-1} \\
&= \sigma^2\mathbf{M}
\end{aligned}
\tag{7.42}
$$

where we have defined $\mathbf{M}$ as in (7.40) to be equal to $(\mathbf{U}^T\mathbf{U})^{-1}$. Thus, the variances and covariances are compactly determined as σ^2 times a matrix whose elements are determined only by $\mathbf{U}$ and not by $\mathbf{y}$. The matrix $\hat{\sigma}^2\mathbf{M}$ can be displayed from a linear regression model menu using the "Display variances" menu item.

The Residual Sum of Squares. Let $\hat{\mathbf{y}} = \mathbf{U}\hat{\eta}$ be the $n \times 1$ vector of fitted values corresponding to the n cases in the data, and similarly, $\hat{\mathbf{e}} = \mathbf{y} - \hat{\mathbf{y}}$ is the vector of residuals. One representation of the residual sum of squares, which is the residual sum of squares function evaluated at $\hat{\eta}$, is

$$
RSS = (\mathbf{y} - \hat{\mathbf{y}})^T(\mathbf{y} - \hat{\mathbf{y}}) = \hat{\mathbf{e}}^T\hat{\mathbf{e}} = \sum_{i=1}^{n} \hat{e}_i^2
$$

which suggests that the residual sum of squares can be computed by squaring the residuals and adding them up. In multiple linear regression, it can also be computed more efficiently based on summary statistics. Using (7.37) and the summary statistics $\mathbf{U}^T\mathbf{U}$, $\mathbf{U}^T\mathbf{y}$, and $\mathbf{y}^T\mathbf{y}$, we write

$$
RSS = RSS(\hat{\eta}) = \mathbf{y}^T\mathbf{y} + \hat{\eta}^T\mathbf{U}^T\mathbf{U}\hat{\eta} - 2\mathbf{y}^T\mathbf{U}\hat{\eta}
$$

We will first show that $\hat{\eta}^T\mathbf{U}^T\mathbf{U}\hat{\eta} = \mathbf{y}^T\mathbf{U}\hat{\eta}$. Substituting for one of the $\hat{\eta}$s, we get

$$
\hat{\eta}^T\mathbf{U}^T\mathbf{U}(\mathbf{U}^T\mathbf{U})^{-1}\mathbf{U}^T\mathbf{y} = \hat{\eta}^T\mathbf{U}^T\mathbf{y} = \mathbf{y}^T\mathbf{U}\hat{\eta}
$$

the last result following because taking the transpose of a 1×1 matrix does not change its value. The residual sum of squares function can now be rewritten as

$$RSS = \mathbf{y}^T\mathbf{y} - \hat{\boldsymbol{\eta}}^T\mathbf{U}^T\mathbf{U}\hat{\boldsymbol{\eta}}$$
$$= \mathbf{y}^T\mathbf{y} - \hat{\mathbf{y}}^T\hat{\mathbf{y}}$$

where $\hat{\mathbf{y}} = \mathbf{U}\hat{\boldsymbol{\eta}}$ are the fitted values. The residual sum of squares is the difference in the squares of the lengths of the two vectors $\mathbf{y}$ and $\hat{\mathbf{y}}$. Another useful form for the residual sum of squares is

$$RSS = SYY(1 - R^2)$$

where R^2 is the square of the sample correlation between $\hat{\mathbf{y}}$ and $\mathbf{y}$. This result generalizes the simple regression relationship, $RSS = SYY(1 - r_{xy}^2)$, to multiple regression, so R^2 can be interpreted as the proportion of variability in the response "explained" by the regression on $\mathbf{u}$ (see also Problem 7.6).

Estimate of Variance. Under the assumption of constant variance, the estimate of σ^2 is

$$\hat{\sigma}^2 = \frac{RSS}{d} \tag{7.43}$$

with d df, where d is equal to the number of cases n minus the number of terms with estimated coefficients in the model. If the matrix $\mathbf{U}$ is of full rank, then $d = n - k$, but the number of estimated coefficients will be less than k if $\mathbf{U}$ is not of full rank.

The Hat Matrix. The leverages in Section 7.6.3 are the diagonal elements of a matrix called the *Hat matrix*, defined by

$$\mathbf{H} = \mathbf{U}(\mathbf{U}^T\mathbf{U})^{-1}\mathbf{U}^T \tag{7.44}$$

assuming that $\mathbf{U}^T\mathbf{U}$ has full column rank. $\mathbf{H}$ is an $n \times n$ symmetric matrix. The fitted values are given by $\hat{\mathbf{y}} = \mathbf{Hy}$, and the residuals are given by $\hat{\mathbf{e}} = \mathbf{y} - \mathbf{Hy} = (\mathbf{I} - \mathbf{H})\mathbf{y}$. The hat matrix has many other useful properties that are easy to verify directly from the definition in (7.44):

$$\mathbf{HU} = \mathbf{U} \tag{7.45}$$

$$(\mathbf{I} - \mathbf{H})\mathbf{U} = \mathbf{0} \tag{7.46}$$

$$\mathbf{H}^2 = \mathbf{H} \tag{7.47}$$

$$\mathbf{H}(\mathbf{I} - \mathbf{H}) = \mathbf{0} \tag{7.48}$$

Equations (7.45) and (7.46) show that the covariance between the residuals $(\mathbf{I} - \mathbf{H})\mathbf{y}$ and any linear combination of the columns of $\mathbf{U}$ has zero covariance, and hence zero correlation. Equation (7.48) shows that the residuals and the fitted values are uncorrelated. The matrix $\mathbf{H}$ is not of full rank, so it does not have an inverse. Its rank is equal to the number of regression coefficients that can be estimated, usually equal to k. The sum of the leverages, which are the diagonal elements of $\mathbf{H}$, is equal to the rank of $\mathbf{H}$.

$$\leftarrow tr(H) = rank(H)$$

7.9.6   References

The transactions data were provided by Alan Welsh. The data for Problem 7.3 are from Tuddenham and Snyder (1954). The Big Mac data are from Enz (1991). The data for Exercise 7.1 were furnished by Mike Camden. The leaf area data used in Problem 7.7 were provided by Todd Pester.

PROBLEMS

7.1   Data on 56 normal births at a Wellington, New Zealand hospital are given in file `birthwt.lsp`. The response variable is *BirthWt*, birth weight in grams. The three predictors are the mother's age denoted by *Age*, term in weeks denoted by *Term*, and the baby's sex, *Sex* = 0 for girls and 1 for boys. Construct a scatterplot matrix of the variables *BirthWt*, *Age*, *Term*, and *Sex*.

   7.1.1   Describe the relationships in each of the marginal response plots with emphasis on the individual regression functions. Why do the points in the scatterplots involving *Sex* fall in two lines? Since the plots including *Sex* contain little information, the information about this binary variable can be encoded in the plot by using it as a marking variable. Redraw the scatterplot matrix, but use *Sex* as a marking variable.

   7.1.2   Study the marginal response plots *BirthWt* versus *Age* and *BirthWt* versus *Term*, separately for each sex. This can be done by clicking on the "1" at the bottom left of the plot to get the boys only, on the "0" to get the girls only, or on the name of the variable *Sex* to get all the points at once. Is there visual evidence to suggest that the relationship between *BirthWt* and *Age* or *BirthWt* and *Term* depends on the sex of the baby? Describe the visual evidence that leads to your conclusion.

7.2   In the Big Mac data, use a scatterplot matrix to find the cities with the most expensive Big Macs. Find the cities with the least expensive Big Macs. Where are bus fares relatively expensive? You might want to use a name list by selecting the "Display case names" item from the data set menu.

TABLE 7.7   Definitions of Variables in the Berkeley Guidance Study Data[a]

Label	Description
Sex	Sex 0 for boys, 1 for girls
WT2	Age 2 weight, kg
HT2	Age 2 height, cm
WT9	Age 9 weight, kg
HT9	Age 9 height, cm
LG9	Age 9 leg circumference, cm
ST9	Age 9 strength, kg
WT18	Age 18 weight, kg
HT18	Age 18 height, cm
LG18	Age 18 leg circumference, cm
ST18	Age 18 strength, kg
Soma	Somatotype, a 1 to 7 scale of body type
Case	Case numbers from the source

[a]The data for boys are in the file BGSboys.lsp, for girls in BGSgirls.lsp, and combined in BGSall.lsp.

7.3   The Berkeley Guidance Study was a longitudinal monitoring of boys and girls born in Berkeley, California between January 1928 and June 1929, and followed for at least eighteen years. The data we use are described in Table 7.7. There are three data files: BGSboys.lsp for boys, BGSgirls.lsp for girls, and BGSall.lsp for all cases.

7.3.1   For the girls, obtain the usual summary statistics (means, standard deviations, and correlations) of all the variables, except *Case* and *Sex*. Obtain a scatterplot matrix for the age 2 variables, the age 9 variables, and *HT18*. Summarize any information in the plot.

7.3.2   Fit the multiple linear regression model

$$HT18 \mid \mathbf{x} = \eta_0 + \eta_1 WT2 + \eta_2 HT2 + \eta_3 WT9 + \eta_4 HT9$$

$$+ \eta_5 LG9 + \eta_6 ST9 + e \qquad (7.49)$$

assuming that $\text{Var}(HT18 \mid \mathbf{x}) = \sigma^2$, and give estimates and standard errors for all parameters and the value of R^2.

7.3.3   Show numerically that R^2 is the same as the square of the sample correlation coefficient between *HT18* and the fitted values from (7.49). Thus a scatterplot of the response versus the fitted values provides a visual interpretation of the coefficient of determination.

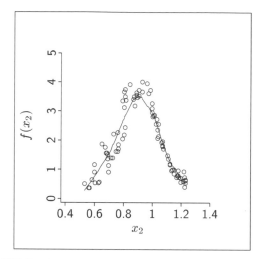

FIGURE 7.10   Curve defining a function of x_2 for Problem.

7.3.4   Give the measurement units of each of the estimated regression coefficients.

7.3.5   Obtain tests that each coefficient is equal to zero, and give the value of the test statistic, its p-value, and a brief summary of the outcome.

7.3.6   Test the hypothesis that the age 2 predictors are not needed in the mean function; that is, test NH: $\eta_1 = \eta_2 = 0$ versus the general alternative.

7.3.7   Write a sentence that summarizes the information in the overall analysis of variance table for (7.49).

7.3.8   Give 95% confidence intervals for η_4 and for $\eta_4 - \eta_2$ in (7.49).

7.4   Consider a regression with two predictors, x_1 and x_2. For each of the following five mean functions, (a) decide if the mean function specifies a multiple linear regression model, and (b) if so, give the number k of terms in the mean function and the definition of each of the terms.

a. $E(y \mid x) = \eta_0 + \eta_1 x_1 + \eta_2 x_2 + \eta_3 x_1 x_2 + \eta_4 (x_1/x_2)$

b. $E(y \mid x) = \eta_0 + \log(\eta_1 x_1 + \eta_2 x_2)$

c. $E(y \mid x) = \eta_0 + \eta_1 \exp(x_1/(x_1 + x_2))$

d. $E(y \mid x) = 1/(1 + \exp\{-[\eta_0 + \eta_1 x_1 + \eta_2 x_2]\})$

e. $E(y \mid x) = \eta_0 + \eta_1 x_1 + \eta_2 f(x_2)$, where $f(x_2)$ is defined by the curve shown in Figure 7.10. The curve was found by smoothing the scatterplot of the points shown in the figure, so there is no explicit formula for computing $f(x_2)$ from x_2.

7.5 This is a demonstration to show that the value of R^2 in a regression depends on the sampling distribution. Load the file demo-r2.lsp. You will get a dialog to generate a bivariate normal sample with means and variances you select. You also get the select the population correlation and the sample size.

7.5.1 Set the means to be 0, the standard deviations to be 1, the population correlation to be 0.8, and $n = 250$. Assuming that (x, y) has a bivariate normal distribution, what is the distribution of $y \mid x$?

7.5.2 Fit the OLS regression of y on x. According to theory, the OLS estimates are unbiased for the population values, and R^2 estimates $\rho^2 = 0.64$. Are the estimates of the regression parameters close to the true values? Is the estimate of R^2 close to the true value?

7.5.3 Draw the plot of y versus x, and use the mouse to select about 30% of the points with x in the middle of its range, and then use the "Case deletions" plot control to remove these points. For this reduced data set, does the true mean function change? Why or why not? Will the OLS estimates of regression coefficients remain unbiased? Why or why not? Select the item "Display estimates" to redisplay the regression computed without the deleted cases. Have the estimates of regression coefficients changed very much? The value of R^2, however, is now much larger; why?

7.5.4 Restore all the points, and then repeat problem 7.5.3 but this time delete about 15% of the points with the largest and smallest values of x. the value of R^2 will now decrease; why?

7.5.5 Repeat this problem a few times, varying the correlation to see that what you observed in this problem will always occur. The same results will also occur with many predictors, where R^2 is the square of the correlation between y and $\hat{y}$. It is a clearly meaningful number when the plot of y versus $\hat{y}$ looks like a bivariate normal sample, but its value depends both on the distribution of points and on the linearity of the mean function between them.

7.6 Let R^2 be the square of the sample correlation between the observed response y and the fitted values $\hat{y}$ in a multiple linear regression. Show that $RSS = SYY(1 - R^2)$, and hence R^2 is the proportion of variability in y explained by the linear regression. (*Hint*: The problem is simplified if $\bar{y} = 0$. This can always be achieved by replacing y by $y - \bar{y}$. Show that the correlation you need is unchanged by this substitution.)

7.7 Study of the growth of plants can be a crucial element in understanding how plants compete for resources. For example, soybean varieties with large early leaves may be desirable in no-till farming because larger

TABLE 7.8   Definitions of Variables in the Leaf Area Data[a]

Label	Description
Year	Year of experiment, 1994 or 1995
DAP	Days after planting
Length	Length of terminal leaflet, cm
Width	Width of terminal leaflet, cm
Area	Actual terminal leaflet area, cm^2

[a]The data are in the file `heifeng.lsp`.

leaves can shade the ground and inhibit weed growth. In a study comparing soybean varieties on leaf growth, it was necessary to develop an inexpensive method for predicting leaf area from measurements that can be taken without removing the leaf from the plant.

Table 7.8 describes variables measured on each of $n = 148$ leaves of the soybean variety Heifeng-25. The goal is to predict *Area*, which is measured using a destructive method in the laboratory, using nondestructive measures. The data are available in file `heifeng.lsp`.

7.7.1 Construct a scatterplot matrix of the three variables, *Area*, *Length*, and *Width*. Briefly describe the information in the plot.

7.7.2 The area of a rectangle is equal to the length times the width, or in log scale, log(*Area*) = log(*Length*) + log(*Width*). This suggests postulating the mean function

$$E(\log(Area) \mid Length, Width) = \eta_0 + \eta_1 \log(Length) + \eta_2 \log(Width)$$

$$(7.50)$$

What is the response? What are the predictors? What are the terms? If leaves were exactly rectangles, what values should we expect to find for η_0, η_1, and η_2?

7.7.3 Fit the multiple linear regression model with mean function (7.50), assuming constant variance. Provide a brief interpretation of the results of the F-test in the summary analysis of variance table.

7.7.4 Obtain the estimated covariance matrix of the estimates, and show numerically that the standard errors in the regression summary are the square roots of the diagonal entries of the estimated covariance matrix.

7.7.5 Assuming normal errors and constant variance, test the null hypothesis that the intercept is zero against the general alternative (7.50).

7.7.6 Assuming normal errors and constant variance, test the null hypothesis that the area of a leaf can be well approximated by the area of a rectangle. That is, test the null hypothesis

$$E(\log(Area) \mid Length, Width) = \log(Length) + \log(Width)$$

against the alternative (7.50).

7.7.7 Assuming normal errors and (7.50), provide a 95% confidence interval for the difference $\eta_1 - \eta_2$.

7.7.8 Assuming normal errors and (7.50), obtain 90% prediction intervals for (a) $\log(Area)$ and (b) $Area$ for a leaf with length 7 cm and width 4 cm.

7.7.9 Draw a plot of $\log(Length)$ versus $\log(Width)$, and from this graph identify the five cases that are likely to have the highest leverage, and explain how you selected these cases. Then, draw the plot of leverage versus case number, and verify that the points you selected from the first plot do in fact have large leverage.

7.7.10 Suppose you wanted to predict $\log(Area)$ using (7.50) for a leaf that has $Length = Width$. Should this prediction be called an extrapolation? Does your answer depend on the value of $Length$?

7.8 Continuing with the leaf area example in Problem 7.7, consider the model with mean function

$$E(Area \mid Width) = \eta_0 + \eta_1 Width + \eta_2 Width^2 \qquad (7.51)$$

7.8.1 Identify all the terms in this model.

7.8.2 Give a short justification of this mean function based on geometric considerations.

7.8.3 Fit the model, and perform a test to compare the fit of the simple linear regression of $Area$ on $Width$ to (7.51). Summarize your results.

7.9 Suppose we have a $k \times k$ symmetric matrix $\mathbf{A}$, partitioned as

$$\mathbf{A} = \begin{pmatrix} \mathbf{A}_{11} & \mathbf{A}_{12} \\ \mathbf{A}_{21} & \mathbf{A}_{22} \end{pmatrix}$$

where $\mathbf{A}_{11}$ is $k_1 \times k_1$, $\mathbf{A}_{22}$ is $k_2 \times k_2$ and $\mathbf{A}_{21} = \mathbf{A}_{12}^T$, and $k_1 + k_2 = k$.

7.9.1 Show by direct multiplication that if $\mathbf{A}$ is of full rank, meaning that $\mathbf{A}^{-1}$ exists, then

$$\mathbf{A}^{-1} = \begin{pmatrix} \mathbf{A}_{11}^{-1} + \mathbf{B}\mathbf{D}^{-1}\mathbf{B}^T & -\mathbf{B}\mathbf{D}^{-1} \\ -\mathbf{D}^{-1}\mathbf{B}^T & \mathbf{D}^{-1} \end{pmatrix}$$

where

$$\mathbf{B} = \mathbf{A}_{11}^{-1}\mathbf{A}_{12}$$

$$\mathbf{D} = \mathbf{A}_{22} - \mathbf{A}_{21}\mathbf{A}_{11}^{-1}\mathbf{A}_{12}$$

7.9.2 Suppose we partition the matrix $\mathbf{U}$ given at (7.34) into $\mathbf{U} = (\mathbf{U}_0, \mathbf{U}_1)$, where $\mathbf{U}_0$ is the $n \times 1$ vector of 1's, and $\mathbf{U}_1$ is the remaining $k - 1$ columns of $\mathbf{U}$. Show that $\mathbf{U}^T\mathbf{U}$ in partitioned form is

$$\mathbf{U}^T\mathbf{U} = \begin{pmatrix} \mathbf{U}_0^T\mathbf{U}_0 & \mathbf{U}_0^T\mathbf{U}_1 \\ \mathbf{U}_1^T\mathbf{U}_0 & \mathbf{U}_1^T\mathbf{U}_1 \end{pmatrix} = \begin{pmatrix} n & n\bar{\mathbf{u}}^T \\ n\bar{\mathbf{u}} & \mathbf{U}_1^T\mathbf{U}_1 \end{pmatrix}$$

where $\bar{\mathbf{u}}$ is the $(k - 1) \times 1$ vector of means of the terms, and $\mathbf{U}_1^T\mathbf{U}_1$ is the $(k - 1) \times (k - 1)$ matrix of uncorrected sums of squares and cross-products. Show also that for the matrices of Problem 7.9.1, $\mathbf{B} = \bar{\mathbf{u}}^T$, and $\mathbf{D}$ is the matrix of *corrected sums of squares and cross products*. This means that if $\mathbf{C}$ is the sample covariance matrix of $u_1, \ldots, u_{k-1}$, then $\mathbf{C} = \mathbf{D}/(n - 1)$. (*Hint*: To obtain this last result, use a matrix generalization of the result that $\sum(a_i - \bar{a})^2 = \sum a_i^2 - n\bar{a}^2$.)

7.9.3 Give an expression for $(\mathbf{U}^T\mathbf{U})^{-1}$ in terms of n, $\bar{\mathbf{u}}$, and the matrix $\mathbf{D}$. If we partition the parameter vector $\eta^T = (\eta_0, \eta_1^T)$, separating out the intercept, give expressions for $\mathrm{Var}(\eta_0)$ and $\mathrm{Var}(\eta_1)$. This will show that $\mathrm{Var}(\eta_1)$ depends only on σ^2 and the inverse of $\mathbf{D}$.

7.9.4 Let $\mathbf{U}_1$ be the matrix $\mathbf{U}_1$, but with column means subtracted off, so $\mathbf{U}_1 = \mathbf{U}_1 - \mathbf{1}\bar{\mathbf{u}}^T$, and also $\mathbf{D} = \mathbf{U}_1^T\mathbf{U}_1$. By direct multiplication based on the partitioned form of $(\mathbf{U}^T\mathbf{U})^{-1}$, show that

$$\hat{\eta}_1 = \mathbf{D}^{-1}\mathbf{U}_1^T\mathbf{y}$$

$$\hat{\eta}_0 = \bar{y} - \bar{\mathbf{u}}^T\hat{\eta}_1$$

7.9.5 If $\mathbf{y}$ is replaced by centering, $\mathbf{y} = \mathbf{y} - \bar{y}\mathbf{1}$, how will the estimates of η_0 and η_1 change?

7.10 The formula for the ith leverage value is $h_i = \mathbf{u}_i^T(\mathbf{U}^T\mathbf{U})^{-1}\mathbf{u}_i$. Show that, using centered matrices as defined in the last problem,

$$h_i = \frac{1}{n} + (\mathbf{u}_1 - \bar{\mathbf{u}})^T(\mathbf{U}_1^T\mathbf{U}_1)^{-1}(\mathbf{u}_1 - \bar{\mathbf{u}})$$

which is the equation of an ellipsoid centered at $\bar{\mathbf{u}}$ with contours determined by $(\mathbf{U}_1^T\mathbf{U}_1)^{-1}$, as shown in Figure 7.7.

CHAPTER 8

Three-Dimensional Plots

In this chapter we return to the theme of using graphs to study regression and introduce a graphical method that uses motion to allow the user to see three dimensions. These plots can have great value in regressions with two predictors, since we can then see all the data at once, and in regressions with many predictors, as discussed in Chapter 20. We also illustrate important ideas in regression introduced in the last two chapters.

8.1 VIEWING A THREE-DIMENSIONAL PLOT

A three-dimensional (3D) plot has three axes, a vertical axis we will generally label V, and a pair of horizontal axes we will label H for the axis that can initially be seen in the flat 2D computer screen, and O for the out-of-page axis that is not visible initially. Figure 8.1 shows how these three axes look in three dimensions. Three-dimensional plots are difficult to represent accurately

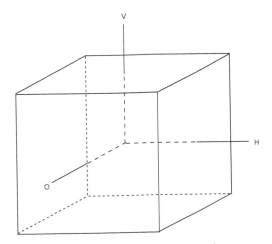

FIGURE 8.1 The 3D plotting region.

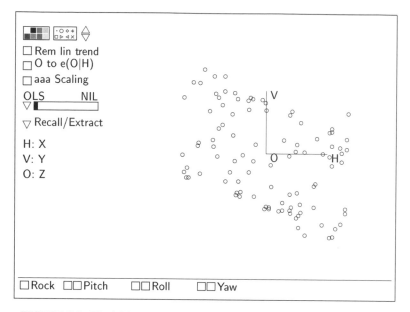

FIGURE 8.2   The initial "Home" view of the "View a surface" demonstration.

on a two-dimensional page, so using the computer while reading this chapter may be particularly useful.

Figure 8.2 is a 3D plot created with *Arc*. It can be duplicated by loading the file demo-3d.lsp, and then selecting "View a surface" from the "Demos:3D" menu. The main difference between Figure 8.2 and the plot you produce is that the background and foreground colors are different: the motion of white points on a dark background is easier to see than the usual black on white. If you want to change the background color, first select "Options" from the plot's menu and then push the "White background" button and finally push "OK." With Mac OS and Windows, the plot's menu is called "3Dplot." The out-of-page dimension is not visible in Figure 8.2 because this dimension is perpendicular to the page of this book. To see the out-of-page dimension, it is necessary to rotate the plot.

In *Arc*, the plot's axes are always labeled with the letters H, V, and O. The names of the variables plotted on the axes are shown to the left of the plot; in Figure 8.2, these are x, y, and z, respectively. We will refer to this as a plot of y versus (x, z), meaning that y is on the vertical V axis; and x and z are on the horizontal axes, with x displayed on the H axis and z on the O axis.

The *vertical screen axis* and the *horizontal screen axis* are the fixed vertical and horizontal directions on the computer screen. When in the initial "Home" position, the H and V axes of a 3D plot are the same as the horizontal and vertical screen axes. But after a 3D plot is rotated, its H and V axes may no longer align with the screen axes. The plot axes and the screen axes will both play a role in the interpretation of 3D plots.

8.1.1 Rotation Control

The basic tools for rotation control are three pairs of buttons at the bottom of the plot. Pushing any of these buttons will cause the points to rotate. The two "Yaw" buttons cause the plot to rotate either to the left or right about the vertical screen axis. The "Roll" buttons cause rotation about the direction perpendicular to the computer screen, while "Pitch" rotates about the horizontal screen axis. Holding down the shift key while pushing a control button causes the plot to rotate continuously until one of the six control buttons is pushed again. In many of the applications of 3D plots in regression graphics, rotation will be about the V-axis only, using the Yaw buttons.

The rate of rotation is changed by selecting the "Faster" or "Slower" item in the plot's menu. These items can be selected more than once, and each selection will result in a slightly faster or slower rate of rotation.

Select the item "Mouse mode" from the plot's menu, and then select "Hand rotate" in the resulting dialog. This changes the pointer into a hand; as you hold down the mouse button you can use the hand to push the point cloud in various directions, much like you might push on the surface of a basketball to start it rotating. Pushing near the outside of the point cloud will result in relatively fast motion, while pushing near the center of the point cloud will result in relatively slow motion. Effective pushing takes some practice.

8.1.2 Recalling Views

The item "Recall home" from the "Recall/Extract" pop-up menu restores the plot to its original orientation. The item "Remember view" will put an internal marker at the current view. You can recall this view by selecting the item "Recall view." The other items in this menu will be described later.

Before rotation in Figure 8.2, the dominant feature is a linear trend; while rotating, distinct curvature emerges. From the rotating plot, one can clearly recognize shapes. Relating the shapes to a specific functional form is quite a different matter, however. Shapes will generally be less obvious in real-world data analysis problems.

8.1.3 Rocking

During rotation a striking pattern may be seen, but it may be visible only while rotating. The impression of three dimensions can be maintained by stopping rotation near the view where the pattern is visible and then holding down the "Rock" button at the bottom left of the plot. As long as the mouse button is down, the plot will rock back and forth, allowing an impression of three dimensions to be maintained while staying near the interesting view.

8.1.4 Show Axes

Occasionally the axes in a 3D plot may be distracting. They can be removed by selecting "Show axes" from the plot's menu. Repeating this operation will restore the axes to the plot.

8.1.5 Depth Cuing

To create an appearance of depth, points in the back of the point cloud are plotted with a smaller symbol than are points in the front. As the points rotate, the symbol changes from small to large or vice versa. If some of the points are marked with a special symbol, either chosen from the symbol palette or obtained using point marking, then depth cuing is turned off. To turn depth cuing off by hand, select "Depth cuing" from the plot's menu. Selecting this item again will turn depth cuing back on.

Depth cuing has been turned off in all the 3D plots shown in this book.

8.1.6 Zooming

The two adjacent triangles just to the right of the symbol palette control the distance that the points are from the observer. Holding down the mouse in the top triangle causes the points to move closer to the observer, while holding the mouse in the bottom triangle causes the points to move away. A plot can be returned to its normal state by using the item "Zoom to normal size" in the "Recall/Extract" menu. Zooming can be useful when there are many points in a plot and increased resolution is needed to see structure.

8.2 ADDING A POLYNOMIAL SURFACE

8.2.1 Parametric Smoother Slidebar

The slidebar on 3D plots initially marked "OLS" is the equivalent of the parametric smoother slidebar on 2D plots. In 3D plots this slidebar is used to fit quadratic polynomial surfaces by using least squares to estimate quadratic models, like the full quadratic,

$$\hat{E}(V \mid H, O) = b_0 + b_1 H + b_2 O + b_{11} H^2 + b_{22} O^2 + b_{12} H O$$

where V is the quantity on the vertical axis, and H and O are on the horizontal axes. As the slider is moved to the right, the coefficients for the linear terms H and O are always estimated, but different combinations of the other b's are set to zero. You can fit a mean function that is a plane ($b_{11} = b_{22} = b_{12} = 0$), fit with one or both of the quadratic terms, fit with only the interaction term, or fit with all terms (the "Full Quad" mean function). Here is a partial list showing how the brief phrases above the slidebar describe the form of the fitted model:

plane: $\hat{E}(V \mid H,O) = b_0 + b_1 H + b_2 O$

pln+H^2: $\hat{E}(V \mid H,O) = b_0 + b_1 H + b_2 O + b_{11} H^2$

pln+HO: $\hat{E}(V \mid H,O) = b_0 + b_1 H + b_2 O + b_{12} HO$

pln+quad: $\hat{E}(V \mid H,O) = b_0 + b_1 H + b_2 O + b_{11} H^2 + b_{22} O^2$

Start the plot shown in Figure 8.2 rotating by shift-clicking in the left "Yaw" button, and then click in the parametric smoother slidebar to see how the various polynomial mean functions match the data. None of these polynomial mean functions match the data very well, since the data were not generated in this way. The parametric smoother slidebar is useful for understanding how the various models can fit data.

The last option on the slidebar is to fit a "1D-Quad" model, which estimates a mean function of the form

$$\hat{E}(V \mid H,O) = a_0 + a_1(b_1 H + b_2 O) + a_2(b_1 H + b_2 O)^2$$

The computations for estimation of the coefficients in this model may take longer, because this mean function is not in the family of linear models. This mean function is called the *1D quadratic model*, and it will be discussed in Section 18.3.2.

8.2.2   Extracting Fitted Values

The parametric smoother slidebar has a menu with several options. The item "Extract mean" creates a new variable with the name you specify that contains the values of $\hat{V}_i = \hat{E}(V \mid H_i, O_i)$, the fitted values for the fit to the data.

8.2.3   Adding a Function

The item "Add arbitrary function" on the parametric smoother slidebar menu allows adding a function to the plot. If you select this item, and in the resulting dialog type

```
y = sin(x)+2 * cos(z)+sin(x) * cos(x)
```

the graph of this function will be added to the plot. It matches the data very closely, as it should since the data were generated using this as the mean function.

8.2.4   Residuals

Recall that the residuals from any of the fits possible with the parametric slidebar are defined as $\hat{e}_i = V_i - \hat{V}_i$, for $i = 1,\ldots,n$. If the model is correct, then the residuals should appear to be independent of H_i and O_i.

The residuals from any of the slidebar fits can be emphasized in a 3D plot by selecting the item "Show residuals" from the slidebar menu. This causes line segments parallel to the V-axis to appear between each point in the plot and the fitted surface. The length of a line segment give the magnitude of the corresponding residual. The line segments for the positive residuals above the surface have a different color than the line segments for the negative residuals below the surface.

Use the parametric smoother slidebar to place the fitted model

$$\hat{E}(V \mid H, O) = b_0 + b_1 H + b_2 O$$

on the 3D plot for the "View a surface" demonstration, and then select "Show residuals" from the slidebar menu. While rotating the plot about the V-axis, the positive residuals all appear at the edges of the fitted plane, while the negative residuals appear in the center. This indicates that the residuals are not independent of (H, O) and thus that the model is wrong.

8.3   SCALING AND CENTERING

The *plotting region* shown in Figure 8.1 is the interior of a cube centered at the origin, with sides running from -1 to 1. The data are *centered* and *scaled* to fit into this region. This can influence the interpretation of a 3D plot.

Suppose a quantity v is to be plotted on one of the axes. In *Arc* and in most other computer programs, centering and scaling are based on the range $r_v = \max(v) - \min(v)$ and the mid-range $m_v = (\max(v) + \min(v))/2$. The quantity actually plotted is $2(v - m_v)/r_v$, which has minimum value -1 and maximum value 1. The centered and scaled variable fills the plotting region along the axis assigned to v. If the program is given instructions to construct the plot of y versus (x, z), what will really be produced is the plot

$$a(y - m_y) \quad \text{versus} \quad (b(x - m_x), c(z - m_z)) \tag{8.1}$$

where

$$a = 2/r_y, \ b = 2/r_x \quad \text{and} \quad c = 2/r_z$$

Centering usually has no effect on the interpretation of the plot. Scaling, however, can have an effect and we refer to the operation leading to (8.1) as *abc-scaling*. Because of the scale factors a, b, and c in (8.1), we will not be able to assess relative size in plots produced with *abc*-scaling: A plot of $100y$ versus $(10x, 1000z)$ will look identical to the plot of y versus (x, z).

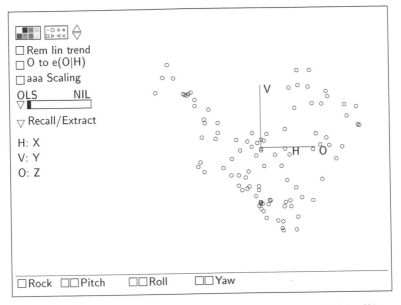

FIGURE 8.3   The same data as in Figure, rotated about the vertical by 60°.

When relative size is important, it may be better to use *aaa-scaling* in which the three scale factors in (8.1) are all replaced by the minimum scale factor, $\min(a,b,c)$. With *aaa*-scaling, the data on one of the axes will fill the plotting region, but the data on the other two axes may not fill their axes. A plot of $100y$ versus $(10x, 1000z)$ will appear quite different from the plot y versus (x, z) when *aaa*-scaling is used. When the plotted variables all have the same units, *aaa*-scaling of a plot can give additional useful information about the relative size and variation of the three plotted variables. In *Arc*, *aaa*-scaling is obtained by pushing the "aaa Scaling" button. Repeating this operation will return to *abc*-scaling.

8.4   2D PLOTS FROM 3D PLOTS

Use the "Recall/Extract" menu to return Figure 8.2 to the "Home" position and then rotate about the vertical axis by using the right "Yaw" button through roughly 60°. The plot should now resemble Figure 8.3. The variable on the vertical screen axis is still y, but the variable on the horizontal screen axis is some linear combination of x and z. Which linear combination is it? After using the right "Yaw" button to rotate about the vertical axis through an angle of θ, the variable on the horizontal screen axis is

$$\text{horizontal screen variable} = d + h = d + b(\cos\theta)x + c(\sin\theta)z \quad (8.2)$$

Here b and c are the scale factors used in (8.1), d is a constant that depends on the centering constants m_x and m_z and on the scale factors b and c, and h is the linear combination of x and z that appears on the horizontal screen axis. Equation (8.2) might be understood by thinking of a circle in the horizontal plane. Each point on the circle determines a linear combination h of x and z. If rotation had been about *both* the vertical axis using a "Yaw" button and about another axis using another button, then the linear combination on the horizontal screen axis would depend on two rotation angles, and it would be a linear combination of x, z and y.

Since horizontal screen variables play an important role in statistical applications of 3D plots, *Arc* allows you to display the values of d and h, and also to save h as a variable for future calculations. Select the item "Display screen coordinates" from the "Recall/Extract" pop-up menu. This will display the linear combination of the quantities plotted on the horizontal and vertical screen axes. For the view in Figure 8.3, the displayed output looks like this:

```
>Linear combinations on screen axes in 3D plot.
Horizontal: - 1.399 + 0.1640 H + 0.2840 O
Vertical: - .00883 + 0.3291 V
```

The H, V, and O refer to the quantities plotted on the horizontal, vertical and out-of-page axes of the plot, which for the example are x, y, and z, respectively. The quantity on the horizontal axis is $h \approx 0.16x + 0.28z$ with the constant $d \approx -1.40$. Apart from a constant, the quantity on the vertical axis is proportional to y. Your output may be slightly different because the rotation angle you use is not likely to be exactly $60°$. You can rotate to exactly $60°$ by selecting the item "Move to horizontal" from the "Recall/Extract" menu. Since $\cos 60° = 0.5$ and $\sin 60° = 0.866$, enter the multipliers $a_1 = 0.5$, $a_2 = 0$, $a_3 = 0.866$ in the resultant dialog.

8.4.1   Saving a Linear Combination

When rotation is stopped, we see a 2D scatterplot of y versus h. The linear combination h that is visible on the horizontal screen axis can be saved for future calculations. Select the "Extract axis" item from the "Recall/Extract" pop-up menu. You will be presented with a dialog to enter a name for this new quantity, and to choose either the horizontal screen variable or the vertical screen variable; you will usually want the default, the horizontal screen variable. Push "OK" after you have chosen an axis and typed a name for the extracted variable. Further rotation of the plot will not change the value of the saved variable.

8.4.2   Rotation in 2D

The static view shown in Figure 8.3 is a *projection* of the points in the full 3D plot onto the plane formed by the vertical screen axis and the horizontal

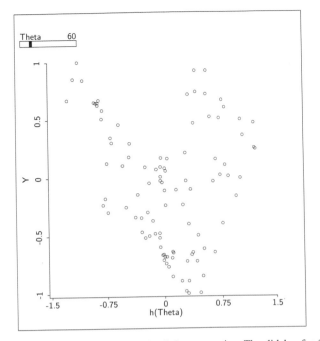

FIGURE 8.4 A 2D view of the "View a surface" demonstration. The slidebar for θ adds rotation to a 2D plot.

screen axis defined by (8.2). Since this is really a 2D plot, it could be viewed in a 2D scatterplot, as in Figure 8.4. We can use this figure to explain how rotation is done. When rotation is about the V axis, we write

$$h = h(\theta) = b(\cos\theta)x + c(\sin\theta)z$$

to recognize the dependence of h on the angle θ of rotation. Imagine changing θ by a small amount; this will change $h(\theta)$, but will not change the variable plotted on the vertical axis. The computer screen can be refreshed by deleting the current points and then redrawing y versus $h(\theta)$ for the updated value of θ. Repeating this process over and over gives the illusion of rotation. For a full 3D rotating plot, depth cuing and updated axes are all that need to be added.

Select the item "Rotation in 2D" from the "Demos:3D" menu to reproduce Figure 8.4, except that the rotation angle is initially set to $\theta = 0$. A slidebar marked "Theta" has been added to this plot. As you hold down the mouse in this slidebar, θ and $h(\theta)$ are changed, and for each new θ the plot is redrawn. The slidebar changes θ in 10° increments, which is a much larger change than is used in the built-in rotating plot. As you change θ, the full 3D plot is visible. If you reverse the colors using the "Options" item in the plot's menu, the rotation is a bit clearer.

We now see that 3D rotation about the vertical axis is nothing more than rapidly updating the 2D plot y versus $h(\theta)$ as θ is incremented in small steps. Rotating a 3D plot once about the vertical axis by using one of the "Yaw" buttons corresponds to incrementing θ between $0°$ and $360°$. During rotation, plots of y versus all possible linear combinations of x and z are visible.

8.4.3 Extracting a 2D Plot

Any 2D view of a 3D plot can be put into its own window by selecting the "Extract 2D plot" item from the "Recall/Extract" menu. The new 2D plot has the usual plot controls, including smoothers.

8.4.4 Summary

The ideas of this section form a basis for viewing regression data in 3D plots, and so a brief summary is in order. The 3D plot is of y versus (x,z), with y on the vertical axis and x and z on the horizontal axes. While rotating this plot once around the vertical axis, we will see 2D plots of *all* possible linear combinations of x and z on the horizontal screen axis, with y on the vertical axis. When the rotation is stopped, we see a 2D plot. The variable (8.2) on the horizontal screen axis can be extracted by using the "Extract axis" item in the "Recall/Extract" menu. This variable will correspond to some particular linear combination of x and z.

8.5 REMOVING A LINEAR TREND IN 3D PLOTS

A strong linear trend in a 3D plot like the one apparent in Figure 8.5 may visually mask other interesting features, particularly nonlinearities. Figure 8.5 can be reproduced by selecting the "Detect a small nonlinearity" item from the "Demos:3D" menu and then rotating the plot. This problem can be overcome by removing the linear trend, leaving any nonlinear effects behind. In *Arc*, a plot is *detrended* in this way by pushing the "Rem lin trend" button on a 3D plot. This will replace the variable on the vertical axis with the residuals $\hat{e}(V \mid HO)$ from the fitted plane

$$\hat{E}(V \mid H,O) = b_0 + b_1 H + b_2 O$$

Also, the label on the vertical axis becomes e(V|HO) as a reminder that we are now plotting residuals. A detrended 3D plot is thus of $\hat{e}(V \mid HO)$ versus (H,O).

Figure 8.6 is obtained from Figure 8.5 by clicking the "Rem lin trend" button. The systematic pattern to the residuals in Figure 8.6 is nearly invisi-

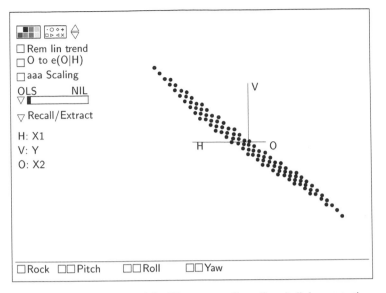

FIGURE 8.5   A 2D view of the "Detect a small nonlinearity" demonstration.

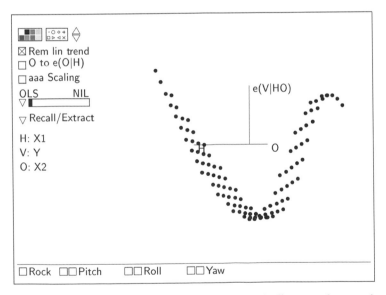

FIGURE 8.6   The same 2D view as in Figure, with the linear trend removed.

ble in Figure 8.5, even while rotating. These data were generated by taking 100 points on a 10×10 grid for x_1 and x_2, and defining $y = z + \exp(-z)/(1 + \exp(-z))$, where $z = 0.909x_1 - 0.416x_2$ is a linear combination of the predictors.

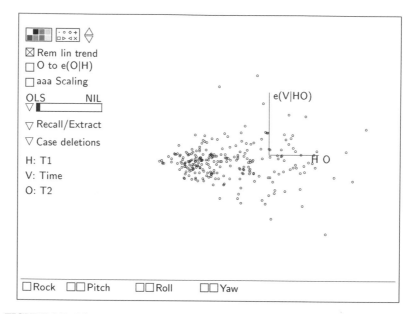

FIGURE 8.7   View of the 3D plot of residual versus (T_1, T_2) from the transaction data.

The illustrative analysis of the transaction data in Chapter 7 was based on the model

$$\text{E}(Time \mid T_1, T_2) = \eta_0 + \eta_1 T_1 + \eta_2 T_2 \quad \text{with} \quad \text{Var}(Time \mid T_1, T_2) = \sigma^2$$

If this model were correct, then we would expect the residuals $\hat{e}$ to be independent of (T_1, T_2). Shown in Figure 8.7 is one view of the 3D plot $\hat{e}_i$ versus (T_{i1}, T_{i2}). The variability of the residuals increases along the horizontal screen axis, indicating that the residuals are not independent of (T_1, T_2) and thus that the data contradict the illustrative model.

8.6   USING UNCORRELATED VARIABLES

Select the item "Collinearity hiding a curve" from the "Demos:3D" menu and rotate the resulting 3D plot. What do you observe? The points in the plot appear to fall close to a rotating vertical sheet of paper. One static view of this plot is given in Figure 8.8. This figure and the full rotating plot are of little help in finding structure because the predictors plotted on the H and O axes are highly correlated. To see this clearly, return the plot to the "Home" position and then use the "Pitch" control to rotate to the 2D plot O versus H. The high correlation between the variables should now be apparent, as all the points fall very close to a line. Finding structure in 3D plots is likely

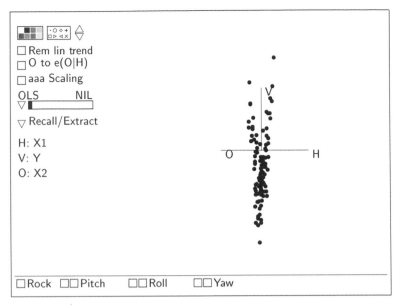

FIGURE 8.8   A 2D view of the "Collinearity hiding a curve" demonstration.

to be difficult when the variables on the H and O axes are even moderately correlated.

With highly correlated variables, some 2D views of the 3D plot lack sufficient resolution for any structure to be clearly visible. This is the case in Figure 8.8, where the values of the horizontal screen variable are tightly clustered about the origin of the plot. To ensure good resolution in a 3D plot, we would like the values of the horizontal screen variable to be well spread in every 2D view. This can be accomplished by replacing the original variables on the H and O axes with an equivalent pair of *uncorrelated variables*. The new plot created in this way will be of the form V versus (H, O_{new}), so we will need to change only the variable on the out-of-page axis. The variable O_{new} is just the residuals from the simple linear regression of O on H, including an intercept. These residuals are represented as $\hat{e}(O \mid H)$. Since the sample correlation between the residuals $O_{new} = \hat{e}(O \mid H)$ and H is zero, the variables on the horizontal and out-of-page axes of the new plot will be uncorrelated, and the values of the horizontal screen variable should be well spread in any 2D view.

Beginning with any 3D plot, a new plot with uncorrelated variables on the horizontal axes is obtained by pushing the "O to e(O|H)" button. Pushing the button again will restore the plot to its original state.

Figure 8.9 is obtained from Figure 8.8 by changing to uncorrelated variables on the horizontal axes. A curved trend is now plainly visible. In this example, x_1 and x_2 have correlation close to 0.99, and the mean function is a full quadratic in the predictors x_1 and x_2.

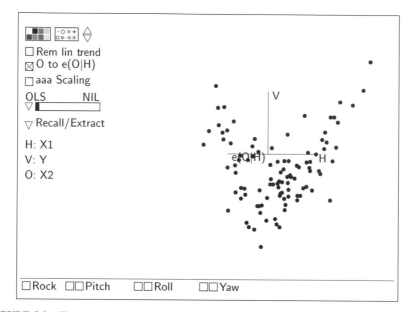

FIGURE 8.9   The same 2D plot as in Figure 8.8, but with the "O to e(O | H)" button pushed.

We gain resolution when changing to uncorrelated horizontal variables in a 3D plot, but do we lose information? Since the residuals $\hat{e}(O \mid H) = O_{new}$ from the regression of O on H are computed as a linear combination of O and H, rotating the new plot of V versus $(H, \hat{e}(O \mid H))$ will still display 2D plots of all possible linear combinations of O and H, just as happens when rotating the original plot V versus (H, O). No information is lost when changing to uncorrelated variables.

8.7   COMPLEMENTS

The data in Figures 8.5 and 8.6 were generated by setting x and z each to be 100 uniform random numbers between $0°$ and $360°$ and then computing $y = \sin(x) + 2\cos(z) + \sin(x)\cos(x)$.

Plot rotation was introduced in the statistics literature by Fisherkeller, Friedman, and Tukey (1974). Many early papers on this subject are reprinted in Cleveland and McGill (1988). Tierney (1990, Section 9.1.3) provides a useful reference for many of the details of plot rotation. The need for uncorrelated variables in a rotating plot was presented by Cook and Weisberg (1989, 1990a). In the latter paper, an optimality property of this procedure is derived. The discussion of rotation in Section 8.4 is limited to rotation about the vertical axis, but it is easily generalized.

A more complete study of 3D plots suitable for readers with Masters level background in statistics is available from Cook (1998b).

The RANDU generator discussed in Exercise 8.2 is taken from Tierney (1990, p. 41). Exercise 8.3 is based on a demonstration included with *Xlisp-Stat*, also by Luke Tierney. The data in Problem 8.7 are discussed in Mielke, Anderson, Berry, Mielke, Chaney and Leech (1983).

PROBLEMS

8.1 In Figure 8.4, the tick-marks on the horizontal axis cover the range from −1.5 to 1.5, even though each of the variables displayed in the plot are scaled to have values between −1 and 1. Why is this larger range necessary? *Hint*: Think of rotating the cube in Figure 8.1.

8.2 Many statistical methods use a sequence of numbers that behave as if they were a random sample from some specified distribution. The most important case is generating a sample from the uniform distribution, so each draw from the distribution is equally likely to be any number between zero and one. Most random number generators are deterministic, so if you start the generator in the same place, it will always give the same sequence of numbers. Consequently, generators must be tested to see if the deterministic sequences they produce behave as if they really are a random sample from a distribution.

A well-known generator of uniform random numbers from the early days of computing is called RANDU. Load the file randu.lsp. This will give you a 3D plot obtained by taking 600 consecutive draws from RANDU, defining x_1 from the first 200 draws, x_2 from the second 200 draws, and x_3 from the third 200 draws, and then plotting x_2 versus (x_1, x_3).

8.2.1 As you spin the plot, the points fall in a cube. Is this expected, or is it evidence that RANDU is not generating numbers that behave like a uniform sample?

8.2.2 If the numbers all behaved as if they were independent draws from a uniform distribution, what would you expect to see in every 2D view of this plot? Spin the plot, and see if you can find a 2D view that is not random. You may need to spin slowly to find anything. What do you conclude?

8.3 Load the file spheres.lsp. This will automatically produce two 3D plots. Move one so both are visible at the same time. Start both rotating by using the left "Yaw" button while pressing down the shift key. What is the difference between Plot 1 and Plot 2? If you are having trouble finding any difference, select a slice in each plot, and then examine the slice as the plots rotate.

8.4 This problem uses the haystack data in the data file `haystack.lsp`.

    8.4.1 Inspect the 3D plot *Vol* versus (*C, Over*) while rotating about the vertical axis. Write a two- or three-sentence description of what you see as the main features of the plot.

    8.4.2 Change to *aaa*-scaling in the 3D plot *Vol* versus (*C, Over*). Write a brief description of the result and why it might have been expected based on the nature of the measurements involved.

    8.4.3 Remove the linear trend from the plot *Vol* versus (*C, Over*) by clicking the "Rem Lin Trend" button. Describe the quantity on the vertical axis of the detrended plot. Write a two- or three-sentence description of the main features of the detrended plot.

8.5 This problem uses data from the Berkeley Guidance Study for girls, in the file `BGSgirls.lsp`.

    8.5.1 Draw the plot of *WT*18 versus (*WT*2, *WT*9). Inspect this plot while rotating about the vertical axis. Write a two- or three-sentence description of what you see.

    8.5.2 Change to *aaa*-scaling in this 3D plot. Explain what happened. Does the scaling convey additional information? What is the information? Remember that all three variables are measured in the same units.

8.6 Data on 56 normal births at a Wellington, New Zealand hospital are given in file `birthwt.lsp`. The response variable for this problem is *BirthWt*, birth weight in grams. The three predictors are the mother's age denoted by *Age*, term in weeks denoted by *Term*, and the baby's sex, *Sex* = 0 for girls and 1 for boys. Construct the 3D plot *BirthWt* versus (*Age, Term*). Inspect the plot using any of the plot controls discussed in this chapter and write a brief description of your impressions.

8.7 It has long been recognized that environmental lead can pose serious health problems, particularly when ingested by young children. Although leaded paint and gasoline have been subject to strict controls for some time, leaded paint in old homes located in inner-city neighborhoods is still a recognized hazard.

    A study was conducted in Baltimore to assess the concentration of lead in the soil throughout the metropolitan area. The lead concentration (ppm) was determined in each of 424 soil samples from the Baltimore area. The data file `baltlead.lsp` contains the lead concentration *Lead*, the coordinates (*X, Y*) in kilometers of the sampling location with the center of Baltimore as the origin, and the distance $D = \sqrt{X^2 + Y^2}$ of the sampling location from the center of Baltimore. The investigators expected that the highest lead concentrations would be found near the center of Baltimore, and that the lead concentration would decrease with the distance D from the city center.

8.7.1 Fit the mean function

$$E(Lead \mid X,Y) = \eta_0 + \eta_1 X + \eta_2 Y$$

using ordinary least squares. Based only on the usual summary statistics from this fit, what do you conclude about the distribution of lead concentration in the metropolitan area?

8.7.2 Construct a 3D plot with *Lead* on the vertical axis and the co-ordinates (X,Y) on the other two axes. Describe what you see while rotating the plot around the vertical axis. Do your impressions agree with the experimenter's prior opinions? In view of this plot, comment on the usefulness of the results from the fit in the previous problem.

8.7.3 As a plot of *Lead* versus D shows, a simple linear regression of *Lead* on D certainly isn't appropriate. Use the power transformation sliders to estimate visually powers λ_L and λ_D that induce a relatively simple relationship in the regression of $Lead^{\lambda_L}$ on D^{λ_D}. Power transformation sliders can be placed on a plot of *Lead* versus D by checking the "Transform sliders" box at the bottom of the plot's "Options" menu.

8.7.4 Construct a *lowess* estimate of the mean function for the regression of $\log(Lead)$ on $\log(D)$. What does the smooth suggest about the distribution of soil lead in the metropolitan area? Does a linear mean function seem appropriate?

8.7.5 Is there any evidence in the data to indicate that the coordinates (X,Y) contain information on $\log(Lead)$ beyond that furnished by $\log(D)$? More specifically, is there evidence in the data to indicate that the distribution of $\log(Lead) \mid (X,Y)$ is different from the distribution of $\log(Lead) \mid D$? *Hint*: This is a challenging problem. Consider transforming (X,Y) to polar coordinates using the "Add a variate" item and the functions for polar coordinates described in Appendix Section A.12.

CHAPTER 9

Weights and Lack-of-Fit

In this chapter we examine two related aspects of the multiple linear regression model: (1) checking for lack-of-fit of a mean function and (2) allowing the variance function $\text{Var}(y \mid x)$ to be nonconstant.

9.1 SNOW GEESE

9.1.1 Visually Assessing Lack-of-Fit

An example on counting snow geese was introduced in Section 3.6. Figure 9.1 provides a scatterplot of the data, along with the OLS fit. Recall that the response $y = Photo$ is the true size of a flock of geese as determined from a photograph, and the single predictor $x = Obs2$ is the visually estimated flock size by the second observer. As in the discussion of Section 3.6, we removed the four cases with the largest values of $Obs2$ to improve visual clarity.

The fitted line in Figure 9.1 seems to give a reasonable characterization of $E(y \mid x)$ for the larger values of x, but it overestimates the size of all flocks with $x \leq 30$. The residuals $\hat{e}_i = y_i - \hat{y}_i$ become important when assessing a fitted mean function against the data. The conclusion that the fitted line in Figure 9.1 overestimates the size of all flocks with $x \leq 30$ comes about because all of the residuals are negative for $x \leq 30$. Similarly, the model seems to fit better for larger values of x because the associated residuals have no apparent pattern. Plotting the residuals versus the predictor can help in assessing the fit of a model since this focuses on deviations from a fit by removing the linear trend from the plot. The linear trend can be removed from any 2D plot produced by *Arc* by clicking the "Rem lin trend" control box. This command causes *Arc* to compute the residuals from the simple linear regression of the y-axis variable on the x-axis variable, and then replace the y-axis variable with the residuals.

Beginning with the plot in Figure 9.1, remove the fitted line, activate the "Rem lin trend" control, and then click the box for "Zero line." The resulting plot of $\hat{e}$ versus $Obs2$ is shown in Figure 9.2. If the multiple linear regression

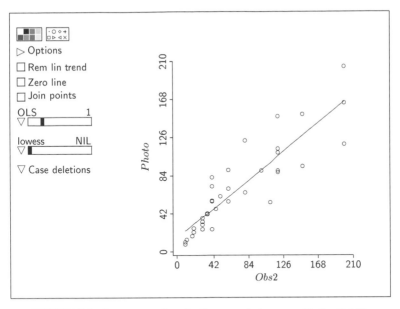

FIGURE 9.1 Snow geese data for the second observer with the OLS fit.

model is correct, then a plot of the residuals versus the predictor should appear as if the mean and variance functions are constant with $E(\hat{e} \mid x) = 0$. The horizontal line at zero in Figure 9.2 is thus the mean function when the model is correct. If either the residual mean or variance function is not constant, then we have evidence that the multiple linear regression model is incorrect. More generally, if the multiple linear regression model is correct, then the plot should appear as if the residuals and predictor are independent. Any evidence of dependence is evidence against the model.

As long as an intercept is included in the mean function, the average of the residuals from an OLS fit of the multiple linear regression model is always zero, $\sum_{i=1}^{n} \hat{e}_i = 0$, and the residuals are always uncorrelated with the predictor, $\mathrm{Cov}(\hat{e}, x) = 0$. This means that the OLS fit of the residuals $\hat{e}_i$ on the predictor x_i will always have slope and intercept equal to zero. The zero line in Figure 9.2 is therefore the fitted line for the simple linear regression of $\hat{e}_i$ on x_i.

The overestimation for smaller values of x in Figures 9.1 and 9.2 is evidence against the simple linear regression model

$$E(y \mid x) = \eta_0 + \eta_1 x \qquad \text{and} \qquad \mathrm{Var}(y \mid x) = \sigma^2 \qquad (9.1)$$

It could be that the constant variance function is not appropriate for the snow geese data. Or it could be that the linear mean function is not appropriate. Perhaps both are inappropriate. We consider nonconstant variance functions and tests for the mean function in this chapter.

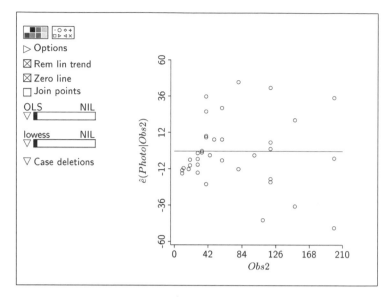

FIGURE 9.2   Scatterplot of $\hat{e}$ versus *Obs2* for the snow geese data.

9.1.2   Nonconstant Variances

The aerial observers could count the number of geese in a small flock, but they visually estimated the size of large flocks. For a very small flock of about ten birds, the counting error will most likely be zero. For a very large flock of about 100 birds, an error of about 20% might be anticipated. Counts for small flocks will therefore be more precise than counts for large ones, and the variance function $\text{Var}(y \mid x)$ should increase with the value of x, as previously discussed in connection with Figure 3.11. When fitting a mean function in a regression with a nonconstant variance function, we should give more weight to cases with small variances. For the regression of Figure 9.1, the weight should be a decreasing function of *Obs2* because the variance function increases with *Obs2*.

The OLS criterion introduced in Chapter 6 gives equal weight to all cases. In the next section we adapt the OLS objective function to allow for differential weighting when the variance function is not constant.

9.2   WEIGHTED LEAST SQUARES

Additional information about the variance function can be incorporated into the multiple linear regression model by writing

$$E(y_i \mid \mathbf{x}_i) = \boldsymbol{\eta}^T \mathbf{u}_i \qquad \text{and} \qquad \text{Var}(y_i \mid \mathbf{x}_i) = \sigma^2/w_i, \qquad i = 1,\ldots,n \quad (9.2)$$

where the *weights* $w_i > 0$ are *known, positive* constants. Figure 4.3, page 60, provides a schematic representation of this regression when the variances are not constant.

The model given by (9.2) can be given in a single equation as

$$y_i \mid \mathbf{x}_i = \eta^T \mathbf{u}_i + e_i / \sqrt{w_i} \qquad \text{for} \quad i = 1, \ldots, n \qquad (9.3)$$

where the random variable e_i has zero mean $E(e_i) = 0$ and constant variance, $\text{Var}(e_i) = \sigma^2$. Taking the mean and variance of both sides of (9.3) gives the results in (9.2).

The interpretation of the constant σ^2 depends on the choice for the weights. Generally, σ^2 is the variance of the response in the subpopulation with weight $w_i = 1$. This subpopulation serves as a reference point for the variances in all other subpopulations. The variance in the subpopulation with $w_i = 2$ is half that in the reference population, for example. Weights are always proportional to reciprocals of variances.

Here are a few ways we might get nonconstant variance with known weights. If y_i is the average of m_i independent observations, each with variance σ^2, then $\text{Var}(y_i \mid x_i) = \sigma^2/m_i$ and the weights are $w_i = m_i$. If y_i is the sum of m_i independent observations, $\text{Var}(y_i \mid x_i) = m_i \sigma^2$, and the weights are $w_i = 1/m_i$. If the variance is a known positive function of a predictor x, say $\text{Var}(y_i \mid x_i) = x_i^2 \sigma^2$, then $w_i = 1/x_i^2$. In addition, weights could be based on a plausible assumption about the variance function. In the snow geese data for example, the assumption that $\text{Var}(Photo \mid Obs2) = \sigma^2 \times Obs2$ may give a reasonable approximation of the variance function. This implies that $w_i = 1/Obs2$.

The known weights are incorporated into the least squares criterion to give the *weighted least squares* criterion. Starting with the multiple linear regression mean function $E(y \mid \mathbf{x}) = \eta^T \mathbf{u}$, we choose estimates to minimize the weighted residual sum of squares,

$$RSS(\mathbf{h}) = \sum_{i=1}^{n} w_i (y_i - \mathbf{h}^T \mathbf{u}_i)^2 \qquad (9.4)$$

The value of $\mathbf{h}$ that minimizes (9.4) is called $\hat{\eta}$, the weighted least squares, or WLS, estimate. The same notation is used for WLS and OLS estimates because the OLS estimates are a special case of WLS obtained by setting all the weights to be equal.

The weighted residual sum of squares pays more attention to squared differences $(y_i - \mathbf{h}^T \mathbf{u}_i)^2$ with larger weights than to differences with smaller weights. Observations with larger weights have smaller variances, and therefore they contribute more information for estimation. On one extreme, letting one weight w_i grow very large forces the variance $\text{Var}(y_i \mid x_i) = \sigma^2/w_i$ to approach zero and also forces the fitted line to pass through the corresponding point (y_i, x_i). On the other extreme, letting a weight w_i approach zero is equivalent to deleting the case from the data so it is ignored in the fitting.

Weighted least squares estimates depend only on the relative magnitudes of the weights. Multiplying all the weights by the same constant $c > 0$ will not change $\hat{\eta}$, but it will change the meaning of σ^2, since it is the variance in the reference population with weight one. In some regressions, scaling the weights to have maximum value one can be helpful because σ^2 is then the largest variance for the data.

The WLS estimates can also be obtained from an ordinary least squares fit of a slightly modified set of data. To see this, rewrite (9.4) as

$$RSS(\mathbf{h}) = \sum_{i=1}^{n} (\sqrt{w_i} y_i - \mathbf{h}^T(\sqrt{w_i} \mathbf{u}_i))^2$$

This is the residual sum of squares function for the regression

$$E(\sqrt{w_i} y_i \mid x_i) = \eta^T(\sqrt{w_i} \mathbf{u}_i) \qquad \text{and} \qquad \text{Var}(\sqrt{w_i} y_i \mid x_i) = \sigma^2$$

This shows that the weighted least squares estimates can be obtained as the OLS estimates with response $\sqrt{w_i} y_i$ and terms obtained by multiplying each of the elements of $\mathbf{u}_i$ by $\sqrt{w_i}$, including the term for the intercept.

In *Arc* and most other regression programs, you can do weighted least squares by declaring a column of data to serve as weights and then specifying the unweighted response and predictor terms as usual. The computer program solves the weighted least squares analysis by solving the equivalent OLS regression.

With WLS, the fitted values, residuals, residual sum of squares, and estimate of σ^2 are, respectively:

$$i\text{th fitted value} = \hat{y}_i = \hat{\eta}^T \mathbf{u}_i$$

$$i\text{th residual} = \hat{e}_i = \sqrt{w_i}(y_i - \hat{y}_i)$$

$$RSS = \sum_{i=1}^{n} w_i(y_i - \hat{y}_i)^2 = \sum_{i=1}^{n} \hat{e}_i^2 \tag{9.5}$$

$$\hat{\sigma}^2 = \frac{RSS}{n - k}$$

These differ from the similar expressions given in (7.15) only in that (1) residuals are multiplied by the square root of the weight and (2) the quantities that depend on residuals also implicitly include the weights.

9.2.1 Particle Physics Example

As an example, we consider an experiment in particle physics. A beam of particles was aimed at a target containing protons. Outgoing particles were

TABLE 9.1 The Physics Data from File
physics.lsp

x	y	S
0.345	367	17
0.287	311	9
0.251	295	9
0.225	268	7
0.207	253	7
0.186	239	6
0.161	220	6
0.132	213	6
0.084	193	5
0.060	192	5

detected and counted. Our goal is to study the total output of particles as a function of the input energy. Physicists measure the total output using the *scattering cross section*, y, which is a function of the number of particles and the density and size of the target, and is measured as a rate per second. Theoretical considerations suggest that

$$E(y \mid x) \approx \eta_0 + \eta_1 x \qquad (9.6)$$

where x is the inverse of the total input energy. Table 9.1, which was constructed from the file physics.lsp, gives the data from the experiment. For each row in this table, the experimenter set x_i and then observed y_i. The third column of the table gives the known population standard deviations $S_i = \sqrt{\mathrm{Var}(y_i \mid x_i)}$ for each value of x_i.

It may seem strange that an experimenter would know the standard deviations but not $E(y_i \mid x_i)$, but this can happen. For example, imagine we are using a sensitive scale to weigh a light object. Each time the object is placed on the scale, a different reading for its weight is obtained. The variance in the readings depends on the accuracy of the scale and not the object itself. After weighing the object a large number of times, we calculate the average weight which is essentially the object's true weight, assuming that our scale is unbiased. Further, the standard deviation of the weights is essentially the true standard deviation for the scale. Now suppose we put a different object on the same scale and take a single reading. We don't know the true weight of the second object, but we do know the true standard deviation because it's the same as the scale's known standard deviation obtained from the many weighings of the first object. Our physics experiment is like ten scales, corresponding to the ten values of x, with known standard deviations.

Returning to the physics experiment, we see that the standard deviations in the third column of Table 9.1 are not the same. This means that weighted least squares should be used when fitting (9.6). According to the general form (9.2), we must therefore have

$$\text{Var}(y_i \mid x_i) = \sigma^2/w_i, \qquad i = 1,\ldots,10 \tag{9.7}$$

How should we pick the weights and what does σ^2 mean? Remembering that weights are proportional to reciprocals of variances and that the positive proportionality constant doesn't matter except for the interpretation of σ^2, we can take the simplest route and set

$$w_i = \frac{1}{\text{Var}(y_i \mid x_i)}$$

Substituting this choice into (9.7) and solving for σ^2 gives $\sigma^2 = 1$. This agrees with the interpretation of σ^2 given previously: σ^2 is the variance in the reference population with weight equal to 1.

The weights should be used to fit the mean function (9.6). They should also be used when smoothing plots that have the response on the vertical axis. First, load the file physics.lsp. To create the weights, you need to compute $1/\text{Var}(y_i \mid x_i)$, and this can be done using the "Transform" menu item by selecting S from the candidate list, and using the power $p = -2$. Next, from the Graph&Fit menu, select "Plot of," and then choose x for the horizontal axis and y for the vertical axis. Finally, select the variable S^{-2} to be used as weights in the plot: click once on the name of this variable, click once in the empty box labeled "Weights/trials," and then click on "OK." The resulting plot is shown in Figure 9.3. The line shown on the plot is the weighted least squares line, computed by clicking in the parametric smoother slidebar to get the number "1" above the slidebar. The title of this slidebar has changed to "WLS" to reflect that the fitted line uses weighted least squares, not ordinary least squares.

To obtain the estimates, select the item "Fit linear LS" from the Graph&Fit menu; the resulting dialog is shown in Figure 9.4. The mean function is specified as usual; the only new feature is setting S^{-2} to be weights. The output is shown in Table 9.2.

The output from a weighted fit is nearly identical to the output for OLS. In particular, the analysis of variance can be used to test NH: $E(y \mid x) = \eta_0$ against the alternative AH: $E(y \mid x) = \eta_0 + \eta_1 x$. The p-value in Table 9.2 is zero to four decimals, so y is certainly not independent of x. The fitted line is $\hat{E}(y \mid x) = 148.5 + 530.8x$, and $\hat{\sigma} = 1.7$. The numerical summaries in Table 9.2 appear to indicate a good fit of the straight-line mean function; the standard errors of the estimates are relatively small, the overall F-value is very large, and $R^2 = 0.94$ is very large. All of these *numerical* summaries conflict with the *visual* impression in Figure 9.3, where the fitted line seems to miss the systematic curvature that can be seen in the graph. We need methodology to supplement the visual comparison of the data to the fitted mean function. In

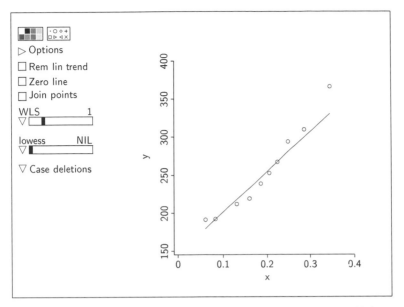

FIGURE 9.3 The physics data with the weighted least squares line added.

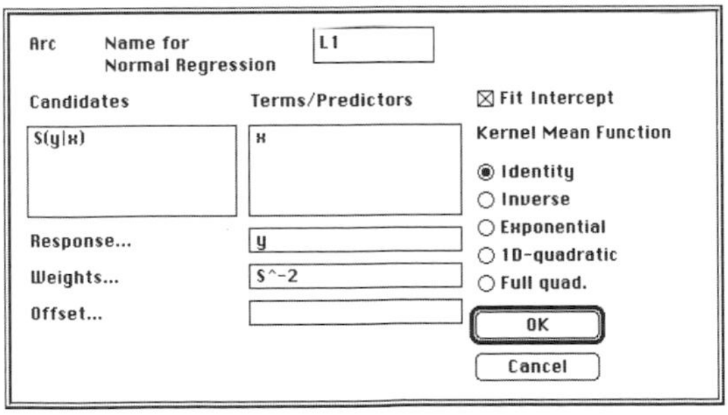

FIGURE 9.4 The regression dialog, with weights specified.

particular, the lack-of-fit tests in Section 9.3 are based on a comparison of $\sigma^2 = 1$, which is an implication of our model, with the estimate $\hat{\sigma}^2 = 1.7^2$ that comes from the data.

9.2.2 Predictions

Predictions for multiple linear regression were discussed in Section 7.6.2. When weighted least squares is used, the methodology for getting the value of the prediction is unchanged, but the standard error of prediction must be modified to include a weight. The revised standard error of prediction, from

TABLE 9.2   Weighted Least Squares Estimates for the Physics Data

```
Data set = Physics, Name of Fit = L1
Normal Regression
Kernel mean function = Identity
Response      = y
Terms         = (x)
Weights       = S^-2
Coefficient Estimates
Label          Estimate      Std. Error     t-value
Constant        148.473        8.07865       18.378
x               530.835       47.5500        11.164

R Squared:                     0.939681
Sigma hat:                     1.65653
Number of cases:                 10
Degrees of freedom:               8

Summary Analysis of Variance Table
Source       df  SS           MS             F            p-value
Regression    1  341.991      341.991        124.63       0.0000
Residual      8  21.9526      2.74408
```

(7.23), is

$$\mathrm{se}(y_{pred}\mid \mathbf{x}) = \left(\hat{\sigma}^2/w_x + \mathrm{se}(\hat{\boldsymbol{\eta}}^T\mathbf{u})^2\right)^{1/2} \tag{9.8}$$

where w_x is the weight attached to the future value to be observed at $\mathbf{x}$. In the physics experiment, the weights are given only at the values of x in the data. To construct the standard error of prediction, w_x must be known for the value of $\mathbf{x}$ used in the prediction.

Standard errors of prediction for weighted fits can be obtained by using the regression model's menu item "Prediction." The appropriate weight should be entered in the corresponding text area "Weight for prediction" of the dialog shown in Figure 6.8, page 116. For example, suppose as part of the physics study that we want a prediction at $x = 0.2$, and we know the variance of the future observation is $50\sigma^2$. Consequently, assuming $\sigma^2 = 1$, the weight for the future observation is $1/50 = 0.02$. Entering 0.2 and 0.02 into the appropriate text areas of the prediction dialog and then clicking "OK" gives the prediction $y_{pred}\mid (x = 0.2) = 254.64$ and its standard error $\mathrm{se}(y_{pred}\mid x = 0.2) = 12.4037$.

9.3   LACK-OF-FIT METHODS

In this section we present methods that can be used to decide if the form of the mean function used to obtain a fitted model is reasonable in a particular

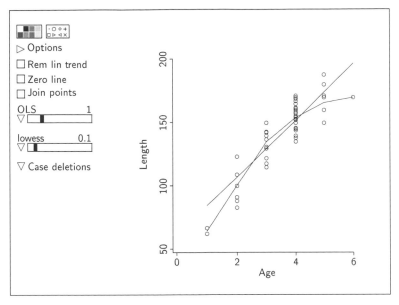

FIGURE 9.5 The Lake Mary bluegill data.

regression. This is a continuing theme of this book: Model checking is an essential part of regression analysis.

9.3.1 Visual Lack-of-Fit with Smooths

We have seen two examples in this chapter where, by visual assessment, a fitted model does not agree with the data. Smooths of the data can be used as well. For illustration we return to the Lake Mary bluegill data, introduced in Section 1.2, page 5, and discussed more fully in Sections 2.1 and 2.2, pages 28–31. We are interested in characterizing the dependence of *Length* of fish on *Age* using a sample of 78 fish. A scatterplot of the data is shown in Figure 9.5. We would like to decide if there is information in the data to contradict the mean function of the simple linear regression model. The straight line shown on Figure 9.5 is the estimate of E(*Length* | *Age*) assuming that the simple linear regression model holds. A curved smooth is also shown on the graph. Recall from Chapter 3 that this smooth is another estimate of E(*Length* | *Age*), one that is not restricted to be linear. If the simple linear regression model were correct for these data, we should expect that these two estimates of the mean function would be approximately the same. Since the two estimates don't seem to match very well, we have a visual clue that simple linear regression may not be appropriate.

The comparison between an OLS fit and a smooth can depend on the smoothing parameter. In Figure 9.5 the smoothing parameter was chosen to

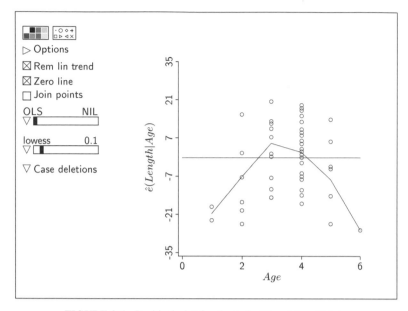

FIGURE 9.6   Residual plot for the Lake Mary bluegill data.

be small so that the smooth joins the average value for each of the ages. This gives a summary of the mean function that does not depend on a model. If a larger smoothing parameter was chosen, then the difference between the two fits would be reduced. Consequently, one might consider a range of smoothing parameters in making such a comparison.

The ideas discussed above can be applied equally to a plot of the residual versus the predictor, as shown in Figure 9.6. Here we have used a smooth after removing the linear trend and drawing the zero line. This smooth is an estimate of the residual mean function $E(\hat{e} \mid x)$. Since the smooth does not seem to sustain the possibility that the residual mean function is constant, we again have visual evidence against the simple linear regression model.

Visual comparisons of two curves can be formalized by using *lack-of-fit tests*. In the context of the multiple linear regression model, the lack-of-fit tests discussed in this chapter are all tests of the hypotheses

$$\text{NH: } E(y \mid \mathbf{x}) = \eta^T \mathbf{u} \quad \text{versus} \quad \text{AH: } E(y \mid \mathbf{x}) \neq \eta^T \mathbf{u} \tag{9.9}$$

given that $\text{Var}(y \mid x) = \sigma^2/w$ where the weights are known.

9.3.2   Lack-of-Fit Based on Variance

The first type of lack-of-fit test is based on comparing two fitted curves visually, as we saw in the last section. We now turn to numerical methods based on variances. Suppose we fit a model with some mean function $E(y \mid x)$ and variance function $\text{Var}(y \mid x) = \sigma^2/w$. The variance function must be correctly specified for any of the lack-of-fit tests discussed in this chapter. If the mean function is correctly specified, then the residual mean square $\hat{\sigma}^2$ from the weighted least squares fit provides an unbiased estimate of σ^2. If the mean function used in the fit is wrong, then $\hat{\sigma}^2$ will be too large on the average, because its size will be increased by systematic biases due to fitting the wrong mean function. The differences between the fitted values for the two curves in Figure 9.5 are estimates of the systematic biases due to fitting a linear mean function.

All of the lack-of-fit tests discussed in this chapter are based on the following general idea. Suppose we can obtain a second estimate of σ^2 that is unbiased regardless of the correctness of the mean function. This second estimate will be used as a baseline for judging the lack-of-fit of the mean function. If the baseline estimate is about the same size as $\hat{\sigma}^2$, then we have information to sustain the mean function. On the other hand, if the baseline estimate is significantly smaller than $\hat{\sigma}^2$, then we have information to contradict the mean function.

9.3.3 Variance Known

In the physics data introduced in Section 9.2, the value of σ^2 is known and equal to one. This is the baseline estimate. From Table 9.2, we have the model-based estimate $\hat{\sigma}^2 = 2.744$ with 8 df. Evidence against the null hypothesis in (9.9) will be obtained if we judge $\hat{\sigma}^2$ large relative to $\sigma^2 = 1$. The test statistic is $(n-k)\hat{\sigma}^2/\sigma^2$. Here, $n-k$ is the df for $\hat{\sigma}^2$, so k is the number of regression coefficients estimated in the mean function; $k = 2$ for simple linear regression. To compute a p-value, we need to know the distribution of this test statistic when the null hypothesis is true. If the distribution of $y \mid x$ is normal and the mean function is correct, then

$$\frac{(n-k)\hat{\sigma}^2}{\sigma^2} \sim \chi^2_{n-k} \tag{9.10}$$

Thus, we should use the Chi-squared distribution with $n-k$ df to get a p-value for this test.

For the physics data we have $(n-2)\hat{\sigma}^2/\sigma^2 = 8(2.744)/1 = 21.95$. Using the "Calculate probability" item, we get a p-value of about 0.005, giving clear evidence against the null hypothesis in (9.9), which agrees with the visual impression of Figure 9.3.

Given that the simple linear regression model is inappropriate for these data, we might consider alternatives. The physical theory suggests fitting a quadratic polynomial in place of simple linear regression. The lack-of-fit test applied to the quadratic polynomial suggests that this alternative model may be adequate (see Problem 9.6).

TABLE 9.3   "Table Data" Output for the Lake Mary Data

Data set = LakeMary, Tabled Values			
Col. 1 = Age			
Col. 2 = Count			
Col. 3 = Length[Mean]			
Col. 4 = Length[SD]			
1	2	64.5000	3.53553
2	7	100.429	14.1640
3	19	134.895	11.4741
4	41	153.829	9.90177
5	8	166.125	13.6532
6	1	170.000	0.000000

9.3.4   External Estimates of Variation

In some regressions, we might have an estimate of σ^2 that is external to the current experiment. For example, imagine an industrial process whose output has variability σ^2. If the process has been running for a while, an estimate of σ^2 might be available. If an experiment that may alter means but not variances is done on the process, we could compare the external baseline estimate of σ^2, say $\hat{\sigma}^2_{\text{ext}}$, to the estimate $\hat{\sigma}^2$ from the experiment. In the last section we discussed the case in which the baseline variation is known. Here the situation is similar, except the baseline variation is estimated and not known.

The test statistic is just the ratio of the two estimates, $F = \hat{\sigma}^2/\hat{\sigma}^2_{\text{ext}}$. Under the null hypothesis that the mean function is correct, this statistic has an $F_{n-k,d}$ distribution, where d is the df for $\hat{\sigma}^2_{\text{ext}}$.

9.3.5   Replicate Observations

Suppose we have observed several responses at the same value, $\tilde{x}$ say, of the predictor $\mathbf{x}$, so we have several observations from the distribution of $y \mid (\mathbf{x} = \tilde{x})$, called *replicates* because they are taken under the same experimental conditions, and thus come from the same subpopulation. In the Lake Mary experiment, multiple bluegills were observed at all ages except age six. The lengths of the five-year-old bluegills, for example, are replicates because they come from the same subpopulation. The standard deviation of the replicate values of y at a fixed value of $\mathbf{x}$ provides a baseline estimate of variance that does not depend on the mean function, and this in turn provides the basis for another test of lack-of-fit.

Consider the data from Lake Mary summarized in Table 9.3. There are 78 observations, but only six different values of $x = Age$. The five nonzero standard deviations in Table 9.3 estimate the corresponding subpopulation standard deviations. Our first task in developing a test for lack-of-fit is to combine the individual estimates of σ into one overall estimate. A little notation will help describe how this is done.

Suppose there are g unique values of $\mathbf{x}$ corresponding to g subpopulations, so $g \leq n$, the total number of observations. Let m_i be the number of observations from the ith subpopulation, for $i = 1,\ldots,g$, and let y_{ij} denote the jth observation from the ith subpopulation, $j = 1,\ldots,m_i$. In the Lake Mary data, we have $g = 6$ and $m_2 = 7$, for example. Next, let sd_i be the sample standard deviation of the responses from the ith subpopulation. For example, in the Lake Mary data, $sd_2 = 14.1640$, which is computed from the $m_2 = 7$ observations $y_{2,1},\ldots,y_{2,7}$. We assume that the variance in the ith subpopulation is σ^2/w_i. The estimates of the variances, sd_i^2, are then combined into a baseline estimate, given by

$$\hat{\sigma}_{pe}^2 = \frac{\sum_{i=1}^{g} w_i(m_i - 1)sd_i^2}{\sum_{i=1}^{g}(m_i - 1)} = \frac{SS_{pe}}{df_{pe}} \tag{9.11}$$

$$= \sum_{i=1}^{g} c_i sd_i^2 \tag{9.12}$$

where

$$c_i = \frac{w_i(m_i - 1)}{\sum_{i=1}^{g}(m_i - 1)} = \frac{w_i(m_i - 1)}{n - g}$$

We have used the notation $\hat{\sigma}_{pe}^2$ for the baseline estimate of σ^2 to denote a *pure error* estimate that imposes no constraints on the mean function. The estimate is given by the ratio of $SS_{pe} = \sum w_i(m_i - 1)sd_i^2$, the *sum of squares for pure error*, divided by its df, $df_{pe} = \sum(m_i - 1) = n - g$.

A second form of $\hat{\sigma}_{pe}^2$ is given in (9.12) to show that it is a weighted average of the subpopulation estimates sd_i^2. The subpopulation estimate sd_i^2 has $m_i - 1$ df, so the weight c_i for sd_i^2 is its df divided by the total df for $\hat{\sigma}_{pe}^2$.

For the Lake Mary data all the $w_i = 1$ and,

$$\hat{\sigma}_{pe}^2 = \frac{1 \times (3.53553)^2 + \cdots + 7 \times (13.6532)^2}{1 + 6 + 18 + 40 + 7} = \frac{8812.68}{72} = 122.398$$

with $\sum(m_i - 1) = 72$ df.

We now use the pure error estimate of σ^2 to construct the test for lack-of-fit. First, the rationale: If the mean function does hold, then the residual mean square from that fit will also estimate σ^2; but if the mean function does not hold, then the residual mean square will be too large. Comparison of these two estimates of variance is the essence of the test for lack-of-fit.

TABLE 9.4   Simple Linear Regression Summary for the Lake Mary Data

```
Data set = LakeMary, Name of Fit = L1
Normal Regression
Kernel mean function = Identity
Response        = Length
Terms           = (Age)
Coefficient Estimates
Label           Estimate     Std. Error     t-value
Constant        62.6490         5.75454      10.887
Age             22.3123         1.53726      14.514

R Squared:                    0.734882
Sigma hat:                     12.5094
Number of cases:                  78
Degrees of freedom:               76

Summary Analysis of Variance Table
Source          df      SS          MS          F        p-value
Regression       1    32965.8     32965.8     210.66     0.0000
Residual        76    11892.8     156.485
Lack of fit      4     3080.15    770.038       6.29     0.0002
Pure Error      72     8812.68    122.398
```

Comparison of these two estimates of variance is done via an F-test. The F-statistic is given by

$$F = \frac{(RSS - SS_{pe})/(n - k - df_{pe})}{\hat{\sigma}^2_{pe}} = \frac{SS_{lof}/df_{lof}}{\hat{\sigma}^2_{pe}} \qquad (9.13)$$

where RSS is the residual sum of squares for the fitted model, k is the number of regression coefficients (terms) in the model, $SS_{lof} = RSS - SS_{pe}$, and $df_{lof} = n - k - df_{pe}$. We have defined two new quantities in this equation, the *sum of squares for lack-of-fit*, SS_{lof}, and its df, df_{lof}. Under the normality assumption for $y \mid x$ and assuming that the mean function holds, $F \sim F_{n-k-df_{pe},df_{pe}}$ and large values of F, corresponding to small p-values, provide evidence of lack-of-fit.

The computations needed for this test can be obtained using *Arc* in a variety of ways. First, Table 9.3 could be used to find the sum of squares for pure error as demonstrated above. The residual sum of squares and its df can be obtained in the usual way. These can then be substituted into (9.13) to get the test statistic.

Arc provides a simpler way to do this test. For the Lake Mary data, fit the simple linear regression of *Length* on *Age*; the output is shown in Table 9.4. In the summary analysis of variance table, the *RSS* is 11892.8 with 76 df. *Arc* checks to see if the data has replicate observations. If it does, then *Arc* computes the sum of squares for pure error and its df, which for this regression are 8812.68 and 72, respectively. The difference between the residual sum of squares and the SS_{pe}, $11892.8 - 8812.68 = 3080.15$, is labeled as the sum of squares for lack-of-fit, with $76 - 72 = 4$ df. These values are used in the numerator of the lack-of-fit test; the denominator is $\hat{\sigma}^2_{pe} = 122.398$. The F-statistic for lack-of-fit and corresponding p-value are given in the output. Since $F = 6.29$ and $p = 0.0002$, we have strong evidence that the null hypothesis (9.9) is false. If we want to continue with parametric modeling for the mean function, we need to seek a more flexible functional form than the straight line, or else consider using a transformation of x or y in the fitting procedure. One possibility would be to use a quadratic mean function, which we tried in Section 7.7. The quadratic fit is shown in Figure 7.4, page 148, and the output for fitting the quadratic model was given in Table 7.6. The p-value for lack-of-fit of the quadratic model is 0.83, suggesting that the quadratic fit gives an adequate description of the mean function over the range of ages available in the data.

9.3.6 Subsampling

A warning is in order here: Observations taken at the same value of **x** do not always supply replication. For example, in the haystack data, suppose the volume of each haystack had been measured twice, rather than once, giving a total of 240 measurements. The variability between the two repeated measurements is due to variability in the measuring method, not the variability in volume between two different haystacks with the same values of C and *Over*. This is an example of *subsampling*, in which the same experimental unit is measured more than once. To get replication, different experimental units must be measured. For example, in a study of fish health as a function of lake characteristics, measurements of fish in different lakes can provide replication, but measurement of fish in the same lake will generally provide only subsampling.

Subsampling is a special case of *variance components*. Methodology for variance components problems is beyond the scope of this book. However, problems with subsampling can be handled using the methods we describe in Section 9.4 by taking as the response variable the mean of the subsamples within an experimental unit.

9.4 FITTING WITH SUBPOPULATION AVERAGES

When replicates are present, the data are often summarized in a table of the unique values of **x**, the subpopulation response averages $\bar{y}$, standard deviations sd, and sample sizes m. We have seen how to get data summarized like this using the "Table data" item on the data set menu in Section 2.1. For the Lake Mary data, this table is given in Table 9.3.

Consider conducting an analysis based just on the subpopulation averages. In particular, if

$$E(y_{ij} \mid \mathbf{x}_i) = \eta^T \mathbf{u}_i \quad \text{and} \quad \text{Var}(y_{ij} \mid \mathbf{x}_i) = \sigma^2 \tag{9.14}$$

then

$$E(\bar{y}_i \mid \mathbf{x}_i) = \eta^T \mathbf{u}_i \quad \text{and} \quad \text{Var}(\bar{y}_i \mid \mathbf{x}_i) = \sigma^2/m_i \tag{9.15}$$

In the second model (9.15), the ith response is the subpopulation average $\bar{y}_i$. Thus, when using subpopulation averages as the response, fitting must be done by weighted least squares with weights $w_i = m_i$ because $\text{Var}(\bar{y}_i \mid \mathbf{x}_i) = \sigma^2/m_i$. This differs from the physics data because here we assume that the variance per observation is constant, while in the physics data the variance is different for each subpopulation.

There are close connections between the fits of (9.14) and (9.15), and the lack-of-fit test statistic (9.13):

- The OLS estimates of the regression coefficients from (9.14) are identical to the weighted least squares estimates of the regression coefficients from (9.15).
- The residual sum of squares from the weighted fit of the mean model (9.15) is the SS_{lof} needed for computation of the F lack-of-fit test statistic (9.13). Weighted least squares must be used in (9.15) for this to hold, even if the m_i are all equal. Let

$$\hat{\sigma}^2_{\text{lof}} = \text{SS}_{\text{lof}}/\text{df}_{\text{lof}}$$

  Then the numerator $\hat{\sigma}^2_{\text{lof}}$ of F is the estimate of σ^2 coming from the weighted fit of (9.15), while the denominator is the model-free, pure error estimate of σ^2.
- The RSS needed for computation of (9.13) is the residual sum of squares from the OLS fit of (9.14).
- The estimate $\hat{\sigma}^2$ from the OLS fit to the full data (9.14) is a weighted average of the estimate $\hat{\sigma}^2_{\text{lof}}$ from the subpopulation mean model (9.15) and the pure error estimate $\hat{\sigma}^2_{\text{pe}}$:

$$\hat{\sigma}^2 = \frac{\text{df}_{\text{lof}} \times \hat{\sigma}^2_{\text{lof}} + \text{df}_{\text{pe}} \times \hat{\sigma}^2_{\text{pe}}}{\text{df}_{\text{lof}} + \text{df}_{\text{pe}}} \tag{9.16}$$

Equivalently,

$$RSS = \text{SS}_{\text{lof}} + \text{SS}_{\text{pe}}$$

The summary of the analysis for the Lake Mary data is shown in Table 9.5. To reproduce this table, load the data file lakemary.lsp, and select "Table

TABLE 9.5 Weighted Least Squares Estimates for the Data Shown in Table 9.3

```
Data set = LakeMary1, Name of Fit = L1
Normal Regression
Kernel mean function = Identity
Response      = Length[Mean]
Terms         = (Age)
Weights       = Count
Coefficient Estimates
Label       Estimate    Std. Error    t-value
Constant    62.6490     12.7653       4.908
Age         22.3123      3.41011      6.543

R Squared:            0.914549
Sigma hat:           27.7496
Number of cases:           6
Degrees of freedom:        4

Summary Analysis of Variance Table
Source        df      SS         MS         F        p-value
Regression     1    32965.8    32965.8    42.81     0.0028
Residual       4     3080.15     770.038
```

data" from the data set menu. Set up the resulting dialog as shown in Figure 2.3, page 30, except that the item "Make new data set" should be activated. This will create a new data set, with values as shown in Table 9.3. The variable *Count* contains the sample sizes m_i, the variable *Length[Mean]* is the average length, $\bar{y}_i$, *Age* is the predictor x_i, and *Length[SD]* gives the standard deviations of the response at each value of the predictor. After you push the "OK" button, the data will be displayed in the text window, and a new data set menu will appear for the summarized data. From the menus for the newly created data set, fit weighted least squares to get the estimates shown in Table 9.5.

Compare the fit in Table 9.5 based on the averages to the fit in Table 9.4 based on the original data. The estimates of regression coefficients are identical in each, but all the other summary statistics differ. The cause of this difference is easy to find: In the fit to the original data in Table 9.4, the residual sum of squares has 76 df, and $RSS = SS_{lof} + SS_{pe}$.

9.5 COMPLEMENTS

9.5.1 Weighted Least Squares

Using matrix notation, Section 7.9.1, suppose we have that

$$\text{Var}(\mathbf{y} \mid \mathbf{X}) = \sigma^2 \mathbf{W}^{-1}$$

where $\mathbf{W}$ is a diagonal matrix whose diagonal elements w_i are known positive weights. In addition, suppose we have a mean function

$$E(\mathbf{y} \mid \mathbf{X}) = \mathbf{U}\eta$$

for some model matrix $\mathbf{U}$ derived from $\mathbf{X}$. The weighted least squares estimate $\hat{\eta}$ is the value of $\mathbf{h}$ that minimizes

$$RSS(\mathbf{h}) = (\mathbf{y} - \mathbf{U}\mathbf{h})^T \mathbf{W}(\mathbf{y} - \mathbf{U}\mathbf{h})$$

which is equivalent to (9.4). As before, we can convert this into an OLS regression. Let $\mathbf{W}^{1/2}$ be the diagonal matrix with elements $w_i^{1/2}$ on the diagonal. We can then write

$$RSS(\mathbf{h}) = (\mathbf{W}^{1/2}\mathbf{y} - \mathbf{W}^{1/2}\mathbf{U}\mathbf{h})^T (\mathbf{W}^{1/2}\mathbf{y} - \mathbf{W}^{1/2}\mathbf{U}\mathbf{h})$$

and we need only solve the OLS regression with response $\mathbf{W}^{1/2}\mathbf{y}$ and model matrix $\mathbf{W}^{1/2}\mathbf{U}$.

9.5.2   The *lowess* Smoother

The *lowess* smoother is a *locally weighted scatterplot smoother*. For a 2D scatterplot of y versus x, a fitted value $\hat{y}_\ell$ at a particular point x_ℓ is obtained as follows. (1) Select a value for a smoothing parameter f, a number between zero and one, for example, set $f = 0.6$. (2) Find the fn closest points to x_ℓ, for example, if $n = 100$, find the $fn = 60$ closest points. (3) Among the fn *nearest neighbors* to x_ℓ, compute the WLS estimates for the regression of y on x, with weights determined so that points close to x_ℓ have the highest weight, and the weights decline toward zero for points farther from x_ℓ. We use a triangular weight function that linearly decreases from a maximum value at x_ℓ to zero at the edge of the neighborhood. (4) Return the fitted value at x_ℓ. (5) Repeat 1–4 for many values of x_ℓ, and join the points.

The *lowess* smooth was first suggested by Cleveland (1979), and it is also given as the first step in Algorithm 6.1.1 by Härdle (1990, p. 192).

9.5.3   References

The physics data were taken from Weisberg *et al.* (1978). The apple shoot data were collected by Bland (1978). Galton's data on sweet peas used in Problem 9.5 were given by Pearson (1930). Jevons' data in Problem 9.9 were provided by Stephen Stigler. A recent discussion of lack-of-fit tests based on smoothing is given by Bowman and Azzalini (1997), who provide Splus computer code to get numerical tests for comparing parametric and nonparametric curves.

Variance component models mentioned in Section 9.3.6 are discussed in virtually any book on experimental designs and analysis of variance, such as Kuehl (1994, Chap. 5).

PROBLEMS

9.1 An experiment was conducted to study the effect on turkey growth of supplemental methionine in the diet. Ten diets were formulated to contain $0.05\%, 0.10\%, \ldots, 0.5\%$ methionine M, and each diet was randomly assigned to two turkeys. The turkeys, which were all of the same age, were weighed at the beginning of the experiment and again four weeks later. The response G is the weight gain, the final weight minus the initial weight.

    9.1.1 Suppose it is known that the variance of gain is proportional to the square of methionine, $\text{Var}(G \mid M) = c^2 \times M^2$, where c is an unknown constant. Describe how to estimate the regression coefficients in the mean function

$$E(G \mid M) = \eta_0 + \eta_1 M$$

    9.1.2 Describe how to perform a lack-of-fit test for the mean function in Problem 9.1.1.

9.2 For the Camp Lake data in file `camplake.lsp`, Problem 2.1, fit the simple linear regression of *Length* on *Age*, and examine for lack-of-fit. The results do not seem to be consistent with the results for the Lake Mary data, even though the data sets were collected in the same way in the same year by the same person using the same methodology.

    Summarize the differences between the dependence of *Length* on *Age* for the two lakes, and give at least two possible explanations or theories what might cause the differences in the mean functions. To aid in this, you may want to use the file `bluegill.lsp` that contains the data for both lakes. You can draw a plot of *Length* versus *Age*, and use *Lake* as a marking variable.

9.3 Continuing with the Camp Lake data, suppose that the length increment g_{ij} in the jth year of life for the ith fish is approximately normally distributed:

$$g_{ij} \sim N(\mu_j, \sigma^2), \qquad j = 1, 2, \ldots$$

    9.3.1 For the ith fish of age a at capture, let $L_{ia} = \sum_{j=1}^{a} g_{ij}$ be the total length at capture. What is the distribution of $L_{ia} \mid a$?

9.3.2 Suppose that to a reasonable approximation for the range of ages of fish in the data, we have

$$\sum_{j=1}^{a} \mu_j = \eta_0 + \eta_1 a + \eta_2 a^2$$

Given these assumptions, describe $E(L_{ia} \mid a)$ and $Var(L_{ia} \mid a)$, and how you would fit this model.

9.3.3 Do you think the assumption that $Var(g_{ij}) = \sigma^2$ is reasonable? If not, what assumption do you think might be better? How would this change your method of fitting in the last part of this problem?

9.4 Continuing the discussion of the haystack data in file `haystack.lsp` started in Problem 4.6, let's approximate a haystack with a hemisphere having circumference $C_1 = (C + 2Over)/2$.

9.4.1 Construct a scatterplot of *Vol* versus C_1^3. Does your visual impression lead you to think a simple linear model would be appropriate for the regression of *Vol* on C_1^3? Why?

9.4.2 Give the mean square for pure error for the model

$$E(Vol \mid C_1) = \eta_0 + \eta_1 C_1^3$$

with $Var(Vol \mid C_1) = \sigma^2$. State in your own words what the mean square for pure error is estimating in terms of these data, and describe how it is computed.

9.4.3 Give the mean square for pure error for the model with mean function

$$E(Vol \mid C_1) = \frac{C_1^3}{12\pi^2}$$

9.4.4 Give the lack-of-fit mean square for the model of Problem 9.4.2. In terms of these data, describe what this mean square is estimating and how it is constructed.

9.4.5 Construct a lack-of-fit test for the simple linear regression model of Problem 9.4.2 based on the estimate of pure error. Do the results of the test agree with your visual impressions indicated above?

9.4.6 Construct the lack-of-fit test based on pure error for the model of Problem 9.4.3.

9.4.7 Construct a new data set by conditioning on C_1^3 using "Table data" in the data set menu. The *i*th row of the data set will consist of three values: the *i*th unique value C_{1i}^3 of C_1^3, the number m_i of observations in the original data set with that value of C_{1i}^3,

TABLE 9.6   Galton's Sweet Pea Data[a]

Diameter of Parent Peas (hundredths of an inch)	Average Diameter of Progeny Peas (hundredths of an inch)	Standard Deviation
21	17.26	1.988
20	17.07	1.938
19	16.37	1.896
18	16.40	2.037
17	16.13	1.654
16	16.17	1.594
15	15.98	1.763

[a]The data are also given in the file `galton.lsp`.
Source: Pearson (1930).

and the mean $\overline{Vol}_i$ of the volumes for these m_i haystacks. Should the simple linear regression of $\overline{Vol}_i$ on C_{1i}^3 be fit with ordinary or weighted least squares? If weights are needed, what are the weights? Describe how the correct least squares fit is related to the lack-of-fit mean square in Problem 9.4.4. Carry out the necessary calculations to check your answer.

9.5 Many of the ideas of regression first appeared in the work of Sir Francis Galton on the inheritance of characteristics from one generation to the next. In a paper on "Typical Laws of Heredity" delivered to the Royal Institution on February 9, 1877, Galton discussed his experiments with sweet peas. By comparing the sweet peas produced by parent plants to those produced by offspring plants, he could observe inheritance from one generation to the next. Galton categorized parent plants according to the typical diameter of the peas they produced. For seven size classes, which are essentially slices, from 0.15 to 0.21 inches, he arranged for each of nine friends to grow ten plants from seed in each size class; however, two of the crops were total failures. A summary of Galton's data was later published by Karl Pearson (1930), as given in Table 9.6. Only average diameter and standard deviation of the offspring peas were given by Pearson. For the purpose of this exercise, assume that each average diameter is based on 90 observations, the number of friends times the number of plants. The data are available in the file `galton.lsp`.

9.5.1 Draw the scatterplot of y = average progeny diameter versus x = parent diameter. Does a straight line seem plausible for the mean function? Why or why not?

9.5.2   Assuming that the standard deviations (sd's) given are population values, compute the appropriate weighted regression of y on x, and add the fitted line to your graph.

9.5.3   Galton wanted to know if characteristics of the parent plant such as size were passed on to the offspring plants. In fitting the regression a parameter value of $\eta_1 = 1$ would correspond to perfect inheritance, while $\eta_1 < 1$ would suggest that the offspring are "reverting" toward "what may be roughly and perhaps fairly described as the average ancestral type." (The substitution of "regression" for "reversion" was probably due to Galton in 1885.) Test the hypothesis that $\eta_1 = 1$ against the alternative that $\eta_1 < 1$.

9.5.4   Obtain a test for lack-of-fit of the mean function $E(y \mid x) = \eta_0 + \eta_1 x$ assuming that $Var(y \mid x) = \sigma^2$.

9.6   Fit a quadratic polynomial to the physics data in file `physics.lsp`, Section 9.2, and test for lack-of-fit of the quadratic polynomial.

9.7   Several particle physics experiments like those described in Section 9.2.1 were done. The experiments differed only by the choice of the input particle a and the measured output particle c. The input a could be either the π^- meson or its anti-particle π^+, and similarly c could be either π^- or π^+, giving four possible choices for (a,c) combinations. All four are given in the data file `physics2.lsp`. After loading the data file, obtain a listing of the data using the item "Display data" from the data set menu.

   Draw the plot of y versus x, setting the weight variable appropriately and using *Expt* as a marking variable (click once on *Expt* and then click again on the text area next to "Mark by"). The plot will then show the points from the four experiments using different colors and symbols. Add four weighted least squares lines, one for each experiment. This is done by selecting the item "Fit by marks–general" from the parametric smoother slidebar (which should be labeled WLS) and then using the slider to add the linear fits to the plot. Is there visual evidence that a straight line mean function is appropriate for any of the groups? Why or why not?

9.8   Many types of trees produce two types of morphologically different shoots. Some branches remain vegetative year after year and contribute considerably to the size of the tree. Called long shoots or leaders, they may grow as much as 15 or 20 cm over a single growing season.

   On the other hand, some shoots will seldom exceed 1 cm in total length. Called short, dwarf, or spur shoots, these usually produce flowers from which fruit may arise. To complicate the issue further, long shoots occasionally change to short in a new growing season and vice versa. The mechanism that the tree uses to control the long and short shoots is not well understood.

The data in this exercise come from a descriptive study of the difference between long and short shoots of McIntosh apple trees. Using healthy trees of clonal stock planted in 1933 and 1934, samples of long and short shoots were taken from the trees every few days throughout the 1971 Minnesota growing season, about 106 days. The shoots sampled are presumed to be a random sample of the available shoots at the sampling dates. The sampled shoots were removed from the tree, marked, and measured in a laboratory, and the number of stem units in each shoot was counted. The long and short shoots could differ because of the number of stem units, the average size of stem units, or both. A summary of the data is given in Table 9.7. The data file shoots.1sp has six columns including the day number *Day*; the number of replicate observations *m*; the average length *Ave.len* = $\bar{y}$, which is the average of the *m* measurements on *y* = number of stem units; and *sd*, the standard deviation of the *m* values of *y* observed on that day. The remaining two columns both indicate the type: *Type.num* is 1 for long shoots and 0 for short shoots, and *Type* is equal to "Long" or "Short."

9.8.1 Draw a scatterplot of *Ave.len* versus *Day*, using *Type* as a marking variable. Delete all the points for long shoots both from the plot and from the computations. An easy way to do this is to draw the histogram of the variable *Type.num*, select all the points with *Type.num* = 1, which are the long shoots, and then select the item "Delete selection from data set" from the "Case deletions" pop-up menu. These points will then not be used in any succeeding regression calculation.

9.8.2 Draw a plot of the within-day standard deviations versus day. Is constant variance Var(*y* | *Day*) = σ^2 plausible? If it is, then the within-day standard deviations can be used to obtain a mean square of pure error, and hence lack-of-fit tests will be possible.

9.8.3 Computing the pure error is a little harder in this problem. The sum of squares for pure error is given by $\sum(m-1)sd^2$, where the sum is over the short shoot observations. The first step in this computation is to create a new variable. Select "Add a variate," and then type z = (m-1)*sd^2 in the text area. Next, select the item "Table data," and choose *Type* to be the conditioning variable, and *z*, the newly created variable, to be the variate. Push the buttons for "Total," and for "Table all cases," and then push "OK." A small table will be displayed that gives the value of $\sum(m-1)sd^2$ separately for long and short shoots.

9.8.4 From the plot of *Ave.len* versus *Day* and assuming constant within-day variance, a simple linear regression model seems

TABLE 9.7 **McIntosh Apple Shoot Data**[a]

Long Shoots				Short Shoots			
Day	m	$\bar{y}$	sd	Day	m	$\bar{y}$	sd
0	5	10.20	0.83	0	5	10.00	0.00
3	5	10.40	0.54	6	5	11.00	0.72
7	5	10.60	0.54	9	5	10.00	0.72
13	6	12.50	0.83	19	11	13.36	1.03
18	5	12.00	1.41	27	7	14.29	0.95
24	4	15.00	0.82	30	8	14.50	1.19
25	6	15.17	0.76	32	8	15.38	0.51
32	5	17.00	0.72	34	5	16.60	0.89
38	7	18.71	0.74	36	6	15.50	0.54
42	9	19.22	0.84	38	7	16.86	1.35
44	10	20.00	1.26	40	4	17.50	0.58
49	19	20.32	1.00	42	3	17.33	1.52
52	14	22.07	1.20	44	8	18.00	0.76
55	11	22.64	1.76	48	22	18.46	0.75
58	9	22.78	0.84	50	7	17.71	0.95
61	14	23.93	1.16	55	24	19.42	0.78
69	10	25.50	0.98	58	15	20.60	0.62
73	12	25.08	1.94	61	12	21.00	0.73
76	9	26.67	1.23	64	15	22.33	0.89
88	7	28.00	1.01	67	10	22.20	0.79
100	10	31.67	1.42	75	14	23.86	1.09
106	7	32.14	2.28	79	12	24.42	1.00
				82	19	24.79	0.52
				85	5	25.00	1.01
				88	27	26.04	0.99
				91	5	26.60	0.54
				94	16	27.12	1.16
				97	12	26.83	0.59
				100	10	28.70	0.47
				106	15	29.13	1.74

[a]The data are given in the file shoots.lsp.

plausible for short shoots. To obtain the estimates of parameters, weighted least squares should be used since the responses are averages. Fit the simple linear regression model with weighted least squares, and obtain a test for lack-of-fit of the straight-line mean function. The p-value for the lack-of-fit test provides evidence against the straight-line mean function. However, an F-test with so many df is very powerful and will detect very small deviations from the straight-line mean function. Thus, while the result here may be statistically significant, it may not be scientifically important, and for purposes of describing the

TABLE 9.8 Jevons' Gold Coin Data[a]

Ave, x, Decades	Sample Size = m	Average Weight = $\bar{y}$	sd	Minimum Weight	Maximum Weight
1	123	7.9725	0.01409	7.900	7.999
2	78	7.9503	0.02272	7.892	7.993
3	32	7.9276	0.03426	7.848	7.984
4	17	7.8962	0.04057	7.827	7.965
5	24	7.8730	0.05353	7.757	7.961

[a]The data are also given in the file jevons.lsp.

growth of these apple shoots, the straight-line mean function may be adequate.

9.8.5 Repeat the earlier sections of this problem, but for the long shoots rather than the short shoots. When you repeat Problem 9.8.2, you will discover that a constant within-day variance assumption is not very plausible; be sure to summarize the evidence for this conclusion. Justify the empirical choice of $Var(y \mid Day) = (Day + 1)\sigma^2$ for the variance function, based on examining a graph. Complete the remaining sections of this problem assuming this variance function rather than constant within-day variance. (*Hint*: If $Var(y \mid Day) = (Day + 1)\sigma^2$, then sd_i^2 estimates $(Day_i + 1)\sigma^2$, so $sd_i^2/(Day_i + 1)$ estimates σ^2. Thus, the mean square for pure error that estimates σ^2 is

$$\hat{\sigma}_{pe}^2 = \frac{\sum_{i=1}^{n}\left[(m_i - 1)sd_i^2/(Day_i + 1)\right]}{\sum_{i=1}^{n}(m_i - 1)}$$

with $\sum(m_i - 1)$ df.)

9.9 The data in this example are deduced from a diagram in a paper written by W. Stanley Jevons in 1868. In a study of coinage, Jevons weighted 274 gold sovereigns that he had collected from circulation in Manchester, England. He cleaned each coin and then recorded the weight to the nearest 0.001 gram, along with the date of issue. Table 9.8 and the data file jevons.lsp list the average, minimum, and maximum weight for each of five age classes. The age classes are coded 1 to 5, roughly corresponding to the age of the coins in decades. The standard weight of a gold sovereign was supposed to be 7.9876 grams; the minimum legal weight was 7.9379 grams.

9.9.1 Let x be the coded age and $\bar{y}$ be the average weight. Draw a scatterplot of $\bar{y}$ versus x, and comment on the applicability of the usual assumptions of the simple linear regression model. Also draw a scatterplot of sd versus x, and summarize the information in this plot.

9.9.2   Assuming that the population variances are equal to the squares of the sample values given in Table 9.8, $\text{Var}(\bar{y} \mid X) = sd^2/m$, compute the appropriate weighted regression of $\bar{y}$ on x.

9.9.3   Compute a lack-of-fit test for the linear regression model, and summarize your results.

9.9.4   Based on these data and the fitted model, obtain a 95% confidence interval for the mean weight of new coins (that is, coins with $x = 0$). Give a brief written description of why you think this interval estimate is useful, or why you think it might give misleading results.

9.9.5

    a. For previously unsampled coins of age $x = 1$ and $x = 5$, estimate the probability that the weight of a coin is less than the legal minimum. For these calculations, use the assumption that the subpopulation variance for a coin of age x decades is the known value sd^2. Hence, predicted values will have normal, not t, distributions.

    b. Estimate the age at which the mean weight of coins is equal to the legal minimum. The point estimate of this value can be obtained by setting $\hat{E}(y \mid x) = 7.9379$ and solving the fitted regression equation for x. This problem is called *calibration* and is discussed at length by Brown (1993).

9.10   Repeat the analysis of Section 9.4 for the Lake Mary data in file `lakemary.lsp`, except this time fit the quadratic mean function rather than the simple linear regression mean function. Compare the results you get to those in Table 7.6.

9.11   This problem continues the analysis of the transactions data in file `transact.lsp`.

    9.11.1   Figure 7.9, page 166, gave a summary plot for the OLS fit to the transactions data. In the discussion of that plot, we argued that variability increased from the lower left to the upper right of that graph. Use a residual plot to verify that variance seems to be increasing from left to right.

    9.11.2   Let $S = T_1 + T_2$ equal the total number of transactions in a given branch. If all transaction times are equally variable, then a reasonable form for the variance function might be $\text{Var}(Time \mid T_1, T_2) = \sigma^2 S$. Fit the weighted least squares regression of *Time* on T_1 and T_2 using the implied weights. Compare the resulting output to Table 7.3. Are the coefficient estimates very different?

Are the standard errors very different? Compare and interpret the estimates of σ^2.

9.11.3 Obtain a prediction and prediction standard error at $T_1 = 500$, $T_2 = 1000$, and compare to Section 7.6.2.

9.11.4 Construct a 95% confidence interval for the difference $\eta_1 - \eta_2$ and compare to the results in Section 7.6.4. Gilstein and Leamer (1983) present a comprehensive discussion of weighted least squares estimates that can be obtained as the weights are changed, and Bloomfield and Watson (1975) discuss the inefficiency of fitting with the wrong weights. The OLS estimates and the weights used in this problem are only two of the possible weightings.

CHAPTER 10

Understanding Coefficients

One of the primary advantages of fitting a parametric mean function is that the regression can then be characterized by just a few numbers like $\hat{\sigma}$ and the estimates of the regression coefficients. Consequently, understanding what the regression coefficients mean is of primary importance, and in this chapter we discuss how parameters in linear regression can be interpreted and used, as well as limitations on their use.

10.1 INTERPRETING COEFFICIENTS

10.1.1 Rescaling

Measurements generally have units attached to them. In the transactions data described in Section 7.3.3, the response *Time* is measured in minutes, and the terms T_1 and T_2 are counts of the number of transactions of type one and two. In the mean function

$$E(Time \mid T_1, T_2) = \eta_0 + \eta_1 T_1 + \eta_2 T_2 \qquad (10.1)$$

the regression coefficient η_1 has the units of a rate, minutes per type one transaction, that converts the term T_1 to the units of the response *Time*. Similarly, the units of η_2 are minutes per type two transaction. The intercept has the same units as the response; it is measured in minutes in this example. The variance function $Var(Time \mid T_1, T_2)$ is measured in minutes squared, so the standard deviation has the same units as the response. The only commonly encountered unit-free statistics in regression are test statistics, correlation coefficients, and R^2.

What happens in a regression if the units change? Suppose, for example, that the response *Time* had been measured in hours rather than minutes. One would hope that the analysis would be unchanged when a variable or a term is multiplied by a constant, and in a sense this is true. Unit-free quantities like tests and correlations are unchanged, but regression coefficients will change

230

in predictable ways. For example, if *Time* was replaced by *Hours* = *Time*/60, all the regression coefficient estimates using *Time* would be divided by 60 because

$$E(Hours \mid T_1, T_2) = E\left(\frac{1}{60} Time \mid T_1, T_2\right)$$

$$= \frac{1}{60} E(Time \mid T_1, T_2)$$

$$= \frac{\eta_0}{60} + \frac{\eta_1}{60} T_1 + \frac{\eta_2}{60} T_2$$

The variance would be divided by 60^2,

$$Var(Hours \mid T_1, T_2) = Var\left(\frac{1}{60} Time \mid T_1, T_2\right)$$

$$= \frac{1}{60^2} Var(Time \mid T_1, T_2)$$

so the standard deviation would be divided by 60.

If T_1 was replaced by $T_1^* = T_1/100$, which is the number of hundreds of transactions, then in the regression of *Time* on T_1^* and T_2, the regression coefficient estimate for η_1 would be multiplied by 100. The key in understanding this operation is to remember that the product of a coefficient times a term in a model is always in the units of the response regardless of the units of the term. If we *divide* a term by a constant, then we must *multiply* its coefficient by the same constant to keep the product unchanged. For example,

$$\eta_1 T_1 = (100 \times \eta_1)\frac{T_1}{100} = (100 \times \eta_1)T_1^* = \eta_1^* T_1^*$$

where η_1^* is the coefficient of T_1^* in the regression of *Time* on T_1^* and T_2. If a constant is added to any of the terms or to the response, the intercept changes but not the other regression coefficients or the standard deviation.

10.1.2   Rate of Change

A regression coefficient can be interpreted as the change in the expected response given a change in the corresponding term by one unit, assuming that the other terms are held fixed. For example, the estimated mean function for the transactions data assuming that the variance function is constant was found in Section 7.5 to be

$$\hat{E}(Time \mid T_1, T_2) = 144.37 + 5.46 T_1 + 2.03 T_2$$

If in a particular branch the number of T_1 transactions was increased by one, then the expected increase in *Time* is estimated to be $\hat{\eta}_1$ minutes, or 5.46

minutes. In some problems, this simple description of a regression coefficient cannot be used, particularly if changing one term while holding others fixed doesn't make sense. For example, in the polynomial mean function

$$E(y \mid x) = \eta_0 + \eta_1 x + \eta_2 x^2$$

changing x by one unit while holding x^2 fixed is generally impossible. For polynomials, the plot of the fitted curve is likely to be more informative than the values of the parameters. In other mean functions, the relationship between terms may be more subtle and may require knowledge external to the data at hand. In a study of the effects of economic policies on quality of life, for example, changing one term like the prime interest rate may necessarily cause a change in other possible terms like the unemployment rate. In situations like these, interpretation of coefficients can be difficult.

10.1.3   Reparameterization

In general, we could consider replacing terms by linear transformations of them. For example, suppose we define $A = (T_1 + T_2)/2$ to be the average number of transactions of the two types and define $D = T_1 - T_2$ to be the difference in the number of transactions. Consider fitting the mean function with three terms, including the intercept:

$$E(Time \mid T_1, T_2) = \eta_0 + \eta_3 A + \eta_4 D \tag{10.2}$$

One might hope that fitting with this reparameterized mean function will give the same essential information as fitting (10.1). To examine this, Table 10.1 summarizes the OLS fit of these two mean functions along with two others:

$$E(Time \mid T_1, T_2) = \eta_0 + \eta_2 T_2 + \eta_4 D \tag{10.3}$$

$$E(Time \mid T_1, T_2) = \eta_0 + \eta_1 T_1 + \eta_2 T_2 + \eta_3 A + \eta_4 D \tag{10.4}$$

All four mean functions give the same overall summaries—the same df, the same value for $\hat{\sigma}$ and R^2, and the same value for the overall F statistic. All four mean functions also give the same fitted values and residuals. They all have the same values for the estimated intercept.

Compare the coefficient estimates for (10.3) to those for (10.1) and (10.2). It is not an accident that the coefficient estimate for the term T_2 in (10.3) is identical to the coefficient estimate for A in (10.2), and the estimate for D in (10.3) is identical to the estimate for T_1 in (10.1). In mean function (10.3), if T_2 is increased by one unit, then the only way that D could be held fixed is if T_1 is also increased by one unit. Thus an increase in T_2 of one unit in (10.3) corresponds to an increase of one unit in both T_1 and T_2 in (10.1), so the coefficient for T_2 should equal the sum of the coefficients for T_1 and T_2 in (10.1). Apart from rounding error, this is what we see in Table 10.1. Similarly,

TABLE 10.1   Coefficient Estimates for Four Linear Transformations of the Transactions Data[a]

| | Mean Function for | | | |
Term	Equation (10.1)	Equation (10.2)	Equation (10.3)	Equation (10.4)
Constant	144.37	144.37	144.37	144.37
T_1	5.46			5.46
T_2	2.03		7.50	2.03
A		7.50		Aliased
D		1.71	5.46	Aliased

$\hat{\sigma} = 1142.56, R^2 = 0.909$

[a]All four mean functions produce the same fitted values and residuals.

increasing D by a unit with T_2 fixed can occur only if T_1 is increased by a unit, so the coefficient for D in (10.3) must be the same as the coefficient for T_1 in (10.1). The moral of this story is that the value of the coefficient estimate for a term depends on the other terms in the model, and trying to interpret a coefficient without reference to the rest of the mean function will often lead to incorrect conclusions. We can see the same thing using a little algebra:

$$\mathrm{E}(Time \mid T_1, T_2) = \eta_0 + \eta_1 T_1 + \eta_2 T_2$$
$$= \eta_0 + \eta_1(T_1 - T_2 + T_2) + \eta_2 T_2$$
$$= \eta_0 + \eta_1 D + (\eta_1 + \eta_2)T_2$$
$$= \mathrm{E}(Time \mid D, T_2)$$

The coefficient of D in the regression of *Time* on (D, T_2) is the same as the coefficient of T_1 in the regression of *Time* on (T_1, T_2). The coefficient of T_2 is different in the two forms of the regression: It equals η_2 in the regression of *Time* on (T_1, T_2), but it equals $(\eta_1 + \eta_2)$ in the regression of *Time* on (D, T_2). Again, we see that the value and meaning of a coefficient depend on the other terms in the mean function.

The last of the four mean functions (10.4) is different from the other three in that it nominally includes four terms beyond the intercept, terms for T_1 and T_2, and also for their average and their difference. But, given T_1 and T_2, we can calculate the terms A and D, so they contain no additional information. *Coefficients cannot be estimated for redundant terms which are exact linear combinations of other terms in the mean function.* Arc checks for such redundancy and, when found, notes it in the output using the word *aliased* in place of an estimated value. If the regression had been specified in another order—possibly A, then D, then T_1, and finally T_2—then the aliased terms would be the last two specified.

The property of *invariance*—giving essentially the same summary of the data after a linear transformation of the terms or the response—is an important property shared by least squares estimation and all other estimation methods described in this book.

10.1.4   Nonlinear Functions of Terms

While fitted regressions are invariant under *linear* transformations of terms, they are not invariant under *nonlinear* transformations. For example, if in the transactions data we replaced T_1 and T_2 by $R_1 = T_1/(T_1 + T_2)$ and $R_2 = \log(T_2)$, we will get different results. In particular, if the regression of *Time* on T_1 and T_2 has a linear mean function, then the regression of *Time* on R_1 and R_2 will not have a linear mean function. Later in this book, we often seek transformations for which the mean function has a useful form.

10.1.5   Variances of Coefficient Estimates

The standard error of a coefficient estimate $\hat{\eta}_j$ depends on the other terms in the model. These standard errors can be computed from a rather simple and revealing formula.

Define R_j^2 to be the value of R^2 for the linear regression of the jth term u_j on all other terms in the mean function. This quantity tells us how closely u_j can be approximated by some linear combination of the other terms; if R_j^2 is close to one, then u_j is nearly redundant given the other terms, while if R_j^2 is close to zero, then u_j has the potential for containing information about the response that is not available from the other terms in the mean function. The standard error of $\hat{\eta}_j$ can be written as

$$\text{se}(\hat{\eta}_j) = \frac{\hat{\sigma}}{\text{sd}(u_j)\sqrt{n-1}}\sqrt{\frac{1}{1-R_j^2}} \tag{10.5}$$

where n is the total sample size. Only the last factor on the right side of (10.5) depends on the other terms in the mean function. If R_j^2 is close to its maximum value of one, then u_j is nearly equal to a linear combination of the other terms and $1/(1 - R_j^2)$ is large. Consequently, the standard error of $\hat{\eta}_j$ is then large. If R_j^2 has its minimum value of zero, then $\text{se}(\hat{\eta}_j)$ has its minimum value of $\hat{\sigma}/((n-1)^{1/2}\text{sd}(u_j))$, which is essentially the same as the standard error of the regression coefficient in simple regression. If terms are added to a mean function, the value of R_j^2 for the jth term can either increase or stay the same, but it never decreases. Consequently, adding variables generally leads to larger standard errors for coefficient estimates. Adding terms also changes the value and meaning of all regression coefficients.

There is another lesson in (10.5) that tells us about the likely consequences of collecting more data. As we increase the sample size n, the standard error of a regression coefficient decreases at approximately the same rate as $1/\sqrt{n}$. For example, if we start with 100 cases and would like to collect additional data to reduce the coefficient standard errors by $1/2$, then we must take an additional 300 cases, so the total is 400, four times the number of cases we started with. The other quantities in (10.5) will also change with the addition of data, but these changes are likely to be relatively inconsequential if the same experimental protocol is used.

10.1.6 Standardization of Terms

Terms are often rescaled by subtracting their sample averages and then dividing by their sample standard deviations, so for $j = 1,\ldots,k-1$, the jth term u_j is replaced by the standardized terms $u_j^* = (u_j - \bar{u}_j)/\mathrm{sd}(u_j)$. When we fit the multiple linear regression model using these standardized terms, the estimate of the intercept is $\hat{\eta}_0^* = \bar{y}$, and the estimate of η_j becomes

$$\hat{\eta}_j^* = \mathrm{sd}(u_j)\hat{\eta}_j$$

One unit in the standardized scale is equivalent to one standard deviation, so $\hat{\eta}_j^*$ is the estimated increase in the expected response per standard deviation increase in the term.

Some investigators compare standardized coefficient estimates for different terms. Under this logic, terms with larger standardized coefficients are more important. Comparing standardized coefficient estimates can be more informative than comparing estimates in different scales, but it is also possible to be misled. For example, if two analysts collect data on the same predictors, one collecting data over a small range and the other over a larger range, they may come to completely different conclusions about the relative magnitudes of the standardized coefficients because of the different standard deviations.

10.2 THE MULTIVARIATE NORMAL DISTRIBUTION

Another very useful interpretation of regression coefficients is possible when the response and the terms have a joint distribution that is multivariate normal. We will write

$$\begin{pmatrix} y \\ \mathbf{u}_1 \end{pmatrix} \sim \mathrm{N}\left(\begin{pmatrix} \mu_y \\ \mu_u \end{pmatrix}, \begin{pmatrix} \sigma_y^2 & \Sigma_{yu}^T \\ \Sigma_{yu} & \Sigma_u \end{pmatrix} \right) \tag{10.6}$$

where y is the response and $\mathbf{u}_1$ is the $(k-1) \times 1$ vector of terms excluding the term for the intercept. In this compact notation, y is normally distributed with mean μ_y and variance σ_y^2. The terms $\mathbf{u}_1$ are also normally distributed. Each element has mean given by the corresponding element of the $(k-1) \times 1$ vector μ_u, and variance given by the corresponding diagonal element of Σ_u. The off-diagonal elements of Σ_u are the covariances between the elements of $\mathbf{u}_1$. Finally, the elements of Σ_{yu} are the covariances between y and the elements of $\mathbf{u}_1$. (Some readers may wish to consult Section 7.9.4 for more on this notation.)

With multivariate normal data, we get a generalization of results summarized in Section 4.2 for the bivariate normal. In particular, the conditional distribution $y \mid \mathbf{u}_1$ is another normal distribution,

$$y \mid \mathbf{u}_1 \sim \mathrm{N}(\eta_0 + \boldsymbol{\eta}_1^T \mathbf{u}_1, \sigma^2) \tag{10.7}$$

which is just the multiple linear regression model with normal errors and constant variance. The unknown parameters in (10.7) depend only on the parameters of the multivariate normal,

$$\eta_0 = \mu_y - \boldsymbol{\eta}_1^T \boldsymbol{\mu}_u, \qquad \boldsymbol{\eta}_1 = \Sigma_u^{-1} \Sigma_{yu}, \qquad \sigma^2 = \sigma_y^2 (1 - \mathcal{R}^2)$$

and $\mathcal{R}^2$ is the population squared multiple correlation coefficient, defined to be the square of the correlation between y and $\boldsymbol{\eta}_1^T \mathbf{u}_1$. With one term in $\mathbf{u}_1$, this reduces to the results of Section 4.2 for the bivariate normal. For a sample from a multivariate normal population, the regression parameters are therefore determined by parameters of the normal, and the conditional distribution of $y \mid \mathbf{u}_1$ can always be studied in terms of the linear regression model with normal errors. The multivariate normal provides a helpful target when using graphical methods to help build regression models.

The multivariate normal has several other important characteristics that we will use in later sections:

Selection of the Response. If the role of the response y is reversed with any of the terms, say u_j, we still get a linear regression model with normal errors for study of the distribution of u_j given y and the remaining terms.

Normal Subsets. If y and $\mathbf{u}_1$ are multivariate normal, then y and any subset of $\mathbf{u}_1$ are also multivariate normal. Consequently, we will have a linear regression model with constant variance for the distribution of y given any subset of the terms.

Joint Distribution of Pairs of Variables. All pairs of variables will have a bivariate normal distribution. In particular, all 2D scatterplots must exhibit linear mean functions and constant variance functions. Although linear mean function and constant variance function for all pairs of variables does not guarantee multivariate normality, in many problems transforming predictors to achieve this goal can provide a good starting point for regression modeling; we return to this in Chapter 13.

Partial Correlation. Let u_2 be a single term in $\mathbf{u}_1$ and collect the remaining terms in $\mathbf{u}_1$ into the vector $\mathbf{u}_3$ so that

$$\boldsymbol{\eta}_1^T \mathbf{u}_1 = \eta_2 u_2 + \boldsymbol{\eta}_3^T \mathbf{u}_3 \tag{10.8}$$

Then the joint conditional distribution of (y, u_2) given $\mathbf{u}_3$ is bivariate normal. The mean of this bivariate normal distribution can depend on the value of $\mathbf{u}_3$ but the variances, $\text{Var}(y \mid \mathbf{u}_3)$ and $\text{Var}(u_2 \mid \mathbf{u}_3)$, and the covariance $\text{Cov}(y, u_2 \mid \mathbf{u}_3)$ do not depend on the value of $\mathbf{u}_3$. Consequently, the correlation between y and u_2 given $\mathbf{u}_3$

$$\rho(y, u_2 \mid \mathbf{u}_3) = \frac{\text{Cov}(y, u_2 \mid \mathbf{u}_3)}{\sqrt{\text{Var}(y \mid \mathbf{u}_3)\text{Var}(u_2 \mid \mathbf{u}_3)}} \tag{10.9}$$

also does not depend on the value of $\mathbf{u}_3$. This correlation is called the population *partial correlation* between y and u_2 given $\mathbf{u}_3$. This is the same as the usual correlation between y and u_2 computed in a bivariate normal subpopulation in which $\mathbf{u}_3$ is held fixed at some value, the particular value being irrelevant.

There is a close connection between the regression coefficient η_2 of u_2 in (10.8) and the partial correlation (10.9):

$$\eta_2 = \rho(y, u_2 \mid \mathbf{u}_3)\frac{\sqrt{\text{Var}(y \mid \mathbf{u}_3)}}{\sqrt{\text{Var}(u_2 \mid \mathbf{u}_3)}} \tag{10.10}$$

If there is only one u-term other than the intercept, then (y, u_2) follows a bivariate normal distribution and, from (4.15),

$$\eta_2 = \rho(y, u_2)\frac{\sqrt{\text{Var}(y)}}{\sqrt{\text{Var}(u_2)}} \tag{10.11}$$

Thus we see that the coefficient (10.11) in a simple linear regression model has the same interpretation as the coefficient η_2 in a multiple linear regression model relative to a subpopulation in which $\mathbf{u}_3$ is held fixed.

10.3 SAMPLING DISTRIBUTIONS

The usefulness of coefficients and inferences about them and about the regression depends on the *sampling distribution* of the data. Consider again the haystack data. If the haystacks we have are a random sample of haystacks in the area under study at the time of the study, then any model estimated from these data would be relevant to all other unmeasured haystacks of that time and place. The model could be used to predict volumes, for example, of the other haystacks. Similarly, in the Lake Mary data, if we view the fish in the sample as a random sample of fish in the lake, a model fit to these data can be applied to all the fish in that lake. The fish in the sample were probably not a random sample of fish in the lake because some fish are harder to catch than others, and the fishing gear used may be size selective, so fisheries managers are often concerned about potential biases introduced by the way data are collected.

In many designed experiments the investigator assigns the values of the predictors to experimental units at random. For the generic response y and predictors $\mathbf{x}$, we often assume that the observed values are obtained from

$$y = E(y \mid \mathbf{x}) + e$$

where the error e is treated as if it were a random draw from a distribution with mean zero and often constant variance σ^2. The distribution of the errors then provides the basis for understanding coefficients and for making inferences and predictions. Random assignment of values of $\mathbf{x}$ to experimental units is enough to make this paradigm useful.

This paradigm can be used with Forbes' data. Although Forbes chose a convenience sample of locations to take measurments, the division of the observation into an expectation plus an error is reasonable, and so we expect that a model estimated from Forbes' data could be used to predict pressure, and hence altitude, at other locations. The Baltimore lead study described in Problem 8.7 is similar, except the fitted model would be thought to be appropriate only for the geographical area studied, the area around Baltimore, with no suggestion in the data that the same model could be applied to other areas.

In many problems, inference from a convenience sample to a population may not make any sense. In the Big Mac regression, the data were collected for a list of 45 world cities, mostly in Europe and the United States. These are hardly representative of a population of cities, and so the regression provides a description of the data, but a fitted model should not be used for inference outside of the data without either additional information or additional assumptions. As a second example, imagine a hypothetical study of the relationship between birth weight and mother's weight, with data collected from all births in a particular clinic. If we take the population of interest to be women served by that clinic, the data provide a census for a particular time period. We may be able to justify using a fitted model to predict for the next woman in that clinic by assuming that she is from the same population as the previous data. Generalizing to women in a particular geographic area, or to other clinics, or to women in general, is not possible without additional information. If, for example, this particular clinic serves poor women, then models may not apply to clinics that serve the wealthy.

10.4   CORRELATION VERSUS CAUSATION AND THE SLEEP DATA

In the transactions data, it is easy to imagine a *causal* relationship between the number of transactions and the total time: Increasing the number of transactions should increase the total time spent on transactions. In many regressions, we need to be much more cautious about interpreting an observed association in data as implying a causal relationship. We illustrate this by example.

TABLE 10.2 Definitions of Variables in the Sleep Example

Name	Description of variable
Label	Species of animal
SWS	Slow wave nondreaming sleep, hrs/day
PS	Paradoxical dreaming sleep, hrs/day
TS	Total sleep, hrs/day
BW	Body weight in kg
BrW	Brain weight in g
Life	Maximum life span, years
GP	Gestation time, days
D_1	Danger index, 0 = relatively low danger from other animals, 1 = relatively high danger from other animals

All known mammalian species spend at least part of each day sleeping, some species sleeping more than others. Sleep must serve some biological function, but why do sleep requirements vary so much from species to species? One approach to getting information that might help understand this question is to study the dependence of hours of sleep on species characteristics. In the file sleep.lsp, we have data on 62 species, covering at least 13 different orders of mammals. This is surely not a random sample of species, so some care must be taken in generalizing any results to all mammal species. The variables defined in Table 10.2 are species averages. The problem of interest is to understand the dependence of hours of sleep on the other species characteristics recorded in the data.

These data are more complicated than most of the data we have encountered so far. First, the number of variables is larger. Also, three possible response variables are given: *SWS*, the hours of slow-wave or deep, nondreaming sleep; *PS*, the hours of paradoxical or dreaming sleep, and their sum $TS = SWS + PS$. Using any of these three as the response could be meaningful. We will use the total hours of sleep, *TS*, as the response variable. The predictor D_1 is really an indicator variable that equals one for species that face danger from other animals, and equals zero for species that face less danger from other animals.

10.4.1 Missing Data

A problem we haven't encountered before is of missing data: Not all variables are measured on each species. You can verify this by loading the data file and selecting the item "Display data" from the Sleep menu. *Arc* can accommodate missing data if the missing data code ? is used in place of a value in the data file. *Arc* follows a simple rule with missing data: Any given calculation is based only on fully observed cases. You can change both the missing value code and the way missing data are handled, using the "Settings" item in the Arc menu (see Section A.5.9). When using *TS* as the response, the data include only 51 fully observed species, and all computations are based on these. Using

SWS as the response would give fewer complete cases; only 39 of the species are observed on all the variables. Further discussion of missing data is given in the Complements to this chapter.

10.4.2   The Mean Function

We can define a multiple linear regression mean function that might be used to describe $E(TS \mid \mathbf{x})$ as a function of the predictors. The mean function will consist of six terms: an intercept, D_1, and the logarithms of *BW*, *BrW*, *Life*, and *GP*. The need for logarithmic transformations of these variables is indicated by the general rule for ranges discussed in Section 5.4 that positive predictors that have the ratio between their largest and smallest values equal to 10 and preferably 100 or more should very likely be transformed to logs. We have used logs to the base two in this example.

From the scatterplot matrix in Figure 10.1 we see that the pairwise relationships between the continuous terms generally have linear mean functions. We also see that at least some of the terms are associated with the response; for example, the conditional mean of *TS* given $\log_2(BW)$ is clearly decreasing: Larger species sleep less. Each scatterplot in the bottom row appears to have a mean function that is decreasing with the term. The plots that include D_1 are hard to analyze because D_1 has only two values, zero or one.

We now turn to fitting the mean function

$$E(TS \mid \mathbf{x}) = \eta_0 + \eta_1 \log_2(BW) + \eta_2 \log_2(BrW) + \eta_3 \log_2(Life)$$
$$+ \eta_4 \log_2(GP) + \eta_5 D_1 \tag{10.12}$$

where $\mathbf{x}$ is the 5×1 vector of predictors.

10.4.3   The Danger Indicator

Model (10.12) is our first example of a mean function that contains an indicator. To understand the role of D_1, think of the two subpopulations of species defined by the danger index, $D_1 = 0$ for species in low danger and $D_1 = 1$ for species in high danger. For notational convenience, let $\mathbf{x}^*$ denote the 4×1 vector of predictors excluding D_1. The mean function (10.12) defines two other mean functions, one for each of the subpopulations defined by D_1. We can write these subpopulation mean functions by substituting $D_1 = 0$ and $D_1 = 1$ into (10.12):

$$E(TS \mid \mathbf{x}^*, D_1 = 0) = \eta_0 + \eta_1 \log_2(BW) + \eta_2 \log_2(BrW)$$
$$+ \eta_3 \log_2(Life) + \eta_4 \log_2(GP) \tag{10.13}$$
$$E(TS \mid \mathbf{x}^*, D_1 = 1) = (\eta_0 + \eta_5) + \eta_1 \log_2(BW) + \eta_2 \log_2(BrW)$$
$$+ \eta_3 \log_2(Life) + \eta_4 \log_2(GP) \tag{10.14}$$

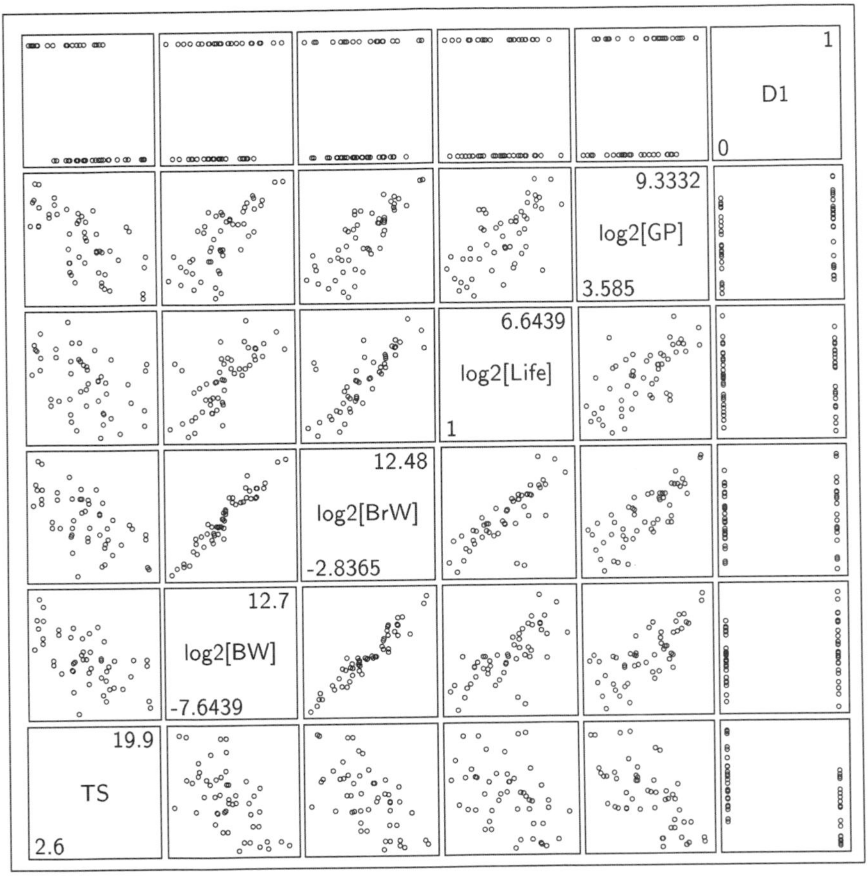

FIGURE 10.1 Scatterplot matrix for the sleep data.

These two subpopulation mean functions are identical except for their intercepts. The intercept in the low danger subpopulation is η_0 just as it is in the overall model (10.12), but the intercept in the high danger subpopulation is $\eta_0 + \eta_5$, where η_5 is the coefficient of D_1 in (10.12). The subpopulation mean functions are parallel hyperplanes in the four logarithmic terms, the constant distance between the hyperplanes being η_5. In Chapter 12 we will encounter subpopulation mean functions that are nonparallel planes.

To illustrate the idea of separate planes arising from an indicator, let's take a brief detour and consider the regression of TS on $\mathbf{w} = (\log_2(Life), \log_2(GP), D_1)^T$. Construct a 3D plot of TS versus $(\log_2(Life), \log_2(GP))$, setting D_1 to be the marking variable in the "Plot of" dialog. In the resulting plot, select the item "Fit by marks" in the pop-up menu for the parametric smoother slidebar. The parametric slidebar can now be used to fit separate planes, one for $D_1 = 1$ and one for $D_1 = 0$. The planes are very nearly parallel for the sleep data, suggesting that the model

$$E(TS \mid \mathbf{w}) = \eta_0 + \eta_1 \log_2(GP) + \eta_2 \log_2(Life) + \eta_3 D_1$$

may be appropriate.

10.4.4   Interpretation

Returning to mean function (10.12), we now add the assumption of constant variance, $\text{Var}(TS \mid \mathbf{x}) = \sigma^2$, and estimate the mean function via OLS. The resulting output is shown in Table 10.3. Before examining the fit of this model to the data, let's assume that the mean function (10.12) is correct, and for the sake of discussion let's suppose that the coefficient η_4 for log gestation period is negative. Can we infer that longer gestation periods *cause* less sleep? Here the causal link seems tenuous at best. More likely is the possibility that *both* longer gestation period and fewer hours of sleep are caused by some other factors; and what we see in the data is an association, but not a causation. In observational studies like this one, regression models generally tell us about association, but not about causation.

We turn now to the output in Table 10.3. Look first at the coefficient estimates and their standard errors. The estimate for D_1 is negative and large relative to its standard error, indicating that animals with $D_1 = 0$ sleep more than animals with $D_1 = 1$. The value of the estimate, -3.8, gives us an idea of how much more: If we could find two species with all characteristics the same, except one has $D_1 = 0$ and the other $D_1 = 1$, we would expect the species with $D_1 = 1$ to sleep about 3.8 hours less than the species with $D_1 = 0$. The coefficient for $\log_2(GP)$ is estimated to be -0.97. Since we used logs to the base two, this means that every time the gestation period doubles, so $\log_2(GP)$ increases by one unit, the average amount of sleep decreases by 0.97 hours.

Two of the remaining coefficient estimates are positive, in contradiction to our expectation from examination of the last row of the scatterplot matrix. In addition, even though $\log_2(BW)$ and $\log_2(BrW)$ appear to have the strongest relationship with TS in Figure 10.1, they both have coefficient estimates that are small relative to their standard errors; indeed the p-value for testing the coefficient for $\log_2(BW)$ equal to zero is larger than 0.9, so $\log_2(BW)$ does not appear to be an important predictor given the other terms in the model. Furthermore, the signs of the coefficient estimates for these two terms are opposite, one positive and one negative.

There is nothing wrong with the regression calculations shown here, what is wrong is trying to infer about individual coefficients without reference to the other terms in the mean function. We have already seen that the standard error of a coefficient estimate depends on the relationship between the corresponding term and the other terms in the mean function; the value of the coefficient estimate also depends on this relationship. The plot of the response against a

TABLE 10.3   Regression Summary Mean Function (10.12) in the Sleep Data

```
Data set = Sleep, Name of Fit = L1
20 cases are missing at least one value.
Normal Regression
Kernel mean function = Identity
Response    = TS
Terms       = (log2[BW] log2[BrW] log2[Life] log2[GP] D1)
Cases not used and missing at least one value are:
(Arctic ground squirrel Desert hedgehog Genet Giant armadillo
Giraffe Kangaroo Mole rat Mountain beaver Okapi
Star-nosed mole Yellow-bellied marmot)
Coefficient Estimates
Label         Estimate        Std. Error      t-value
Constant      19.3091         2.66184         7.254
log2[BW]       0.0225154      0.343342        0.066
log2[BrW]     -0.550752       0.526717       -1.046
log2[Life]     0.425713       0.534618        0.796
log2[GP]      -0.971283       0.486174       -1.998
D1            -3.81995        0.907142       -4.211

R Squared:              0.631445
Sigma hat:              2.99451
Number of cases:        62
Number of cases used:   51
Degrees of freedom:     45

Summary Analysis of Variance Table
Source      df     SS       MS       F       p-value
Regression  5      691.349  138.27   15.42   0.0000
Residual    45     403.519  8.96709
```

single term is not directly relevant to an interpretation of its estimated regression coefficient. However, we *can* find a plot that allows visualization of the effect of each term in a mean function. This plot is called an *added-variable plot*.

10.5   2D ADDED-VARIABLE PLOTS

The added-variable plot is a graphical object that (1) always provides visual information on the numerical calculation of the coefficient of a term, and in some situations provides (2) diagnostic information on the model and (3) a visual assessment of the *net effect* of a predictor or term: the effect of a predictor in a subpopulation in which all other predictors are held fixed.

We begin by describing added-variable plots for mean functions with two terms in addition to an intercept, and then look at the general case.

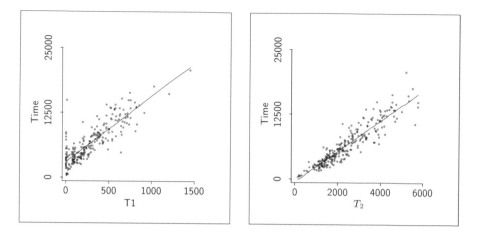

a. T_1 as predictor.   b. T_2 as predictor.

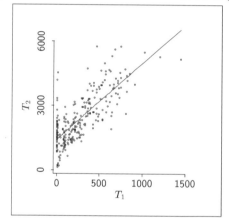

c. Predictors.

FIGURE 10.2   Three 2D plots for the transactions data.

10.5.1   Adding a Predictor to Simple Regression

Let's return to the transactions data. In (10.1), the mean function depends on a linear combination of T_1 and T_2, and the regression is the study of the conditional distribution of *Time* given both T_1 and T_2. Let's suppose that we have already examined the conditional distribution of *Time* given only T_1. We now study what happens when a second term T_2 is added to the mean function. The three plots in Figure 10.2 are relevant here. Figure 10.2a is a summary of the marginal relationship between *Time* and T_1. The fitted OLS line on the plot seems to match the data quite well, except perhaps at $T_1 = 0$, reflecting some branches with no T_1 transactions but many T_2 transactions. Figure 10.2b, which is the plot for the simple regression of *Time* on T_2, tells a very similar story: Ignoring T_1, *Time* and T_2 are linearly related.

The third plot in Figure 10.2c is a 2D scatterplot of the predictors, T_2 versus T_1. This plot allows us to visualize the mean function $E(T_2 \mid T_1)$. While the relationship between T_2 and T_1 is not of direct interest to us, this plot is important because it shows that $E(T_2 \mid T_1)$ depends on T_1 linearly. To put it another way, if we know the value of T_1, we have some information about T_2. If we contemplate using T_2 as a second predictor of *Time* after using T_1, we should expect that the effect of T_2 would be adjusted for the part of T_2 that can be explained by T_1. In other words, only the part of T_2 that constitutes new information beyond that already furnished by T_1 should be used. Multiple linear regression does an appropriate adjustment.

Use the "Fit linear LS" item in the Graph&Fit menu to estimate the mean function for *Time* $\mid T_1$, assuming the simple linear regression model. The estimated mean function is

$$\hat{E}(Time \mid T_1) = 3044 + 12.67T_1 \tag{10.15}$$

with $\hat{\sigma}(Time \mid T_1) = 1909$, $R^2(Time \mid T_1) = 0.75$. We added the "$(Time \mid T_1)$" to these symbols to remind us explicitly that these statistics have been computed from the regression of *Time* on T_1 only. Notice also that the coefficient estimate for T_1, about 12.7 minutes per transaction, is quite different from the 5.5 minutes per transaction we found when using both T_1 and T_2. We can get a similar fitted equation for the mean function in Figure 10.2b by fitting the regression of *Time* on T_2,

$$\hat{E}(Time \mid T_2) = -542 + 2.95T_2 \tag{10.16}$$

with $\hat{\sigma}(Time \mid T_2) = 1450$ and $R^2(Time \mid T_2) = 0.85$. The variable T_2 alone gives a somewhat larger value of R^2 than fitting T_1 alone.

We can view the fitted mean function $\hat{E}(Time \mid T_1)$ as dividing each observation *Time* into two pieces, the part of *Time* explained by T_1 and the part of *Time* not explained by T_1, so

$$Time = \text{Explained} + \text{Unexplained}$$

$$= \hat{E}(Time \mid T_1) + (Time - \hat{E}(Time \mid T_1))$$

$$= \hat{E}(Time \mid T_1) + \hat{e}(Time \mid T_1)$$

where $\hat{e}(Time \mid T_1)$ are the residuals from the regression of *Time* on T_1. Let's think about adding T_2 to a model that already includes T_1. We want T_2 to explain the part of *Time* not explained by T_1. This suggests fitting a regression

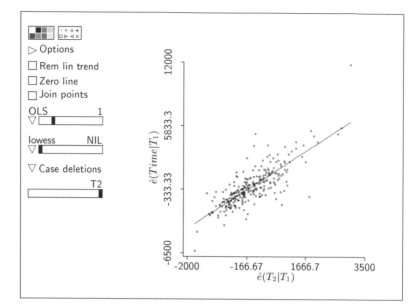

FIGURE 10.3   Added-variable plot for T_2 in the transactions data. The axis labels have been modified to correspond to the notation in this book. In particular, there are no "hats" on the e's in the plots produced by *Arc*.

model, not with *Time* as the response but with the residuals $\hat{e}(Time \mid T_1)$ as the response. However, this won't quite work because T_1 and T_2 are related to each other, as shown in Figure 10.2c, and so some of the information in T_2 is redundant. The solution is to fit the linear regression of T_2 on T_1 and get the residuals $\hat{e}(T_2 \mid T_1)$ from that regression. This is the part of T_2 that is not explained by T_1. Then, fit the regression of $\hat{e}(Time \mid T_1)$ on $\hat{e}(T_2 \mid T_1)$ to study the effect of adding T_2 to a model that already includes T_1. The slope estimate from the simple linear regression of $\hat{e}(Time \mid T_1)$ on $\hat{e}(T_2 \mid T_1)$ will give the change in *Time* for a unit change in T_2, adjusted for fitting T_1 first, and this is exactly what we want. The plot of $\hat{e}(Time \mid T_1)$ versus $\hat{e}(T_2 \mid T_1)$ is an *added-variable plot*.

Arc can be used for these calculations. We need the residuals from the regressions of *Time* on T_1 and of T_2 on T_1. In the transactions example, use the Graph&Fit menu to fit these OLS regressions. Then, use the Graph&Fit menu to plot the residuals from these two regressions against each other. This plot, shown in Figure 10.3, can be studied like any other 2D plot, checking for strength of linear relationship or examining for points that are separated from all the other points. In Figure 10.3, the points are clustered about the OLS regression line shown on the plot, apart from one separated point. This is a graphical indication that T_2 is a useful additional predictor when T_1 has already been used as a predictor.

10.5.2 Added-Variable Plots in *Arc*

The example we have worked here for added-variable plots has only two terms, but exactly the same methodology can apply no matter how many terms we have. Suppose we have a mean function

$$E(y \mid \mathbf{x}) = \eta^T \mathbf{u} = \eta_1^T \mathbf{u}_1 + \eta_2 u_2 \qquad (10.17)$$

where we have divided the $k \times 1$ vector of terms $\mathbf{u}$ into two pieces, $\mathbf{u}_1$ with $k - 1$ terms and u_2 with the remaining term. The added-variable plot for u_2 is a plot of $\hat{e}(y \mid \mathbf{u}_1)$ versus $\hat{e}(u_2 \mid \mathbf{u}_1)$.

While the procedure outlined in the last section can be carried out to get an added-variable plot, there is a shortcut available in *Arc*. Fit the model with mean function (10.17), and then select the item "AVP–All 2D" from the model's menu. You will get a 2D scatterplot that initially shows the added-variable plot for the first term. You can obtain the added-variable plot for any other term by using the new slidebar on the left of the plot to select the desired term.

10.6 PROPERTIES OF 2D ADDED-VARIABLE PLOTS

Added-variable plots are closely related to regression calculations. Consider fitting a line by OLS to the points in an added-variable plot, as shown in Figure 10.3. To do this, the two sets of residuals must be saved as variables, as described in Appendix A, Section A.11.

10.6.1 Intercept

As long as the intercept is in the mean function, the estimated intercept in the added-variable plot will be zero.

10.6.2 Slope

The estimated slope in an added-variable plot will always be the same as the estimated coefficient $\hat{\eta}_2$ in the OLS fit of (10.17), so fitting the simple linear mean function by OLS to an added-variable plot gives

$$\hat{E}(\hat{e}(y \mid \mathbf{u}_1) \mid \hat{e}(u_2 \mid \mathbf{u}_1)) = 0 + \hat{\eta}_2 \hat{e}(u_2 \mid \mathbf{u}_1)$$

In this way the added-variable plot always provides visual information on the numerical calculation of the coefficient of a term.

For the transactions data, the fitted mean function for the added-variable plot is

$$\hat{E}(\hat{e}(Time \mid T_1) \mid \hat{e}(T_2 \mid T_1)) = 0 + 2.03455\hat{e}(T_2 \mid T_1)$$

and the estimated regression coefficient for $\hat{e}(T_2 \mid T_1)$ is the coefficient estimate for η_2 in the OLS fit to (10.1).

10.6.3   Residuals

The residuals in an added-variable plot are identical to the residuals $\hat{e}$ from the OLS fit of the full model (10.17),

$$\hat{e} = \hat{e}(y \mid \mathbf{u}_1, u_2) = \hat{e}(y \mid \mathbf{u}_1) - \hat{\eta}_2 \times \hat{e}(u_2 \mid \mathbf{u}_1) \qquad (10.18)$$

since the added-variable plot has "response" $\hat{e}(y \mid \mathbf{u}_1)$ and "fitted values" $\hat{\eta}_2 \times \hat{e}(u_2 \mid \mathbf{u}_1)$. For the transactions data this relationship simplifies to

$$\hat{e} = \hat{e}(Time \mid T_1, T_2) = \hat{e}(Time \mid T_1) - \hat{\eta}_2 \times \hat{e}(T_2 \mid T_1)$$

This can be verified numerically by plotting the residuals from the model fit to the points in the added-variable plot versus the residuals obtained from the fit of the full model.

10.6.4   Sample Partial Correlation

Under multivariate normality (10.6), the sample correlation between $\hat{e}(y \mid \mathbf{u}_1)$ and $\hat{e}(u_2 \mid \mathbf{u}_1)$ is the sample partial correlation between y and u_2 given $\mathbf{u}_1$:

$$\hat{\rho}(y, u_2 \mid \mathbf{u}_1) = r(\hat{e}(y \mid \mathbf{u}_1), \hat{e}(u_2 \mid \mathbf{u}_1))$$

Thus, an added-variable plot for u_2 after $\mathbf{u}_1$ allows us to visualize the sample partial correlation in the same way as a scatterplot of y versus u_2 permits us to visualize the marginal correlation between y and u_2. The use and interpretation of the sample partial correlation is subject to the same limitations discussed in Section 4.2.2 for the ordinary correlation.

10.6.5   *t*-Statistics

If k is small relative to n, the t-statistic for testing the hypothesis that the slope equals 0 in an added-variable is approximately the same as the t-statistic for testing $\eta_2 = 0$ in the full mean function (see Problem 10.3). Thus, the term u_2 is important if its added-variable plot is strongly linear, with small residuals, indicating that the t-statistic for testing $\eta_2 = 0$ is large. This is the situation we have observed in the transactions data.

10.6.6   Three Extreme Cases

We next discuss three extreme cases of added-variable plots. While these cases rarely occur in practice, they may provide additional understanding of 2D

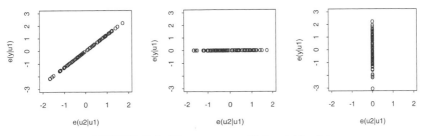

FIGURE 10.4   Three extreme added-variable plots.

added-variable plots. The three cases are depicted in Figure 10.4 as the added-variable plot for u_2 in the generic regression model (10.17).

Points on a Diagonal Line: $\hat{e}(y \mid \mathbf{u}) = 0$. In the left frame of this plot, all the points lie exactly on a straight line with nonzero slope. What does this tell us about the fit of (10.17)? Since the points fall exactly on a line, the residuals from the added-variable plot regression are all zero. This tells us that the residuals from (10.17) must all be zero and thus all the data fall exactly on a plane in the k terms. Adding u_2 to the model gives a perfect fit, so u_2 is useful even after the contributions of the terms in $\mathbf{u}_1$.

Points on Horizontal Line: $\hat{e}(y \mid \mathbf{u}_1) = 0$. Suppose next that the points in an added-variable plot lie exactly on a straight line with zero slope, as in the second frame of the plot. In this case we must have $\hat{e}(y \mid \mathbf{u}_1) = 0$ for every case in the data. There is no reason to consider adding u_2 because the regression of y on $\mathbf{u}_1$ explains all the variation in the response. In practice, scatter about a horizontal line indicates that a term will have a coefficient close to zero in a fitted mean function.

Points on a Vertical Line: $\hat{e}(u_2 \mid \mathbf{u}_1) = 0$. The third frame illustrates the extreme case with $\hat{e}(u_2 \mid \mathbf{u}_1) = 0$. This means that u_2 is an exact linear function of $\mathbf{u}_1$. In the context of the multiple linear regression model, all of the information about the response available from u_2 is already available from $\mathbf{u}_1$, so u_2 is an entirely redundant term. Most computer programs will give an error message and refuse to include terms that are an exact or nearly exact linear combination of other terms in the model. This situation is often identified by the word *collinearity*. We have seen an example of exact collinearity already in the attempt to fit mean function (10.4) to the transactions data, where the program recognized that the two additional terms were of no value after including the first two terms T_1 and T_2, and characterized these terms as aliased. Approximate collinearity, where one of the predictors is almost an exact linear combination of the others, is quite common in some areas of application. It can be diagnosed if the range of the values of $\hat{e}(u_2 \mid \mathbf{u}_1)$ is tiny relative to the range of the values of u_2.

10.7   3D ADDED-VARIABLE PLOTS

Two-dimensional added-variable plots can be generalized to three dimensions straightforwardly. Consider the mean function

$$E(y \mid \mathbf{x}) = \boldsymbol{\eta}^T \mathbf{u} = \boldsymbol{\eta}_1^T \mathbf{u}_1 + \eta_2 u_2 + \eta_3 u_3 \qquad (10.19)$$

where we have divided the $k \times 1$ vector of terms $\mathbf{u}$ into three pieces, $\mathbf{u}_1$ with $k - 2$ terms and the individual terms u_2 and u_3. The 3D added-variable plot for (u_2, u_3) is a plot of $\hat{e}(y \mid \mathbf{u}_1)$ versus $(\hat{e}(u_2 \mid \mathbf{u}_1), \hat{e}(u_3 \mid \mathbf{u}_1))$.

The item "AVP–3D" in an *Arc* model menu can be used to construct a 3D added-variable plot, and the parametric smoother slidebar can then be used to superimpose the fitted OLS plane, as discussed in Section 8.2. The properties of a 3D added-variable plot constructed in this way are the same as the properties of a 2D added-variable plot. In particular,

- the intercept of the fitted plane will be zero if the full model contains an intercept η_0,
- the estimated coefficients for the fitted plane are $\hat{\eta}_2$ and $\hat{\eta}_3$, and
- the residuals from the fitted plane are the same as those from the full model.

10.8   CONFIDENCE REGIONS

Load the file BGSboys.lsp. This file contains the data on boys from the Berkeley Guidance Study. For this example, we will study how weight at age 18, *WT18*, depends on measurements of height, weight, leg circumference, and strength at age nine, via the mean function:

$$E(WT18 \mid \mathbf{x}) = \eta_0 + \eta_1 HT9 + \eta_2 WT9 + \eta_3 LG9 + \eta_4 ST9 \qquad (10.20)$$

with $\mathrm{Var}(WT18 \mid \mathbf{x}) = \sigma^2$.

In Chapter 7 we learned how to make confidence statements concerning regression coefficients. A confidence statement for a coefficient η_2, for example, paid no attention to the value of any of the other coefficients. However, coefficient estimates can be correlated, and thus joint confidence statements can be more informative than marginal ones.

To construct a joint 95% confidence region for η_2 and η_4 in mean function (10.20), select "Confidence regions" from the model's menu. In the resulting dialog, move *WT9* and *ST9* from the left list to the right list. The result is shown in Figure 10.5a. The cross hairs mark off marginal 95% confidence intervals

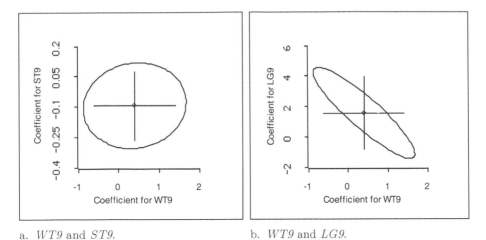

a. *WT9* and *ST9*.   b. *WT9* and *LG9*.

FIGURE 10.5 Joint and marginal 95% confidence regions for the Berkeley Guidance Study for boys.

on the coordinate axes for η_2 and η_4. For example, the 95% confidence interval for η_2 runs from about -0.6 to 1.4.

The elliptical region in Figure 10.5a is a joint 95% confidence region for η_2 and η_4. Confidence regions for two coefficients in a multiple linear regression model assuming that the errors are normally distributed are always elliptical. The ellipse in Figure 10.5a is nearly a circle, but this characteristic of the shape depends on the aspect ratio in the plot and on the correlations between the estimated coefficients. The cross hairs in Figure 10.5a sit well within the joint confidence region, so the range allowed for either coefficient in the joint region is larger than the corresponding range of either marginal confidence interval.

Construct a joint confidence region for the coefficients of *WT9* and *LG9*, as shown in Figure 10.5b. Figures 10.5a and 10.5b are qualitatively different. The major axis of the ellipse in Figure 10.5b has a negative slope, so $\hat{\eta}_2$ and $\hat{\eta}_3$ are negatively correlated. When the scaling in the plot is the default scaling like in these figures, elongation of the ellipse indicates large correlation. The cross hairs in Figure 10.5b extend outside of the joint confidence region because of the high correlation. In Figure 10.5a the near circularity suggests that the correlation between the coefficients of *WT9* and *ST9* is small.

10.8.1 Confidence Regions for Two Coefficient Estimates

Consider a joint confidence region for the regression coefficients (η_1, η_2) in the mean function

$$E(y \mid \mathbf{x}) = \eta_0 + \eta_1 u_1 + \eta_2 u_2$$

This mean function has three terms: The intercept and two nonconstant terms u_1 and u_2. We assume further that errors are normally distributed with mean zero and constant variance function $\text{Var}(y \mid \mathbf{x}) = \sigma^2$. Let $\eta = (\eta_1, \eta_2)^T$, with OLS estimate $\hat{\eta}$. From (7.18), we know that the estimated covariance matrix for $\hat{\eta}$ is of the form $\hat{\sigma}^2 \mathbf{M}$, or

$$\widehat{\text{Var}}(\hat{\eta}) = \hat{\sigma}^2 \begin{pmatrix} m_{11} & m_{12} \\ m_{12} & m_{22} \end{pmatrix}$$

The elements m_{ij} of $\mathbf{M}$ depend on u_1 and u_2 only through their standard deviations $\text{sd}(u_1)$ and $\text{sd}(u_2)$ and through their sample correlation r_{12}.

A joint $(1 - \alpha) \times 100$ percent confidence region for η is the set of all values of the 2×1 vector $\mathbf{h}$ that satisfies the inequality

$$(\mathbf{h} - \hat{\eta})^T [\widehat{\text{Var}}(\hat{\eta})]^{-1} (\mathbf{h} - \hat{\eta}) \le 2Q(F_{2,n-3}, 1 - \alpha) \qquad (10.21)$$

where $Q(F_{2,n-3}, 1 - \alpha)$ is the $1 - \alpha$ quantile of the $F_{2,n-3}$ distribution, and $[\widehat{\text{Var}}(\hat{\eta})]^{-1}$ is the *inverse* of the 2×2 covariance matrix for $\hat{\eta}$ (see Section 7.9.1). The points satisfying this inequality fall inside an ellipse. The center of the ellipse is at $\hat{\eta}$. The size of the ellipse is partially controlled by the choice of the level $1 - \alpha$. As we require more confidence, $1 - \alpha$ becomes bigger, $Q(F_{2,n-3}, 1 - \alpha)$ becomes bigger, and the area of the ellipse increases.

The element of $\widehat{\text{Var}}(\hat{\eta})$ that corresponds to $\widehat{\text{Var}}(\hat{\eta}_1)$ is

$$\widehat{\text{Var}}(\hat{\eta}_1) = \frac{\hat{\sigma}^2}{(n-1)\text{sd}^2(u_1)} \times \frac{1}{(1 - r_{12}^2)} \qquad (10.22)$$

where r_{12} is the sample correlation between u_1 and u_2. This is the same as the variance estimate we encountered when discussing the standard error of coefficient estimates in Section 10.1.5. The standard errors of the coefficient estimates determine the lengths of marginal intervals, and these are strongly determined by r_{12}^2. The length of a confidence interval increases as r_{12}^2 increases, so high correlation between the predictors gives relatively long confidence intervals.

The shape of a joint confidence interval depends on $\rho(\hat{\eta}_1, \hat{\eta}_2)$, the correlation between $\hat{\eta}_1$ and $\hat{\eta}_2$, and on scale factors. Holding the scale factors fixed, the joint confidence region becomes elongated as $|\rho(\hat{\eta}_1, \hat{\eta}_2)|$ increases. What will make $|\rho(\hat{\eta}_1, \hat{\eta}_2)|$ large? There is a close connection between this correlation and r_{12}:

$$\rho(\hat{\eta}_1, \hat{\eta}_2) = -r_{12} \qquad (10.23)$$

Thus, the correlation between the OLS estimates of η_1 and η_2 in a regression model with just two nonconstant terms is the negative of the sample correlation between these terms. To gain some intuition about why this is so, imagine a

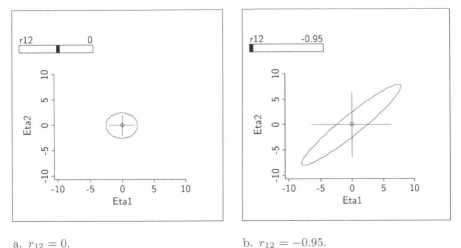

a. $r_{12} = 0.$ b. $r_{12} = -0.95.$

FIGURE 10.6   Confidence region demonstration.

3D plot of y versus (u_1, u_2) in which the points occur at the tips of the slats in a picket fence that runs up a hill. Now imagine trying to balance (fit) a sheet of plywood (a regression plane) on the picket fence. The plywood will be stable and easily balanced in the direction of the fence, but it will be unstable and hard to balance in the direction perpendicular to the fence. Similarly, the regression coefficients are well-estimated in the direction of the fence, but not so in the perpendicular direction. The uncertainty in the regression coefficients runs primarily in the direction perpendicular to the fence. Along the fence, there is little variability, so the confidence region along the fence will be narrow, while perpendicular to the fence variability is high, producing wide confidence regions. These combine to explain why a high positive correlation between u_1 and u_2 results in coefficient estimates that are highly negatively correlated.

Load the demonstration `demo-cr.lsp`. A brief description of the demonstration will be displayed, and a 2D plot with a slider labeled "r12" will appear on the computer screen: "r12" is the sample correlation r_{12} between u_1 and u_2. This demonstration is for a regression with u_1 and u_2 scaled to have mean zero and equal standard deviations, sd(u_1) = sd(u_2). In this scaling and with an aspect ratio of one, the shape of a confidence region depends only on the correlation r_{12}. The initial plot shown in Figure 10.6a is the joint 95% confidence region for (η_1, η_2) when $r_{12} = 0$. The cross hairs on the plot give marginal intervals, assuming that $\hat{\eta} = (0,0)^T$ and that $\hat{\sigma}^2 = 1$. The slidebar controls the sample correlation r_{12}. As the slidebar is moved, the confidence regions are redrawn.

Figure 10.6a with $r_{12} = 0$ is similar to Figure 10.5a. Before moving the slider, consider what will happen as it is moved to the right. What will happen to the cross hairs? The demonstration has been constructed to keep the center at the origin, so only lengths and orientation can change. What will happen to the ellipse? How will its orientation change?

As the slider is moved to the right, the marginal intervals depicted by the cross hairs get longer. Was this predicted by the previous discussion? Also, the slope of the major axis of the ellipse is now negative when r_{12} is positive. Was this predicted by the previous discussion? The plot when the sample correlation $r_{12} = -0.95$ is shown in Figure 10.6b.

10.8.2 Bivariate Confidence Regions When the Mean Function Has Many Terms

With minor modifications, we can apply our understanding of confidence regions to the general model,

$$E(y \mid \mathbf{x}) = \eta_0 + \eta_1 u_1 + \eta_2 u_2 + \eta_3^T \mathbf{u}_3 \qquad (10.24)$$

where u_1 and u_2 are single terms and all remaining terms are collected in the vector $\mathbf{u}_3$. This problem differs from that discussed in Section 10.8.1 because it has more than two terms beyond the intercept.

How do we construct a joint confidence region for (η_1, η_2)? Aside from the presence of $\mathbf{u}_3$, this is the same problem considered in Section 10.8.1. The required confidence region can be constructed by first getting $\hat{e}(u_1 \mid \mathbf{u}_3)$, the residuals from the OLS fit of u_1 on $\mathbf{u}_3$, and $\hat{e}(u_2 \mid \mathbf{u}_3)$, the residuals from the OLS fit of u_2 on $\mathbf{u}_3$. Both of these regressions should include an intercept term because there is an intercept term in (10.24). Now go back to Section 10.8.1 and replace u_1 with $\hat{e}(u_1 \mid \mathbf{u}_3)$ and u_2 with $\hat{e}(u_2 \mid \mathbf{u}_3)$. Except for a change in df, the discussion of the previous section now applies verbatim to a confidence region for (η_1, η_2) in (10.24). To apply equation (10.21), replace $Q(F_{2,n-3}, 1 - \alpha)$ with $Q(F_{2,n-k}, 1 - \alpha)$, where k is the number of terms including the intercept.

The shape of confidence regions in regressions with two terms beyond the intercept is determined by the sample correlation r_{12}. In the many-term generalization, the shape of confidence regions for two terms is determined by the sample correlation between $\hat{e}(u_1 \mid \mathbf{u}_3)$ and $\hat{e}(u_2 \mid \mathbf{u}_3)$. This correlation is the sample partial correlation between u_1 and u_2 adjusted for $\mathbf{u}_3$. It is the sample correlation between u_1 and u_2 after removing any linear association with $\mathbf{u}_3$, and therefore the sample partial correlation between u_1 and u_2 is just the correlation for the scatterplot of $\hat{e}(u_1 \mid \mathbf{u}_3)$ versus $\hat{e}(u_2 \mid \mathbf{u}_3)$. Recall that a 3D added-variable plot for (u_1, u_2) is the 3D plot of $\hat{e}(y \mid \mathbf{u}_3)$ versus $(\hat{e}(u_1 \mid \mathbf{u}_3), \hat{e}(u_2 \mid \mathbf{u}_3))$. The correlation between the variables in the horizontal plane of a 3D added-variable plot is thus the same as the partial correlation between u_1 and u_2. In addition, a 3D added-variable plot for two terms in a many-term model can be used to understand their coefficients just as a 3D plot of y versus (u_1, u_2) is used when there are only two nonconstant terms in the

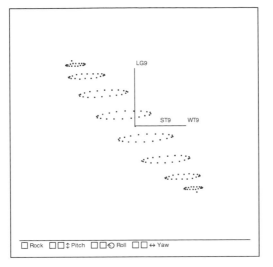

FIGURE 10.7   A three-dimensional confidence region.

model. In particular, the picket fence analogy for the behavior of confidence regions for (η_1, η_2) in the model $E(y \mid \mathbf{x}) = \eta_0 + \eta_1 u_1 + \eta_2 u_2$ applies also to a 3D added-variable plot for (u_1, u_2) in model (10.24).

10.8.3   General Confidence Regions

Joint confidence regions for more than two coefficients follow the same general ideas. In particular, letting η_S denote a subset of q coefficients in a multiple regression model, a $(1 - \alpha) \times 100\%$ confidence region for η_S is the set of all values of the $q \times 1$ vector $\mathbf{h}$ that satisfy the following inequality:

$$(\mathbf{h} - \hat{\eta}_S)^T [\widehat{\mathrm{Var}}(\hat{\eta}_S)]^{-1} (\mathbf{h} - \hat{\eta}_S) \leq q Q(F_{q,n-k}, 1 - \alpha) \qquad (10.25)$$

where $\hat{\eta}_S$ is the OLS estimate of η_S, and $[\widehat{\mathrm{Var}}(\hat{\eta}_S)]^{-1}$ is the inverse of the $q \times q$ covariance matrix for $\hat{\eta}_S$ (see Section 7.9.1). This specifies a q-dimensional ellipsoid. To get a 3D ellipsoid using *Arc*, select the "Confidence regions" item from the model menu, and choose three terms. All the points in this plot will be on the surface of a 3D ellipsoid giving a confidence region. A view of this plot for the three variables *WT9*, *LG9*, and *ST9* is shown in Figure 10.7. Since the 3D plot is automatically centered and scaled, about all the user can see in this plot is orientation and elongation. When rotating this plot, one learns that the coefficient estimates for *WT9* and *LG9* are strongly and negatively correlated; the coefficient estimates for *LG9* and *ST9* are nearly uncorrelated, as are the coefficient estimates for *WT9* and *ST9*.

10.9   COMPLEMENTS

10.9.1   Missing Data

The practical expedient of deleting cases with some values of the predictor or the response missing is but one method of dealing with missing values. This can be a reasonable approach in many problems, as long as the cause of the missing values is independent of the values themselves. For example, in the sleep data if *SWS* is unobserved only because its value is very small and therefore difficult to observe, simply deleting cases with *SWS* unobserved will lose information and can result in biases. In general, dealing with missing data is a hard problem, and each problem may call for different solutions. Little and Rubin (1987) provide a summary of the literature on analysis of problems with missing data.

In *Arc*, you can force the use of only the same fully observed cases on all computations by adding the line delete-missing = t to the data file, as discussed in Appendix A, Section A.5.9.

10.9.2   Causation, Association, and Experimental Designs

When can we infer causation? A useful distinction is between *observational studies* and *randomized designs*. In an observational study, we collect a sample of observed variables, as in the sleep data presented in Section 10.4. In observational studies we can rarely declare causation because other factors may be the causes of both the response and the predictors. This is not always the case: The transaction data are also observational, but inferring causation is logically acceptable.

In a randomized study, we set the values of some of the predictors by assigning treatments to units using a randomization scheme. On average, treated and untreated units differ only because of the added treatment effect, so a regression coefficient attached to a treatment effect can generally be interpreted as a causal effect. For example, consider a cloud seeding experiment in which days are randomly assigned to be seeded or not. If the conditional distribution of rainfall given the seeding indicator and other predictors depends on the seeding indicator, we might be able to infer that seeding causes a change in rainfall.

Berk and Freedman (1995) provide an interesting discussion of when statistical models fit to a data set can be generalized to a larger population, particularly in the context of social justice data.

10.9.3   Net Effects Plots

Added-variable plots will sometimes display the *net effects* of adding a term to a regression equation; discussion is given by Cook (1998b, Chapter 13).

10.9.4 References

The sleep data are from Allison and Cicchetti (1976). The data in Problem 10.1 are from Hald (1960). The data in Problem 10.11 were provided by Paul Weibel. The data in Problem 10.12 are from Samaniego and Watnik (1997).

PROBLEMS

10.1 The data in the file `hald.lsp` give values of the hardness y of $n = 13$ Portland cement mixtures consisting of varying amounts—x_1, x_2, x_3, and x_4—of four chemicals. Define several additional terms, as follows:

$$S_1 = x_1 + x_3$$
$$D_1 = x_1 - x_3$$
$$S_2 = x_2 + x_4$$
$$D_2 = x_2 - x_4$$

10.1.1 Draw the scatterplot matrix for y and the four predictors x_1, x_2, x_3, and x_4, and present a qualitative summary of the information it contains.

10.1.2 Using OLS, fit four multiple linear regression models, all with y as the response variable and with predictors given in the order specified by

$$L1: \quad x_1, x_2, x_3, x_4$$
$$L2: \quad S_1, S_2, D_1, D_2$$
$$L3: \quad D_1, D_2, x_3, x_4, x_1, x_2$$
$$L4: \quad D_1, D_2, x_1, x_2, x_3, x_4$$

  a. In the fit of L3, some of the coefficient estimates are labeled as *aliased*. Explain what this means.

  b. What aspects of the fitted regressions are all the same? What aspects are different?

  c. Why is the coefficient estimate for x_2 in L1 equal to the coefficient estimate for D_2 in L3?

  d. Show how each of the coefficient estimates in L2 could be obtained from the coefficient estimates in L1.

  e. Describe how an added-variable plot for x_1 in L3 would look.

10.2   Consider a multiple linear regression with mean function

$$E(y \mid \mathbf{x}) = \eta_0 + \eta_1 u_1 + \eta_2 u_2$$

with terms $u_1 = x_1$ and $u_2 = x_2$.

10.2.1   Suppose we fit the regression, but with terms given by u_1 and $u_3 = \hat{e}(x_2 \mid x_1)$, the residuals from the simple linear regression of x_2 on x_1, so the mean function is

$$E(y \mid \mathbf{x}) = \gamma_0 + \gamma_1 u_1 + \gamma_3 u_3$$

Find expressions for γ_0 and γ_1 in terms of η_0 and η_1.

10.2.2   Show that the sample correlation between u_1 and u_3 is zero and hence the OLS estimates of η_2 and γ_3 are identical. (*Hint*: This is very similar to Problem 6.15.2.) This also shows that $\hat{e}(y \mid x_1) = y - \hat{\gamma}_0 - \hat{\gamma}_1 x_1$.

10.2.3   The added-variable plot for u_2 is the plot of $\hat{e}(y \mid u_1)$ versus $\hat{e}(u_2 \mid u_1)$. For the regression of $\hat{e}(y \mid u_1)$ on $\hat{e}(u_2 \mid u_1)$, use the results of Problems 10.2.2 and 10.2.3 to show that (1) the intercept is zero; (2) the slope is $\hat{\eta}_2$, and the residuals are the same as $\hat{e}(y \mid u_1, u_2)$.

10.3

10.3.1   Verify numerically that in the transactions data, the estimated slope in the added-variable plot for T_2 is the same as $\hat{\eta}_2$, the coefficient for T_2 in the multiple linear regression model. Also, verify that the intercept in this plot is zero.

10.3.2   Let t^* be the t-statistic for testing the hypothesis that the slope in the added-variable plot is zero, and let t_2 be the t-statistic for testing $\eta_2 = 0$ in the original regression model. Show that the values of these two statistics are different. Then, show that $t_2 = t^* \sqrt{(n-k)/(n-2)}$, where k is the number of terms in the mean function. The constant corrects for df in the estimates of variance.

10.4   This is a continuation of Problem 7.3 using data from the Berkeley Guidance Study for girls, in the file BGSgirls.lsp.

10.4.1   Describe the population for which these data are relevant. Do you think the results from analysis of these data can be applied to girls born in Berkeley in 1998–1999? Why or why not?

10.4.2   Consider the following linear regression model:

$$HT18 \mid \mathbf{x} = \eta_0 + \eta_1 WT2 + \eta_2 HT2 + \eta_3 WT9$$

$$+ \eta_4 HT9 + \eta_5 LG9 + \eta_6 ST9 + e \qquad (10.26)$$

with $\text{Var}(e \mid \mathbf{x})$ assumed to be constant. Give estimates and standard errors for all coefficients, and the value of the proportion of variability explained. Give the *units* of each of the estimates.

10.4.3 If *HT9* were changed from cm to inches, how would $\hat{\eta}_4$ and $\text{se}(\hat{\eta}_4)$ change? How would R^2, $\hat{\sigma}^2$, the other elements of $\hat{\eta}$, and the overall F statistic change?

10.4.4 If *HT18* were changed from cm to inches, how would $\hat{\eta}_4$ and $\text{se}(\hat{\eta}_4)$ change? How would R^2, $\hat{\sigma}^2$, the other elements of $\hat{\eta}$, and the overall F statistic change?

10.4.5 If each predictor were centered to have average zero, what would be the estimate of η?

10.4.6 Construct added variable plots for *HT2* and *HT9*. Briefly describe the information these plots contain. The sample correlations between the variables on the axes of these plots are sample partial correlations. What are these sample partial correlations measuring?

10.5 Continuing with the data from the Berkeley Guidance Study for girls, obtain the OLS fit of the model

$$HT18 \mid \mathbf{x} = \eta_0 + \eta_1 WT9 + \eta_2 HT9 + e$$

and notice that $\hat{\eta}_1$ is negative. On the other hand, the sample correlation coefficient between *HT18* and *WT9* is positive, which seems to agree with intuition but is contradicted by $\hat{\eta}_1$. Recall that η_1 can be interpreted as the change in $\text{E}(HT18 \mid \mathbf{x})$ when *HT9* is fixed and *WT9* is increased by one unit.

Construct two plots, a scatterplot of *HT18* versus *WT9* and a histogram of *HT9*. The scatterplot shows a positive correlation as mentioned before. Now, use the histogram to slice *HT9* and observe the corresponding highlighted points in the scatterplot. Repeat this operation for several different slices.

Explain how this demonstration helps us understand the estimate of η_1 and the partial correlation between *HT18* and *WT9* adjusted for *HT9*.

10.6 Refer to model (10.26).

10.6.1 Construct the sample partial correlation coefficient between *LG9* and *WT9* given fixed values of the remaining terms in (10.26). Describe what this partial correlation means and how it differs from the correlation between *LG9* and *WT9*.

10.6.2 What is the relationship between the sample partial correlation coefficient between *LG9* and *WT9* and a joint confidence region for their coefficients in model (10.26)?

10.6.3   What is the relationship between the sample partial correlation coefficient between *LG9* and *WT9* and a 3D added-variable plot for $(LG9, WT9)$?

10.7   In this problem we continue with the data from the Berkeley Guidance Study for girls, but now change the response to somatotype, *Soma*. Somatotype is an index of body type on a seven-point scale, with value one for the most slender body type and the value seven for the least slender. This measurement was taken from photographs at age 18.

10.7.1   Obtain the OLS fit of the model

$$Soma \mid \mathbf{x} = \eta_0 + \eta_1 WT2 + \eta_2 WT9 + \eta_3 WT18 + e$$

The *p*-value for the coefficient of *WT2* is about 0.065, which is at least suggestive of a significant result. Because the regression coefficient for *WT2* is negative, the results seem to indicate that an overweight two year old will tend to be a relatively slender teenager, a conclusion that doesn't make much sense.

10.7.2   Refit the model after reparameterizing to the three predictors *WT2*, $DW9 = WT9 - WT2$, and $DW18 = WT18 - WT9$. Do the results in terms of the new predictors (weight changes) make more sense than the results in terms of the original predictors? Why?

10.8   This problem uses the sleep data in the file `sleep.lsp`.

10.8.1   Obtain the added-variable plot for $\log_2(Life)$ using mean function (10.12). Three points appear at the extreme right of this plot. To what species do these correspond? How does the added-variable plot change if (1) all three are removed from the data or (2) only the two most extreme cases are removed from the data?

10.8.2   Construct a 3D added-variable plot for $\log_2(BW)$ and $\log_2(BrW)$. Give a brief description of what you think is important in the plot.

10.9   In Problem 7.7 we considered predicting the area of a soybean leaf from its length and width using the model

$$\log(Area) \mid (Length, Width) = \eta_0 + \eta_1 \log(Length) + \eta_2 \log(Width) + e$$

$$(10.27)$$

10.9.1   Construct a 95% confidence region for (η_1, η_2), and give a careful interpretation of its meaning.

10.9.2 Based only on your confidence region for (η_1, η_2), can you tell if log(*Length*) and log(*Width*) are positively correlated or negatively correlated? Why?

10.9.3 Based on the description of the data collection given in Problem 7.7, can the results of this experiment be applied to a larger population? What is that population and why?

10.10 Run the demo-cr.lsp demonstration as described in Section 10.8.1. Then, change the aspect ratio in the plot by making the window twice as long as it is high, and run the demonstration again. Describe any qualitative differences. Since the data have not changed, there are no *quantitative* differences, but perception of the plot can change.

10.11 The file rat.lsp contains data from an experiment on 19 rats that were given a dose of a drug, the dose being roughly in proportion to the body weight of the animal. At the end of the experiment, each rat was sacrificed, its liver was weighed and the amount of drug in the liver was determined. The experimenters believed that the amount recovered, *y*, should be independent of the three predictors, *Dose*, *BodyWt*, and *LiverWt*.

10.11.1 Fit the OLS regression of *y* on the three predictors. Two of the predictors have relatively large *t*-values. What does this indicate? Get the correlation between these two estimated coefficients by selecting "Display variances" from the regression model menu. Guess the shape of the joint confidence region for these two coefficients.

10.11.2 Draw the joint confidence region for the two coefficients with large *t*-values. Is the shape as you expected it to be?

10.11.3 Without removing the plot of the confidence region, draw a 3D added-variable plot for the two predictors identified. As you rotate this plot, examine it for any features that may help understand the plot of the confidence region. Using the "O to e(O|H)" plot control may be helpful.

10.11.4 In the 3D added-variable plot, one point should have been identified as different from the others. How is it different? Select this point, and only this point, and then delete it by using the "Case deletions" plot control. Not only is this plot updated, but the plot of the confidence region is updated as well. On the confidence region plot, the new ellipsoid and cross hairs are drawn in a different color, but the old ones are not removed. How does the confidence region change? What does this tell you about the case that was deleted?

10.12   The data in the file `baseball.lsp` give *Wins*, the number of games won, *Payroll*, the total payroll for all players, *Pitchpay*, the payroll for pitchers, and *Hitpay*, the payroll for other players, for all major league baseball teams in the 1995 baseball season.

10.12.1   Examine the 3D plot of *Wins* versus (*Pitchpay*,*Hitpay*), and summarize the information in this plot. Which teams appear to have done well (that is, won more games than might be expected for their payroll)? Which did poorly (that is, won fewer games than might be expected for their payroll)?

10.12.2   Consider fitting regression models, all with constant variance function and the three mean functions:

$$E(Wins \mid \mathbf{x}) = \eta_0 + \eta_1 Payroll \qquad (10.28)$$

$$E(Wins \mid \mathbf{x}) = \eta_0 + \eta_2 Pitchpay \qquad (10.29)$$

$$E(Wins \mid \mathbf{x}) = \eta_0 + \eta_2 Pitchpay + \eta_3 Hitpay \qquad (10.30)$$

Under what conditions will the fit of (10.28) give the same R^2 as the fit of (10.30)?

10.12.3   Given that *Payroll* = *Pitchpay* + *Hitpay*, without doing any calculations, is it possible for (10.29) to fit better than (10.28)? Why or why not? Fit both models to confirm your answer.

10.12.4   In the fit of (10.30), examine the added-variable plots for *Pitchpay* and *Hitpay* and summarize the information they contain.

10.12.5   In the fit of (10.29), interpret the value of $\hat{\eta}_2$.

10.12.6   These models provide a description of the dependence of wins on baseball salaries, but probably cannot be used for prediction of future values. (That is, if the Minnesota Twins were to increase their pitcher's salaries by 10 million dollars, this model could not be used to predict the additional number of games won.) Explain why this is so.

CHAPTER 11

Relating Mean Functions

In this chapter we continue to assume that the multiple linear regression model with mean function $E(y \mid \mathbf{x}) = \eta^T \mathbf{u}$ is appropriate for a particular problem. But now we begin to think about comparing this mean function to a mean function obtained by constraining some of the coefficients in η to equal specified values, or to equal one another. Models formed in this way are called *submodels* because they are formed by imposing constraints on the regression coefficients in the *full model* $E(y \mid \mathbf{x}) = \eta^T \mathbf{u}$.

For example, here are three ways in which the coefficients $\eta = (\eta_0, \eta_1, \eta_2)^T$ in a regression with three terms could be constrained to yield a submodel mean function:

$$\text{submodel } \eta = \begin{pmatrix} \eta_0 \\ \eta_1 \\ 0 \end{pmatrix} \quad \text{or} \quad \begin{pmatrix} \eta_0 \\ \eta_1 \\ \eta_1 \end{pmatrix} \quad \text{or} \quad \begin{pmatrix} 5 \\ \eta_1 \\ \eta_1 \end{pmatrix}$$

In the first case, η_2 is constrained to equal zero. This is the same as deleting the last term from the full mean function. In the second case, the last two coefficients are set equal. This is equivalent to replacing u_1 and u_2 in the full mean function with their sum,

$$\eta_0 + \eta_1 u_1 + \eta_1 u_2 = \eta_0 + \eta_1 (u_1 + u_2)$$

In the final case, the first coefficient is set equal to five and the last two coefficients are again constrained to be equal.

We begin our study of submodel mean functions in the next section, focusing on the removal of terms from the full model.

11.1 REMOVING TERMS

Suppose we have a linear regression model with mean function

$$E(y \mid \mathbf{u}) = \eta_1^T \mathbf{u}_1 + \eta_2 u_2 \tag{11.1}$$

where $\mathbf{u}_1$ consists of $k - 1$ terms, and u_2 is just one term. The k terms $(\mathbf{u}_1, u_2)$ are functions of the p predictors $\mathbf{x}$; but for the calculations of this section, conditioning on $\mathbf{u}$ is sufficient.

Assume that the variance function $\text{Var}(y \mid \mathbf{x}) = \text{Var}(y \mid \mathbf{u}) = \sigma^2$ is constant. What, if anything, can be said about the regression of y on $\mathbf{u}_1$, the regression in which u_2 is dropped from consideration?

This question can be answered with the assistance of two fundamental results that we previewed in Section 2.5 for a special case. Here we state the results in general.

11.1.1   Marginal Mean Functions

Suppose we know that the mean function for the regression of y on $(\mathbf{u}_1, u_2)$ is given by (11.1). Then the mean function for the regression of y on $\mathbf{u}_1$ alone is given by

$$E(y \mid \mathbf{u}_1) = E[E(y \mid \mathbf{u}_1, u_2) \mid \mathbf{u}_1] \tag{11.2}$$

$$= E[(\eta_1^T \mathbf{u}_1 + \eta_2 u_2) \mid \mathbf{u}_1]$$

$$= \eta_1^T \mathbf{u}_1 + \eta_2 E(u_2 \mid \mathbf{u}_1) \tag{11.3}$$

Equation (11.2) is the general result that shows how to get a mean function based on fewer terms. Equation (11.3) is the specific result for the multiple linear regression model: We replace u_2 by the mean function from the regression of u_2 on $\mathbf{u}_1$.

For example, suppose that the true mean function for the regression of a response y on two predictors x_1 and x_2 is

$$E(y \mid \mathbf{x}) = 1 + 2x_1 + 3x_2 \qquad \text{with} \quad \text{Var}(y \mid \mathbf{x}) = \sigma^2$$

What can we say about the regression of y on x_1 alone? From (11.3), the mean function is

$$E(y \mid x_1) = 1 + 2x_1 + 3E(x_2 \mid x_1) \tag{11.4}$$

and this depends on the regression of x_2 on x_1. There are a number of important special cases. If x_2 and x_1 are independent, then $E(x_2 \mid x_1) = E(x_2)$ which is a constant. Substituting this into (11.4) gives

$$E(y \mid x_1) = (1 + 3E(x_2)) + 2x_1$$

This is the mean function for the simple regression model with terms $u_0 = 1$ and $u_1 = x_1$. The regression coefficient for x_1 is the same whether x_2 is included or not, but the value of the intercept changes from 1 to $1 + 3E(x_2)$.

Suppose next that the mean function for the regression of x_2 on x_1 is linear, $E(x_2 \mid x_1) = \alpha_0 + \alpha_1 x_1$. Substituting this into (11.4) gives

$$E(y \mid x_1) = (1 + 3\alpha_0) + (2 + 3\alpha_1)x_1$$

so we again get a simple linear regression mean function with terms $u_0 = 1$ and $u_1 = x_1$, but the values of both the intercept and the slope have changed. If α_1 was a sufficiently large negative number, the sign of the coefficient for x_1 could even become negative.

Finally, suppose the mean function for the regression of x_2 on x_1 is a nonlinear function, say $E(x_2 \mid x_1) = \alpha_0 + \alpha_1 \exp(\alpha_2 x_1)$. The mean function for the regression of y on x_1 becomes

$$E(y \mid x_1) = (1 + 3\alpha_0) + 2x_1 + 3\alpha_1 \exp(\alpha_2 x_1)$$

Because the parameters α_1 and α_2 do not enter this mean function linearly, this is a *nonlinear* mean function and not a *linear* mean function.

A regression model based on a subset of terms depends on the distribution of the terms, which in turn depends on the distribution of the predictors. If the terms are independent or linearly related, then submodel mean functions generally have the same form as the full mean function. If the terms are dependent but not linearly related, anything can happen, and the regression for the full model may tell us little about the regression for a submodel. These comments apply whenever the term deleted has a nonzero regression coefficient η_2 in (11.1). If the coefficient is zero, then deleting the term has no effect on the mean function.

11.1.2 Marginal Variance Functions

A formula that parallels (2.3) tells us how the marginal variance function for the regression of y on $\mathbf{u}_1$ is related to the variance function for the regression of y on both $\mathbf{u}_1$ and u_2:

$$\mathrm{Var}(y \mid \mathbf{u}_1) = E[\mathrm{Var}(y \mid \mathbf{u}_1, u_2) \mid \mathbf{u}_1] + \mathrm{Var}[E(y \mid \mathbf{u}_1, u_2) \mid \mathbf{u}_1] \qquad (11.5)$$

$$= E[\sigma^2 \mid \mathbf{u}_1] + \mathrm{Var}[(\eta_1^T \mathbf{u}_1 + \eta_2 u_2) \mid \mathbf{u}_1] \qquad (11.6)$$

$$= \sigma^2 + \eta_2^2 \mathrm{Var}(u_2 \mid \mathbf{u}_1) \qquad (11.7)$$

Equation (11.5) is the general result. The second equality (11.6) is obtained by substituting $E(y \mid \mathbf{u})$ and $\mathrm{Var}(y \mid \mathbf{u})$ as specified by the model. Dropping a term u_2 may result in a regression with a nonconstant variance function if $\mathrm{Var}(u_2 \mid \mathbf{u}_1)$ is not constant, even if the variance function in the full model is constant.

11.1.3 Example

Imagine a regression intended to describe the dependence of salary of faculty members in a college on their sex (0 = male and 1 = female) and on the number of years of service. Suppose we have the following two mean functions:

$$E(Salary \mid Sex, Years) = 18065 + 201 Sex + 759 Years$$

$$E(Salary \mid Sex) = 24697 - 3340 Sex$$

In the mean function with two predictors, we see that female faculty members earn about \$201 per year more on average than male faculty members with the same number of years of service. However, if we remove *Years* from the problem, then the coefficient for *Sex* is negative, indicating that females earn \$3340 less than men, on the average. How can both these mean functions be true at the same time?

The results of this section give us the answer: If the mean function for the regression of *Years* on *Sex* is linear, with a negative slope, we can get both the mean functions. In particular, if

$$E(Years \mid Sex) = (6632/759) - (3541/759)Sex$$

then from (11.3) both equations can be seen to hold. The change in sign of the coefficient is due to females having fewer years of service on average than males. Data on these three variables and a few others for a small midwestern college are given in the file `salary.lsp`. The mean functions used in this example are essentially the same as those obtained from the actual data.

11.2 TESTS TO COMPARE MODELS

The results of the previous section show that dropping a term from a regression can change both the mean function and the variance function in important ways, unless the value of the regression coefficient for that term is $\eta_2 = 0$. Consequently, testing η_2 equal to zero, or more generally testing any subset of the regression coefficients to be zero, plays a useful role in many analyses. The paradigm for F-tests described in Sections 6.6 and 7.6.5 can be applied to compare any submodel with a full model. This leads to the descriptive hypotheses

NH: Submodel mean function applies with $\mathrm{Var}(y_i \mid \mathbf{x}_i) = \sigma^2/w_i$
AH: Full model mean function applies with $\mathrm{Var}(y_i \mid \mathbf{x}_i) = \sigma^2/w_i$

where the weights w_i are known. Recalling notation from Section 6.6, let RSS_{NH} and RSS_{AH} be the residual sum of squares from the fits of the mean

TABLE 11.1 Highway Accident Data

Predictor	Term	Description
Rate	$\log_2(Rate)$	1973 accident rate per million vehicle miles.
Len	$\log_2(Len)$	Length of segment in miles.
ADT	$\log_2(ADT)$	Average daily traffic count in thousands.
Trks	$\log_2(Trks)$	Truck volume as a percent of total volume.
Slim	*Slim*	Speed limit (in 1973, speeds above 55 mph were permitted).
Sigs	$(Sigs + 1)^{-1}$	Number of signalized interchanges per mile.
Acpt	$\log_2(Acpt)$	Number of access points per mile.

functions, and let df_{NH} and df_{AH} be the df associated with each of the residual sum of squares. Since the residual sum of squares measures lack of fit, we compare the fits by looking at the difference $RSS_{NH} - RSS_{AH}$. If this number is small, then the two mean functions fit about equally well; and if it is significantly large, then the constraints imposed to get the submodel are contradicted by the data. In other words, the submodel fails to describe significant systematic trends in the data.

If $y \mid \mathbf{x}$ is normally distributed, then the scaled value of the difference provides a test statistic. We compute

$$F = \frac{(RSS_{NH} - RSS_{AH})/(df_{NH} - df_{AH})}{\hat{\sigma}^2} \tag{11.8}$$

where $\hat{\sigma}^2$ is the estimate of σ^2 from the fit of the full model. Large values of F provide evidence against the null hypothesis and in favor of the alternative hypothesis. Significance levels for this test are obtained by comparing the observed value of F to the $F_{df_{NH}-df_{AH}, df_{AH}}$ distribution.

11.3 HIGHWAY ACCIDENT DATA

As an example, we consider data relating automobile accident rates, in accidents per million vehicle miles, to six potential predictors, as described in Table 11.1, for 39 segments of Minnesota highways. The data are available in the file highway.lsp.

A reasonably good full model must be available before considering issues of comparing models. The model we developed has response $y = \log_2(Rate)$ and the six terms shown in Table 11.1 in addition to the intercept. The terms are all power transformations of the predictors. They were chosen, in part, to make the transformed predictors linearly related. According to the results of Section 11.1, the study of mean functions is more straightforward when the predictors are linearly related. To allow for transformation of *Sigs*, which has many

TABLE 11.2   Regression Summary for Highway Accident Data

```
Normal Regression
Kernel mean function = Identity
Response = log2[Rate]
Terms = (log2[Acpt] log2[Trks] log2[ADT] log2[Len] Sigs1^-1
        Slim)
Coefficient Estimates
Label             Estimate        Std. Error       t-value
Constant           5.66197         1.38713           4.082
log2[Acpt]         0.136891        0.109419          1.251
log2[Trks]        -0.258160        0.221716         -1.164
log2[ADT]         -0.0492037       0.0539646        -0.912
log2[Len]         -0.275680        0.0921348        -2.992
Sigs1^-1          -0.539088        0.361623         -1.491
Slim              -0.0346930       0.0169035        -2.052

R Squared:                         0.719392
Sigma hat:                         0.385543
Number of cases:                        39
Degrees of freedom:                     32

Summary Analysis of Variance Table
Source       df        SS           MS            F        p-value
Regression    6     12.1944      2.03241        13.67      0.0000
Residual     32      4.7566      0.148644
```

zero values, the new variate $Sigs1 = Sigs + 1$ was used. As in other problems, we used base-two logarithms to facilitate interpretation. The specific methods and ideas we used in the development of this full model are discussed in Part III of this book.

The fitted OLS mean function for the full model

$$E(\log_2(Rate) \mid \mathbf{u}) = \eta_0 + \eta_1 \log_2(Acpt) + \eta_2 \log_2(Trks) + \eta_3 \log_2(ADT)$$
$$+ \eta_4 \log_2(Len) + \eta_5 Sigs1^{-1} + \eta_6 Slim \quad (11.9)$$

is shown in Table 11.2. The overall F-test indicates that the response is related to the terms.

Let's first test the hypothesis that $\log_2(Rate)$ does not depend on the two terms $Sigs1^{-1}$ and $Slim$. Both the speed limit and the number of signals can be changed by the state highway department, so the effects of these on accident rate might be of special interest. The reduced mean function is

$$E(\log_2(Rate) \mid \mathbf{u}) = \eta_0 + \eta_1 \log_2(Acpt) + \eta_2 \log_2(Trks)$$
$$+ \eta_3 \log_2(ADT) + \eta_4 \log_2(Len) \quad (11.10)$$

which is obtained from (11.9) by setting $\eta_5 = \eta_6 = 0$. The mean function (11.10) is the submodel mean function, and (11.9) is the full model mean function. From Table 11.2, we find $RSS_{AH} = 4.7566$ and $df_{AH} = 32$. After fitting the submodel mean function (11.10), we find $RSS_{NH} = 5.9649$ and $df_{NH} = 34$. The F statistic can now be computed:

$$F = \frac{(5.9649 - 4.7566)/(34 - 32)}{4.7566/32} = 4.06$$

Using the "Calculate probability" item in the Arc menu, we find that the upper-tail p-value is 0.03, providing evidence against the null hypothesis.

11.3.1 Testing Equality of Coefficients

We use the transactions data (transact.lsp) described in Section 7.3.3 to illustrate the procedure for testing equality of regression coefficients. The full mean function is

$$\mathrm{E}(Time \mid T_1, T_2) = \eta_0 + \eta_1 T_1 + \eta_2 T_2 \qquad (11.11)$$

where η_1 is the average time per type one transactions and η_2 is the average time per type two transactions. We assume as in Problem 9.11, page 228, that the variance function is $\mathrm{Var}(Time \mid T_1, T_2) = \sigma^2(T_1 + T_2)$. Consider testing the null hypothesis that the average time per transactions is the same for each of the two types, $\eta_1 = \eta_2$. Substituting for η_2 in (11.11), we get

$$\mathrm{E}(Time \mid T_1, T_2) = \eta_0 + \eta_1 T_1 + \eta_1 T_2$$
$$= \eta_0 + \eta_1(T_1 + T_2) \qquad (11.12)$$

Mean function (11.12) can be estimated via WLS with terms for the intercept and for $S = T_1 + T_2$, and also using $1/S$ as weights. To fit this mean function in *Arc*, use the "Add a variate" item to create the new term S = T1+T2 and to create the weights, Sinv = 1/S. Fit via WLS to get $RSS_{NH} = 129729$ and $df_{NH} = 259$, while under the alternative mean function (11.11), $RSS_{AH} = 107600$ with $df_{AH} = 258$. Consequently, the F-statistic for testing the null hypothesis (11.12) versus the alternative (11.11) is

$$F = \frac{(129729 - 107600)/(259 - 258)}{107600/258} = 53.06$$

which can be compared to the $F_{1,258}$ distribution to get a p-value. To four decimals, $p = 0.0000$, so we have strong evidence against the claim that the number of minutes per transaction is the same for each transaction type.

11.3.2 Offsets

Continuing with the transactions data and assuming model (11.11), suppose we had reason to believe that the value of η_1 should be five minutes per transaction. Using the t-tests described earlier, we could test the hypothesis that $\eta_1 = 5$ against a general alternative, but we could also use an F-test. The advantage of the F-test is that it can be used to test a variety of hypotheses that cannot be tested with a t-test.

Rewrite the alternative model (11.11) so it incorporates the possibility that $\eta_1 = 5$:

$$\begin{aligned}
\mathrm{E}(Time \mid T_1, T_2) &= \eta_0 + \eta_1 T_1 + \eta_2 T_2 \\
&= \eta_0 + (\eta_1 - 5 + 5)T_1 + \eta_2 T_2 \\
&= \eta_0 + \eta_1^* T_1 + 5T_1 + \eta_2 T_2
\end{aligned}$$

where $\eta_1^* = \eta_1 - 5$. Since the term $5T_1$ includes no unknown paramaters, it can be moved to the left side of the equation and used to *offset* the response,

$$\begin{aligned}
\mathrm{E}(Time - 5T_1 \mid T_1, T_2) &= \eta_0 + \eta_1^* T_1 + \eta_2 T_2 \\
\mathrm{E}(Time^* \mid T_1, T_2) &= \eta_0 + \eta_1^* T_1 + \eta_2 T_2
\end{aligned} \tag{11.13}$$

where the offset response $Time^* = Time - 5T_1$. Testing if $\eta_1 = 5$ in model (11.11) is the same as testing if $\eta_1^* = 0$ in model (11.13). Thus, by offsetting the response, we have reformulated the original model hypotheses

$$\begin{aligned}
\text{NH}: \quad & \mathrm{E}(Time \mid T_1, T_2) = \eta_0 + 5T_1 + \eta_2 T_2 \\
\text{AH}: \quad & \mathrm{E}(Time \mid T_1, T_2) = \eta_0 + \eta_1 T_1 + \eta_2 T_2
\end{aligned}$$

in terms of a comparison of equivalent models

$$\begin{aligned}
\text{NH}: \quad & \mathrm{E}(Time^* \mid T_1, T_2) = \eta_0 + \eta_2 T_2 \\
\text{AH}: \quad & \mathrm{E}(Time^* \mid T_1, T_2) = \eta_0 + \eta_1^* T_1 + \eta_2 T_2
\end{aligned}$$

that involves deleting terms. The models of the two null hypotheses are the same, as are the models of the alternative hypotheses.

Arc allows for offsets without explicit reformulation of the mean function. We again use the transactions data for illustration. Use the "Add a variate" menu item to create the new offset variable off=5*T1. Next, to fit the model of the null hypothesis

$$\mathrm{E}(Time \mid T_1, T_2) = \eta_0 + 5T_1 + \eta_2 T_2$$

use the "Fit linear LS" item and specify the response *Time*, the single predictor T_2, move *off* to the Offset box, and use $1/S$ as weights. From this we find that the residual sum of squares is $RSS_{NH} = 108307$ with $\text{df}_{NH} = 259$. The value of the F-statistic, which is now computed in the usual way, is $F = 1.70$, which gives a p-value of 0.19 when compared to the F-distribution with $(1,258)$ df.

In general, an offset will consist of a *known* linear combination of terms. In this example, the offset is $5T_1$.

11.4 SEQUENTIAL FITTING

The sequential analysis of variance table provides a summary of the information needed to compare any two mean functions obtained by successively adding terms to a *base model*. For example, consider the sequence of mean functions for the highway accident data given by

$$E(\log_2(Rate) \mid \mathbf{u}) = \eta_0 + \eta_4 \log_2(Len) \tag{11.14}$$

$$E(\log_2(Rate) \mid \mathbf{u}) = \eta_0 + \eta_4 \log_2(Len) + \eta_6 Slim \tag{11.15}$$

$$E(\log_2(Rate) \mid \mathbf{u}) = \eta_0 + \eta_4 \log_2(Len) + \eta_6 Slim$$
$$+ \eta_1 \log_2(Acpt) \tag{11.16}$$

$$E(\log_2(Rate) \mid \mathbf{u}) = \eta_0 + \eta_4 \log_2(Len) + \eta_6 Slim$$
$$+ \eta_1 \log_2(Acpt) + \eta_2 \log_2(Trks) \tag{11.17}$$

$$E(\log_2(Rate) \mid \mathbf{u}) = \eta_0 + \eta_4 \log_2(Len) + \eta_6 Slim$$
$$+ \eta_1 \log_2(Acpt) + \eta_2 \log_2(Trks)$$
$$+ \eta_3 \log_2(ADT) \tag{11.18}$$

$$E(\log_2(Rate) \mid \mathbf{u}) = \eta_0 + \eta_4 \log_2(Len) + \eta_6 Slim$$
$$+ \eta_1 \log_2(Acpt) + \eta_2 \log_2(Trks)$$
$$+ \eta_3 \log_2(ADT) + \eta_5 Sigs1^{-1} \tag{11.19}$$

The base mean function (11.14) consists of the intercept and $\log_2(Len)$. Each subsequent submodel mean function is obtained by adding one term. Select the item "Examine submodels" from the menu for the full model (11.9). This will give the dialog shown in Figure 11.1. By default, the names of all the terms in the mean function are shown in the right list, and the button for "Fit in specified order" is pushed. Move the term $\log_2(Len)$ to the base model in the left list. Often, the base model will have no terms other than the intercept, in which case no terms need to be moved. You can also change the order of fitting terms by moving the term names between the two lists.

Table 11.3 summarizes the sequential analysis of variance. The two columns under the heading Total give the df and RSS for each of the six mean functions. The labels tell us about the order of the sequential fitting: The first row is

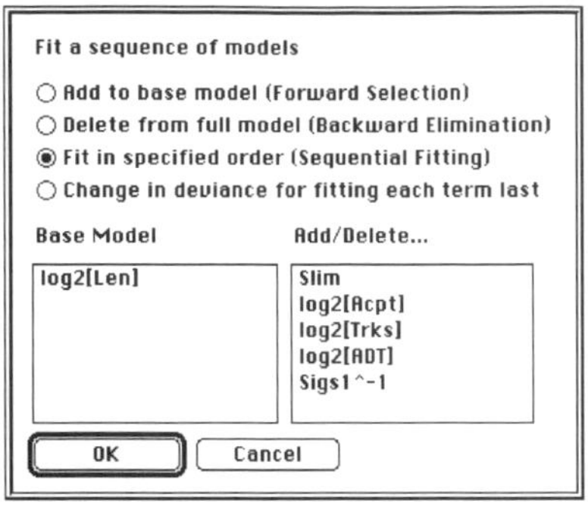

Fit a sequence of models

○ Add to base model (Forward Selection)
○ Delete from full model (Backward Elimination)
◉ Fit in specified order (Sequential Fitting)
○ Change in deviance for fitting each term last

Base Model   **Add/Delete...**

log2[Len]   Slim
                            log2[Acpt]
                            log2[Trks]
                            log2[ADT]
                            Sigs1^-1

[OK]   [Cancel]

FIGURE 11.1 The dialog for fitting a sequence of mean functions based on subsets of terms. The option "Fit in specified order" is used to fit a sequential analysis of variance.

TABLE 11.3   A Sequential Analysis of Variance Table for the Sleep Data

```
Sequential Analysis of Variance
All fits include an intercept.
Base model = (Ones log2[Len])
```

		Total			Change	
Predictor	df	RSS	|	df	RSS	MS
Base Model	37	11.4138	|			
Slim	36	6.11216	|	1	5.30162	5.30162
log2[Acpt]	35	5.49940	|	1	0.612764	0.612764
log2[Trks]	34	5.09933	|	1	0.400072	0.400072
log2[ADT]	33	5.08693	|	1	0.0123937	0.0123937
Sigsl^-1	32	4.75660	|	1	0.330335	0.330335
Residual			|	32	4.7566	0.148644

from fitting the base model, *Slim* was added for the second row, $\log_2(Acpt)$ was added for the third row, and so on. The remaining three columns are derived from these to facilitate some computations: the change in df, *RSS*, and *MS* (mean square) between adjacent mean functions. The *RSS* for the mean function with all the terms is repeated as the last row of the change columns. The information in this table can be used to compare the fits of any two mean functions in (11.14)–(11.19) by using the *F*-statistic in (11.8). Tests for comparing other mean functions can be constructed by changing the order in which terms enter the sequential analysis of variance dialog of Figure 11.1.

11.5   SELECTING TERMS

A standard problem in regression analysis is deciding which terms should appear in the mean function. Sometimes, we will have information to help make decisions. In the transactions data, for example, we used the mean function

$$E(\textit{Time} \mid T_1, T_2) = \eta_0 + \eta_1 T_1 + \eta_2 T_2$$

As long as the average time for a transaction is the same in each branch office, and the fixed time not spent on transactions is the same in each branch, this mean function should be correct. In other problems, we may not be so sure. In the highway accident data, several potential predictors and terms derived from them are available, but we don't have a theory that tells us the functional form of the mean function. The analyst generally faces a number of challenges: (1) What are the relevant predictors? and (2) How should the predictors be combined into terms? These concerns will be addressed in Part III of this book. For now, we consider a more limited issue. Suppose that we have a linear regression model with the usual variance function $\mathrm{Var}(y \mid \mathbf{x}) = \sigma^2/w$ and k terms in the mean function

$$E(y \mid \mathbf{x}) = \eta^T \mathbf{u} \tag{11.20}$$

Throughout this section, *we assume that this mean function is correct.* Our goal is to find which, if any, terms can be deleted from the mean function without important loss of information. A little additional notation will help. Let $\mathcal{I}$ be the subset of term indices that we will keep in the mean function, and let $\mathcal{D}$ be the complementary subset of indices of terms that we will consider deleting from the mean function. The full mean function (11.20) can now be reexpressed as

$$E(y \mid \mathbf{x}) = \eta_{\mathcal{I}}^T \mathbf{u}_{\mathcal{I}} + \eta_{\mathcal{D}}^T \mathbf{u}_{\mathcal{D}}$$

where $\eta_{\mathcal{I}}$ and $\eta_{\mathcal{D}}$ denote the subvectors of η that correspond to the term indices in $\mathcal{I}$ and $\mathcal{D}$. The mean function for the submodel contains just the $\mathcal{I}$ terms,

$$E_{\mathcal{I}}(y \mid \mathbf{x}) = \eta_{\mathcal{I}}^T \mathbf{u}_{\mathcal{I}} \tag{11.21}$$

Here the subscript on $E_{\mathcal{I}}$ is intended as a reminder that this is the hypothesized mean function for the submodel which may or may not be correct. For example, if we start with the five-term mean function

$$E(y \mid \mathbf{x}) = \eta_0 + \eta_1 u_1 + \eta_2 u_2 + \eta_3 u_3 + \eta_4 u_4$$

we might consider deleting u_2 and u_4. In this case, $\mathbf{u}_{\mathcal{D}} = (u_2, u_4)^T$, $\mathbf{u}_{\mathcal{I}} = (1, u_1, u_3)^T$, $\eta_{\mathcal{D}} = (\eta_2, \eta_4)^T$, and $\eta_{\mathcal{I}} = (\eta_0, \eta_1, \eta_3)^T$.

The issue now is how to assess the relative advantages of the full mean function (11.20) and the submodel mean function (11.21). We do this using the appropriate WLS or OLS fitted values from the two models. Let $\hat{y}_i$ and $\hat{y}_{i,\mathcal{I}}$ denote the fitted values from the full model and the submodel, $i = 1, \ldots, n$. Because the full model is assumed to be correct, its fitted values are unbiased in the sense that the expectation of the ith fitted value is equal to the corresponding point on the mean function, $E(\hat{y}_i) = E(y \mid \mathbf{x}_i)$. The same may or may not be true of the submodel. If $\boldsymbol{\eta}_D = \mathbf{0}$, then the fitted values from the submodel are also unbiased, $E(\hat{y}_{i,\mathcal{I}}) = E(y \mid \mathbf{x}_i)$. But if $\boldsymbol{\eta}_D \neq \mathbf{0}$, the fitted values from the submodel will be biased, $E(\hat{y}_{i,\mathcal{I}}) \neq E(y \mid \mathbf{x}_i)$. Bias may be tolerable if it is not too large. To judge the worth of the submodel, we investigate the *mean squared error* (MSE) of a fitted value from the submodel

$$\text{MSE}(\hat{y}_{i,\mathcal{I}}) = E\{(\hat{y}_{i,\mathcal{I}} - E(y_i \mid \mathbf{x}_i))^2\}$$

This measures how close the ith fitted value from the submodel is to the corresponding point on the true mean function, on the average. We would like this number to be small. To see what controls the size of $\text{MSE}(\hat{y}_{i,\mathcal{I}})$ consider the decomposition,

$$\text{MSE}(\hat{y}_{i,\mathcal{I}}) = \text{Var}(\hat{y}_{i,\mathcal{I}}) + \{E(\hat{y}_{i,\mathcal{I}}) - E(y \mid \mathbf{x}_i)\}^2$$

$$= \text{variance} + \{\text{bias}\}^2$$

The mean squared error of $\hat{y}_{i,\mathcal{I}}$ is the sum of its variance and its squared bias. We could make $\text{MSE}(\hat{y}_{i,\mathcal{I}})$ small if we could reduce its bias and its variance. Using the results in Section 10.1.5, deleting terms from a model reduces the variances of the fitted values; the more terms we delete, the more we reduce variance. Deleting terms from a model with nonzero coefficients increases the bias of the fitted values; the more terms we delete, the more we increase bias. Good submodels compromise between deleting terms to reduce variability and including terms to reduce bias.

Dropping a term can be particularly helpful when it is nearly a linear combination of other terms in the model. We have seen previously in (10.5) that the variance of a coefficient estimate from the full model is

$$\text{Var}(\hat{\eta}_j) = \frac{\sigma^2}{(n-1)\text{sd}_j^2} \times \frac{1}{1 - R_j^2}$$

where sd_j is the sample standard deviation of the jth term, and R_j^2 is the R^2 value for the linear regression of the jth term on all the other terms. If R_j^2 is close to one, then $\text{Var}(\hat{\eta}_j)$ will be large, and $\text{Var}(\hat{y}_i)$ will also be large because it is a function of the variances of coefficient estimates (see Section 7.6.4). Consequently, including highly correlated terms in a regression model can inflate the variances of estimates, fitted values and of predictions.

11.5.1 Criteria for Selecting Submodels

Most subset selection methods try to select a submodel that results in relatively small values of the sum of the mean squared errors. For a submodel $\mathcal{I}$, this *total mean squared error* can be expressed as

$$J_{\mathcal{I}} = \sum_{i=1}^{n} \mathrm{MSE}(\hat{y}_{i,\mathcal{I}})$$

$$= \sum_{i=1}^{n} \{ \mathrm{Var}(\hat{y}_{i,\mathcal{I}}) + \mathrm{bias}^2(\hat{y}_{i,\mathcal{I}}) \}$$

The total mean squared error $J_{\mathcal{I}}$ has a variance component and a bias component, and the selected subset must balance these.

The parameter $J_{\mathcal{I}}$ is unknown and therefore must be estimated for each $\mathcal{I}$. One useful estimate of $J_{\mathcal{I}}$, which is called *Mallows' $C_{\mathcal{I}}$* statistic, is given by

$$C_{\mathcal{I}} = \frac{RSS_{\mathcal{I}}}{\hat{\sigma}^2} + 2k_{\mathcal{I}} - n \tag{11.22}$$

$$= \frac{RSS_{\mathcal{I}} - RSS}{\hat{\sigma}^2} + k_{\mathcal{I}} - (k - k_{\mathcal{I}}) \tag{11.23}$$

$$= (k - k_{\mathcal{I}})(F_{\mathcal{D}} - 1) + k_{\mathcal{I}} \tag{11.24}$$

where the subscript $\mathcal{I}$ refers to statistics computed from the submodel, while statistics without subscripts are computed from the full model. Also, $k_{\mathcal{I}}$ is the number of terms in the submodel (which is the same as the number of indices in $\mathcal{I}$), k is the number of terms in the full model, and $F_{\mathcal{D}}$ is the F-statistic for testing

$$\mathrm{NH}: \qquad \mathrm{E}(y \,|\, \mathbf{x}) = \eta_{\mathcal{I}}^T \mathbf{u}_{\mathcal{I}}$$

$$\mathrm{AH}: \qquad \mathrm{E}(y \,|\, \mathbf{x}) = \eta^T \mathbf{u} = \eta_{\mathcal{I}}^T \mathbf{u}_{\mathcal{I}} + \eta_{\mathcal{D}}^T \mathbf{u}_{\mathcal{D}}$$

Good candidates for submodel mean functions will have $C_{\mathcal{I}}$ nearly equal to or less than $k_{\mathcal{I}}$. When applied to the full model, $C_{\mathcal{I}} = k_{\mathcal{I}} = k$. Thus the full model is always judged to be a good candidate by this general rule. Use of $C_{\mathcal{I}}$ will often lead to a number of candidate models for further study rather than one obvious best choice. In many problems, it is best to use $C_{\mathcal{I}}$ to screen out the worst models, rather than to find a single best model.

Equation (11.24) shows that $C_{\mathcal{I}} \le k_{\mathcal{I}}$ if and only if $F_{\mathcal{D}} \le 1$. Thus, it is generally good to delete sets of terms $\mathcal{D}$ when the F-statistic $F_{\mathcal{D}}$ is less than one.

11.5.2 Stepwise Methods

The $C_{\mathcal{I}}$ statistic is easy to use when there are only a few subsets $\mathcal{D}$ to consider for deletion. When there are no obvious first candidate subsets for deletion, we might attempt to find better models by using an automated computer search of all possible subsets. With k terms, there are 2^k possible submodels. If k is small, say ten or less, then this is a manageable problem, with at most

$2^{10} = 1024$ subsets to consider. If k is large, say 20 or more, then considering $2^{20} \approx 1,000,000$ or more subsets can be computationally expensive and time-consuming. Three basic approaches to the computational problem have been developed. The first is brute-force computation, in which C_I is computed for all possible submodels. The second approach uses a clever computational algorithm that can eliminate most submodels without actually fitting them. The third possibility uses *stepwise regression*, in which only a small fraction of submodels is examined. Stepwise regression doesn't guarantee finding optimal subsets, although the results obtained with this approach are often useful in practice. Simple stepwise methods are implemented in *Arc*.

Arc includes two simple algorithms for stepwise regression, using *forward selection*, in which terms are sequentially added to a base mean function, and *backward elimination*, in which terms are removed from a mean function. We discuss these procedures in the context of an example.

11.5.3　Highway Accident Data

The fitted OLS mean function for the full model for the highway accident data was given in Table 11.2. The overall F-test indicates that the response is related to the terms, but most of the terms have small t-values. We may be able to reduce the total mean squared error by deleting some of the terms.

For example, suppose we wish to contrast the full model with the submodel obtained by deleting $\log_2(Acpt)$ and $\log_2(Trks)$. To calculate C_I we need to fit the submodel to get $k_I = 5$ and the residual sum of squares $RSS_I = 5.2463$. Combining this with the results of Table 11.2, we compute

$$
\begin{aligned}
C_I &= \frac{RSS_I}{\hat{\sigma}^2} + 2k_I - n \\
&= \frac{5.2463}{0.1486} + 2 \times 5 - 39 \approx 6.29
\end{aligned}
$$

This is somewhat larger than $k_I = 5$. Although this submodel may be reasonable, we would hope to do better, perhaps by using stepwise methods.

Our general goal is to find a subset of the predictors that can be removed from the mean function without significant loss of information. In examining subsets, the term $\log_2(Len)$ will be considered to be part of the base model that is included in every subset. If we assume that accidents tend to occur at a few dangerous points rather than uniformly over a highway segment, then if the length of a segment were increased, the number of miles driven would increase, but the number of accidents might stay constant unless another dangerous point were added. Lengthening a segment will lower the accident rate, so accident rate and length should be negatively related. Including $\log_2(Len)$ in all models essentially adjusts for this artifact.

Forward Selection. Stepwise methods can be accessed in *Arc* by selecting the item "Examine submodels" from the model menu to get the dialog shown

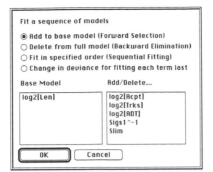

FIGURE 11.2 Dialog for model selection.

in Figure 11.2. We have seen this dialog before when getting sequential analysis of variance tables, but we now use the radio buttons to select the item for forward selection. As shown in the dialog, the variable $\log_2(Len)$ will be included in all mean functions. The other terms were left in the list of terms to add. The results are shown in Table 11.4.

Here is a description of the output. At the first stage, the *base mean function* consists of the intercept and the term $\log_2(Len)$. Arc then fits each mean function obtained by adding one of the remaining terms, so there are five regressions, each consisting of the base mean function plus one other term. The output consists of the df, the residual sum of squares, $C_{\mathcal{I}}$, and $k_{\mathcal{I}}$ for each of the five mean functions. The mean functions are ordered according to the value of $C_{\mathcal{I}}$, from smallest to largest. The term associated with the smallest $C_{\mathcal{I}}$, *Slim*, is now added to the base mean function, and we proceed to the next stage.

At the second stage, the base terms are $\log_2(Len)$ and *Slim*. The process of adding a term is then repeated, adding one of the remaining terms to the two already selected. This whole process is then repeated until all the terms are in the mean function.

Examining Table 11.4, we see that adding $\log_2(Acpt)$ and then $Sigs1^{-1}$ decreases $C_{\mathcal{I}}$ to a value close to the number of terms $k_{\mathcal{I}}$ in the submodel. Thus the four-term model with $\log_2(Len), Slim, \log_2(Acpt)$, and $Sigs1^{-1}$ is a reasonable starting point for further analysis.

Backward Elimination. Backward elimination is similar to forward selection, except at each step a term is removed from the current mean function. The output for backward elimination, again forcing $\log_2(Len)$ into each mean function, is shown in Table 11.5. Use the same dialog as in Figure 11.2, except the button for backward elimination should be pushed.

At the first stage, all six-term submodels obtained by dropping one of the terms other than $\log_2(Len)$ are considered. Summary statistics are reported, again ordering the mean functions based on $C_{\mathcal{I}}$. The term whose removal

TABLE 11.4   Forward Selection Output for the Highway Accident Data

```
Data set = Highway, Name of Fit = L1
Normal Regression
Kernel mean function = Identity
Response = log2[Rate]
Terms = (log2[Acpt] log2[Trks] log2[ADT] log2[Len]
        Sigs1^-1 Slim)
Forward Selection: Sequentially add terms
that minimize the value of CI.
All fits include an intercept.

Base terms: (log2[Len])
                    df  RSS          |   k   CI
Add: Slim           36  6.11216      |   3   8.120
Add: log2[Acpt]     36  6.43817      |   3   10.313
Add: Sigs1^-1       36  9.05288      |   3   27.903
Add: log2[Trks]     36  9.8983       |   3   33.591
Add: log2[ADT]      36  10.5218      |   3   37.785

Base terms: (log2[Len] Slim)
                    df  RSS          |   k   CI
Add: log2[Acpt]     35  5.4994       |   4   5.997
Add: log2[Trks]     35  5.5644       |   4   6.434
Add: Sigs1^-1       35  5.58155      |   4   6.550
Add: log2[ADT]      35  6.05881      |   4   9.761

Base terms: (log2[Len] Slim log2[Acpt])
                    df  RSS          |   k   CI
Add: Sigs1^-1       34  5.07855      |   5   5.166
Add: log2[Trks]     34  5.09933      |   5   5.306
Add: log2[ADT]      34  5.49931      |   5   7.997

Base terms: (log2[Len] Slim log2[Acpt] Sigs1^-1)
                    df  RSS          |   k   CI
Add: log2[Trks]     33  4.88017      |   6   5.831
Add: log2[ADT]      33  4.95813      |   6   6.356

Base terms: (log2[Len] Slim log2[Acpt] Sigs1^-1 log2[Trks])
                    df  RSS          |   k   CI
Add: log2[ADT]      32  4.7566       |   7   7.000
```

corresponds to the smallest C_T is then removed, and we proceed to the next stage. This is continued until all the terms apart from $\log_2(Len)$ are removed from the mean function. The submodels with C_T close to k_T are a reasonable starting point for further analysis. The results from forward selection and backward elimination agree in this example, but there is no guarantee that

TABLE 11.5 Backward Elimination Output for the Highway Accident Data

```
Data set = Highway, Name of Fit = L1
Normal Regression
Kernel mean function = Identity
Response = log2[Rate]
Terms = (log2[Acpt] log2[Trks] log2[ADT] log2[Len]
        Sigs1^-1 Slim)
Backward Elimination: Sequentially remove terms
that give the smallest change in CI.
All fits include an intercept.
```

Current terms: (log2[Len] log2[Acpt] log2[Trks] log2[ADT]
                Sigs1^-1 Slim)

	df	RSS		k	CI
Delete: log2[ADT]	33	4.88017	|	6	5.831
Delete: log2[Trks]	33	4.95813	|	6	6.356
Delete: log2[Acpt]	33	4.98925	|	6	6.565
Delete: Sigs1^-1	33	5.08693	|	6	7.222
Delete: Slim	33	5.38275	|	6	9.212

Current terms: (log2[Len] log2[Acpt] log2[Trks] Sigs1^-1
                Slim)

	df	RSS		k	CI
Delete: log2[Trks]	34	5.07855	|	5	5.166
Delete: Sigs1^-1	34	5.09933	|	5	5.306
Delete: log2[Acpt]	34	5.30504	|	5	6.690
Delete: Slim	34	5.57975	|	5	8.538

Current terms: (log2[Len] log2[Acpt] Sigs1^-1 Slim)

	df	RSS		k	CI
Delete: Sigs1^-1	35	5.4994	|	4	5.997
Delete: log2[Acpt]	35	5.58155	|	4	6.550
Delete: Slim	35	5.76437	|	4	7.780

Current terms: (log2[Len] log2[Acpt] Slim)

	df	RSS		k	CI
Delete: log2[Acpt]	36	6.11216	|	3	8.120
Delete: Slim	36	6.43817	|	3	10.313

Current terms: (log2[Len] Slim)

	df	RSS		k	CI
Delete: Slim	37	11.4138	|	2	41.786

this will always be so. The mean function for the submodel suggested for the highway data is summarized in Table 11.6.

Other Criteria. There is nothing we have said so far that gives a clear indication about how to choose a single "best" submodel. At each stage, terms

TABLE 11.6   **Regression Summary for the Submodel Mean Function for the Highway Data**

```
Data set = Highway, Name of Fit = L2
Normal Regression
Kernel mean function = Identity
Response    = log2[Rate]
Terms       = (log2[Acpt] log2[Len] Sigsl^-1 Slim)
Coefficient Estimates
Label        Estimate      Std. Error    t-value
Constant     4.57452       1.17644        3.888
log2[Acpt]   0.186577      0.101673       1.835
log2[Len]   -0.280324      0.0780369     -3.592
Sigsl^-1    -0.517296      0.308179      -1.679
Slim        -0.0360693     0.0168330     -2.143

R Squared:                 0.700399
Sigma hat:                 0.386483
Number of cases:              39
Degrees of freedom:           34

Summary Analysis of Variance Table
Source        df       SS          MS         F       p-Value
Regression     4      11.8725     2.96813    19.87    0.0000
Residual      34       5.07855    0.149369
```

are added or deleted based on the value of C_I until the list of terms is exhausted. The results may suggest a number of possible mean functions that could be investigated further by using other methods and relevant contextual information.

The stepwise procedures implemented in other computer programs may use different criteria, including specific rules for selecting a "best" model. For example, with forward selection, terms may be added until the t-statistic for the next term is less than a user-selected cutoff called t-in. With backward elimination, terms are deleted until the t-statistic for the next term to be deleted is greater than the cutoff. We do not recommend such stopping rules for routine use since they can reject perfectly reasonable submodels from further consideration. Stepwise procedures are easy to explain, inexpensive to compute, and widely used. The comparative simplicity of the results from stepwise regression with model selection rules appeals to many analysts. But, such algorithmic model selection methods must be used with caution. We conclude with a discussion of two simulated examples that illustrate the problems that can arise when using stepwise procedures to select a "best" model.

Algorithmic Stepwise Model Selection Can Overstate Significance. A data set of $n = 100$ cases with a response y and 50 predictors was generated using standard normal random numbers, so that y is independent of the predictors

TABLE 11.7 Results of a Simulated Example with
$n = 100$ **and 50 Predictors**

Method	k	R^2	Overall F p-Value	Number of Terms with p-Value $\leq$	
				0.25	0.05
No selection	50	0.59	0.13	16	6
t-in $= \sqrt{2}$	16	0.48	< 0.001	16	11
t-in $= 2$	4	0.46	< 0.001	4	4

and thus all 50 predictors have regression coefficients of zero. Table 11.7 summarizes several regressions. The first line is the summary for the regression of y on all 50 predictors. The value of $R^2 = 0.59$ may seem surprisingly large considering that all the data are independent random numbers. The overall F-test gives a p-value of 0.13 for these data. In repeated simulations like this one, the p-value of the test is equally likely to be any value between zero and one. The final two columns give the number of terms in the mean function with p-value less than or equal to 0.25 and 0.05. The second line of the table summarizes the mean function obtained using forward selection, stopping when the t-test for the next term to be added was less than the cutoff t-in $= \sqrt{2}$. Sixteen predictors were added, R^2 is a little lower than for the full model, and the corresponding overall F-test now has a p-value less than 0.001, suggesting that y is indeed related to these terms. We know of course that this is not the case.

We changed the stopping criterion in the third line, continuing to add terms until the t-statistic for the next term to be added was less than the cutoff t-in $= 2$, resulting in only four predictors being added. The value of R^2 remains fairly large, even with just four terms, and the t-values for each of the terms are now all larger than two in absolute value. Similar results would be found repeating this simulation, except of course the terms chosen as important by the algorithm would be different.

This example demonstrates many lessons. First, stepwise selection of terms can have important effects on the apparent significance of results. The coefficients for the terms left in the mean function will generally be too large in absolute value and will appear much more important than they really are. Second, when the response and the terms are independent,

$$E(R^2) = (k - 1)/(n - 1)$$

so R^2 can be large even if the response is independent of the terms. In our simulated example, $n = 100$ and $k = 51$ so $E(R^2) \approx 0.51$ which agrees well with the observed value $R^2 = 0.59$. Since selection retains the terms that make R^2 large, the value of R^2 for the submodel mean function will be close to the value for the full mean function.

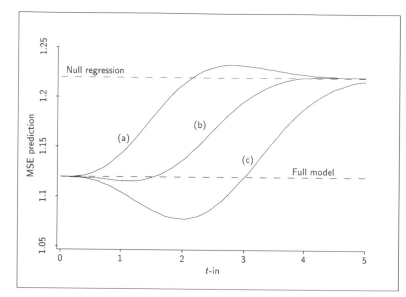

FIGURE 11.3   Plot of mean square error against t-in for three choices of η : $\eta_a = (\sqrt{2}, \sqrt{2}, \sqrt{2},$ $\sqrt{2}, \sqrt{2})^T$; $\eta_b = (\sqrt{5}, \sqrt{5}, 0, 0, 0)^T$; $\eta_c = (\sqrt{10}, 0, 0, 0, 0)^T$. The lower dashed line is the mean square error without selection, while the upper line is the mean square error when all predictors are ignored.

Do Algorithmic Stepwise Methods Reduce Mean Squared Error? An important question is whether a subset $\mathcal{I}$ selected by a stepwise method will actually have lower mean squared error than the mean function without selection. In some cases, exact computations are possible; one of these is depicted in Figure 11.3. This figure was constructed for regressions with $n = 100$, five predictors and $E(y \mid \mathbf{x}) = \eta_0 + \eta_{(\cdot)}^T \mathbf{x}$, where the subscript on $\eta_{(\cdot)}$ will be changed to indicate different coefficient vectors. The predictors are taken to be uncorrelated and with equal variances. Three cases were considered with different values of $\eta_{(\cdot)}$.

- *Case a.* $\eta_a = (\sqrt{2}, \sqrt{2}, \sqrt{2}, \sqrt{2}, \sqrt{2})^T$, and subset selection is probably not desirable because all predictors contribute equally.
- *Case b.* In this case, $\eta_b = (\sqrt{5}, \sqrt{5}, 0, 0, 0)^T$, so it is certainly desirable to delete the last three predictors.
- *Case c.* Here, $\eta_c = (\sqrt{10}, 0, 0, 0, 0)^T$, so we would like to delete all but the first predictor.

Figure 11.3 shows the total mean squared error as a function of t-in for these three choices of η. In Case a, we see that the mean square error for some values of t-in (in the range from 2 to 5) is even larger than the mean square error for the null model, the model $E(y \mid \mathbf{x}) = \eta_0$. Here subset selection is worse than ignoring the predictors altogether!

In Case b, the mean square error of prediction is usually worse and never much better than the expected mean squared error for the full model, so subset selection still does not help.

In Case c, where we expect selection to be the most help, we do get some improvement in mean squared error, but only if t-in is small enough.

11.6 COMPLEMENTS

Formulas (11.2) and (11.5) are proved, for example, in Casella and Berger (1990), Theorem 4.4.1 and Theorem 4.4.2. The simulated example in Section 11.5.3 is similar to examples in Freedman (1983) and in Rencher and Pun (1980). The second example in Section 11.5.3 is from Copas (1983). Mallows' C_T was described by Mallows (1973).

Computationally intensive approaches to subset selection have been proposed. These essentially require computing a simulation for each mean function and estimating a criterion function from the simulation. See, for example, Breiman and Spector (1992) for an example of this approach. Book-length treatments of variable selection are given by Miller (1990), by Linhart and Zucchini (1986) and by McQuarrie and Tsai (1998). Bayesian ideas for mean function selection were introduced by George and McCulloch (1993). Furnival and Wilson (1974) provided a computational algorithm that can find the best few submodel mean functions based on a function like C_T.

In most presentations of regression, an analysis of variance table consists only of the last three columns of Table 11.3.

The salary data are from Weisberg (1985). The black cherry tree data in Problem 11.1 are from Ryan, Joiner, and Ryan (1976). Problem 11.3 is modeled after results in Mantel (1970). The species data in Problem 11.5 are from Johnson and Raven (1973).

PROBLEMS

11.1 The data in the file `bcherry.lsp` provide measurements on 31 black cherry trees from the Allegheny National Forest, Pennsylvania. The goal of the study is to obtain a prediction equation for the volume of marketable timber in a given tree, and perhaps in a whole forest, from measurements taken on diameter D and height Ht. For the sample of trees, the volume Vol was also measured.

   If a tree were shaped like a cylinder, we would have $Vol \approx (\pi/4)D^2 Ht$; if it were another conic section like a cone, the constant multiplier would change but not the exponents on the predictors. Taking logs of this equation, we get a starting point for study of these data:

$$E(\log(Vol) \mid D, Ht) = \eta_0 + \eta_1 \log(D) + \eta_2 \log(Ht) \qquad (11.25)$$

where $\eta_0 = \log(\pi/4)$, $\eta_1 = 2$ and $\eta_2 = 1$. This mean function is likely to be at best an approximation because trees are not cylinders and the log of a mean is not exactly the same as the mean of a logarithm. Use mean function (11.25) as a starting point for this problem, along with the assumption that $Var(\log(Vol) \mid D, Ht) = \sigma^2$. Use natural logarithms, although the results would be the same for any base you select.

11.1.1   Identify the predictors in the problem and the terms in (11.25).

11.1.2   Obtain a test of the hypothesis that $\eta_1 = 2$ against the alternative that η_1 is arbitrary. Do the test two ways, first based on the t-test and then using offsets. Show that the two tests give identical inference (that is $t^2 = F$).

11.1.3   Obtain a test of the null hypothesis that $\eta_1 = 2, \eta_2 = 1$ versus the general alternative in the fit of (11.25). (*Hint:* You will need to fit a mean function that has no additional parameters beyond the intercept. In *Arc*, create a new variable whose value is one for all 31 observations using the "Add a variate" item and typing Const=(repeat 1 31). Then, fit a model *with no intercept* and with the single predictor *Const.*)

11.1.4   Foresters would prefer a prediction equation that uses only $\log(D)$ as a predictor of *Vol* because it is much easier to measure in the field. Starting with the mean function (11.25), under what conditions is the mean function for the regression of $\log(Vol)$ on $\log(D)$ a simple linear regression model? Are those conditions satisfied with these data? Under what conditions is the variance constant in the submodel? Are those conditions satisfied with these data? How much information is lost by predicting volume from diameter alone?

11.2   In the highway data in the file highway.lsp, perform the following hypothesis tests:

NH: eq. (11.14)   versus   AH: eq. (11.15)
NH: eq. (11.14)   versus   AH: eq. (11.16)
NH: eq. (11.14)   versus   AH: eq. (11.17)
NH: eq. (11.14)   versus   AH: eq. (11.18)
NH: eq. (11.15)   versus   AH: eq. (11.16)
NH: eq. (11.15)   versus   AH: eq. (11.17)
NH: eq. (11.16)   versus   AH: eq. (11.17)
NH: eq. (11.17)   versus   AH: eq. (11.18)

TABLE 11.8   Data for Problem 11.3

y	x_1	x_2	x_3
5	1	1004	6
6	200	806	7.3
8	−50	1058	11
9	909	100	13
11	506	505	13.1

All the information needed for these tests is given in Table 11.3.

11.3 Using the simulated data in Table 11.8 in the file `select.lsp`, fit via OLS with the mean function $E(y \mid x) = \eta_0 + \eta_1 x_1 + \eta_2 x_2 + \eta_3 x_3$. Apply both the forward selection and backward elimination algorithms, and compare results. What is the "correct" mean function?

11.4 In this problem, we will generate some random data to show that the fitted coefficients for unnecessary terms need not be zero. First, generate x_1 to be a sample of 100 from a distribution that is Uniform on the interval $[0, 1]$. Then generate $x_2 = 100 x_1 +$ Uniform random number, so x_2 is a rescaling of x_1 plus a small random amount. Finally, compute $y = x_2 +$ Uniform random number on the interval 0 to 5. These computations can be done in *Arc*, and then put into a data set, with the following commands

```
>(def x1 (uniform-rand 100))
>(def x2 (+ (* 100 x1) (uniform-rand 100)))
>(def y (+ x2 (* 5 (uniform-rand 100))))
>(make-dataset :data (list x1 x2 y)
        :data-names '("x1" "x2" "y") :name "Sim")
```

We can clearly view y as depending on either x_1 or x_2, since these differ only in the third significant digit. Fit the OLS regression of y on x_1 and x_2. What are the coefficient estimates for η_1 and η_2? Repeat several times (you can do this using the "Add a variate" menu item to create a new y for the given x_1 and x_2 by typing the expression y1 = (+ x2 (* 5 (uniform-rand 100))) into the dialog and then using y_1 as the response. Summarize your findings.

11.5 The Galápagos Islands off the coast of Ecuador provide an excellent laboratory for studying the factors that influence the development and survival of species. The file `species.lsp` gives data for each of 29 islands. For each island, the total number of animal species was recorded, along with the number of endemic species (those that were not introduced from elsewhere), and several physical characteristics of the island, such as its area, maximum elevation, distance to and size of the nearest island, and distance from Santa Cruz Island. A few elevation measurements are missing for very small islands. Examination of large-scale maps suggest that none of these elevations exceed 200 m.

Use these data to find variables that seem to influence diversity, as measured by some function of the number of species and the number of endemic species, and summarize your results.

11.6   We continue the sleep regression in the file `sleep.lsp` discussed in Section 10.4, using the mean function

$$E(TS \mid \mathbf{x}) = \eta_0 + \eta_1 \log_2(BW) + \eta_2 \log_2(BrW) + \eta_3 \log_2(Life)$$
$$+ \eta_4 \log_2(GP) + \eta_5 D_1$$

A summary of the fit of this function is given in Table 10.3.

11.6.1   Test the hypothesis that $E(TS \mid \mathbf{x})$ does not depend on the size terms, $\log_2(BW)$ and $\log_2(BrW)$, and summarize results.

11.6.2   Next, test the hypothesis that TS does not depend on $\log_2(BrW)$, with $\log_2(BW)$ included in the mean function. Do the results of this test appear to contradict the last result? Compare the coefficient estimate for $\log_2(BW)$ in the full model and in the model without $\log_2(BrW)$. Can you explain the change in sign?

11.6.3   Based on the full model, examine a 3D added-variable plot for $\log_2(BrW)$ and $\log_2(BW)$, and summarize your findings.

11.6.4   Assuming that the full model provides a useful approximation for the mean function for TS given the predictors, use both forward selection and backward elimination to find a subset of terms that can describe the mean function without important loss of information.

11.6.5   Construct the $C_{\mathcal{I}}$-statistic for comparing the full model to the model with mean function

$$E(TS \mid \mathbf{x}) = \eta_0 + \eta_3 \log_2(Life) + \eta_4 \log_2(GP) + \eta_5 D_1$$

Based on $C_{\mathcal{I}}$, what do you conclude about the two models?

11.6.6   Compute the $C_{\mathcal{I}}$ statistic for comparing the full model to (a) the model without $\log_2(BrW)$, and (b) the model without $\log_2(BrW)$ and D_1. In each case, provide an interpretation of $C_{\mathcal{I}}$.

11.6.7   Consider the mean function without the term $\log_2(BW)$. An investigator has a theory that predicts that the coefficient of $\log_2(GP)$ should be twice the coefficient of $\log_2(BrW)$ in this reduced mean function. Test the hypothesis that this theory is correct against the alternative that it is not correct.

CHAPTER 12

Factors and Interactions

Allowing categorical predictors increases the range of questions that can be studied using regression methodology. In this chapter we discuss the use of *factors*, which convert categorical predictors into numerical ones for use in regression models.

12.1 FACTORS

Categorical predictors have two or more *levels*. Sex is an example of a categorical predictor with two levels, male and female. Treatment status, with the levels treated and control, is another example of a categorical predictor with two levels. Other categorical predictors can have many levels, such as age, grouped as young, not-so-young and old; auto type, with levels sedan, van, and truck; or location, with one level for each location where data were collected. Categorical predictors serve to distinguish populations on qualitative rather than quantitative variables.

To use categorical predictors in regression models we must have a way to indicate their levels using numerical variables. This is done by creating *indicator variables* and sets of indicator variables called *factors*.

12.1.1 Two Levels

Suppose we wanted to include a predictor *Action* in a mean function, where *Action* has two levels, either *Treatment* or *Control*, indicating whether an experimental unit is given the treatment or given the control. We can obtain an indicator for the levels of *Action* by defining the variable v_1 to have the value 1 if *Action = Treatment*, and the value 0 if *Action = Control*. We know that if $v_1 = 1$, then the experimental unit received the treatment, and if $v_1 = 0$, then it received the control. In the data file, v_1 would appear as a column of zeroes and ones. Any two different numbers could be used in place of zero and one to indicate the levels of *Action*, but zero and one are used because the resulting models are the easiest to interpret. The variable v_1 is called an *indicator vari-*

able, or sometimes a *dummy variable*. Only one indicator variable is required to describe a two-level categorical predictor.

Suppose further that we have a sample of n_1 independent treated experimental units and an additional n_2 independent control units, and we wish to test the hypothesis that the expected response for the control population is the same as that for the treated population. The usual two-sample t-test is often appropriate for this test. It arises naturally in the context of the regression model

$$E(y \mid v_1) = \eta_0 + \eta_1 v_1 \quad \text{with} \quad \text{Var}(y \mid v_1) = \sigma^2$$

where v_1 is the indicator for *Action* as described above. The mean for the control group is $E(y \mid v_1 = 0) = \eta_0$, while the mean for the treated group is $E(y \mid v_1 = 1) = \eta_0 + \eta_1$. Thus, η_1 is the difference between the population means for the treatment population and the control population. The t-statistic for the hypothesis that $\eta_1 = 0$ is the same as the usual two-sample t-statistic for comparing the means of two populations with the same variance.

12.1.2 Many Levels

Two indicator variables v_2 and v_3 are required to describe a categorical predictor with 3 levels:

$$v_2 = 0 \quad \text{and} \quad v_3 = 0 \quad \text{for level one}$$

$$v_2 = 1 \quad \text{and} \quad v_3 = 0 \quad \text{for level two}$$

$$v_2 = 0 \quad \text{and} \quad v_3 = 1 \quad \text{for level three}$$

The variable v_2 is the indicator for level two, and v_3 is the indicator for level three. By convention, the subscript on the variable is the same as the level it is indicating. If the level is neither two nor three, we know it must be level one, the only remaining level.

A categorical predictor C with ℓ levels requires $\ell - 1$ indicator variables $v_2, \ldots, v_\ell$. For any observation, at most one of these indicator variables is equal to one, and all the rest are equal to zero. If they are all zero, then C is equal to its first level. If $v_j = 1$, then C is equal to its jth level.

The set of indicator variables used to describe the levels of a categorical predictor is called a *factor*. In *Arc*, a categorical predictor can be converted into a factor using the "Make factors" item in the data set menu. We will illustrate the use of this dialog shortly.

12.2 TWIN DATA

Known particularly for his work on the heritability of IQ, Sir Cyril Burt (1883–1971) was among England's most honored psychologists during his lifetime. Shortly after his death, scientists began to question statistical peculiarities in his published work, leading to the headline "Crucial data was faked by eminent

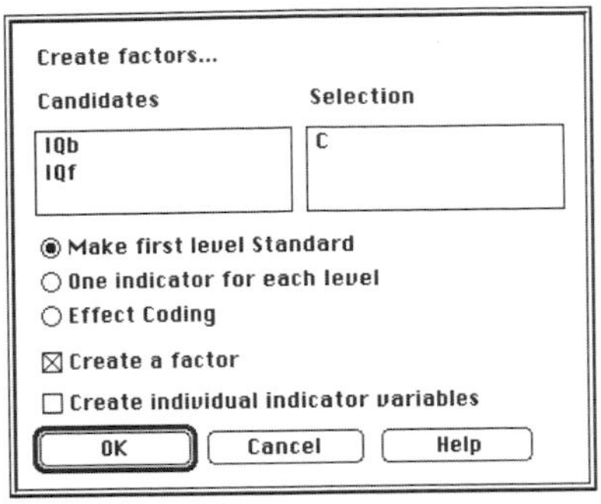

FIGURE 12.1   The "Make factors" dialog.

psychologist" on the front page of the *London Sunday Times* in October 1976. Burt's work fell into disrepute for some time thereafter. However, it has been argued recently that Burt did not get a fair hearing at the time and that he is in fact innocent of the charges that he "faked the data."

The data file twins.1sp contains one of Burt's data sets that was published in 1966. The data set contains the IQ scores for 27 pairs of identical twins, one reared in a foster home and the other reared by their biological parents. In addition, there is a categorical predictor C which indicates the social class of the biological parents, with $C = 1$ for upper class, $C = 2$ for middle class, and $C = 3$ for lower class. We begin by considering the regression of IQb, the IQ for the twin reared by the biological parents, on social class.

The first step is to convert C into a factor, a set of two indicator variables that indicate the levels of social class. To create factors in *Arc*, select the item "Make factors" from the data set menu. This results in the dialog shown in Figure 12.1. Select the variables from which you wish to create factors and then select the way you want the factors to be computed. The default is to use the first level as the baseline for measuring the effects of the other variables. This is the way we defined indicator variables in Section 12.1. This option creates a factor with the same name as the variable, but with the prefix $\{F\}$. The factor $\{F\}C$ represents the collection of indicator variables describing the categorical predictor C. The other two choices for computing factors are to create one indicator variable for each level, or to use effect coding. These options are described in the complements to this chapter. For now, we will use only the default.

To fit the model

$$\mathrm{E}(IQb \,|\, C) = \eta_0 + \eta_2 v_2 + \eta_3 v_3 \quad \text{with} \quad \mathrm{Var}(IQb \,|\, C) = \sigma^2 \quad (12.1)$$

in *Arc*, specify $\{F\}C$ as the predictor and IQb as the response. The regression output is shown in Table 12.1. The indicator variables v_2 and v_3 are called $\{F\}C[2]$ and $\{F\}C[3]$ in the regression output, the number in square brackets indicating the level of C. The first level of C is used as the baseline and thus does not appear in the output.

TABLE 12.1 Fit of the Regression Model (12.1)

```
Data set = Twins, Name of Fit = L1
Normal Regression
Kernel mean function = Identity
Response       = IQb
Terms          = ({F}C)
Coefficient Estimates
Label             Estimate    Std. Error    t-value
Constant          107.571      5.46176       19.695
FC[2]             -17.0714     8.03949       -2.123
FC[3]             -16.3571     6.68926       -2.445

R Squared:                   0.221519
Sigma hat:                   14.4504
Number of cases:             27
Degrees of freedom:          24

Summary Analysis of Variance Table
Source        df       SS          MS         F      p-value
Regression    2       1426.06      713.029    3.41   0.0495
Residual      24      5011.57      208.815
Pure Error    24      5011.57      208.815
```

According to the regression coefficient for $\{F\}C[2]$, the mean difference $E(IQb \mid C = 2) - E(IQb \mid C = 1)$ is estimated to be about $\hat{\eta}_1 = -17.1$ points, while the difference $E(IQb \mid C = 3) - E(IQb \mid C = 1)$ is estimated to be about $\hat{\eta}_3 = -16.4$ points. Note also from the analysis of variance table that the residual sum of squares is the same as the pure error sum of squares; can you explain why this happens?

The results of our analysis so far can be visualized by constructing boxplots of IQb conditioning on C, as shown in Figure 12.2. The boxplots confirm the results of the numerical analysis that the population mean for the upper social class $E(IQb \mid C = 1)$ is greater that the population means for the other two classes. The plots suggest also that the variance in IQ for the upper social class could be larger than the variance in the other two populations. Methods of testing for nonconstant variance are discussed in Chapter 14, and the task of testing for nonconstant variance in this regression is set as Problem 14.6.

12.3 ONE-WAY ANALYSIS OF VARIANCE

Let C be a categorical predictor with ℓ levels, and consider the regression model

$$E(y \mid C) = \eta_0 + \eta_2 v_2 + \cdots + \eta_\ell v_\ell \quad \text{with} \quad \text{Var}(y \mid C) = \sigma^2 \quad (12.2)$$

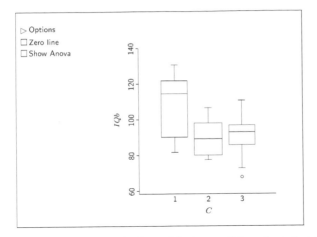

FIGURE 12.2   Boxplots of *IQb* conditioning on *C* from the twin data.

The mean for the population associated with the first level of the categorical predictor is just η_0 because this population is indicated by $v_2 = \cdots = v_\ell = 0$. For $j > 1$, the mean for the population with $v_j = 1$ is $E(y \mid C = j) = \eta_0 + \eta_j$. The means of the ℓ populations determined by the levels of C are all equal if and only if $\eta_j = 0$ for $j = 2, \ldots, \ell$. Thus, we can test the hypothesis that the population means are equal by using the F-statistic to compare the null hypothesis mean function $E(y \mid C) = \eta_0$ with the mean function (12.2) for the alternative hypothesis:

$$\text{NH}: E(y \mid C = 1) = \cdots = E(y \mid C = \ell) = \eta_0$$

$$\text{AH}: \text{NH not true; that is, at least one } \eta_j \neq 0, \qquad j = 2, \ldots, \ell.$$

This procedure is called a *one-way analysis of variance*. Despite its name, the one-way analysis of variance is a procedure to analyze differences in means. When fitting (12.2) in *Arc*, the analysis of variance output corresponds to a one-way analysis of variance. For example, the F-statistic in the summary analysis of variance table of Table 12.1 can be used to test the hypothesis that the population mean IQ for children reared by their biological parents does not depend on social class. The "Show ANOVA" button that appears on boxplots will cause the one-way anova table for comparing the subpopulation means to appear on the plot.

We can test the hypothesis that two population means are equal by formulating equivalent tests based on the regression coefficients in (12.2). The test depends on whether the first (standard) level of C is one of the two means compared. The hypothesis $E(y \mid C = 1) = E(y \mid C = j)$ is equivalent to the hypothesis that $\eta_j = 0$. This test can be carried out using a t-test from the fit of (12.2) in the usual way.

On the other hand, for $j > 1$ and $k > 1$, the hypothesis

$$E(y \mid C = j) = E(y \mid C = k)$$

is equivalent to the hypothesis that $\eta_j = \eta_k$ discussed in Section 11.3.1. We can also use an equivalent t-test based on the test statistic

$$t = \frac{\hat{\eta}_j - \hat{\eta}_k}{se(\hat{\eta}_j - \hat{\eta}_k)} = \frac{\bar{y}_j - \bar{y}_k}{se(\bar{y}_j - \bar{y}_k)}$$

The standard error of a linear combination of coefficient estimates is discussed in Section 7.6.4, but for the special case of one-way analysis of variance the standard error is just

$$se(\bar{y}_j - \bar{y}_k) = [\hat{\sigma}^2(1/n_j + 1/n_k)]^{1/2}$$

with $\hat{\sigma}^2$ the residual mean square from the one-way analysis of variance table, and $\bar{y}_j$ is the mean response for the jth level of the factor. Under the hypothesis of equal population means, this statistic has a $t_{n-\ell}$ distribution.

12.4   MODELS WITH CATEGORICAL AND CONTINUOUS PREDICTORS

Continuing with the twin data, consider the regression of $y = IQf$, the IQ of the twin reared by foster parents, on IQb and C. There is now one continuous predictor IQb and one categorical predictor, social class C. Let's think about how social class can affect the mean function. Suppose that for each fixed level of C the mean function for the regression of IQf on IQb can be expressed using a simple linear regression model:

$$E(y \mid IQb, C = c) = \eta_{0c} + \eta_{1c}IQb, \qquad \text{with} \quad Var(y \mid IQb, C) = \sigma^2$$

$$(12.3)$$

The additional subscript c on the η's allows for the possibility that *the effect of social class could be reflected in the intercept, the slope, or both.*

We distinguish four cases for (12.3).

Model 1: Unrelated Regression Lines. In this, the most general case, the intercept and slope parameters for each social class are different, and we have a circumstance like that in Figure 12.3a. The social class with the highest average y may depend on the value of IQb: For some values of IQb one social class is best, and for others a different social class is best. We say that there is *an interaction between C and IQb.*

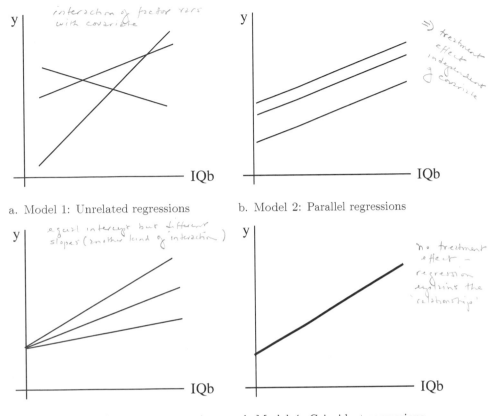

a. Model 1: Unrelated regressions   b. Model 2: Parallel regressions

c. Model 3: Equal intercept regressions   d. Model 4: Coincident regressions.

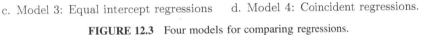

FIGURE 12.3   Four models for comparing regressions.

Model 2: Parallel Regression Lines. In this model, the slopes are equal, $\eta_{11} = \eta_{12} = \eta_{13}$, but the intercepts may differ. The mean functions shown in Figure 12.3b are three parallel lines. This is a very important special case, because the effect of social class does not depend on the value of *IQb*, and we can speak unambiguously of a social class effect as the difference between the intercepts. In this case we may also say that *there is no interaction between C and IQb.*

Model 3: Equal Intercept Regression Lines. In this model, intercepts are equal, $\eta_{01} = \eta_{02} = \eta_{03}$, but the slopes may differ, as shown in Figure 12.3c. The lines never cross for *IQb* > 0, so the ordering of the social classes is always the same, but the size of the difference changes with *IQb*. In other experiments, the lines may cross at some point other than on the *y*-axis.

Model 4: Coincident Regression Lines. Here, all lines are the same, so all the intercepts are the same and all the slopes are the same. This case is

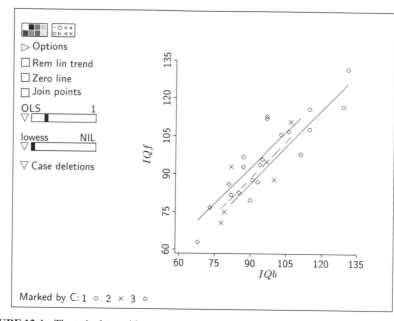

FIGURE 12.4   The twin data, with a separate regression line shown for each level of C, according to general model 1. The solid line with the largest intercept is for $C = 3$, the dashed line is for $C = 2$, and the remaining line is for $C = 1$.

illustrated in Figure 12.3d, where there is only one regression line that is common to all social classes. In this case, social class has no effect.

To examine these four possibilities for the twin data, we begin by looking at a plot of the data, shown in Figure 12.4. In creating this plot you should specify C as a marking variable. This will cause points to have a different symbol and color for each social class. Ignore the lines on the plot for now; they will be discussed shortly.

12.4.1   Fitting

We now discuss how *Arc* can be used to fit each of the four models. We start with Model 4 and then progress to the more complicated models.

Fitting Model 4: Coincident Regression Lines.   To fit Model 4, ignore social class and use OLS to fit the usual simple linear regression of y on IQb,

$$E(y \mid IQb) = \eta_0 + \eta_1 IQb \qquad \text{with} \quad Var(y \mid IQb) = \sigma^2$$

You can superimpose the fitted version of this model on the plot of Figure 12.4 in the usual way using the parametric smoother slidebar.

TABLE 12.2 Summary of Four Models for the Twin Data

Name	Model	df	RSS	Residual MS
Unrelated	1	21	1317.47	62.74
Parallel	2	23	1318.40	57.32
Equal intercepts	3	23	1326.46	57.67
Coincident	4	25	1493.53	59.74

Fitting Model 2: Parallel Regression Lines. Model 2 has three intercept parameters and one slope parameter. If the two terms making up the factor $\{F\}C$ are called v_2 and v_3, so $v_2 = 1$ when $C = 2$ and $v_3 = 1$ when $C = 3$, the mean function for parallel regressions is just

$$E(y \mid IQb,C) = \eta_0 + \eta_{02}v_2 + \eta_{03}v_3 + \eta_1 IQb \qquad (12.4)$$

In this mean function, the common slope is η_1, and the intercepts are η_0 if $C = 1$, $\eta_0 + \eta_{02}$ if $C = 2$, and $\eta_0 + \eta_{03}$ if $C = 3$. The subpopulation with the first level, $C = 1$, has been used as a standard and the intercepts for the other two levels are measured relative to it. Thus, η_{02} is the change in the intercept when passing from the standard level to level 2, and η_{03} is the change in the intercept when passing from the standard level to level 3. For this reason it is often useful to construct data files so the first level of any factor corresponds to the experimental control, or another meaningful standard.

You can fit model (12.4) in *Arc* using "Fit linear LS," specifying the response y along with terms IQb and $\{F\}C$. To compare models using F-tests, we need only the residual sum of squares and residual degrees of freedom; these are given in Table 12.2.

Assuming you have used C as a marking variable, you can superimpose the fitted version of model (12.4) on Figure 12.4 by selecting the item "Fit by marks—parallel" from the pop-up menu for the parametric smoother slidebar.

Fitting Model 3: Equal Intercept Regression Lines. To fit Model 3, we need to have a way of specifying a separate slope for IQb for each level of C. This is done by creating an *interaction* between IQb and the factor $\{F\}C$, using the "Make interactions" item in the data set menu. The interaction dialog creates a new variable called either $IQb\{F\}C$ or $\{F\}CIQb$, depending on the order you specify the variates, consisting of the two terms $v_2 \times IQb$ and $v_3 \times IQb$. These are used in the coincident lines mean function:

$$E(y \mid IQb,C) = \eta_0 + \eta_1 IQb + \eta_{12}v_2 IQb + \eta_{13}v_3 IQb \qquad (12.5)$$

The common intercept is η_0 and the three slopes are η_1 if $C = 1$, $\eta_1 + \eta_{12}$ if $C = 2$, and $\eta_1 + \eta_{13}$ if $C = 3$. Again, the first level of C is used as a standard, and the slopes for the other two levels are measured relative to it. For example, η_{12} is the change in the slope when passing from the standard level to level 2.

Fit model (12.5) using y as the response, along with terms IQb and $IQb\{F\}C$. The results are summarized in Table 12.2. You can add these regression lines to your version of Figure 12.4 by selecting "Fit by marks—equal intercept" from the parametric smoother slidebar pop-up menu.

Fitting Model 1: Unrelated Regression Lines. The most general model, Model 1, requires fitting the regression of y on $\{F\}C$, IQb, and $IQb\{F\}C$ to give separate slopes and intercepts for each level of C:

$$E(y \mid IQb, C) = \eta_0 + \eta_{02}v_2 + \eta_{03}v_3 + \eta_1 IQb + \eta_{12}v_2 IQb + \eta_{13}v_3 IQb$$

$$(12.6)$$

The results are also summarized in Table 12.2. You can add these regression lines to your plot by selecting "Fit by marks—general" from the parametric smoother slidebar pop-up menu. These are the lines we added to Figure 12.4. Although the lines are allowed to have different slopes, it appears from the plot that the estimated slopes are very similar, suggesting that Model 2 may be appropriate.

12.4.2 Tests

Most tests concerning the slopes and intercepts of different regression lines will use the general model, Model 1, as the alternative model. The usual F-statistic for testing Model 2, 3, or 4 against Model 1 is

$$F_\ell = \frac{(RSS_\ell - RSS_1)/(\mathrm{df}_\ell - \mathrm{df}_1)}{\hat{\sigma}_1^2}$$

$$(12.7)$$

where $\ell = 2, 3, 4$ represents the model being tested. In (12.7), RSS_ℓ is the residual sum of squares for model number ℓ, with df_ℓ degrees of freedom, and $\hat{\sigma}_1^2$ is the estimate of σ^2 from Model 1. To get a p-value, compare F_ℓ to the F distribution with $(\mathrm{df}_\ell - \mathrm{df}_1, \mathrm{df}_1)$ df.

For the twin data,

$$F_2 = \frac{(1318.40 - 1317.47)/(23 - 21)}{62.74} = 0.007$$

The very large p-value, about 0.99 agrees with our visual impression of Figure 12.4. This p-value was obtained by comparing the value 0.007 with the $F_{2,21}$ distribution.

Other comparisons are possible, provided the model of the null hypothesis is nested within the model of the alternative hypothesis. For example, we could test NH: Model 4 versus AH: Model 2, but we cannot use the F-test for NH: Model 3 versus AH: Model 2.

TABLE 12.3   The Turkey Growth Data

S = Source	D = Dose	m = Reps	y = AveGain	WS = WithinSS
Control	0	10	623.0	3408.0
1	0.04	5	680.2	206.8
1	0.10	5	721.4	1841.2
1	0.16	5	750.4	1223.2
1	0.28	5	789.4	861.2
2	0.04	5	672.2	2810.8
2	0.10	5	709.2	860.8
2	0.16	5	731.2	592.8
2	0.28	5	778.2	2642.8
3	0.04	5	668.4	2399.2
3	0.10	5	715.6	327.2
3	0.16	5	732.0	1254.0
3	0.28	5	794.0	1488.0

12.5   TURKEY DIETS

Methionine is an amino acid that is essential for normal growth in turkeys. Depending on the ingredients in the feed, it can be necessary for turkey producers to add supplemental methionine for a proper diet. Too much methionine could be toxic. Too little methionine could result in malnourished birds.

An experiment was conducted to study the effects on turkey growth of different doses D of methionine, ranging from 0.04% to 0.28% of the total diet, from three different commercial sources S. Methionine from each of the three sources was added to turkey feed in each of four doses, resulting in 12 treatment combinations. In addition, the experiment included a control diet in which no methionine was added. The total experiment thus consisted of the 13 treatment combinations shown in the first two columns of Table 12.3. Each of the 12 methionine treatments was randomly assigned to five pens of turkeys, and the control was randomly assigned to 10 pens. The total experiment consisted of $(5 \times 12) + 10 = 70$ pens, each containing the same number of birds.

Pen weights were obtained at the beginning and the end of the experiment three weeks later. The response variable is the average weight gain in grams per pen. Reported in Table 12.3 are all treatment combinations, the number m of pens that received each treatment, the average y of the weight gains per treatment, and WS, the sum of squares between the pen weight gains within a treatment. These data are available in file `turkey.lsp`.

The within-treatment sum of squares WS can be used to compute a sum of squares for pure error, to be used in testing lack of fit; see Section 9.3.5, page 214, to refresh your memory about data summarized in this way. The sum of squares for pure error is just the sum of the WS, 19916, with (4×12)

$+9 = 57$ df. Thus, $\hat{\sigma}_{\text{pe}}^2 = 19916/57 = 349.404$ is a model-free estimate of pen to pen variation with 57 df.

12.5.1 The Zero Dose

The results of this experiment are similar to those for the twin data because there is one categorical predictor S and one continuous predictor D. We can think of within-source models just as we thought of models within social class in the twin data. There are two important differences between the two data sets, however. First, in this experiment we do not have the data on individual pens, and the number of replicate pens for the control diet is different from that for the methionine diets. This means that we will need to use weighted least squares when fitting models, where the number of replicates is used as the weight. Second, there is a zero dose that complicates matters a bit. Since a zero dose of one source is the same as a zero dose of any other source, the within-source mean functions must agree at $D = 0$, so that the intercepts for the within-source mean functions must be equal. This means that only coincident and equal-intercept models are reasonable for this experiment.

How should the zero dose be included in the indicator variables? Actually, it doesn't matter for fitting mean functions with a common intercept for all sources. In the data file, we have coded the zero dose to have $S = 1$.

12.5.2 Adapting to Curvature

Figure 12.5 is a plot of the data obtained by using S as a marking variable and the number of replicates as weights. The plot appears curved, suggesting that straight-line within-source mean functions may not be adequate for these data. The curvature can be seen more clearly by removing the linear trend. To confirm, fit the weighted least squares regression of y on D and $D\{F\}S$, fitting the mean function for equal intercepts and different slopes. For the fit of this mean function, the residual mean square is 1391.90 with 9 df, and the F-test for lack of fit is $F = 1391.90/349.404 = 3.98$, which, when compared to the $F_{9,57}$ distribution, gives a p-value less than 0.001. Fitting with straight lines doesn't make sense here. Some reflection on the problem may convince you that this finding is appropriate. If weight gain increased linearly with the amount of methionine, we could keep adding the supplement to get bigger and bigger turkeys. We might reasonably expect that, above some value for D, additional supplementation has no effect, or a toxic effect. This suggests that the within-source mean functions should be curved, and flatten out as D increases.

As discussed in Chapter 7, quadratic polynomials might be useful for describing the response to dose over the range in this experiment. In particular, the quadratic model

$$E(y \mid D,S) = \eta_0 + \eta_{1s}D + \eta_{2s}D^2$$

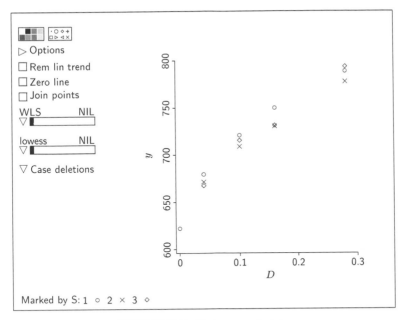

FIGURE 12.5   Turkey growth data.

allows separate regression coefficients of D and D^2 for each level of S, but requires a common intercept. To fit this mean function, use weighted least squares with response y and terms D, D^2, $D\{F\}S$, and $D^2\{F\}S$. The final term $D^2\{F\}S$ provides for interaction terms between D^2 and $\{F\}S$. We leave it as an exercise to (a) show that this quadratic mean function matches the data well and (b) compare it to the coincident regression mean function,

$$E(y \mid D, S) = \eta_0 + \eta_1 D + \eta_2 D^2$$

12.6   *CASUARINA* DATA

Seeds from a tropical tree called *Casuarina cunninghamii* were collected from two seed sources. Six plants from each seed source were grown with fertilizer and another six without. After four years, diameter was measured at 65 cm above ground. The trees were then cut down and weighed. The goal for this example is to study the dependence of the response, W = weight in kg, on the three predictors, diameter D in cm, the use of fertilizer F, and seed source S. The data are in the file `casuarin.lsp`.

The variables S and F are categorical predictors, so to use them in regression models we need to form factors. However, S is coded to have the values 0 for one source and 1 for the other, and similarly F is coded 0 for trees grown without fertilizer and 1 for trees grown with fertilizer. The factors we create from these categorical predictors would in this case be identical to the variable itself; consequently, in this special case the variables are already factors. If there were more than two sources, or if S were coded as a text variable with values "S0" and "S1," for example, then factors would have to be created.

We begin by thinking about how W might depend on D for fixed values of S and F. Elementary physics and geometric considerations suggest that

$$\text{Weight} = \text{Volume of wood} \times \text{Wood density per unit volume}$$

Furthermore, if a tree were a perfect cylinder, then

$$\text{Volume of wood} = (\pi/4) \times \text{height} \times D^2$$

Putting these two together, we get

$$\text{E}(W \mid D, \text{ height, density}) \approx \pi/4 \times \text{density} \times \text{height} \times D^2 \qquad (12.8)$$

This relationship depends on height and density, which are unknown. However, we can use the results of Section 11.1 to get marginal relationships. First, take logarithms,

$$\text{E}(\log(W) \mid D, \text{height}, \text{density}) \approx \log(\pi/4) + \log(\text{density})$$
$$+ \log(\text{height}) + 2\log(D)$$

This equation is an approximation because (12.8) is approximate and because the logarithm of an expectation is only approximately the same as the expectation of a logarithm. We now average over the predictors we have not observed to get a mean function involving only D. Using equation (11.2), we find

$$\text{E}(\log(W) \mid D) \approx \log(\pi/4) + \text{E}(\log(\text{density}) \mid D)$$
$$+ 2\log(D)$$
$$+ \text{E}(\log(\text{height}) \mid D)$$

This equation depends on $\log(D)$ and on the mean functions for the regressions of $\log(\text{density})$ on D and $\log(\text{height})$ on D. We have no data to estimate these mean functions, so we must make an informed guess about them. Suppose that wood density is independent of diameter of the tree, so $\text{E}(\log(\text{density}) \mid D)$ is constant for all trees in a seed source/fertilizer combination. Call this constant d. We can also hypothesize about the relationship between height and diameter. Suppose

$$\text{E}(\text{height} \mid D) = m_0 D^{m_1}$$

If $m_1 = 1$, then height and diameter are linearly related. If $m_1 = 0$, then height and diameter are unrelated. Other values of m_1 provide nonlinear relationships between height and diameter. We can then write $E(\log(\text{height}) \mid D) \approx \log(m_0) + m_1 \log(D)$. Combining these terms gives a mean function for $\log(W)$:

$$E(\log(W) \mid D) \approx \log(\pi/4) + d + \log(m_0) + (2 + m_1)\log(D)$$

$$= \eta_0 + \eta_1 \log(D) \tag{12.9}$$

The intercept is $\eta_0 = \log(\pi/4) + d + \log(m_0)$, and the slope is $\eta_1 = 2 + m_1$. We can estimate η_0 and η_1 as usual. An estimate of m_1 will be given by $\hat{\eta}_1 - 2$, but we cannot estimate d or m_0.

The discussion so far is for fixed values of S and F. The effect of either of these categorical variables, if there is one, can be through the intercept, the slope, or both. There are four possible populations formed by the two levels of S and the two levels of F.

12.6.1 Effect Through the Intercept

The model

$$E(\log(W) \mid D,S,F) = \eta_0 + \eta_2 S + \eta_3 F + \eta_4 SF + \eta_1 \log(D) \tag{12.10}$$

allows for separate intercepts, but forces a common slope. Terms of the form SF are called *two-factor interactions* because they allow for interactions between two factors, in this case S and F. This mean function has four separate intercepts, as follows:

$S = 0,$	$F = 0:$	η_0
$S = 0,$	$F = 1:$	$\eta_0 + \eta_3$
$S = 1,$	$F = 0:$	$\eta_0 + \eta_2$
$S = 1,$	$F = 1:$	$\eta_0 + \eta_2 + \eta_3 + \eta_4$

The population with $S = 0$ and $F = 0$ has intercept η_0 and is used as the standard. η_3 is the effect of passing from ($S = 0$, $F = 0$) to ($S = 0$, $F = 1$), while η_2 is the effect of passing from ($S = 0$, $F = 0$) to ($S = 1$, $F = 0$). What happens if we change both seed source and fertilizer? Would we get the benefits of both changes, or might there be some unique effect? The effects of S and F are said to be *additive* if η_4, the coefficient of the interaction term SF, is zero. Otherwise, when $\eta_4 \neq 0$, there is an interaction between S and F and the effects are said to be *nonadditive*. The effects could be nonadditive if the fertilizer was effective with one seed source but not with the other seed source. Additive

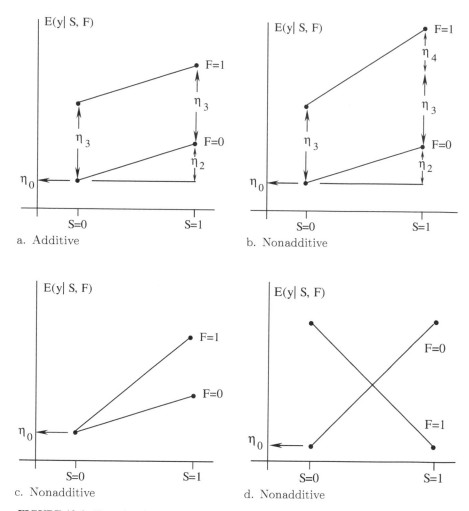

FIGURE 12.6   Four situations illustrating additive and nonadditive effects in model (12.11).

effects are often much easier to understand than nonadditive effects, so testing $\eta_4 = 0$ is usually of interest.

   To gain additional insight into the difference between additive and nonadditive effects, let's momentarily consider a simplified version of model (12.10) without the term for D:

$$E(y \mid S, F) = \eta_0 + \eta_2 S + \eta_3 F + \eta_4 S F \qquad (12.11)$$

where $y = \log(W)$.

   Figure 12.6a illustrates additive effects when $\eta_4 = 0$. The plot consists of four points, the four values of the mean function $E(y \mid S, F)$ for model (12.11).

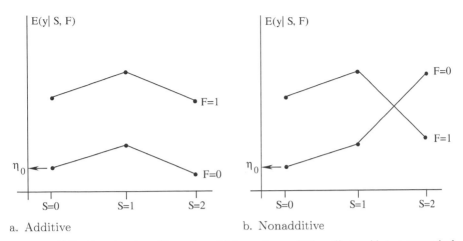

a. Additive   b. Nonadditive

FIGURE 12.7 Two situations illustrating additive and nonadditive effects with two categorical predictors with 2 and 3 levels.

The two lines connecting the pairs of points for $F = 0$ and $F = 1$ are for reference and visual clarity and are not part of the mean function. η_0 marks the mean of the standard reference population $(S = 0, F = 0)$, and all other effects arc measured relative to it. The two reference lines are parallel because $\eta_4 = 0$. The effect of fertilizing is the same for the two seed sources, and the effect of changing from seed source zero to seed source one is the same whether we fertilize or not.

Figures 12.6b–d illustrate different types of nonadditive effects. In Figure 12.6b, fertilizing is beneficial for both seed sources, but the effect is greater for seed source one than for seed source zero. Figure 12.6c is similar to Figure 12.6b, except that $\eta_3 = 0$ so there is no benefit to fertilizing seed from source zero, but there is still a benefit for seed source one. An extreme interaction is illustrated in Figure 12.6d.

Figure 12.7 illustrates additive and nonadditive effects when S has three levels. Again, the line segments connect the population means for $F = 0$ and $F = 1$. These segments are not part of the mean function, but they help us recognize additive effects: The parallel line segments in Figure 12.7a indicate additive effects, while the nonparallel line segments in Figure 12.7b indicate nonadditive effects. Also, in Figure 12.7b there is an additive effect between $S = 0$ and $S = 1$, but not between $S = 1$ and $S = 2$. Overall, S has a nonadditive effect because we still need the interaction term SF in the mean function. Graphs such as these could be used in practice by replacing population means with sample means. Their appearance will depend on the order in which the categories are represented on the horizontal axis, but reordering cannot change additive effects into nonadditive effects.

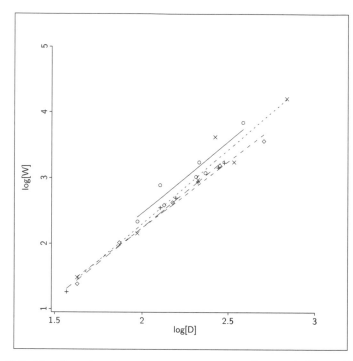

FIGURE 12.8 The *Casuarina* data, with a separate OLS line fit to each *SF* combination. For $S = F = 0$, the points are circles, and the fitted line is the solid line. For $S = F = 1$, points are +, and the fitted line has long dashes. For $S = 0$, $F = 1$, points are ×, and the fitted line has short dashes. For $S = 1$, $F = 0$, points are diamonds, and the fitted line has intermediate dashes.

12.6.2   Effect Through Intercept and Slope

Returning to the regression including S, F and $\log(D)$, the most general mean function with separate slopes and intercepts is given by

$$E(\log(W) \mid D, S, F) = \eta_0 + \eta_2 S + \eta_3 F + \eta_4 SF + \eta_1 \log(D)$$
$$+ \eta_5 S \log(D) + \eta_6 F \log(D)$$
$$+ \eta_7 SF \log(D) \tag{12.12}$$

This mean function allows a separate slope and a separate intercept for each combination of S and F. It requires fitting with several *two-variable interactions* and a *three-variable interaction* $SF \log(D)$.

Figure 12.8 is a plot of $\log(W)$ versus $\log(D)$, with points marked separately for the four groups defined by combinations of S and F. The marking variable was created using the "Add a variate" item and then typing SF1=10*S+F in the text area of the dialog. $SF1 = 11$ if $S = F = 1$, 10 if $S = 1$, $F = 0$, 1 if $S = 0$, $F = 1$, and 0 if $S = F = 0$. Also shown on the plot are the OLS regression lines fit to each group separately. Although formal testing may be de-

TABLE 12.4 Regression Summary for the *Casuarina* Data

```
Data set = casuarina, Name of Fit = L1
Normal Regression
Kernel mean function = Identity
Response       = log[W]
Terms          = (log[D] S F)
Coefficient Estimates
Label      Estimate     Std. Error    t-value
Constant   -1.88002     0.221641      -8.482
log[D]      2.15514     0.0946287     22.775
S          -0.168918    0.0614421     -2.749
F          -0.0558921   0.0605592     -0.923

R Squared:            0.965764
Sigma hat:            0.148293
Number of cases:      24
Degrees of freedom:   20

Summary Analysis of Variance Table
Source       df      SS          MS          F        p-value
Regression   3       12.4068     4.1356      188.06   0.0000
Residual     20      0.439819    0.0219909
```

sirable, from the plot it appears that a common slope model may be adequate, and for exposition here we will assume that the effects of F and S are additive.

This suggests fitting a mean function with $\log(W)$ as the response and with terms S, F, and $\log(D)$, as summarized in Table 12.4. Of the two categorical variables, only source S seems to be important, and F can probably be dropped from the mean function without important loss of information. We see that the estimate for the coefficient of $\log(D)$ is 2.155 with a standard error of 0.095. From this we can compute the t-statistic for testing $m_1 = 0$ as $(2.155 - 2)$ $/0.095 = 1.63$. Comparing this to the t_{20} distribution gives a two-tailed p-value of about 0.12, so these results are reasonably consistent with $m_1 = 0$. At least for these four-year-old trees, $E(\text{height} \mid D)$ appears to be independent of D = diameter, trees of seed source zero weigh more on the average than trees of seed source one, and fertilizer does not appear to have an effect on weight.

12.7 FACTORIAL EXPERIMENTS

In some experiments, data are collected at all combinations of levels of a few factors. For example, a $2 \times 3 \times 4$ factorial experiment has three factors, and the first factor has two levels, the second has three levels, and the third

TABLE 12.5   The Wool Data

Variable	Definition
Len	Length of test specimen (250, 300, 350 mm)
Amp	Amplitude of loading cycle (8, 9, 10 mm)
Load	Load put on the specimen (40, 45, 50 g)
Cycles	Number of cycles until the specimen fails

factor has four levels. In a single replicate of a complete factorial experiment, one observation is taken at each possible combination of the factors, so in a $2 \times 3 \times 4$ experiment there will be $n = 24$ observations. Experiments are often done with more than one replication; and in experiments with many factors, fractional replication, in which some combinations of the factor levels are not used, may be required. Most of the methodology described in this book can be applied to the analysis of factorial experiments. However, there is a huge literature on designing and analyzing experiments with factors, including many topics that will not be covered here; the complements include a few references.

As an example, the data in the file wool.lsp are from a small experiment to understand the strength of wool as a function of three factors that were under the control of the experimenter. The variables are summarized in Table 12.5. We will use the logarithm of the number of cycles to failure $\log_2(Cycles)$, as the response. This data set has three factors, each at three levels, and would be called a 3^3 design. It has a single replication, so $n = 3^3 = 27$.

As a first analysis of these data, consider fitting a model with just *main effects*, which correspond to fitting indicator variables for the three factors without any interaction terms. Create three factors from the three variables and then fit the OLS regression of $\log_2(Cycles)$ on $\{F\}Len$, $\{F\}Amp$, and $\{F\}Load$, as summarized in the top part of Table 12.6. The coefficient estimates in Table 12.6 can be interpreted similarly to the coefficient estimates for factors in the earlier examples in this chapter. For example, the estimate for $\{F\}Len[300]$ is the estimated difference between the response when the load is 300 and when the load is 250, all other factors held fixed. The overall F-test confirms that the response is not independent of the predictors.

The analysis of experimental data like these is often summarized in a sequential analysis of variance table. Select the item "Examine submodels" from the regression menu. You need to select an order for entering the terms, but in balanced factorial experiments like this one, the order doesn't matter because all the terms are *orthogonal*. This means that the estimated coefficient for a term is the same with or without any of the other terms in the mean function. If the experiment is just slightly unbalanced—for example, if one of the 27 observations were missing—then orthogonality does not hold. Select any order you like. The result is in the bottom part of Table 12.6.

TABLE 12.6 Main Effects Only Model for the Wool Data

```
Normal Regression
Kernel mean function = Identity
Response      = log2[Cycles]
Terms         = ({F}Amp {F}Len {F}Load)
Coefficient Estimates
Label           Estimate    Std. Error    t-value
Constant         9.35281      0.139128     67.225
FAmp[9]         -0.945268     0.128807     -7.339
FAmp[10]        -1.82029      0.128807    -14.132
FLen[300]        1.32487      0.128807     10.286
FLen[350]        2.40175      0.128807     18.646
FLoad[45]       -0.469294     0.128807     -3.643
FLoad[50]       -1.13286      0.128807     -8.795

R Squared:              0.96908
Sigma hat:              0.273242
Number of cases:        27
Degrees of freedom:     20

Summary Analysis of Variance Table
Source       df       SS          MS          F       p-value
Regression    6      46.7999     7.79999    104.47    0.0000
Residual     20       1.49322    0.074661
```

```
Terms         = ({F}Amp {F}Len {F}Load)
Sequential Analysis of Variance
All fits include an intercept.
                    Total                         Change
Predictor   df     RSS        |   df     RSS          MS
Ones        26    48.2932     |
FAmp        24    33.3751     |    2    14.918       7.45901
FLen        22     7.32501    |    2    26.0501     13.0251
FLoad       20     1.49322    |    2     5.83179     2.9159
Residual                      |   20     1.49322     0.074661
```

The usual summary for this analysis consists of the last three columns of the bottom part of Table 12.6. We see that each of the changes for adding a term has two df, because each main effect has two df. Because of the orthogonality of the design, the ratio of each of the changes in sum of squares for fitting a main effect to the residual mean square is an F-statistic for testing that the regression coefficients for that term are all equal to zero; all three main effects are clearly nonzero in this regression.

The analysis of these data could continue by adding interactions to the mean function to see if they are required to summarize $\log_2(Cycles)$ adequately; we reserve this to Problem 12.7.

12.8   COMPLEMENTS

12.8.1   Alternate Definitions of Factors

In addition to the method described in Section 12.1, the "Make factors" dialog includes two other options for constructing the terms that make up a factor. These alternatives can be useful in some circumstances.

One Indicator for Each Level. Let's return to the turkey growth data. Suppose that we choose to define a factor from S using the option "One indicator for each level," as shown in the sample dialog in Figure 12.1. This would create three indicators: v_1 for level one of S, v_2 for level two of S, and v_3 for level three of S. A factor created in this way is indicated by the prefix $\{T\}$.

Since by construction $v_1 + v_2 + v_3 = 1 = u_0$, the additional term v_1 is in fact redundant information: Given the value of v_2 and v_3, the value of v_1 is determined. However, it can be useful to fit all three indicators to allow alternative interpretations of the parameters. If we fit the regression model with response y and predictors $\{T\}S$ and D *through the origin*, we will be fitting the mean function

$$E(y \mid S,D) = \eta_{01}v_1 + \eta_{02}v_2 + \eta_{03}v_3 + \eta_1 D$$

and the three parameters η_{01}, η_{02} and η_{03} are directly interpretable as the three source intercepts. Similarly, if we fit the regression of y on $\{T\}S$ and $D\{T\}S$, again through the origin and omitting the term D from the mean function, we will get

$$E(y \mid S,D) = \eta_{01}v_1 + \eta_{02}v_2 + \eta_{03}v_3 + \eta_{11}v_1 D + \eta_{12}v_2 D + \eta_{13}v_3 D$$

which give the intercepts and the slopes directly. These are just reparameterizations of the models from Section 12.4, as discussed in Section 10.1.3. Thus the results are really the same if we fit with the factor defined as $\{F\}S$ or $\{T\}S$.

Effect Coding. The final option of *effect coding* is a bit more complicated, although it too will give results that are equivalent to using simple indicator variables. The terms created under effect coding have the values 0, 1, or -1. For a factor S with ℓ levels, the $\ell - 1$ terms $v_1, \ldots, v_{\ell-1}$ are defined as follows:

$$v_j = \begin{cases} 1 & \text{if } S = j \\ -1 & \text{if } S = \ell \\ 0 & \text{otherwise} \end{cases}$$

A factor created using this coding is denoted by the prefix $\{C\}$.

If we fit the regression model with response y and predictors $\{C\}S$ and D, we will be fitting the mean function

$$E(y \mid S, D) = \eta_0 + \eta_{11}v_1 + \eta_{12}v_2 + \eta_2 D$$

In this parameterization, the intercepts for the three sources are $\eta_0 + \eta_{11}$, $\eta_0 + \eta_{12}$, and $\eta_0 - \eta_{11} - \eta_{12}$. If you add the three intercepts together, you will get a multiple of η_0. The quantities η_{11}, η_{12} and $-\eta_{11} - \eta_{12}$ are called *effects* that add to zero. This parameterization is generally used in discussing experimental designs, using the constraint to make the effects of a factor sum to zero. Effect coding is used in some standard computer programs.

12.8.2 Comparing Slopes from Separate Fits

Probably the most common problem in comparing groups is testing for parallel slopes in simple regression with two groups. Since this F-test has 1 df in the numerator, it is equivalent to a t-test. Let $\hat{\eta}_{1j}, \hat{\sigma}_j^2, n_j$, and SXX_j be, respectively, the estimated slope, residual mean square, sample size, and sum of squares of the predictor within group j, for $j = 1, 2$. Then a pooled estimate of σ^2 is

$$\hat{\sigma}_p^2 = \frac{(n_1 - 2)\hat{\sigma}_1^2 + (n_2 - 2)\hat{\sigma}_2^2}{n_1 + n_2 - 4}$$

and the t-test for equality of slopes is

$$t = \frac{\hat{\eta}_{11} - \hat{\eta}_{12}}{\hat{\sigma}_p(1/SXX_1 + 1/SXX_2)^{1/2}}$$

with $n_1 + n_2 - 4$ df. The square of this t-statistic is numerically identical to the corresponding F-statistic.

12.8.3 References

Factorial experiments and designs mentioned briefly in Section 12.7 are very widely used. There are many books devoted to this topic, including Kuehl (1994) and the classic in this area, Cochran and Cox (1957). The twin data were given by Burt (1966). The turkey data are from Noll, Weibel, Cook, and Witmer (1984). The *Casuarina* data were provided by Ross Cunningham. The wool data are taken from Box and Cox (1964). The FACE-1 data in Problem 12.8 are from Woodley, Simpson, Bioindini, and Berkeley (1977). The cathedral data in Problem 12.3 were provided by Stephen Jay Gould.

PROBLEMS

12.1 In the turkey growth data, show that the quadratic regression model discussed in Section 12.5 is adequate, perform an appropriate test comparing the equal intercept and coincident mean functions, and summarize your results.

12.2 Give a complete summary of the differences between long and short shoots in the apple shoots data, Problem 9.8.

12.3 The data in the file gothic.lsp give y = total length and x = nave height, both in feet for medieval English cathedrals. The cathedrals are classified according to their architectural style, either Romanesque for the earlier cathedrals or Gothic for the later cathedrals. Some cathedrals have both a Gothic and a Romanesque part, each of differing height; these cathedrals are included twice.

Use the data to decide if the relationship between length and height is the same for the two architectural styles. If they differ, describe the differences.

12.4 In the twin data, one way to assess the effects of environment on IQ is to investigate the difference in IQ, $IQd = IQb - IQf$. If environment has no effect, then we might expect that

$$\mathrm{E}(IQd \,|\, C) = 0 \quad \text{with} \quad \mathrm{Var}(IQd \,|\, C) = \sigma^2$$

Using an F-test, compare this model to the more general model in which

$$\mathrm{E}(IQd \,|\, C) = \eta_0 + \eta_1 v_2 + \eta_3 v_3 \quad \text{with} \quad \mathrm{Var}(IQd \,|\, C) = \sigma^2$$

12.5 Table 12.7 gives output from fitting Model 4, coincident lines, to the turkey data shown in Table 12.3 with the number of replicates as weights. In forming the output related to pure error, Arc assumes that the three values of y at each value of D are replicates. They are not, however, because each corresponds to a different source. An important point here is that computer programs cannot assess true replication. The three values of y at each value of D would be equivalent to replicates only if the three sources were in fact equivalent.

Construct a test of lack-of-fit of Model 4 using the within treatment sum of squares for an estimate of pure error.

12.6 The file salary.lsp contains data on yearly salary and other characteristics of faculty in a small midwestern college. The data were collected as evidence in a class-action case to judge a claim of salary discrimination against women. All persons represented in the data file held

TABLE 12.7   Fit of the Coincident Regression Model to the Turkey Data

```
Data set = Turkey, Name of Fit = L1
Normal Regression
Kernel mean function = Identity
Response    = y
Terms       = (D)
Weights     = N
Coefficient Estimates
Label         Estimate      Std. Error    t-value
Constant      647.288       6.91625       93.589
D             532.281       43.9435       12.113

R Squared:                  0.930256
Sigma hat:                  35.5023
Number of cases:            13
Degrees of freedom:         11

Summary Analysis of Variance Table
Source        df      SS          MS          F         p-value
Regression    1       184929.     184929.     146.72    0.0000
Residual      11      13864.5     1260.41
Lack of fit   3       11289.2     3763.07     11.69     0.0027
Pure Error    8       2575.33     321.917
```

TABLE 12.8   Variables in the Salary Data

Variable	Description
Sex	Sex, coded 1 for female and 0 for male.
Rank	Rank: 1 = Assistant Professor, 2 = Associate Professor, 3 = Full Professor.
Year	Number of years in current rank.
Degree	Highest degree, coded 1 for Doctorate and 0 for Masters.
YD	Number of years since highest degree was earned.
Salary	Academic year salary in dollars.

tenured or tenure-track positions at the time of the study; temporary faculty were not included in the action. The variables described in Table 12.8 are included in the data file. In all sections of this problem, use $y = \log_2(Salary)$ as the response variable.

12.6.1   Write a mean function for the regression of y on all five predictors that allows for the possibility of an interaction between *Rank* and *Sex*, but forces the effects of the other three predictors to be additive. Describe how to test for an interaction between *Rank* and *Sex* in your model.

12.6.2 Write a mean function for the regression of y on all five predictors that allows for the possibility of an interaction between *Rank* and *Sex* and an interaction between *Year* and *Sex* but forces the effects of the other two predictors to be additive. Describe how to test for an interaction between *Rank* and *Sex* in your model. Describe how to test for an interaction between *Year* and *Sex* in your model.

12.6.3 Construct a plot of y versus *Year* with the points marked by *Rank*. Based on your interpretation of the plot, suggest a mean function for the regression of y on *Year* and *Rank*.

12.6.4 Using all five predictors, give the mean function that contains the two-factor interactions $\{F\}Rank \times Degree$, $\{F\}Rank \times Sex$, and $Sex \times Degree$ and is additive in the other two predictors. Briefly explain the roles of the various terms in this mean function.

12.6.5 If there is no discrimination, the mean function should be independent of *Sex*:

$$E(y \mid Rank, Degree, Year, YD, Sex) = E(y \mid Rank, Degree, Year, YD)$$

Starting with the mean function of Problem 12.6.4, investigate the evidence for discrimination using appropriate tests.

12.6.6 Starting with the mean function of Problem 12.6.4 and neglecting the issue of discrimination, develop a model for predicting y.

12.6.7 Starting with the mean function of Problem 12.6.4, is there any evidence in the data that *Year* and *YD* have nonlinear effects?

12.6.8 Starting with the mean function of Problem 12.6.4, describe when the *three-factor interaction* would be needed. No calculations are necessary.

12.6.9 Salary increases are usually given as a percentage of the current salary. Using the regression of y on *Rank*, estimate the average percentage increase for each rank. Is there evidence that the three ranks receive different percentage increases on the average? Is there evidence that the percentage increase changes systematically with time?

12.7 This problem continues the analysis of the wool data, begun in Section 12.7.

12.7.1 Obtain the sequential analysis of variance table, as in the bottom part of Table 12.6, but do the fitting using a different order. Show that the results are the same.

12.7.2 Since this design is orthogonal, the correlations between the coefficient estimates for the indicator variables should all be

TABLE 12.9   Definitions of Variables in the FACE Experiment on Cloud Seeding

A	Action, coded 1 of the day is seeded with silver iodide, 0 otherwise. The decision to seed or not was made at random for each suitable day.
D	A time trend: The number of days after June 15, 1975.
S	The suitability for seeding score. A score of $S \geq 1.5$ was required for including the day in the study.
C	Cloud cover in the experimental area, measured using radar in Coral Gables, Florida.
P	Prewetness, the amount of rainfall in the hour preceding seeding, in 10^7 cubic meters.
E	Echo motion category, either 1 or 2, a measure of the type of clouds present.
$Rain$	Rainfall in 10^7 cubic meters in the target area following the action.

zero. Obtain the correlation matrix of the estimates and verify that this is in fact true.

12.7.3   Use the "Table of" command to obtain a table of means for $\log_2(Cycles)$, conditioning on *Load*. Show that the estimated coefficient for $\{F\}Load[45]$ is equal to the difference between the mean for $\log_2(Cycles)$ given *Load* = 45 and *Load* = 40. Show that the standard error of this coefficient is given by $\hat{\sigma}(2/9)^{1/2}$, and explain where this formula comes from.

12.7.4   Refit the data, but use a mean function that includes all main effects and all two-factor interactions. Verify that the two-factor interactions are orthogonal to each other and orthogonal to the main effects. Test the hypotheses that each of the two-factor interactions is zero against a general alternative.

12.8   Cloud Seeding. Judging the success or failure of cloud seeding designed to increase rainfall is an important practical problem. Results for cloud seeding experiments have generally been mixed: Sometimes the observed effect of seeding has been to increase rainfall, sometimes to decrease rainfall, and sometimes no effect is observed. During the rainy season, the suitability of each day for seeding is judged, and on suitable days a decision to seed or not is made at random. The total rainfall in a target area during a fixed period of time following the treatment is the response variable. Comparing the rainfall on seeded days to unseeded days provides the basis for deciding if seeding increases rainfall.

Data from the first Florida Area Cumulus Experiment (FACE) from 1975 are given in the file cloud.lsp. In that year, $n = 24$ days were judged to be suitable, and the variables in Table 12.9 were recorded.

12.8.1   In these data, how many days are seeded? How many unseeded? Draw a boxplot of $\log(Rain)$ conditioned on A, and summarize the information in this plot. Given the information in this boxplot, which assumption underlying the test for $\eta_1 = 0$ in the fit the mean function $E(\log(Rain) \mid A) = \eta_0 + \eta_1 A$ with $\text{Var}(\log(Rain) \mid A) = \sigma^2$ is likely to be violated?

A storm, whether seeded or not, may produce rain over a target or it may change course and produce rain elsewhere. It could produce heavy rain in a small area or light rain over a larger area. Consequently, direct comparison of seeded and unseeded days may be too variable to find differences due to seeding because of the natural variability in rainfall. To improve precision, we use other predictors that may account for some of the variability that is not due to the treatment.

12.8.2 Suppose we consider first only a single additional predictor S, and consider the mean function

$$E(\log(Rain) \mid A, S) = \eta_0 + \eta_1 A + \eta_2 S + \eta_3 AS$$

Give hypothetical values of the η's that correspond to the following situations.

a. Rainfall is independent of A.

b. Rainfall is independent of S.

c. Seeding increases $\log(Rain)$ with the same expected increase for all values of S.

d. Seeding increases $\log(Rain)$, but the increase is larger for larger values of S.

e. When S is small, seeding *increases* $\log(Rain)$, but when S is large enough, seeding *decreases* $\log(Rain)$.

12.8.3 Consider next adding the term $\log(P)$ to the mean function,

$$E(\log(Rain) \mid A, S, P) = \eta_0 + \eta_1 A + \eta_2 S + \eta_3 AS$$
$$+ \eta_4 \log(P) + \eta_5 A \log(P)$$

Summarize the meaning of η_4 and η_5 in this mean function.

12.8.4 Draw the scatterplot matrix of $\log(Rain)$ and the predictors (D, S, C, P), using A as a marking variable. Identify any apparently unusual cases, and explain why they are unusual. Transform the predictors as needed to make the mean functions in the 2D frames of the scatterplot matrix linear. What additional information is available from marking the points in two colors?

12.8.5 Use OLS to fit the model with mean function $E(\log(Rain) \mid \mathbf{x})) = \eta^T \mathbf{u}$, with

$$\mathbf{u}^T = (D, S, \log(C), A, AS)$$

and summarize results, particularly with regard to the coefficients for A and AS.

12.8.6 Identify any cases with high leverage in the fit of the model in Problem 12.8.5. What causes the large leverage? Was this obvious from examination of the scatterplot matrix in Problem 12.8.4? Refit the model as in Probelm 12.8.5, after deleting the two cases with largest leverage. Is there much change in the results?

We will continue this analysis in Problem 15.8.

CHAPTER 13

Response Transformations

In the transactions data, mean function (7.9)

$$E(Time \mid T_1, T_2) = \eta_0 + \eta_1 T_1 + \eta_2 T_2$$

makes sense: The expected number of minutes is equal to the number of each type of transaction multiplied by the average transaction time. In the haystack data, we are not certain which form to use for the mean function, but since the volume of a hemisphere is proportional to the cube of its circumference, we might expect to use C^3 and $Over^3$ as terms, or else possibly transform both predictors and the response to log scale. In the Big Mac data, Section 7.3.5, or the sleep data, Section 10.4, we don't have a theory to help us convert predictors to terms or to choose an appropriate transformation for the response. For either of these data sets and for many others, the multiple linear regression model may provide a useful approximation to the conditional distribution of the response y given the predictors $\mathbf{x}$, but only after y and $\mathbf{x}$ are transformed, often using a logarithmic transformation, but in general using various types of transformations.

The general goal of transforming is simply stated: Find a transformation of y and terms $\mathbf{u}$ derived from $\mathbf{x}$, so that the regression of the transformed y on $\mathbf{u}$ is linear and perhaps has other desirable properties as well. In this chapter, we consider transforming y alone, as well as *transformations to normality* where both the response and the predictors may be transformed. Predictor transformations are discussed in Chapter 16.

13.1 RESPONSE TRANSFORMATIONS

We use the notation $T(y)$ to represent a strictly monotonic transformation of y. Examples might be $T(y) = \log(y)$ or $T(y) = \sqrt{y}$. In the first case, $T(y)$ is the logarithmic transformation, and in the second it is the square root transformation. A strictly monotonic function is either always increasing or always decreasing.

316

Transformations of y are chosen to meet a particular goal. For example, we can choose $T(y)$ so that the mean function $E(T(y) \mid \mathbf{x}) = \eta^T \mathbf{u}$ for a fixed set of terms $\mathbf{u}$. We call this the goal of *transforming for linearity*. A second possibility is to select the transformations to make the errors in the transformed scale as close to normally distributed as possible. This is the goal of *transforming for normality*. A third goal of transforming the response might be to stabilize the variance function $\text{Var}(T(y) \mid \mathbf{x})$ by making it constant. In a surprisingly large number of regressions, linearity and variance stabilization can be achieved with the same transformation.

13.1.1 Variance Stabilizing Transformations

Variance stabilizing transformations can be useful when the variance function of a strictly positive response y at $\mathbf{x}$ depends on the value of the mean function at $\mathbf{x}$, or symbolically,

$$\text{Var}(y \mid \mathbf{x}) = \sigma^2 v(E(y \mid \mathbf{x})) \tag{13.1}$$

where $v(\cdot)$ is a function called the *kernel variance function*. For example, if $y \mid \mathbf{x}$ has a Poisson distribution, Section 23.5.1, then the variance function is equal to the mean function, so $v(E(y \mid \mathbf{x})) = E(y \mid \mathbf{x})$. For distributions in which the mean and variance are functionally related as in (13.1), a general theory tells us how to find a transformation $T(y)$ that will stabilize variance. For the Poisson case, for example, the transformed response $T(y) = \sqrt{y}$ has variance that is nearly constant. Table 13.1 summarizes the common variance stabilizing transformations. The first three transformations are used when the relationship between the mean and variance is monotonic, while the last one can be used with binomial data, for which the variance depends on the probability of success, and is largest when the probability is 0.5 and is smallest when the probability is close to zero or one.

An alternative to variance stabilizing transformations is to use *generalized linear models* that use information about the relationship between the mean function and the variance function in estimation. These models are discussed for binomial data in Chapters 21 and 22, and for other types of data in Chapter 23.

13.1.2 Transforming to Linearity with One Predictor

Suppose we have a regression with response y and single predictor x. If the response plot y versus x is clearly curved, then we know that the mean function $E(y \mid x)$ cannot be summarized by the simple linear regression on x. Sometimes a nonlinear relationship can be turned into a linear one by a suitable monotonic transformation $T(y)$ of y. If transformation can linearize the mean function, the regression model in the transformed scale is

$$T(y) \mid x = \eta_0 + \eta_1 x + e \tag{13.2}$$

TABLE 13.1 Common Variance Stabilizing Transformations

$T(y)$	Comments
$\sqrt{y}$ or $\sqrt{y} + \sqrt{y+1}$	Appropriate when $\mathrm{Var}(y \mid \mathbf{x}) \propto E(y \mid \mathbf{x})$, for example for Poisson distributed data. The latter form is called a *Freeman–Tukey deviate*, and it gives better results if the y_i are small or if some $y_i = 0$.
$\log(y)$	Though most commonly used to achieve linearity, this is a variance stabilizing transformation when $\mathrm{Var}(y \mid \mathbf{x}) \propto [E(y \mid \mathbf{x})]^2$. It can be appropriate if the errors are a percentage of the response, like $\pm 10\%$, rather than an absolute deviation, like ± 10 units.
$1/y$	The inverse transformation stabilizes variance when $\mathrm{Var}(y \mid \mathbf{x}) \propto [E(y \mid \mathbf{x})]^4$. It can be appropriate when responses are mostly close to zero, but occasional large values occur.
$\sin^{-1}(\sqrt{y})$	This is usually called the *arcsine square-root* transformation. It stabilizes variance when y is a proportion between zero and one, but it can be used more generally if y has a limited range by first transforming y to the range $(0, 1)$ and then applying the transformation.

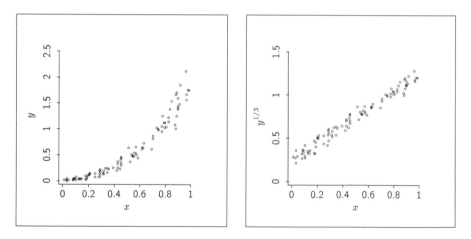

a. Response y. b. Response $y^{1/3}$.

FIGURE 13.1 The mean function for the summary plot (a) is curved. In (b), y has been replaced by $y^{1/3}$, and the mean function is linear. The variance function is also different in the two plots.

Analyzing a model that is linear in x on a transformed scale is often much easier than analyzing a model that is nonlinear in x in the original scale.

Consider the response plot y versus x shown in Figure 13.1a. You can reproduce this figure from the data file `resptran.lsp`. Since the mean function for this plot is nonlinear, we can use simple linear regression only if we transform y, transform x, or transform both. Figure 13.1b is obtained from Figure 13.1a by replacing y with $T(y) = y^{1/3}$. No curvature is apparent in this plot, and so we could proceed by fitting the linear regression model (13.2) with $T(y) = y^{1/3}$. The plot in Figure 13.1b provides a visualization of the regression function $E(T(y) \mid x)$. Transforming the response in this example has linearized the mean function and changed the variance function. In Figure 13.1a the variance increases with x, while in Figure 13.1b the variance function is constant. The transformation thus achieved a linear mean function and a constant variance function.

In the example we knew the appropriate transformation $T(y) = y^{1/3}$, but we will not normally have this information. How can we determine a suitable response transformation with real data? The plot in Figure 13.1a does not help because the mean function for the plot is $E(y \mid x)$. However, the *inverse response plot* of x versus y can help us choose a suitable transformation when the untransformed regression function $E(y \mid x)$ is monotonic, as in Figure 13.1a. If $E(y \mid x)$ is not monotonic, then a response transformation may not be appropriate, and adding terms like a quadratic to the mean function may be needed.

An inverse response plot is useful for choosing a transformation $T(y)$ when $E(y \mid x)$ is monotonic because then

$$E(x \mid y) \approx T(y) \qquad\qquad (13.3)$$

This equation tells us that the mean function for the regression of x on y is approximately the required transformation. The transformation itself can be estimated by fitting a curve to the plot. Once an appropriate estimate is determined, the transformed values can be extracted and then used as the response in further analysis.

This method will work well as long as equation (13.3) holds to a reasonable approximation. Equation (13.3) will be a good approximation if the signal dominates the noise. If the plot x versus y shows a well-determined curve, then (13.3) is a good approximation and the curve can be used to guide the selection of a transformation. If the plot x versus y does not show a well-determined curve, then finding a useful response transformation will likely be problematic.

For the example in Figure 13.1a, the inverse response plot of x versus y is shown in Figure 13.2. The plot shows a power curve obtained by using the "Power curve" menu item from the parametric slidebar pop-up menu, with power 0.33. Since the curve matches the data, the cube-root transformation is suggested. Figure 13.1b shows the response plot in the cube-root scale.

Arc fits power curves in the following way. Let λ denote the power for the scaled power transformation defined in Section 5.2, as shown at the right of the power curve slidebar, and let h and v denote the variables plotted

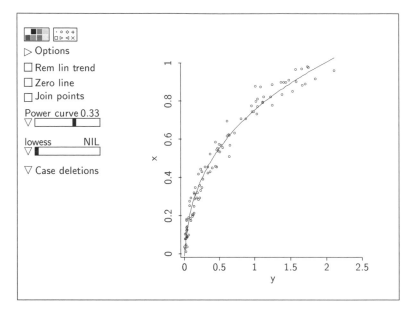

FIGURE 13.2   Inverse response plot corresponding to Figure 13.1a. The power curve added suggests transforming y to $y^{1/3}$.

on the horizontal and vertical axes. For each value of λ, *Arc* first

- constructs the fitted values $\hat{v}_i = \hat{\eta}_0 + \hat{\eta}_1 h_i^{(\lambda)}$ from the OLS fit of the simple linear regression with response v and predictor $h^{(\lambda)}$ and then
- superimposes a plot of $\hat{v}_i$ versus h_i on the plot, connecting adjacent points for visual clarity.

13.1.3   Inverse Fitted Value Plot

The ideas discussed in the last section can be extended to regressions with many predictors. We suppose that there is a transformation $T(y)$ so that

$$\mathrm{E}(T(y) \mid \mathbf{x}) = \eta^T \mathbf{u} \tag{13.4}$$

The analog to the inverse response plot for the one-predictor case is the *inverse fitted value plot*, in which we plot on the vertical axis the fitted values $\hat{y}$ from the OLS fit of y on $\mathbf{u}$ and the response y on the horizontal axis. As with the one-predictor case, a power curve or even a nonparametric curve can be used to estimate the mean function of this plot. The estimated mean function is then our estimate of $T(y)$. If the inverse fitted value plot has a well-determined mean function, then the estimate of that mean function provides an estimate of $T(y)$.

The procedure just described requires a condition that is not needed for the one-predictor case. The condition has to do with all the u-terms in model (13.4) excluding the constant $u_0 = 1$ for the intercept; the condition does not involve the response. Let u_j be one of the nonconstant terms. Then the mean functions $E(u_j \mid \eta^T \mathbf{u})$ should all be linear:

$$E(u_j \mid \eta^T \mathbf{u}) = a_j + b_j \eta^T \mathbf{u} \qquad \text{for} \quad j = 1, \ldots, p \qquad (13.5)$$

This condition enables us to gain information on η in (13.4) without knowing $T(y)$ in the first place. We will identify situations in which (13.5) holds by saying that the *terms are linearly related*. In many situations the terms will be the same as the predictors, in which case we will refer to *linearly related predictors*. The consequences of having linearly related predictors are discussed more fully in Section 19.2.

Condition (13.5) cannot be checked directly because η is unknown. However, it is guaranteed to hold when the terms follow a multivariate normal distribution. Condition (13.5) should also hold to a good approximation if every frame in a scatterplot matrix of the terms has a mean function that is either linear, or at least not noticeably curved. If any frame has a clearly curved mean function, then condition (13.5) may fail. In that case we may attempt to induce linearly related terms by transforming to multivariate normality. Methodology for inducing multivariate normality is outlined in this chapter and expanded in Section 19.1.

As an example, we use the wool data described in Section 12.7. The response variable $y = Cycles$ is the number of loading cycles to failure of worsted yarn. The three predictors are *Len*, *Amp*, and *Load*. The response was measured at all possible combinations of three settings for each predictor, resulting in $3^3 = 27$ observations. Load the file `wool.lsp` and specify *Cycles* as the response and specify the other three variables as terms. The fitted values $\hat{y}$ from the OLS fit of y on the three terms will be used to study the need for a transformation of the response.

Draw the plot $\hat{y}$ versus y as shown in Figure 13.3. If the mean function for Figure 13.3 were a straight line, no transformation would be needed. Since the plot is plainly curved and the mean function is monotonic, a transformation is needed.

The next question is the choice of transformation, and this can be determined visually by adding a fitted curve to the plot. Since the response is strictly positive, we can try power curves. Shown in Figure 13.3 is the curve for $\log(y)$, which seems to match the data quite well, suggesting that $\log(y)$ is appropriate.

13.1.4   Numerical Choice of Transformation

The *Box–Cox method* is a numerical procedure for choosing a response transformation. The use of an inverse fitted value plot requires linear relationships

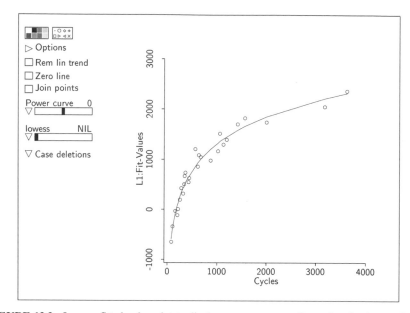

FIGURE 13.3   Inverse fitted value plot to display a response transformation for the wool data.

among the terms, and then a transformation to linearity is selected. The Box-Cox method doesn't require linearly related terms, although it will work better if terms are linearly related. While the inverse fitted value plot selects a transformation for linearity, the Box–Cox method selects a transformation to make the errors as close to normally distributed as possible.

Using the Box–Cox method requires specifying a *family of transformations* indexed by a parameter λ, so choice of transformation is reduced to estimation of λ. We will find both point and interval estimates of this parameter.

We recall from Section 5.2 the definition of a scaled power transformation. If the variable y is positive, we define $y^{(\lambda)}$ from (5.1) by

$$
y^{(\lambda)} = \begin{cases} (y^\lambda - 1)/\lambda & \text{if} \quad \lambda \neq 0 \\ \log(y) & \text{if} \quad \lambda = 0 \end{cases}
$$

We assume that there is a value of λ so that the linear regression model holds in the transformed scale,

$$
y^{(\lambda)} \mid \mathbf{x} = \eta^T \mathbf{u} + e \tag{13.6}
$$

where the errors are normally distributed with mean zero and constant variance. We might consider estimating λ by choosing the value that minimizes the residual sum of squares $\text{RSS}(\lambda)$ from the OLS regression of $y^{(\lambda)}$ on $\mathbf{u}$. While the general idea is right, the details are wrong because the units of $\text{RSS}(\lambda)$ are different for every value of λ. Consequently, we can't compare values of

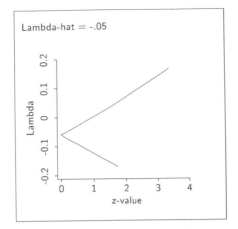

FIGURE 13.4   Confidence curves for choosing a transformation in the wool data.

RSS(λ) for different values of λ. A way out is to adjust the transformation so that the units are always the same. Define a *modified power transformation* $z(\lambda)$ by

$$z(\lambda) = y^{(\lambda)} \mathrm{gm}(y)^{1-\lambda}$$

where $\mathrm{gm}(y)$ is the geometric mean of the observed values of y. The estimate of λ minimizes $\mathrm{RSS}_z(\lambda)$, the residual sum of squares from the regression of $z(\lambda)$ on **x**.

Finding the value $\hat{\lambda}$ that minimizes $\mathrm{RSS}_z(\lambda)$ is now a one-dimensional minimization problem easily solved numerically on the computer. The results can be viewed in a plot using *confidence curves*, which are a plot of

$$\lambda \text{ versus } \{n[\log(\mathrm{RSS}_z(\lambda)) - \log(\mathrm{RSS}_z(\hat{\lambda}))]\}^{1/2}$$

For the wool data, the confidence curves are given in Figure 13.4. The value $\hat{\lambda}$ that minimizes $\mathrm{RSS}_z(\lambda)$ is the point where the curves meet the vertical axis, which is $\hat{\lambda} = -0.05$ for the wool data. The confidence curves can be used to get an interval estimate for λ. The horizontal axis of Figure 13.4 is labeled z-value. Values on this axis correspond to the standard normal distribution. For example, the interval between the two curves at 1.96 on the horizontal axis, which is approximately -0.18 to $+0.06$, is a 95% confidence interval for λ because the area under a standard normal curve between -1.96 and $+1.96$ is 95%. The log transformation is suggested because $\hat{\lambda}$ is close to zero, in agreement with the transformation chosen using the inverse fitted value plot.

To draw Figure 13.4, select the item "Choose response transform" from the regression menu. You will get a dialog to choose the transformation family, and whether or not a constant is to be added to the response before transformation.

The defaults are appropriate here, so just push the "OK" button. The other choices are discussed in the complements.

The estimate of λ will generally fall in the range between about -2 and $+2$. Estimates outside this range often, but not always, indicate that transformation of the response will not aid in understanding a regression.

We now have two methods for choosing a response transformation: (1) the inverse fitted value plot described in Section 13.1.3 and (2) the Box–Cox method. For the wool data, these two gave the same transformation, but they need not always agree. The inverse fitted value plot chooses transformations to linearize the response function, while the Box–Cox method tries to make the errors in the transformed scale as close to normally distributed as possible. For example, suppose that $E(y \mid x)$ was independent of x, but with errors that have a skewed distribution. The inverse fitted value plot will suggest no transformation, while the numerical method will choose a transformation to make y more nearly normally distributed.

13.2 TRANSFORMATIONS TO NORMALITY

The multivariate normal distribution was introduced in Section 10.2. If the pair (y, x) has a multivariate normal distribution, then according to (10.7), the regression of y on x follows a linear regression model with constant variance function. This brings up an interesting idea: Can we find a transformation $T(y)$ of y and a set of terms $\mathbf{u}$ that are element-wise transformations of x, such that the pair $(T(y), \mathbf{u})$ has a multivariate normal distribution, excluding the constant $u_0 = 1$? If so, we then know that the multiple linear regression model is appropriate for the conditional distribution of $T(y) \mid \mathbf{u}$, and this model can provide a basis for understanding. We know this cannot be done in all data sets—for example, a factor cannot be transformed to be approximately normal—but this idea can be very useful in problems with continuous predictors.

13.2.1 Visual Choice of Transformation

Return to the Big Mac data in file `big-mac.lsp` described in Section 7.1, and focus on the conditional distribution of $y = BigMac$ given $x = (TeachSal, EngSal)$. These three variables are shown in the scatterplot matrix in Figure 13.5. If these three variables had a joint normal distribution, then every frame of this scatterplot matrix would resemble a plot from a bivariate normal distribution: All mean functions would be linear, and all variance functions would be constant. In examining Figure 13.5, we see that the mean functions $E(TeachSal \mid EngSal)$ and $E(EngSal \mid TeachSal)$ are approximately linear.

Figure 13.6a gives the plot of *EngSal* versus *TeachSal*, but with the linear trend removed and with a zero line and *lowess* smooth added. The plot suggests that there is some nonlinearity in the mean function, so perhaps linearity, and therefore normality, can be improved by transformation. Since both

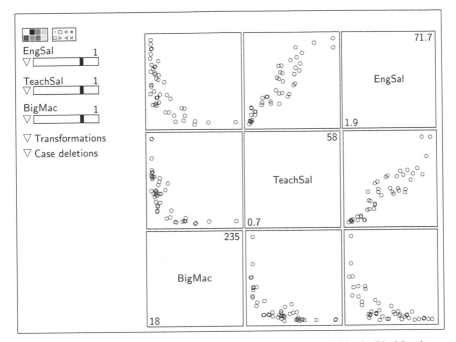

FIGURE 13.5 Scatterplot matrix for *BigMac*, *TeachSal*, and *EngSal* in the Big Mac data.

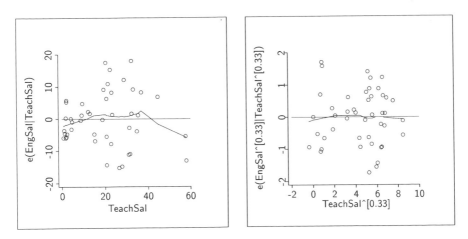

a. No transformations.   b. Cube root scale.

FIGURE 13.6 Residuals from the regression of *EngSal* on *TeachSal* with the linear trend removed.

EngSal and *TeachSal* are measured in the same units, we might require using the same transformation for both variables. Figure 13.6b shows the resulting plot with both variables in cube root scale. Linearity seems apparent here.

The plots in Figure 13.5 that include *BigMac* are clearly curved, and once again transforming may help. After a bit of exploration, using $BigMac^{-1/3}$ seems to linearize the mean function in all frames of the scatterplot matrix.

We have found that $(BigMac^{-1/3}, EngSal^{1/3}, TeachSal^{1/3})$ has an approximately linear mean function for each pair of variables, with no clear evidence of nonconstant variance. We know that if the three variables have a multivariate normal distribution, then all 2D plots have a linear mean function with constant variance. Finding linear mean functions and constant variance in the frames of a scatterplot matrix does not guarantee multivariate normality, but it is usually a good indicator. With only three variables, we can explore multivariate normality using a 3D plot, by plotting the three variables and then removing all linear trends (push the "Rem lin trend" and "O to e(O|H)" buttons). The plot should then resemble a spherical cloud of points with constant mean and variance functions in every 2D view. Draw this plot to verify that this actually happens in this data set. Treating $(BigMac^{-1/3}, EngSal^{1/3}, TeachSal^{1/3})$ as if it were normally distributed is therefore reasonable, and this immediately implies that the regression of $BigMac^{-1/3}$ on $(EngSal^{1/3}, TeachSal^{1/3})$ can be well-approximated by a linear model.

The units of the transformed variables are an interesting feature of these results. Both *EngSal* and *TeachSal* are in units of dollars per year, or in generic terms, items per unit time. *BigMac*, on the other hand, is in units of minutes of labor to buy a hamburger and French fries, or more generally in units of time per item. In the transformed scale, all three variables are in units of the cube root of items per unit time.

13.2.2 Automatic Choice of Transformations

Arc provides automatic tools that can be used to select a transformation so that the transformed data behave as much like a normally distributed random sample as possible. Consider transforming *TeachSal*. The goal is to find a scaled power transformation parameter λ so that $TeachSal^{(\lambda)}$ is as close to normally distributed as possible. This is actually just an application of the Box–Cox method introduced in Section 13.1.4.

Select the item "Find normalizing transformation" from the pop-up menu next to the slidebar for *TeachSal* in the scatterplot matrix in Figure 13.5. After a short calculation, the estimate from the Box–Cox method is returned as $\hat{\lambda}_{TeachSal} = 0.48$, not far from the value of $1/3$ determined visually. This process can be repeated for each of the variables, yielding transformations that make the *marginal* distribution of each variable as normal as possible.

A preferable procedure is to choose all the transformations at the same time to make the *joint* distribution of the variables as normal as possible, using a generalization of the Box–Cox method. Let λ be a vector of transformation

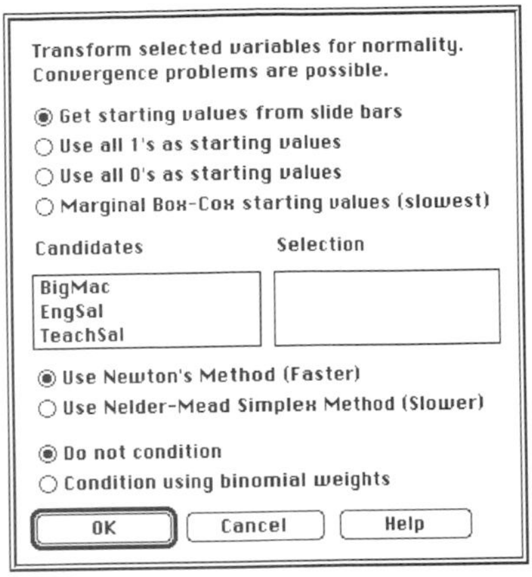

FIGURE 13.7 The transformation dialog from a scatterplot matrix.

parameters, one for the response and one for each of the predictors. The goal is to choose λ to normalize the resulting component-wise transformed variates.

From the "Transformations" plot control on the scatterplot matrix, select the item "Find normalizing transformations." You will get the dialog shown in Figure 13.7. Select the variables to be transformed. You then can choose from among several options that will help the computational algorithm find the transformations that do the best job at achieving joint normality; the defaults are good choices, but you might want to change the defaults if the computational method fails. The buttons at the bottom of the dialog that allow conditioning will be described in a later chapter and are not relevant here. The computation can be very slow with several variables or with large samples, but good starting values can speed up computations. Starting values can be set by first doing visual transformation, and leaving your estimated transformations on the slidebars which are then used by *Arc* as the starting values for the computational algorithms. The Nelder–Mead computational algorithm is slower, but it will sometimes give answers when the standard Newton's method fails.

After pushing the "OK" button in the dialog, the selected variables are transformed in the scatterplot matrix using the estimated transformation parameters. You can judge visually if the transformations succeed in obtaining approximate normality by checking each of the relevant frames in the scatterplot for approximate linearity and constant variance. Also, output is obtained as shown in Table 13.2.

The column headed Lambda-hat in Table 13.2 gives the estimates of the scaled power transformations. The next column marked SE gives the standard errors of these estimates. The next two columns are test statistics for testing the hypotheses that each individual transformation parameter is either zero or one. For example, the test that the transformation parameter for *EngSal* is zero, and

TABLE 13.2   **Output from "Finding Normalizing Transformations"**

```
Summary of Transformations to Normality
Variable         Lambda-hat      SE      Wald test     Wald test
                                         lambda =0     lambda =1
BigMac             -0.396       0.185     -2.14         -7.56
EngSal              0.383       0.116      3.31         -5.32
TeachSal            0.328       0.098      3.34         -6.87

Likelihood ratio tests...
    of all lambda = 0:  18.651 df = 3 p = .000
    of all lambda = 1: 113.561 df = 3 p = .000
```

hence that a log transformation is needed, against the alternative that it is not zero is given by $(0.383 - 0)/0.116 = 3.31$. The test statistic for the hypothesis that no transformation is needed, $\lambda = 1$, is $(0.383 - 1)/0.116 = -5.32$. These Wald test statistics can be compared to a normal distribution to get approximate p-values. The p-values for all three variables for the log-transformation are very small, as are the p-values for no transformation, suggesting that neither choice is acceptable.

At the foot of Table 13.2 are two more test statistics. The first of these is for the *simultaneous* test that *all* the optimal transformations are logarithms, while the second is for the test that all the optimal transformations are $\lambda = 1$ for untransformed values. These likelihood ratio test (*LRT*) statistics have approximate χ^2 distributions, with degrees of freedom equal to the number of variables, which is three in this example, leading to the p-values of 0.000 to three digits for each of these tests. Neither of these hypotheses is supported by the data.

Other overall tests can be performed as well. For example, using visual fitting, we estimated the transformations to be $\lambda_{\text{Visual}} = (-0.33, 0.33, 0.33)^T$, not too far from the value $\hat{\lambda} = (-0.396, 0.383, 0.328)^T$ chosen by the automatic procedure. We can test that the transformation is equal to λ_{Visual} by selecting the item "Evaluate LRT at" from the "Transformations" pop-up menu. You will then get a dialog, in which you can type the hypothesized values for the three transformation parameters. These must be typed in the same order as the variables listed in the dialog. The resulting output is

```
LRT that lambda = (-0.33 0.33 0.33) for (BigMac EngSal
    TeachSal): 0.413 p-value = .938
```

providing no evidence against this hypothesis. The transformations chosen visually are as good as the transformations chosen by the automatic procedure. In practice, transformations can and should be rounded to convenient or meaningful values, provided that the rounded values are not contradicted by the LRT. If a transformation parameter of 0.33 is as good as one of 0.383, then the simpler 0.33 will generally be preferred to the 0.383, which is less likely to be meaningful in any real-life situation. In this regression, selecting the transformation parameters to be equal for *EngSal* and *TeachSal*, and equal but of opposite sign for *BigMac*, makes the units of the three quantities comparable.

The transformation plot controls have several other options. First, you can add an individual transformed variable to the data set by selecting the item "Add transformed values to data set" from the variable's slidebar. The variables added are the ordinary power transformations, not the scaled power transformations; if the value of the transformation parameter is zero, then the natural log transformation is saved. You can add all transformed variables to the data set by selecting "Add transformed variables to the data set" from the "Transformations" pop-up menu. You can add a value for λ to a slidebar using the item "Add power choice to slidebar." Finally, you can restore to the plot either all untransformed variables, or all their logarithms, using items in the "Transformations" menu.

13.2.3 Possible Routes

There are at least two ways of using transformations of the response and the predictors **x** in an attempt to induce a linear regression model:

1. We could try transforming the response and predictors simultaneously to joint normality using the methods discussed in Section 13.2.
2. Or, we could transform just the predictors to joint normality and then use the inverse fitted value plot discussed in Section 13.1.3 to transform the response as needed. This use of inverse fitted value plots requires linearly related predictors. The transformation of the predictors to joint normality is used to ensure this condition.

In many regressions these two routes will lead to essentially the same answers. Nevertheless, the requirements for transforming y and **x** to joint normality are more stringent. The method based on inverse fitted value plots is more robust and may give reasonable results when the other fails.

The requirement of linearly related predictors needed for the fitted value plot plays a role in methods discussed later in this book, particularly in Chapter 16 and in Part III.

13.3 COMPLEMENTS

13.3.1 The Box–Cox Method

Transformations to normality in the univariate setting were first discussed in a pioneering paper by Box and Cox (1964). They also gave the wool data and various technical details concerning the derivation of the modified power family; see also Cook and Weisberg (1982) and Atkinson (1985). Velilla (1993) proposed the multivariate generalization of the Box–Cox method for simultaneous transformation. Whereas the Box–Cox method minimizes the residual sum of squares in the transformed response after modification to correct for scale changes, the multivariate generalization minimizes a function of the matrix of sums of squares and cross products.

These transformation methods try to make the data appear as normal as possible. This certainly does not guarantee success. Hernandez and Johnson (1980) give several examples where closest-to-normal is hardly normal at all.

13.3.2   Profile Log-Likelihoods and Confidence Curves

The confidences curves derived in Section 13.1.4 are a rescaling of a plot of the *profile log-likelihood*, which is defined by

$$L(\lambda) = -\frac{n}{2} \log(\text{RSS}_z(\lambda))$$

The value of λ that *maximizes* $L(\lambda)$ is the same as the value that *minimizes* $\text{RSS}_z(\lambda)$. The usual plot of the profile log-likelihood is of $L(\lambda)$ versus λ, as shown in the upper left of Figure 13.8. The value $\hat{\lambda}$ maximizes this curve. The second frame of Figure 13.8 is $2(L(\hat{\lambda}) - L(\lambda))$ versus λ, obtained from the first frame by flipping the curve and then changing the values on the vertical axis. In the third frame, the values on the vertical axis are replaced by their square roots. This results in a sharp point at $\hat{\lambda}$. The final frame interchanges the axes, giving the confidence curves.

13.3.3   Transformation Families

Other families of transformations have been suggested in place of the power family, particularly for cases in which the response either is not strictly positive or is bounded on the interval zero to one. In *Arc*, we include two additional families. The *modulus* family can be used when the response is not strictly positive. It was defined by John and Draper (1980) to be

$$y(\lambda) = \begin{cases} \text{sign}(y)((|y| + 1)^{\lambda} - 1)/\lambda, & \lambda \neq 0 \\ \text{sign}(y) \log(|y| + 1), & \lambda = 0 \end{cases}$$

This is like a power transformation, with the same power applied to y and to $-y$. The *folded power* family, Mosteller and Tukey (1977), is defined for data bounded on the interval from zero to one to be

$$y(\lambda) = \begin{cases} (y^{\lambda} - (1 - y)^{\lambda})/\lambda, & \lambda \neq 0 \\ \log(y/(1 - y)), & \lambda = 0 \end{cases}$$

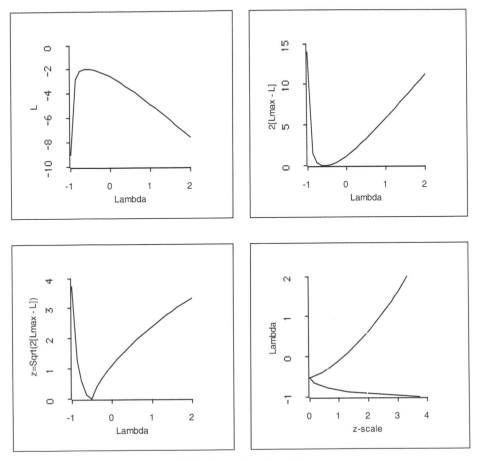

FIGURE 13.8 Derivation of confidence curve.

This family gives an alternative to the arcsin square-root transformation for binomial data. Either of these families can be selected in the "Choose response transform" dialog.

13.3.4 References

Scheffé (1959, Section 10.7) gave the general method of finding variance stabilizing transformations. The inverse fitted value plot discussed in Section 13.1.3 for visualizing a response transformation was proposed by Cook and Weisberg (1994a,b). Confidence curves were proposed by Cook and Weisberg (1990b). They are generally useful for displaying the uncertainty in the estimates of parameters in nonlinear models where confidence intervals can be asymmetric.

The data for Problem 13.1 are available from the UCLA Statistics Web server.

PROBLEMS

13.1 A major source of water in Southern California is the Owens Valley. This water supply is in turn replenished by spring runoff from the Sierra Nevada mountains, to the east of the valley. If runoff could be predicted, engineers, planners, and policy makers could do their jobs more efficiently. The data in the file `water.lsp` contains 43 years worth of precipitation measurements taken at six sites in the mountains and stream runoff volume at a site near Bishop, California. The three sites with name starting with "O" are fairly close to each other, and the three sites starting with "A" are also fairly close to each other.

  13.1.1 Load the datafile, and construct the scatterplot matrix of the six snowfall variables, which are the predictors in this problem. Using the methodology for automatic choice of transformations outlined in Section 13.2.2, find transformations to make the predictors as close to normal as possible. Obtain a test of the hypothesis that all $\lambda_j = 0$ against a general alternative, and summarize your results. Do the transformations you found appear to achieve linearity? How do you know?

  13.1.2 Consider the multiple linear regression model with mean function given by

$$E(\log(y) \mid \mathbf{x}) = \eta_0 + \eta_1 \log(APMAM) + \eta_2 \log(APSAB)$$

$$+ \eta_3 \log(APSLAKE) + \eta_4 \log(OPBPC)$$

$$+ \eta_5 \log(OPRC) + \eta_6 \log(OPSLAKE)$$

  with constant variance function. Estimate the regression coefficients using OLS. You will find that two of the estimates are negative; which are they? Does a negative coefficient make any sense? Why are the coefficient estimates negative?

  13.1.3 Examine the scatterplot matrix of the terms in the mean function (the predictors in log scale). This scatterplot matrix gives guidance on how one might proceed to remove terms from the model. What is the guidance?

  13.1.4 In the OLS fit, the regression coefficient estimates for the three predictors beginning with "O" are approximately equal. Would you expect these coefficients to be equal? Why or why not?

13.2

    13.2.1 In the wool data, we have seen that the model with response *Cycles* and terms given by the three predictors does not match the data well, and that transformation is required. However, we did not check for linearly related predictors. Does the condition of linearly related predictors seem reasonable for these data?

    13.2.2 Instead of transforming the response in the wool data, we might continue to use *Cycles* as the response and consider adding interactions to the mean function. Are any of the two-factor interactions important with response *Cycles*? Are any of the two-factor interactions important with response log(*Cycles*)? The phrase *removable nonadditivity* refers to interactions that can be made insignificant by transforming the response.

13.3 In transforming $(BigMac, TeachSal, EngSal)^T$ for multivariate normality, test the hypothesis that $\lambda = (-1, 1, 1)^T$. This transformation would also make all three variables in the same units of items per time.

13.4 The following questions relate to transforming the response in the Big-Mac data after replacing the four predictors *Bread*, *BusFare*, *TeachSal*, and *TeachTax* by their logarithms.

    13.4.1 Set up the regression using the four predictors in log scale and the response, *BigMac*. Draw the inverse fitted value plot of $\hat{y}$ versus *BigMac*. Does this plot suggest that transformation of *BigMac* may be necessary? Estimate the best power transformation. Check on the adequacy of your estimate by refitting the regression model with the transformed response and then drawing the inverse fitted value plot again. If transformation was successful, this second inverse fitted value plot should have a linear mean function.

    13.4.2 Construct confidence curves for the best Box–Cox power transformation parameter and read an approximate 95% confidence interval from the plot. Do these results agree generally with the graphical results?

    13.4.3 The inverse fitted value plot $\hat{y}$ versus *BigMac* contains four cities with the largest values of *BigMac* that are likely to be very important in determining the transformation. Identify these cities. The "Case deletions" plot control can be used to see how deleting these cities will change the information concerning a transformation. Did deleting these cities change your view of the need to transform *BigMac*?

CHAPTER 14

Diagnostics I: Curvature and Nonconstant Variance

In the last few chapters we studied various statistical methods for the multiple linear regression model,

$$E(y_i \mid \mathbf{x}_i) = \boldsymbol{\eta}^T \mathbf{u}_i \qquad \text{and} \qquad \text{Var}(y_i \mid \mathbf{x}_i) = \sigma^2 / w_i \qquad (14.1)$$

for $i = 1, \ldots, n$. The vector $\mathbf{u}_i$ consists of k terms $u_j(\mathbf{x}_i)$, $j = 1, \ldots, k$, each derived from the p predictors in the predictor vector $\mathbf{x}_i$. The variance function depends on the known positive weights w_i and the unknown positive constant σ^2.

Diagnostic methods are used to help decide if we have information in the data to contradict the model. The need for diagnostic methods can be illustrated by the four artificial data sets in file anscombe.lsp shown in Figure 14.1. Each data set consists of 11 data pairs (x, y) and each produces identical estimates from fitting the simple linear regression model with $u_1 = x$: $\hat{\eta}_0 = 3.0$, $\hat{\eta}_1 = 0.5$, $\hat{\sigma}^2 = 1.53$, and $R^2 = 0.667$. Since the fitted models are the same, one might conclude that the simple linear regression model is equally appropriate for each data set, but this is clearly contradicted by the plots.

The first data set in Figure 14.1a is what we might expect to observe if the simple linear regression model was appropriate. The second data set in Figure 14.1b shows that the simple linear regression model is incorrect and that a quadratic polynomial would likely be a much better choice.

We see in Figure 14.1c that a simple linear regression could be correct, but one data point is unusually far from the other data points. Perhaps that case was recorded incorrectly, and either the actual value of x should have been much larger, or the actual value of y should have been smaller to get the point to match the linear relationship in the rest of the data. Possibly the outlying point should be deleted from the data set, and the regression re-

334

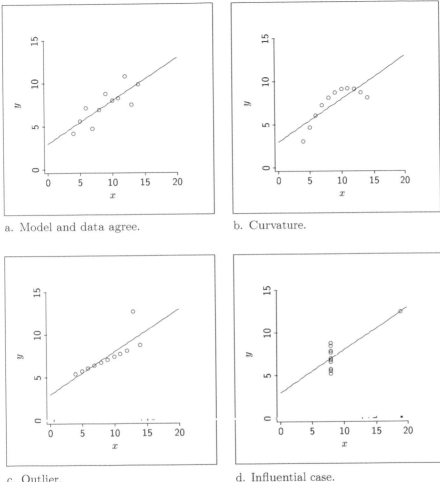

a. Model and data agree.

b. Curvature.

c. Outlier.

d. Influential case.

FIGURE 14.1 Four hypothetical data sets.

computed. Without a clear context for the regression, we cannot judge the outlying point as "correct" or "incorrect." Finding *outliers*, which are cases with unusually large or small values of the response, is a topic in Chapter 15.

For the final data set shown in Figure 14.1d, there is not enough information to make an informed judgment about the correctness of the model. The fitted line must pass through the isolated point with the largest value of the predictor. The leverage for this point is $h_i = 1$; and if it is deleted, we could not estimate the slope. We must be wary of conclusions from an analysis that is so heavily dependent upon a single case. Finding such *influential cases* is also a topic in Chapter 15.

14.1 THE RESIDUALS

14.1.1 Definitions and Rationale

Write the multiple linear regression model in the additive form given at (9.3):

$$y_i \mid \mathbf{x}_i = \eta^T \mathbf{u}_i + e_i / \sqrt{w_i} \qquad \text{for} \quad i = 1,\dots,n \tag{14.2}$$

Regardless of the weights, the unobservable e_i have constant variance, while the variance function depends on the weights, or

$$\text{Var}(y_i \mid \mathbf{x}_i) = \text{Var}((e_i / \sqrt{w_i}) \mid \mathbf{x}_i) = \sigma^2 / w_i \tag{14.3}$$

A final assumption of the multiple linear regression model that we make explicit is

$$e_i \text{ independent of } \mathbf{x}_i \tag{14.4}$$

This independence assumption has two consequences that will be useful when interpreting diagnostic plots:

$$E(e_i \mid \mathbf{x}_i) = E(e_i) = 0 \tag{14.5}$$

$$\text{Var}(e_i \mid \mathbf{x}_i) = \text{Var}(e_i) = \sigma^2 \tag{14.6}$$

Equation (14.5) says that the mean function for the errors must equal zero, while from (14.6) the variance function for the errors must be constant. If we actually observed the errors, diagnostic information about the model could be obtained by plotting the errors versus functions of $\mathbf{x}$. Information to contradict (14.5) or (14.6) suggests that the model is wrong. If (14.5) are (14.6) supported, we have no evidence against the model.

The immediate obstacle to using these ideas in practice is that the errors are unobservable, and so we estimate them using residuals. Let $\hat{y}_i = \hat{\eta}^T \mathbf{u}_i$ denote the ith fitted value from the WLS fit of (14.2). The ith residual $\hat{e}_i$ estimates e_i and is defined as

$$\hat{e}_i = \sqrt{w_i}(y_i - \hat{y}_i)$$

To see how the residuals estimate the errors, substitute $\eta^T \mathbf{u}_i + e_i / \sqrt{w_i}$ for y_i and substitute $\hat{\eta}^T \mathbf{u}_i$ for $\hat{y}_i$:

$$\begin{aligned}
\hat{e}_i &= \sqrt{w_i}(y_i - \hat{y}_i) \\
&= \sqrt{w_i}(\eta^T \mathbf{u}_i + e_i / \sqrt{w_i} - \hat{\eta}^T \mathbf{u}_i) \\
&= e_i + \sqrt{w_i}(\eta^T \mathbf{u}_i - \hat{\eta}^T \mathbf{u}_i) \\
&= e_i + \varepsilon_i
\end{aligned}$$

The ith residual $\hat{e}_i$ equals e_i plus a random deviation $\varepsilon_i = \sqrt{w_i}(\eta - \hat{\eta})^T\mathbf{u}_i$. Whether the model is correct or not, $E(\varepsilon_i \mid \mathbf{x}_i) = 0$ so the mean function for the residuals equals the mean function for the errors $E(\hat{e}_i \mid \mathbf{x}_i) = E(e_i \mid \mathbf{x}_i)$. The mean function estimated from a plot of residuals versus a function of $\mathbf{x}$ estimates the mean function for the unobservable errors versus that function of $\mathbf{x}$, so the plot can be used to find model deficiencies.

Although the mean functions for $e \mid \mathbf{x}$ and $\hat{e} \mid \mathbf{x}$ match, the variance functions do not because of variation due to estimation. If all the leverages are small, then $\text{Var}(\hat{e}_i \mid \mathbf{x}_i) \approx \text{Var}(e_i \mid \mathbf{x}_i)$ whether the model is correct or not. There are notable exceptions to this conclusion that are discussed in Chapter 15. Also, if all the leverages are approximately equal, then the variance function for the residuals will differ from the true variance function only by a multiplicative constant.

14.1.2   Residual Plots

If the model is correct, then in a plot of residuals versus predictors, u-terms, or linear combinations of predictors and u-terms, the residual mean function is constant and the residual variance function is approximately constant. If the data show that either function is clearly not constant, then we have information to contradict the model. For example, Figures 14.2a–d and Figures 14.3a and 14.3b are plots of the residuals versus a single predictor from six different data sets, each with $n = 75$ cases. For each data set, the residuals were constructed from the simple linear regression of the response on the predictor. The interpretations of the plots don't depend on the number of terms k in the model, as long as k is small relative to n. A *lowess* estimate $\hat{E}(\hat{e} \mid x)$ of the residual mean function is shown on each figure; *lowess* estimates of the residual standard deviation function are shown in Figures 14.2a and 14.3a.

The residuals in Figure 14.2a appear to be independent of the predictor, and the *lowess* estimates of the residual mean and standard deviation functions don't vary much relative to the variation in the residuals. Thus, the plot is consistent with our expectations under the model. The *lowess* smooths in Figure 14.2a are not exactly constant, and we should not expect them to be so, even if the model is correct. Anomalies can be found in all residual plots if we look hard enough. Generally, the model underlying a residual plot is sustained when the variation in the estimated residual mean function $\hat{E}(\hat{e} \mid x)$ is dominated by the variation in the residuals about $\hat{E}(\hat{e} \mid x)$. Problem 14.1 describes how to gain intuition on the natural variation in residual plots when the underlying model is correct.

In Figure 14.2b we see clear curvature in $\hat{E}(\hat{e} \mid x)$, and thus the model is not sustained by the data. We reach the same conclusion in Figure 14.2c, except now the estimated residual variance function is clearly not constant. In Figure 14.2d the data seem to sustain the model, except for a single outlying

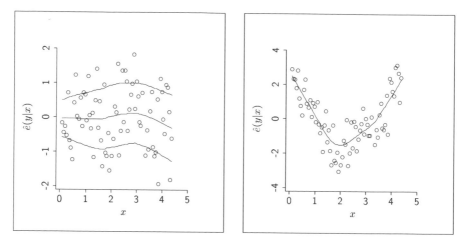

a. Correct model.   b. $E(\hat{e}|x)$ nonconstant.

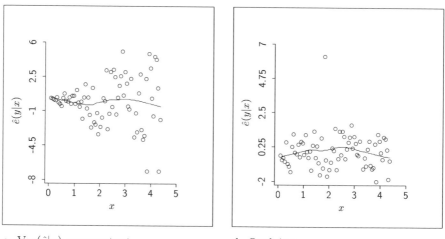

c. $Var(\hat{e}|x)$ nonconstant.   d. Outlying case.

FIGURE 14.2   Plots of residuals versus a single predictor x from four hypothetical data sets.

point. The pattern in Figure 14.3a, which is a combination of the patterns in Figures 14.2b and 14.2c, indicates that the residual mean and variance functions are both nonconstant.

Figures 14.2a–d and Figure 14.3a represent the kinds of patterns to watch for in residual plots: curvature indicating a nonconstant residual mean function, a fan shape indicating a nonconstant residual variance function, a few outlying points that are well-separated from the main point cloud, or a combination of these patterns.

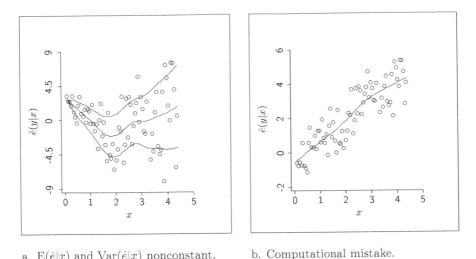

a. $E(\hat{e}|x)$ and $\mathrm{Var}(\hat{e}|x)$ nonconstant.   b. Computational mistake.

FIGURE 14.3   Plots of residuals versus a single predictor x from two hypothetical data sets.

Figure 14.3b illustrates a different type of problem, and we must conclude that there is a mistake some place. Why?

If the weights are constant, all $w_i = 1$, and there is an intercept in the mean function, *the sample correlation coefficient between the residuals and any of the u-terms in the model or any linear combinations of them is zero,* as indicated in (7.45)–(7.48). This also implies that the average of the residuals is zero. Similarly, the OLS regression coefficient of the simple linear regression of $\hat{e}$ on any u-term equals zero.

The sample correlation coefficient in Figure 14.3b is clearly not zero, so either there must be a mistake or the model fit did not include an intercept but probably should have.

14.1.3   Choosing Residual Plots

There are many possible residual plots. Here are the residual plots that have been found to be most useful in practice, roughly in order of their importance: Plot residuals $\hat{e}$ versus

- fitted values $\hat{y}$,
- individual predictors x_k, or pairs of predictors in a 3D plot; equivalently, plot u-terms of the form $u(\mathbf{x}) = x_k$,
- potential predictors that are not represented in the model,
- individual u-terms that do not correspond to individual predictors,
- linear combinations of the u-terms.

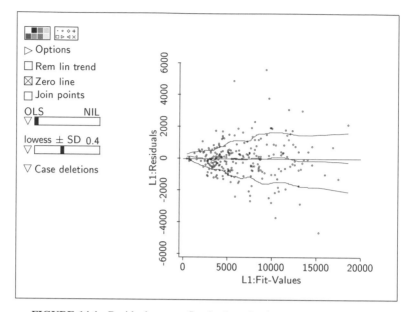

FIGURE 14.4   Residuals versus fitted values for the transaction time data.

Regardless of the quantity on the horizontal axis, Figures 14.2a–d and Figures 14.3a and 14.3b represent the kinds of patterns to watch for in residual plots.

14.1.4   Examples of Residual Plots

As a first example, we return to the transaction time data in file `transact.lsp`. The fitted mean function is, assuming a constant variance function (see Table 7.3, page 156),

$$\widehat{Time} = 144.369 + 5.46206T_1 + 2.03455T_2 \tag{14.7}$$

and the estimate of σ is $\hat{\sigma} = 1142.56$.

As a first step in examining this model for deficiencies, we plot the residuals against the fitted values. After fitting with *Arc*, select the item "Plot of" from the Graph&Fit menu and then select `L1:Residuals` and `L1:Fit-values`, where `L1` is assumed to be the name of the model. The resulting plot with *lowess* estimates of the mean and standard deviation functions is shown in Figure 14.4.

We see in this plot that the estimated residual mean function does not change much across the plot. Since the model includes a constant term, the average value of the residuals is zero. If the mean function is constant, it must equal zero for all values on the horizontal axis. The standard deviation function

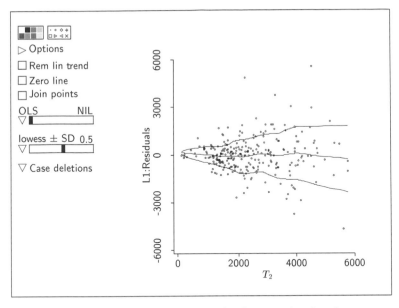

FIGURE 14.5 Residuals versus T_2 for the transaction time data.

does not appear to be constant, however. Judging from the residual plot, the variability in time must be smaller in branches with few transactions than in branches with many transactions.

Shown in Figure 14.5 is a scatterplot of the residuals versus the predictor T_2. The interpretation of this plot is similar to that in Figure 14.4, although the conclusion of nonconstant variance seems a bit stronger here. The same conclusion was reached when we discussed the 3D plot of $\hat{e}$ versus (T_1, T_2) in Section 8.5.

Is the finding that $\text{Var}(Time \mid T_1, T_2)$ is not constant important? This depends on the goal of the analysis. If the only goal is to estimate the number of minutes each transaction takes on the average, then the estimates obtained when using the wrong variance function, while somewhat inefficient, are still unbiased and useful. But any computation that depends on the estimated variance function, such as standard errors, prediction variances, tests, and so on, can be misleading. For example, prediction variances for very small branches will be too large, while prediction variances for very large branches will be too small. In addition, the F-tests in Section 11.3.1 involving the transactions data may not be accurate.

As a second example, Figure 14.6 is a plot of residuals against fitted values for the sleep data in file `sleep.lsp` using the mean function given by (10.12), page 240, with assumed constant variance. The residual mean and variance functions both seem to be constant, so there is no clear information to contradict the model in this plot.

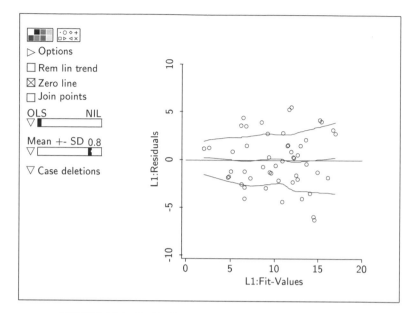

FIGURE 14.6   Residuals versus fitted values for the sleep data.

14.1.5   A Note of Caution

A general residual plot is of residuals against some linear combination $\mathbf{h}^T\mathbf{u}$ of the u-terms from the model. The linear combination could correspond to a single u-term, or to the fitted values when $\mathbf{h} = \hat{\eta}$. Suppose the plot clearly shows that $\text{Var}(\hat{e} \mid \mathbf{h}^T\mathbf{u})$ is *constant* and that $\text{E}(\hat{e} \mid \mathbf{h}^T\mathbf{u})$ is *nonconstant*, as illustrated in Figure 14.2b. Does this imply that the model's mean function is incorrect? Or, suppose that the plot clearly shows that $\text{Var}(\hat{e} \mid \mathbf{h}^T\mathbf{u})$ is *nonconstant* and that $\text{E}(\hat{e} \mid \mathbf{h}^T\mathbf{u})$ is *constant*, as in Figures 14.4 and 14.5. Does this imply that the model's variance function is incorrect?

If we conclude that either $\text{E}(\hat{e} \mid \mathbf{h}^T\mathbf{u})$ or $\text{Var}(\hat{e} \mid \mathbf{h}^T\mathbf{u})$ is not constant, then we must conclude that the model is incorrect *in some respect*, but we may not be able to tell if the problem is due to a misspecified mean function or a misspecified variance function. For example, if a plot of $\hat{e}$ versus x_k suggests that $\text{E}(\hat{e} \mid x_k)$ is nonconstant, the model's mean function could be deficient in the way x_k is used, it could be deficient in the way some other predictor is used, or both. Adding higher-order terms in x_k like x_k^2 can sometimes, but not always, help.

Shown in Figure 14.7 is a plot of residuals versus OLS fitted values from a constructed data set with two predictors and 100 cases. The data are available in file `caution.lsp`. The model used to fit the data and generate the residuals is

$$y \mid (x_1, x_2) = \eta_0 + \eta_1 x_1 + \eta_2 x_2 + e \tag{14.8}$$

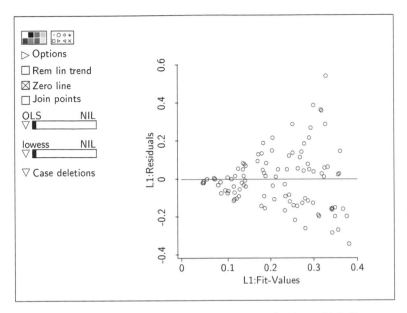

FIGURE 14.7   Plot of residuals versus fitted values for the fit of a multiple linear regression model in a constructed data set with two predictors.

with $\mathrm{Var}(y \mid \mathbf{x}) = \mathrm{Var}(e \mid \mathbf{x}) = \sigma^2$. We conclude from the residual plot in Figure 14.7 that $\mathrm{E}(\hat{e} \mid \mathbf{x})$ is constant and that $\mathrm{Var}(\hat{e} \mid \mathbf{x})$ is nonconstant. Does this imply that $\mathrm{Var}(y \mid \mathbf{x})$ is nonconstant? While nonconstant $\mathrm{Var}(y \mid \mathbf{x})$ may explain a pattern such as that in Figure 14.7, it is not the only explanation. The *true* regression for this example is

$$y \mid \mathbf{x} = \frac{|x_1|}{2 + (1.5 + x_2)^2} + e \qquad (14.9)$$

where e is a standard normal error. Thus, $\mathrm{Var}(y \mid \mathbf{x}) = \mathrm{Var}(e \mid \mathbf{x}) = \sigma^2$ is in fact correct, while the model's mean function (14.8) is incorrect because the true mean function is a nonlinear function of $\mathbf{x}$. A 3D plot of the residuals versus (x_1, x_2) shows the situation quite clearly. 3D plots can be superior to 2D plots because they allow for an additional predictor to be included.

14.2   TESTING FOR CURVATURE

Suppose we plot the residuals versus a linear combination $\mathbf{h}^T \mathbf{u}$ of the u-terms in the linear model and judge through visual inspection that the residual mean function is curved. It is generally useful at this point to quantify the visual impression with a statistical test for curvature. In this section we present two general tests for curvature. The first is applicable when $\mathbf{h}$ is not random, and the second is for the plot of residuals versus fitted values so $\mathbf{h} = \hat{\boldsymbol{\eta}}$.

When $\mathbf{h}$ is fixed, we can quantify our visual impressions of curvature by adding the term $(\mathbf{h}^T\mathbf{u})^2$ to the model and testing whether its coefficient is zero. Specifically, fit the model

$$y_i \mid \mathbf{x}_i = \boldsymbol{\eta}^T\mathbf{u}_i + \delta(\mathbf{h}^T\mathbf{u}_i)^2 + e_i/\sqrt{w_i} \qquad (14.10)$$

and then construct the t-test for the hypothesis that $\delta = 0$. A small p-value is taken as confirmation of our visual impression that the original model (14.2) is incorrect.

When $\mathbf{h} = \hat{\boldsymbol{\eta}}$ we must consider the possibility that the curvature depends on $\boldsymbol{\eta}^T\mathbf{u}$. We test the hypothesis that $\delta = 0$ in the model

$$y_i \mid \mathbf{x}_i = \boldsymbol{\eta}^T\mathbf{u}_i + \delta(\boldsymbol{\eta}^T\mathbf{u}_i)^2 + e_i/\sqrt{w_i} \qquad (14.11)$$

This model is nonlinear in the regression coefficients so the testing procedure for (14.10) cannot be applied straightforwardly. Instead, we use a diagnostic testing procedure called *Tukey's test for nonadditivity*. Tukey's test for $\delta = 0$ is computed by first finding the fitted values $\hat{y}_i = \hat{\boldsymbol{\eta}}^T\mathbf{u}_i$ under the null hypothesis that $\delta = 0$. These fitted values are the same as the fitted values from the WLS fit of (14.2). Next, substitute these fitted values for the nonlinear term of (14.11) to get the approximating linear model

$$y_i \mid \mathbf{x}_i = \boldsymbol{\eta}^T\mathbf{u}_i + \delta(\hat{\boldsymbol{\eta}}^T\mathbf{u}_i)^2 + e_i/\sqrt{w_i} \qquad (14.12)$$

Tukey's test statistic is the same as the usual t-test for $\delta = 0$ in this approximating model but it is compared to standard normal distribution.

We use the haystack data as a first example of detecting curvature in residual plots. Fit the multiple linear regression model

$$Vol \mid \mathbf{x} = \eta_0 + \eta_1 C + \eta_2 Over + e \qquad (14.13)$$

by using OLS. Then select the item "Residual plots" from the model's menu. The resulting dialog allows you to select variables to be plotted on the horizontal axis of a residual plot. The default choices are the fitted values and the u-terms in the model. The scatterplot produced in this way has an extra slider that controls the variable on the horizontal axis, and the appropriate curvature test statistic and p-value are shown at the top of the plot. *Arc* automatically calculates Tukey's test for the plot of residuals versus fitted values.

For example, shown in Figure 14.8 is the residual plot for *Over*. The plot shows definite curvature which is confirmed by the curvature test shown at

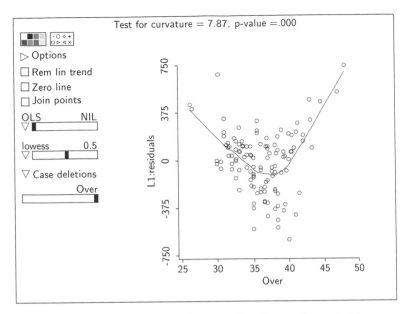

FIGURE 14.8 Plot of residuals versus *Over* from the haystack data.

the top of the plot. The curvature statistic in this case is the usual t-statistic for the coefficient of the added term $Over^2$.

At this point we might try to improve the model by actually adding the term $Over^2$ to obtain the new model

$$Vol \mid \mathbf{x} = \eta_0 + \eta_1 C + \eta_2 Over + \eta_3 Over^2 + e$$

This model could be checked for curvature using the same procedure. However, no test will be reported in the plot of residuals versus *Over* because the term $Over^2$ is already in the model. The curvature test reported on the plot of residuals versus $Over^2$ is constructed by adding the term $(Over^2)^2 = Over^4$ to the model.

Adding quadratic or higher-order terms sometimes removes curvature found in a residual plot, but this is not generally a very satisfying way to build models. For example, we saw in Problem 4.6, equation (4.18), that a reasonable mean function for the haystack data could be

$$E(Vol \mid C, Over) \approx \frac{(C/2 + Over)^3}{12\pi^2}$$

To match this mean function by adding terms to (14.13), we would need to add the four cubic polynomial terms in the expansion of $(C/2 + Over)^3$. If we also added quadratic terms, we would end up with a ten-term model.

14.3   TESTING FOR NONCONSTANT VARIANCE

To study nonconstant variance, it is useful to have a functional form for the variance function. Let $\mathbf{v}$ consist of terms constructed like the u-terms in the mean function. The component terms v_j of $\mathbf{v}$ are functions of the predictor vector $\mathbf{x}$ and will be called v-terms. The v-terms can be individual predictors, $v_j(\mathbf{x}) = x_k$, or they can be the same as some of the u-terms, $v_j(\mathbf{x}) = u_j(\mathbf{x})$,
 We model the variance function as

$$\text{Var}(y \mid \mathbf{x}) = \sigma^2 \exp(\alpha^T \mathbf{v}) \tag{14.14}$$

The parameter σ^2 is the variance of y when $\mathbf{v} = \mathbf{0}$, and α is a vector of parameters. The exponential in (14.14) ensures that the variance function is positive for all values of $\alpha^T \mathbf{v}$, but for the tests described in this section the exact form of the variance function isn't very important. The constant variance function is a special case of (14.14), obtained when $\alpha = \mathbf{0}$.
 Taking the logarithm of (14.14) gives an alternate representation for the variance function that can be interpreted like a mean function:

$$\log(\text{Var}(y \mid \mathbf{x})) = \log(\sigma^2) + \alpha^T \mathbf{v} \tag{14.15}$$

The role of the intercept is taken by $\log(\sigma^2)$, and so a constant term should not be one of the v-terms.
 The variance is often a function of the mean; generally, as the mean increases, so will the variance. This is just an important special case of (14.15) obtained when $\mathbf{v} = \mathbf{u}$, $\alpha^T \mathbf{v} = \gamma \eta^T \mathbf{u} = \gamma E(y \mid \mathbf{x})$, giving

$$\log(\text{Var}(y \mid \mathbf{x})) = \log(\sigma^2) + \gamma \eta^T \mathbf{u}$$
$$= \log(\sigma^2) + \gamma E(y \mid \mathbf{x}) \tag{14.16}$$

in which the log variance is a linear function of the mean. The variance function (14.16) is constant when $\gamma = 0$.
 For (14.14), and equivalently (14.15), a test of $\alpha = \mathbf{0}$ is a test for constant variance against an alternative of nonconstant variance. We use a *score test*, which requires that the mean function $E(y \mid \mathbf{x})$ be correctly specified. To compute the score test, fit model (14.2) via OLS. The squared residuals $\hat{e}^2$ contain information about the variance function, and the test statistic is just the regression sum of squares $SSreg$ from the OLS regression of $\hat{e}^2$ on $\mathbf{v}$, divided by the scale factor $2(\sum \hat{e}_i^2/n)^2$:

$$\text{Test statistic} = \frac{SSreg(\hat{e}^2 \text{ on } \mathbf{v})}{2(\sum \hat{e}_i^2/n)^2}$$

To get a p-value, this statistic should be compared to the χ^2 distribution with df equal to the number of v-terms.

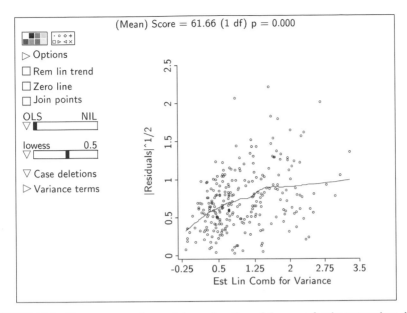

FIGURE 14.9   Nonconstant variance plot as a function of the mean for the transactions data.

To test $\gamma = 0$ in model (14.16), fit the OLS regression of $\hat{e}^2$ on $\hat{y}$, the fitted values from the fit of (14.2), and then compute the statistic in the same way. The resulting statistic is to be compared to a χ^2 distribution with 1 df.

If we knew α, a plot of $\hat{e}$ versus $\alpha^T\mathbf{v}$ should show the typical fan-shaped pattern that characterizes nonconstant variance. The sign of the residuals is not usually relevant information when assessing nonconstant variance. Consequently, visual impressions can be enhanced by folding the fan-shaped pattern in half at zero so the negative residuals overlay the positive residuals, changing the fan shape into a right triangle. *Arc* goes a step further and provides plots of $|\hat{e}|^{1/2}$ versus an estimate of $\alpha^T\mathbf{v}$.

14.3.1   Transactions Data

After fitting the mean function $\mathrm{E}(\textit{Time} \mid T_1, T_2) = \eta_0 + \eta_1 T_1 + \eta_2 T_2$ by using OLS, select the item "Nonconstant variance plot" from the model's menu. This will produce the plot shown in Figure 14.9. The vertical axis in this plot is $|\hat{e}|^{1/2}$. The plot is initially based on model (14.16) with the logarithm of the variance a linear function of the mean, and so the fitted values are plotted on the horizontal axis. The general trend in the plot is increasing to the right, suggesting that variance increases with the mean. This is confirmed by the score test, given at the top of the plot. The value of the statistic is 61.66, with one df and the corresponding *p*-value is zero to three decimal places.

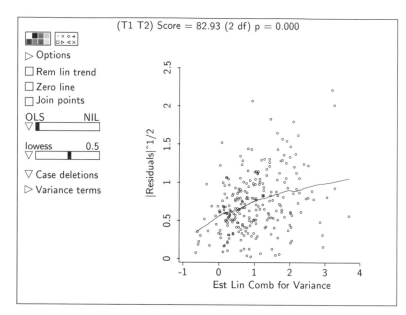

FIGURE 14.10 Nonconstant variance plot as a function of T_1 and T_2 for the transactions data.

An additional plot control called "Variance terms" appears on the nonconstant variance plot. This control is used to change the v-terms in the variance function. The initial plot is obtained by specifying model (14.16), where the variance depends on the mean, or you can specify model (14.15) by selecting v-terms from the list of all the candidates in the data file. In the transactions data we selected T_1 and T_2 as v-terms, allowing the variance function to depend on a linear combination of T_1 and T_2. The resulting plot is shown in Figure 14.10. The horizontal axis in this plot is an estimate of $\alpha^T v$.

The score test shown in Figure 14.10 is equal to 82.93, compared to 61.66 for the variance as a function of the mean. Since the variance model (14.16) is a submodel of the full variance model (14.15), we can subtract the two score statistics, $82.93 - 61.66 = 21.27$ with $2 - 1 = 1$ df, to get an approximate χ^2 test for comparing the two models. For this test, both the null and alternative hypotheses specify that the variance is nonconstant. Under the null hypothesis the variance changes with the mean, while under the alternative the variance changes with a linear combination of the terms, which need not be the one that determines the mean. In the transactions data, the p-value for the test is very small, suggesting that the logarithm of the variance is not just a rescaling of the mean function.

We have discovered in several different ways that the variance function in the transactions data is not constant, but rather is a function of T_1 and T_2. What shall we do next? Quite often it is possible to induce constant variance by transforming the response, but this will not be appropriate for the transactions

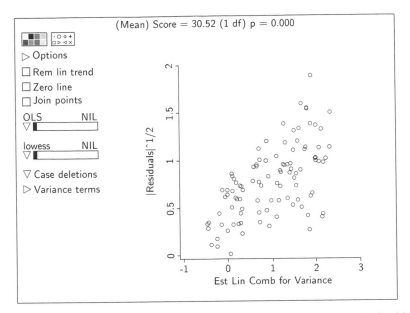

FIGURE 14.11   Score test for nonconstant variance in the constructed data of Section 14.1.5.

data because the mean function is known to be correct, and transforming will change the mean function. If the variance is a function of the mean, we could use *generalized linear models*, in which the relationship between the mean and variance functions is part of the model. We pursue this option in Chapter 23, and in particular for the transactions data in Section 23.4.

Another approach is to use a reasonable approximation for weights and then use weighted least squares. In the transactions data, we might hypothesize that $\text{Var}(Time \mid T_1, T_2) = (T_1 + T_2)\sigma^2$, so each transaction contributes the same variability to the response. We used this approach in Problem 9.11 and in Section 11.3.1.

14.3.2   Caution Data

Consider testing for nonconstant variance in the data (`caution.lsp`) introduced in Section 14.1.5. The test and plot shown in Figure 14.11 are for the linear mean function

$$E(y \mid x_1, x_2) = \eta_0 + \eta_1 x_1 + \eta_2 x_2$$

and the variance function (14.16). The *p*-value is quite small, and the plot seems to confirm the test. However, we know from the way in which the data were generated [see equation (14.9)] that $\text{Var}(y \mid \mathbf{x})$ is constant. The issue here can be traced back to one of the assumptions underlying the test; namely, that the mean function is correct. The mean function used for the test in

Figure 14.11 is incorrect, so we can't be sure what the test is telling us. The same sort of issues can arise in testing for curvature while assuming that the variance function is correctly specified.

14.4   COMPLEMENTS

Cook (1998b) and Cook and Weisberg (1982) provide a more theoretical development of residuals, including additional uses for them.

Tukey (1949) proposed Tukey's test for curvature in the context of two-way tables; see St. Laurent (1990) for a discussion of generalizations of this test. Like the nonconstant variance test, Tukey's test is a score test; Buse (1982) provides a comparison of score tests, which he calls Lagrange multiplier tests, likelihood ratio tests, and Wald tests.

The score test for nonconstant variance was given by Cook and Weisberg (1983). The results of Chen (1983) suggest that the form of the variance function is not very important. Hinkley (1985) stated that the difference of score statistics as used in Section 14.3.1 can be compared to a χ^2 distribution to compare nested models. The use of $|\hat{e}|^{1/2}$ in the nonconstant variance plots is based on results given by Hinkley (1975). Other relevant references on variance modeling and diagnostics concerning the variance function include Carroll and Ruppert (1988) and Verbyla (1993).

The example discussed with Figure 14.1 was first presented by Anscombe (1973). The mussels data in Exercise 14.3 were furnished by Mike Camden. The lettuce data in Exercise 14.4 are taken from Cochran and Cox (1957, p. 348). The data in `caution.lsp` are from Cook (1994).

PROBLEMS

14.1   The demonstration program `demo-bn.lsp` discussed in Section 4.4 can be used to gain intuition on the behavior of residual plots when the model is correct. Load the program and generate a scatterplot of $n = 10$ cases from a standard bivariate normal distribution with $\rho = 0$. Next, remove the linear trend, which replaces the response by the residuals, and then superimpose *lowess* estimates of the residual mean and variance functions. Does the plot leave the impression that the residuals are independent of the predictor? Using a button on the plot, get four new samples and keep the sample that gives the clearest impression that the residuals are *not* independent of the predictor.

Finally, repeat the simulation for $n = 20$, 30, and 50. What do you conclude about the usefulness of residual plots?

14.2   Explore fitting the transactions data as suggested in Section 14.3.1 using weighted least squares, and show that nonconstant variance is no longer a problem. Summarize the analysis of these data.

TABLE 14.1   The Mussels Data Used in Problem 14.3

Variable	Description
M	Mass of the mussel's muscle (g), the edible part of a mussel.
L	The length of the mussel's shell in mm.
W	The width of the mussel's shell in mm.
H	The height of the mussel's shell in mm.
S	The mass of the mussel's shell in g.

14.3   The file `mussels.lsp` contains data on horse mussels sampled from the Marlborough Sounds off the coast of New Zealand. The data were collected as part of a larger ecological study of the mussels. The variables are defined in Table 14.1.

   Load the data and then delete the cases numbered 7 and 47 from the data set. This can be done by drawing any plot, and then clicking on cases 7 and 47 in the window created by selecting "Display case names" from the data set menu. The cases can then be deleted by using the "Case deletions" plot control on any plot. All problems below refer to the data set without cases 7 and 47.

   14.3.1   Fit the mean function

$$E(M \mid H, L, W) = \eta_0 + \eta_1 H + \eta_2 L + \eta_3 W$$

   using OLS. Inspect the residual plots obtained using the item "Residual plots" in the model menu. Write a summary of your conclusions, including justification.

   14.3.2   Does an appropriate Box–Cox power transformation of the response improve the mean function of Problem 14.3.1? Use residual plots to justify your answer.

   14.3.3   Fit the mean function

$$E(M \mid H, L, W, S) = \eta_0 + \eta_1 H + \eta_2 L + \eta_3 W + \eta_4 S$$

   using OLS.

   a. Inspect the residual plots obtained using the item "Residual plots" in the model menu. Is this mean function contradicted by the data? Provide justification for your answer.

   b. The residual plots from this mean function are noticeably different from the residual plots of Problem 14.3.1. Explain why the residual plots changed after adding *S* to the mean function.

   c. Obtain score tests for nonconstant variance assuming that the logarithm of the variance function is (a) a linear function of the terms in the mean function, and (b) a linear function of the mean function. Next, compare these two models using an appropriate test.

14.3.4   Fit the mean function

$$E(\log(M) \mid H,L,W,S) = \eta_0 + \eta_1 H + \eta_2 L + \eta_3 W + \eta_4 \log(S)$$

using OLS. Is there evidence in the data to indicate that this mean function can be improved or that the variance function $\text{Var}((\log(M) \mid H,L,W,S)$ is not constant?

14.4   The data in file `lettuce.lsp` are the results of a *central composite design* on the effects of minor elements copper *Cu*, molybdenum *Mo*, and iron *Fe* on the growth of lettuce in water culture. The response is lettuce yield, *y*. The sample size is only 20, so seeing trends in plots may be a bit difficult.

Set up the regression with *y* as the response and the other three variables as the predictors. Examine the scatterplot matrix of the four variables. From the scatterplot matrix, describe the design. Use the methods described in this book to model *y* as a function of the predictors, and then use appropriate graphical methods to assess the lack of fit. Summarize your results. *Hint:* central composite designs are used to find the maximum or the minimum value of a response surface, so it is likely that $E(y \mid Cu,Mo,Fe)$ is not a monotonic function of the predictors.

14.5   As an alternative approach in the *Casuarina* data in file `casuarin.lsp`, Section 12.6, suppose we used the mean function

$$W \mid \mathbf{x} = \eta_0 + \eta_1 D + \eta_{11} D^2 + \eta_2 S + \eta_3 F + error$$

This mean function specifies additive effects for the two categorical variables and a quadratic effect for *D*.

14.5.1   Graphically show that this is a plausible alternative to the logarithmic model derived in Section 12.6.

14.5.2   After fitting this mean function with constant variance using OLS, draw the plot of $\hat{e}$ versus $\hat{y}$. The points with the five largest absolute residuals correspond to larger fitted values and to the five heaviest trees. This might lead one to suspect that larger trees are more variable. Explore this further, by getting a score test for nonconstant variance as a function of the mean, and summarize both the plot and the results of the test.

14.5.3   Individual points separated from the main trend may determine the value of the score statistic. These can be identified in the plot and deleted with the "Case deletions" item. The plot will then be automatically updated. Five of the 24 points are generally to the right and above the rest of the points. Deletion all five of these cases and summarize how the score test changes. Are these the same five points you found previously? In such a small data set, deleting the five largest trees is undesirable because it limits inference to smaller trees. Restore all the data using the "Restore all" item from the "Case deletions" pop-up menu to continue the analysis.

14.5.4 Obtain a score test for nonconstant variance that is a linear function of the four terms in the mean function, and then a test for comparing this model for the variance to the variance as a function of the mean. Summarize your results.

14.5.5 One common remedy for nonconstant variance is to transform the response to another scale. As we have seen in Section 12.6, the mean function

$$E(\log(W) \mid D) = \eta_0 + \eta_1 \log(D) + \eta_2 F + \eta_3 S$$

provided a good approximation to the data. Does the change to the log scale correct the nonconstant variance? If not, find the simplest summary of nonconstant variance you can, and provide a complete summary of the analysis of these data.

14.6 During our discussion of the twin data in Section 12.2, we conjectured that the variance in IQb, the IQ for the twin reared by biological parents, for the upper social class might be larger than the variance in IQb for the middle and lower social classes. The score test discussed in this chapter can be used to assess that possibility.

After creating a factor $\{F\}C$ for social class, test the hypothesis of constant variance using the variance model

$$\log \operatorname{Var}(IQb \mid C) = \log(\sigma^2) + \alpha_2 v_2 + \alpha_3 v_3$$

where v_2 and v_3 are the indicator variables for the middle and lower social classes.

CHAPTER 15

Diagnostics II: Influence and Outliers

Separated or outlying points can merit special attention for many reasons. They can cause us to miss important trends by reducing visual resolution in plots. Temporarily removing them from the plot often helps, as we did in the analysis of the snow geese data in Section 3.6. Transformations of the data to more suitable scales is another option for dealing with separated points that reduce visual resolution. We saw an example of this during the discussion of the brain weight data near Figure 5.4.

Separated points can also have a large *influence* on the results of the analysis: Deleting them from the data set could produce conclusions quite unlike those based on the full data. For example, we left the analysis of the transaction data in Section 14.3.1 rejecting the variance model

$$\log(\mathrm{Var}(y \mid \mathbf{x})) = \log(\sigma^2) + \gamma \eta^T \mathbf{u} \qquad (15.1)$$

in favor of the more general model

$$\log(\mathrm{Var}(y \mid \mathbf{x})) = \log(\sigma^2) + \alpha^T \mathbf{v} \qquad (15.2)$$

Shown in Figure 15.1 is a scatterplot of the residuals versus the fitted values from the OLS regression of *Time* on $\mathbf{x} = (T_1, T_2)^T$. One point, case 160, on the plot is highlighted. Deleting that point from the data set by using the "Case deletions" plot control and then recomputing the tests for nonconstant variance gives the results shown in Table 15.1. The striking feature of these results is that our conclusion regarding the nature of the nonconstant variance changes completely once case 160 is deleted from the analysis: With case 160 we conclude that the data do not sustain model (15.1), but without case 160 the data provide no evidence to reject that model. Case 160 is an *influential* case, at least for testing for nonconstant variance.

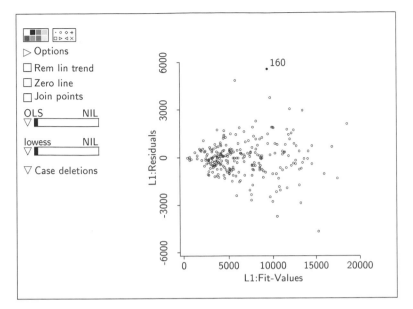

FIGURE 15.1   Scatterplot of OLS residuals versus fitted values for the transaction data. The point with the largest residual is highlighted.

TABLE 15.1   Score Tests for Nonconstant Variance in the Transaction Data with and without Case 160

	$\alpha = 0$	$\gamma = 0$	Difference
All data	82.93	61.66	21.27
Without 160	56.95	56.57	0.38

It is difficult to know what to do about case 160 without knowing more about the issues behind the data, but there are three possibilities that should always be considered. First, case 160 may contain a mistake, perhaps a recording error, that can be identified and corrected by tracing its history. Second, it may be that model (15.1) is correct and that case 160 represents just a relatively rare observation. Rare events will occur with the appropriate frequency no matter how surprised we are that they occur in a particular analysis. Finally, model (15.1) may be incorrect, and case 160 is the only clear evidence of that in the data. More generally, cases corresponding to unusual points may represent new and unexpected information.

Points in scatterplots that stand apart from the main point cloud are always candidates for influential cases. Case 160 lies on the edge of the point cloud in Figure 15.1, as do several other cases. However, in a 3D plot of residuals versus (T_1, T_2), case 160 is clearly separated from the rest of the data. We first

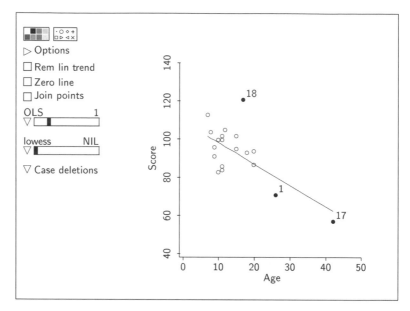

FIGURE 15.2 The adaptive score data.

identified case 160 in this 3D plot. The effects of individual points on an analysis can always be assessed by deleting them and examining the changes in the analysis. The "Case deletions" plot control appears on most plots to make this operation easy.

In this chapter we describe methods that can be used to identify influential cases and to find outlying points in linear regression. Remember that *Arc always starts numbering at zero.* For example, case 160 is the 161st case in the file `transact.lsp`.

15.1 ADAPTIVE SCORE DATA

Load the file `adscore.lsp`. This is a small data set with observations on 21 children, giving their *Age* in months at first spoken word, and a *Score*, which is a measure of the development of the child. The goal is to study the conditional distributions of *Score* given *Age*. Construct the plot of *Score* versus *Age*, as shown in Figure 15.2. Three separated points are identified on the plot. The mean function E(*Score* | *Age*) appears to decrease with *Age*, and case 18 seems to be fit poorly by the linear trend relative to the other data. Cases 1 and 17 are interesting by virtue of their relatively large values of *Age*.

Suppose we fit a simple linear regression model,

$$E(Score \mid Age) = \eta_0 + \eta_1 Age \qquad \text{and} \qquad Var(Score \mid Age) = \sigma^2 \quad (15.3)$$

TABLE 15.2 Adaptive Score Data Estimates with Cases Removed

	Intercept	Slope	SE(slope)	$\hat{\sigma}$	R^2
All data	110	−1.1	0.31	11.0	0.41
Not 1, 17	98	−0.1	0.62	10.5	0.00
Not 1	109	−1.0	0.33	11.1	0.35
Not 17	106	−0.8	0.52	11.1	0.11
Not 18	109	−1.2	0.24	8.6	0.57

to these data, as shown on Figure 15.2. How do you think the fit of the model will change when cases 1 and 17 are deleted? What will happen to the estimates of η_0, η_1, and σ^2? Select those two points and use the "Case deletions" pop-up menu to delete them. Use the "Display fit" item in the regression menu to see statistics for the new fitted model.

Selected summaries from regressions with a few cases deleted are given in Table 15.2. Without cases 1 and 17, the relationship between *Age* and *Score* disappears: A small fraction of the data effectively determines the fitted model. Cases 1 and 17 are clearly *influential*. In this example, the two cases with large values of *Age* are influencing the fit.

15.2 INFLUENTIAL CASES AND COOK'S DISTANCE

When we have more than two terms, finding influential points graphically may not be so easy. However, a diagnostic statistic called *Cook's distance* can be used to summarize essential information about the influence of each case on the estimated regression coefficients. Cook's distance is a mathematical measure of the impact of deleting a case. It is not intended for use as a statistical test.

To assess the influence of the ith case, we compare $\hat{\eta}_{(i)}$, the estimate of η that is computed without case i, to the estimate $\hat{\eta}$ based on all the data. Cook's distance combines the vector of differences $\hat{\eta}_{(i)} - \hat{\eta}$ into a summary number D_i that is a squared distance between $\hat{\eta}_{(i)}$ and $\hat{\eta}$; if this number is sufficiently large, then case i is *influential* for $\hat{\eta}$. For a linear model with mean function $E(y_i \mid x_i) = \eta^T u_i$ and constant variance function $Var(y_i \mid x_i) = \sigma^2$, we get a formula for D_i using the matrix notation developed in Section 7.9. Let

- U denote the $n \times k$ matrix defined at (7.34) to have rows u_i^T, $i = 1, \ldots, n$,
- $\hat{y} = U\hat{\eta}$ denote the $n \times 1$ vector of fitted values for the full data with jth element $\hat{y}_j$, and
- $\hat{y}_{(i)} = U\hat{\eta}_{(i)}$ denote the $n \times 1$ vector of fitted values when estimating η without the ith case, with jth element $\hat{y}_{(i),j}$.

Then Cook's distance can be written as

$$D_i = \frac{(\hat{\boldsymbol{\eta}}_{(i)} - \hat{\boldsymbol{\eta}})^T (\mathbf{U}^T \mathbf{U})(\hat{\boldsymbol{\eta}}_{(i)} - \hat{\boldsymbol{\eta}})}{k\hat{\sigma}^2} \tag{15.4}$$

$$= \frac{(\hat{\mathbf{y}}_{(i)} - \hat{\mathbf{y}})^T (\hat{\mathbf{y}}_{(i)} - \hat{\mathbf{y}})}{k\hat{\sigma}^2} \tag{15.5}$$

A scalar version of this last equation is

$$D_i = \frac{1}{k\hat{\sigma}^2} \sum_{j=1}^{n} (\hat{y}_{(i),j} - \hat{y}_j)^2 \tag{15.6}$$

Since equation (15.4) is the equation of an ellipsoid, like the confidence regions discussed in Section 10.8, contours of constant D_i are elliptical, defining a squared distance from $\hat{\boldsymbol{\eta}}_{(i)}$ to $\hat{\boldsymbol{\eta}}$. Equations (15.5) and (15.6) show that D_i is the squared Euclidean distance between the vectors of fitted values $\hat{\mathbf{y}}$ and $\hat{\mathbf{y}}_{(i)}$, or simply the sum of squares of the differences in fitted values when case i is deleted, divided by a scale factor.

Return to the adaptive score data and restore any deleted cases. Using OLS, fit the linear model with *Score* as the response and *Age* as the predictor; let's suppose the name of this model is L1. From the Graph&Fit menu, select "Plot of" to draw a plot of Cook's distances, which are called *L1:Cooks-D*, versus case numbers. Case 17 has the largest value, $D_{17} = 0.68$. Deletion of case 17 will cause the largest change in estimated coefficients, in agreement with what we found graphically. Case 1 does not have a particularly large value of D_1 when all the data are used. However, delete case 17 from the regression and see how D_1 changes. You will need to rescale the plot by using the item "Rescale Plot" from the plot's menu.

Figure 15.3 shows a scatterplot of Cook's distances versus case number for the transactions data discussed at the beginning of this chapter. There are three relatively large values of D_i, and the largest value is for case 160, the same case that was found to influence tests for nonconstant variance. This plot could now be used as a control for deleting cases and studying their influence on the analysis.

Cook's distance provides an ordering of the cases in terms of their influence on $\hat{\boldsymbol{\eta}}$. We should always study the impact of cases that have a relatively large value of D_i, like case 160 in the transaction data. Otherwise, it is generally useful to study cases that have $D_i > 0.5$ and is always important to study cases with $D_i > 1$. These benchmarks are intended as an aid to finding influential cases, but they do not represent a test. Again, *there is no significance test associated with D_i.*

Shown in Figure 15.4 is an added-variable plot for T_1 in the transactions data with three cases marked. Recall from Chapter 10 that the slope of the OLS line on the plot is the same as $\hat{\eta}_1$. We might expect cases that are influential for $\hat{\eta}_1$ to stand apart from the main point cloud in an added-variable plot. This

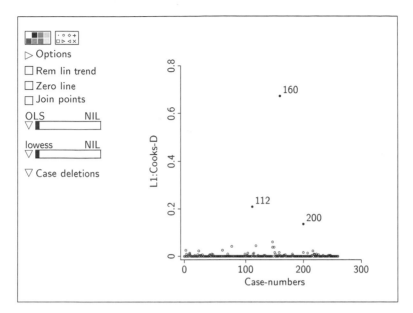

FIGURE 15.3   Scatterplot of Cook's distances versus case number for the transactions data.

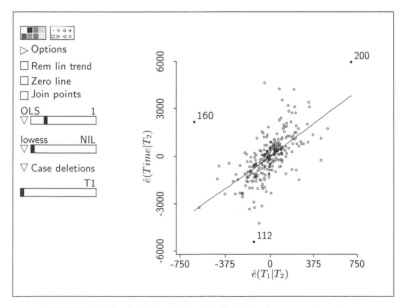

FIGURE 15.4   Added variable plot for T_1 in the transactions data.

expectation turns out to be essentially correct, and added-variable plots are useful diagnostics for identifying cases that influence the estimate of a single coefficient. 3D added-variable plots can be used in the same way, except now we look for cases that could influence two coefficient estimates. In Figure 15.4 we see that the remote cases are the same as those identified by using Cook's distance, although there will not always be such good agreement.

15.3 RESIDUALS

15.3.1 Studentized Residuals

The variance of a residual is, from (7.24), equal to

$$\text{Var}(\hat{e}_i \mid \mathbf{x}_i) = \sigma^2(1 - h_i) \tag{15.7}$$

where h_i is the ith leverage, discussed in Section 7.6.3, page 161. Residuals can be standardized to correct for unequal variance caused by unequal leverages by dividing the residual by an estimate of its standard deviation. The result is called a *Studentized residual*,

$$r_i = \frac{\hat{e}_i}{\hat{\sigma}\sqrt{1 - h_i}} \tag{15.8}$$

Each Studentized residual has mean zero and variance one.

15.3.2 Cook's Distance Again

In the multiple linear regression model, Cook's distance for the ith case can be computed without actually deleting the ith case and refitting. It can be expressed as a function of the Studentized residual and the leverage for that case,

$$D_i = \frac{r_i^2}{k} \times \frac{h_i}{1 - h_i} \tag{15.9}$$

where, as usual, k is the number of u-terms in the model. This representation gives some insight into influence. The factor $h_i/(1 - h_i)$ will be large when the leverage is close to one. Cases with $\mathbf{u}_i$ sufficiently far from $\bar{\mathbf{u}}$ relative to the density contours of the u-terms will have large values of D_i. In particular, the highlighted case in Figure 7.7d, page 162, will likely be influential.

D_i could also be large because the ith Studentized residual is large. Large Studentized residuals are due to outliers, points whose response falls far from the fitted mean function, an idea that will be developed in the next section. Thus, a case can be influential because the response is outlying, $\mathbf{u}_i$ has high leverage, or a combination of the two.

Return to the transactions data and construct a 3D plot of *Time* versus (T_1, T_2) with the fitted plane and the residuals superimposed. We discussed *Arc* controls for doing this in Section 8.2.4, page 190. While spinning the plot, identify the most influential case by visualizing the leverages and the magnitudes of the residuals. We previously identified case 160 as the most influential using Cook's distance. Did you identify the same point visually? 3D added-variable plots can be interpreted in much the same way, by looking for points that influence the estimates of the corresponding coefficients.

15.4 OUTLIERS

We have used the term *outlier* to refer to cases that somehow seem different from the rest of the data, in particular cases with a response that does not fit the pattern in the data. For example, we judged the remote case in Figure 14.1c to be *outlying* because the response seems much too large relative to the trend in the rest of the data. On the other hand, a case will have high *leverage* if its u-vector $\mathbf{u}_i$ falls in a region of low density relative to the sample distribution of the u-terms in the mean function. A high leverage case may or may not be an outlier. In this section we explore the idea of an outlier and give a method of testing for outliers in linear models.

By definition, an outlier must outlie something. Consider the legal case of Hadlum v. Hadlum: 349 days after Mr. Hadlum departed for military service abroad, Mrs. Hadlum gave birth. Mr. Hadlum judged the observation of 349 days to be outlying relative to the distribution of gestation times and therefore filed for divorce. The statistical issue here is how to judge the weight of evidence against the hypothesis that the outlying observation is a valid, albeit extreme, realization from the known distribution of gestation times. Mr. Hadlum could have computed a p-value, the probability that a randomly selected gestation time exceeds 349 days, although we imagine that the legal case was more involved.

During the Cold War the former USSR inadvertently dropped one of their satellites in central Canada. The joint U.S. and Canadian project to locate the debris was based on monitoring the level of radiation while flying over probable locations. Levels of radiation that were high relative to the background were taken as indicators of satellite debris. In this situation, we think of looking for radiation levels that outlie the level of background radiation. Most of the debris was located using this method.

Outliers may indicate important new information. It is now generally recognized that an anomalous reading, which went unnoticed at the time, in Millikan's famous oil drop experiment is the first documented evidence for quarks, a type of elementary particle in physics. Similarly, residuals of large magnitude can indicate that the model is wrong, possibly leading to a better understanding of the regression.

Outliers may or may not be influential. The second phase of the Florida Area Cumulus Experiment (FACE-2) was designed to confirm the first-phase indications that cloud seeding can produce increases in natural rainfall. But the analysis of the second-phase data was complicated by a single outlier, an unseeded day on which the rainfall was four standard deviations above the average. The outlying day was very influential: With the outlier the estimated increase in rainfall due to seeding was about 5%, but without the outlier the estimated increase was 25%.

The concern for outliers is quite old and probably dates from the first attempts to form conclusions from data. The sometimes imprudent practice of discarding outliers was commonplace 200 years ago. Writing in 1887, Frank Edgeworth's informal notion of an outlier is much the same as it is today: "Discordant observations may be defined as those which present the appearance of differing in respect of their law of frequency from other observations with which they are combined."

15.4.1   Testing for a Single Outlier

We now discuss a standard method of testing for a single outlier in the linear model

$$E(y_i \mid \mathbf{x}_i) = \eta^T \mathbf{u}_i \qquad \text{with} \quad \text{Var}(y_i \mid \mathbf{x}_i) = \sigma^2 \qquad (15.10)$$

for $i = 1, \ldots, n$. More specifically, we introduce a test of the hypothesis that the ℓth case does not outlie the mean function of this model versus the alternative that it does. We assume that *the index ℓ is chosen without reference to any quantity that depends on the response variable.* Thus, the choice of the case to test cannot depend on the response, but it can depend on the predictors. The test itself *will* depend on the response variable, however. For example, consider a large university department with 40 male faculty and one female faculty member. We could reasonably ask if the salary of the female faculty member outlies the mean function for the male faculty. The actual test would depend on all salaries.

The first obstacle we face in testing for an outlier in multiple linear regression is that the vector of regression coefficients η is unknown. If the ℓth case is an outlier, the estimate of η based on the full data may be biased. This suggests using estimates computed without the suspected outlier, case ℓ. Use the following procedure:

- Obtain the OLS estimate $\hat{\eta}_{(\ell)}$.
- Predict the response for the ℓth case, $\tilde{y}_\ell = \hat{\eta}_{(\ell)}^T \mathbf{u}_\ell$.
- Compute the test statistic that compares the predicted value $\tilde{y}_\ell$ to the observed value y_ℓ,

$$t_\ell = \frac{y_\ell - \tilde{y}_\ell}{\text{se}(\tilde{y}_\ell)} \qquad (15.11)$$

- Compare t_ℓ to the quantiles of a t-distribution with $n - k - 1$ df.

Here, $\mathrm{se}(\tilde{y}_\ell)$ is the standard error for prediction, as discussed in Section 7.6.2, still based on the reduced data without case ℓ. The tilde on $\tilde{y}_\ell$ is intended to distinguish this prediction, which is based on the reduced data, from the usual prediction $\hat{y}_\ell$ based on all the data.

The test statistic (15.11) can also be obtained by adding another term to the model and fitting with the full data. Define the term $u_i^{(\ell)} = 1$ if $i = \ell$ and $u_i^{(\ell)} = 0$ otherwise. The term $u_i^{(\ell)}$ is an indicator variable for the ℓth case, as defined in Chapter 12. We will call it a *case indicator* to distinguish it from other indicator variables.

The outlier test statistic t_ℓ is the same as the usual t-test statistic based on all the data for the hypothesis that $\delta = 0$ in the expanded linear model

$$\mathrm{E}(y_i \mid \mathbf{x}_i) = \boldsymbol{\eta}^T \mathbf{u}_i + \delta u_i^{(\ell)} \qquad \text{with} \quad \mathrm{Var}(y_i \mid \mathbf{x}_i) = \sigma^2 \qquad (15.12)$$

In this expanded model, the regression coefficient δ allows for the possibility that $\mathrm{E}(y_\ell \mid \mathbf{x}_\ell)$ differs from the mean function for the rest of the data. If it is inferred that $\delta \neq 0$, then we have evidence that $\mathrm{E}(y_\ell \mid \mathbf{x}_\ell)$ is in fact different, implying that case ℓ is an outlier.

For the multiple linear regression model, there are very simple formulas for $\hat{\delta}$ and t_ℓ:

$$\hat{\delta} = \frac{\hat{e}_\ell}{1 - h_\ell} \qquad (15.13)$$

$$t_\ell = r_\ell \left(\frac{n - k - 1}{n - k - r_\ell^2} \right)^{1/2} \qquad (15.14)$$

where $\hat{e}_\ell$ is the ℓth residual, h_ℓ is the ℓth leverage, and r_ℓ is the ℓth Studentized residual as defined in (15.8). These statistics depend on n and k, and on the leverages and on the Studentized residuals from the fit to all the data. The magnitude of t_ℓ increases with the magnitude of r_ℓ and so they contain the same information about the agreement between the ℓth case and the mean function. Thus, the squared Studentized residual r_ℓ^2 in Cook's distance (15.9) can be thought of as measuring the agreement between the mean function for the model and the mean for the ℓth case.

Suppose that $\delta \neq 0$ in (15.12) so the ℓth case is in fact an outlier. The power of the outlier t-test to detect that $\delta \neq 0$ depends on the noncentrality parameter

$$\lambda = \frac{\delta^2 (1 - h_\ell)}{\sigma^2}$$

The power increases as λ increases, although the power itself is a rather complicated function of λ. For fixed values of δ and σ^2, the outlier will be the hardest to find using this test when it occurs at a high leverage case, just where it can do the most damage.

The discussion so far has been under the condition that $\text{Var}(y_i \mid \mathbf{x}_i)$ is constant. When there are known weights, replace the condition $\text{Var}(y_i \mid \mathbf{x}_i) = \sigma^2$ in model (15.12) with $\text{Var}(y_i \mid \mathbf{x}_i) = \sigma^2/w_i$. The t-statistic for $\delta = 0$ from the weighted least squares fit is then the outlier statistic t_ℓ.

15.4.2 Checking Every Case

In the development of the outlier t-statistic t_ℓ we assumed that the index ℓ was chosen without reference to any quantity that depends on the response variable. Often we may want to test for a single outlier without choosing ℓ first. We can test for a single outlier with *unknown index* ℓ by using the maximum absolute value of t_ℓ,

$$t_{\max} = \max_{1 \leq \ell \leq n} |t_\ell|$$

Because $t_{\max}$ is a maximum over n test statistics, its distribution under the null hypothesis of no outliers is no longer a t-distribution. The exact p-value for $t_{\max}$ is difficult to calculate. Instead, we use the *Bonferroni inequality* to compute a bound on the p-value:

$$p\text{-value}(t_{\max}) \leq 2n \times \Pr(t_{n-k-1} > t_{\max})$$
$$= n \times p\text{-value from } t_{n-k-1}$$

These equations tell us that the p-value for $t_{\max}$ is not greater that n times the p-value computed from a t-distribution with $n - k - 1$ df. This bound can be computed in *Arc* by typing the command `outlier-pvalue` with three arguments

$$(\texttt{outlier-pvalue} \quad t_{\max} \quad n-k-1 \quad n)$$

15.4.3 Adaptive Score Data

A plot of the outlier t_ℓ statistics versus case numbers for model (15.3) is shown in Figure 15.5. Case 18 has the largest value of the statistic, $t_{18} = 3.60698$. This value can be obtained from *Arc* by selecting the item "Mouse mode" from the plot's menu, choosing "Show coordinates," and clicking on the plotted point, or, since this is a relatively small data set, listing the outlier t_ℓ statistics, labeled *L1:Outlier-t*, and case numbers using the item "Display data" from the data set menu. If case 18 was identified prior to inspecting the data, the appropriate p-value would be computed from a $t_{n-k-1} = t_{18}$ distribution, giving a two-tailed p-value of 0.002. It is a coincidence that the case number is the same as the df—usually this will not be so.

However, we did not identify case 18 before hand. Instead, we selected case 18 because it has the largest absolute value of t_ℓ. Thus we need to compute

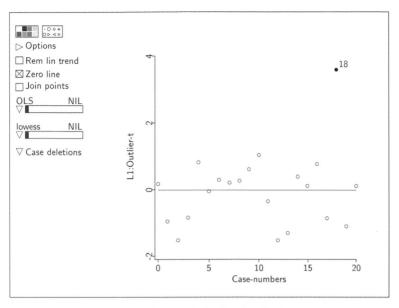

FIGURE 15.5 Outlier t_ℓ statistics for the adaptive score data.

the outlier p-value bound for $t_{\max}$,

```
>(outlier-pvalue 3.60698 18 21)
0.0423288
```

The output from this command is $0.042 = n \times 0.002 = 21 \times 0.002$. The p-value for $t_{\max}$ is thus smaller than its upper bound of 0.042, indicating that case 18 significantly outlies the mean function. There are at least three possibilities to explain this result. Either case 18 is anomalous and should perhaps be discarded, or the mean function is wrong, or the variance function is wrong. If we can rule out the latter two possibilities, then we are left to explain the anomalous case.

Outliers may or may not be influential. From (15.9), if an outlier (high r_i^2) has a low leverage (h_i), then D_i may not be particularly large and case i may not have a large influence on $\hat{\eta}$. For example, in Table 15.2 we see that deleting case 18 has almost no effect on the coefficient estimates, but its deletion reduces the estimate of σ from 11 to 8.6. This case is influential for estimating σ^2, but not for estimating coefficients.

15.5 FUEL DATA

Load the file `fuel90.lsp`. These data consist of several measurements on the states in the United States and the District of Columbia that provide an oppor-

TABLE 15.3   The Fuel Consumption Data

Variable	Description
FUEL/POP	1990 per person fuel consumption.
INC	1990 per capita personal income in 1000s of U.S. dollars.
VEH/POP	1990 registered cars, buses and trucks per person.
TAX	State gas tax in cents per gallon, as of April 1, 1992.
VM/VEH	1989 1000s of vehicle miles of travel per vehicle.

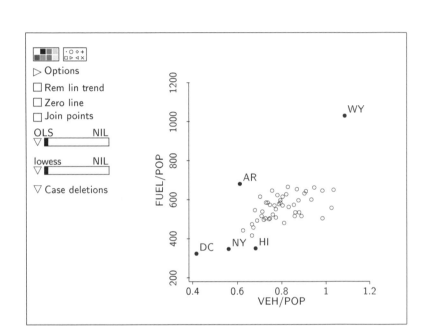

FIGURE 15.6   Marginal response plot for *VEH/POP* for the fuel data.

tunity to understand how statewide fuel consumption varies with characteristics of the state. The variables are described in Table 15.3. Five points appear separated from the main body of the marginal response plot of *FUEL/POP* versus *VEH/POP* shown in Figure 15.6. Three of these states have the lowest per capita fuel consumption, one has the highest fuel consumption, and one state has high fuel consumption given its value of *VEH/POP*. Mark these points with a color or symbol for future reference.

Next, fit the OLS regression of *FUEL/POP* on the four predictors, and construct the plot of residuals versus fitted values as shown in Figure 15.7. This plot does not sustain the mean function for the linear model. The point for Wyoming is clearly outlying, and the test for curvature is significant. The *lowess* estimate of the residual mean function was computed by removing

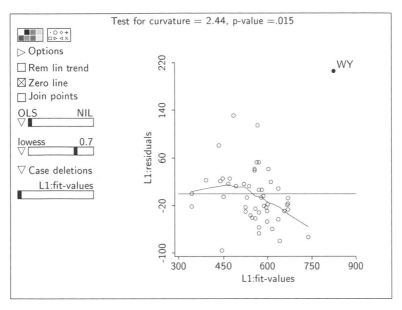

FIGURE 15.7 Residual plot for the fuel data.

(not deleting) the point for Wyoming, placing the *lowess* smooth on the plot, and then restoring the point. The *lowess* estimate suggests that the residual mean function is not constant. We are now faced with a dilemma: Either the data for Wyoming are unusual in some important respect or the model is wrong.

Continuing our graphical exploration of the regression, select the item "AVP–All 2D" from the regression menu. As you cycle through the added-variable plots, note the locations of the points you have marked (you may wish to select the item "Show labels" from the plot's menu). In most of the added-variable plots, Wyoming is separated from the bulk of the data both vertically, suggesting this state might be an outlier, and horizontally, suggesting that this state is a high leverage case. Combining vertical and horizontal separation, Wyoming is probably influential, and its deletion may change conclusions. To confirm this indication, construct the 3D added-variable plot for *VEH/POP* and *TAX*, along with the regression plane and the residuals.

Delete Wyoming and see what happens to the 2D added-variable plots. The most striking change is for *INC*, which changes from a plot with little evidence of a linear trend to one that has a negative trend. This is confirmed by using the "Display fit" item in the model's menu and comparing coefficient estimates with and without Wyoming. Virtually all the coefficients show large changes when Wyoming is deleted. Cook's distance is also very effective in locating this case, as its value is about 1.34, while the second largest value is only 0.18.

15.6   COMPLEMENTS

15.6.1   Updating Formula

Gauss (1821–1826) was the first to give a remarkable formula that can be used to obtain the estimates for a multiple linear regression model when the ith observation is deleted. Using the notation of Section 15.2, the basic formula is

$$(\mathbf{U}_{(i)}^T \mathbf{U}_{(i)})^{-1} = (\mathbf{U}^T \mathbf{U})^{-1} + \frac{(\mathbf{U}^T \mathbf{U})^{-1} \mathbf{u}_i \mathbf{u}_i^T (\mathbf{U}^T \mathbf{U})^{-1}}{1 - h_i} \qquad (15.15)$$

A history of this equation is given by Henderson and Searle (1981). Straightforward application of this formula shows that

$$\hat{\eta}_{(i)} = \hat{\eta} - \frac{(\mathbf{U}^T \mathbf{U})^{-1} \mathbf{u}_i \hat{e}_i}{1 - h_i}$$

$$\hat{\sigma}_{(i)}^2 = \hat{\sigma}^2 \frac{n - k - r_i^2}{n - k - 1}$$

and from these deriving (15.9), (15.13), and (15.14) is straightforward.

15.6.2   Local Influence

The use of added-variable plots to identify influential cases described at the end of Section 15.5 is based in part on a generalized method of influence assessment. The basic idea behind influence analysis is that a regression solution should be stable: Small changes in the data should not produce large changes in the results. Deleting cases is one way of introducing small changes in the data, but there are others as well.

   We might assume that the variance function is not quite constant. Suppose we set $\mathrm{Var}(y_i \mid \mathbf{x}_i) = \sigma^2 / \omega_i$, where ω is the Greek letter omega. Let $\omega = (\omega_1, \ldots, \omega_n)^T$. We can study the change in a single coefficient estimate, say $\hat{\eta}_1$, as ω changes. Let $\hat{\eta}_1(\omega)$ denote the WLS estimate of η_1 with weights ω. If $\omega_i = 1$ for all cases, then $\hat{\eta}_1(\omega) = \hat{\eta}_1$, the OLS estimate, but for other values of ω we can get a different estimate. We can get a *worst* situation by solving the following problem: Find ω_i not too far from 1 for every case such that $\hat{\eta}_1(\omega)$ is as far from $\hat{\eta}_1$ as possible. Cases whose values for ω_i are most different from 1 are potentially influential cases. Finding the worst case is the same as the mathematical problem of maximizing the rate of change in $\hat{\eta}_1(\omega)$ as ω is varied. This maximization problem can be solved, and the potentially influential cases can be identified in the added-variable plot for u_1.

   The added-variable plot for *VEH/POP* in the linear regression model fit to the Fuel data is shown in Figure 15.8. The most separated points are likely to be the most influential. From the plot, modifying the case weight for Wyoming will result in the greatest rate of change in the estimate of the coefficient of *VEH/POP*.

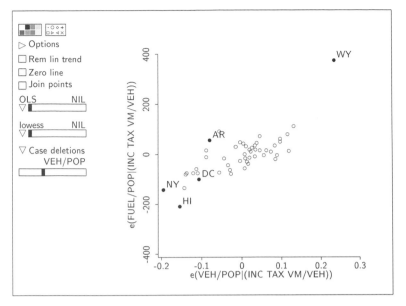

FIGURE 15.8   Added-variable plot for *VEH/POP* in the fuel data.

15.6.3   References

A good elementary introduction to residuals, leverages, outliers, and influence measures is given by Fox (1991). More advanced treatments are given by Cook and Weisberg (1982) and by Atkinson (1985). Deletion influence measures were introduced by Cook (1977). The use of added-variable plots for studying local influence is developed in Cook (1986a). Many other influence measures have been suggested; see Cook, Peña, and Weisberg (1988) and Cook (1986b) for comparisons. Cook and Prescott (1981) discuss the power of tests for outliers and also discuss the accuracy of Bonferroni bounds. The case of Hadlum *vs.* Hadlum is discussed by Barnett (1978). The Russian satellite debris problem is discussed by Beckman and Cook (1983). The FACE-2 experiment is discussed by Kerr (1982); FACE-1 data are presented in Problems 12.8 and 15.8. The quote on page 362 is from Edgeworth (1887).

The adaptive score data first appeared in Mickey, Dunn, and Clark (1967). The muscles data used in Exercise 15.6 are from Cochran and Cox (1957). The fuel consumption data come from the *World Almanac*.

PROBLEMS

15.1   In this problem we use the adaptive score data `adscore.lsp` to verify certain properties of the residuals and fitted values. First, verify that the residuals add to zero. This can be done as follows. Assuming the name of the model is L1, type:

```
(sum (send L1 :residuals))
```

Similarly, replacing :residuals with :leverages, verify that the sum of the leverages is equal to the number of coefficients in the model, including the intercept, which is 2 for these data. Next, verify that $y = \hat{e} + \hat{y}$ by adding both the residuals and the fitted values to the data set, and then plotting their sum versus *Score*. Verify that the regression of $\hat{e}$ on $\hat{y}$ has slope equal to zero, and show that the square of the correlation between y and $\hat{e}$ is equal to $1 - R^2$. These results hold for any OLS fit of a linear model with an intercept.

15.2   The purpose of this problem is to demonstrate numerically that the *t*-statistic for $\delta = 0$ in model (15.12) is the same as the outlier-*t*. We will do this by using the adaptive score data to construct t_{18} according to model (15.12). If all goes well, this should be the same as the value of t_{18} of the variable *Outlier-t* for model (15.10), apart from rounding error. To begin, use the "Add a variate" item and the case-indicator function (Section A.12) by typing the following in the text area in the dialog:

$$u18 = (\text{case-indicator } 18 \ 21)$$

This will create a new variable called *u18* that is of length 21, and all of whose values are equal to 0 except for the element for case number 18, which is equal to 1 (remember again that the numbering of cases always starts with 0, so case number 18 is actually the nineteenth case in the data file). Fit model (15.12) and compare the *t*-statistic for $\delta = 0$ with the corresponding element of Outlier-t for model (15.10).

15.3   It is possible to test for two outliers simultaneously by adding two case indicators to the model. Suppose we wish to test the *j*th and ℓth cases, where the cases are chosen without reference to the response, as in Section 15.4.1. Letting $u_i^{(j)}$ and $u_i^{(\ell)}$ denote the corresponding case indicators, form the linear model

$$E(y_i \mid \mathbf{x}_i) = \eta^T \mathbf{u}_i + \delta_j u_i^{(j)} + \delta_\ell u_i^{(\ell)} \qquad \text{with} \quad \text{Var}(y_i \mid \mathbf{x}_i) = \sigma^2$$

We can now test the hypotheses

$$\text{NH}: \qquad \delta_j = \delta_\ell = 0$$

$$\text{AH}: \qquad \text{either } \delta_j \neq 0 \quad \text{or} \quad \delta_\ell \neq 0 \quad \text{or both are nonzero}$$

by using an *F*-test. Additional cases can be tested simultaneously in the same way by adding additional case indicators.

In the adaptive score data, test the hypothesis that $\delta_{18} = \delta_1 = 0$ against the general alternative. Provide a brief interpretation of the results.

15.4 With Wyoming deleted, is there any clear information in the fuel data fuel90.lsp to contradict the linear model? Curvature, nonconstant variance, outliers, and influential cases are all relevant issues.

15.5 In the Big Mac data, given the regression with response log(*BigMac*) and terms (log(*Bread*), log(*BusFare*), log(*TeachSal*), log(*TeachTax*)) check for influential cases and for outliers. Which cities seem to be different from the rest? How are they different? Do they strongly influence any conclusions?

15.6 The data in file muscles.lsp comprise two replications of a $4 \times 4 \times 3$ factorial experiment on rats to investigate the use of electrical stimulation to prevent deterioration of denervated muscles. The response y is the weight (1 unit $= 0.01$ g) of the denervated muscle at the end of the experiment. Since larger animals tend to have larger muscles, the weight of the untreated muscle, x, on the other side of the rat was used as a covariate. The other factors in the experiment were: *Rep*, the replication number, either zero or one; *TrtTime*, the length of stimulation in minutes, either 1, 2, 3, or 5; *Trt/day*, the number of treatments per day, 1, 3, or 6; and *Trt*, a qualitative factor for the type of current used, 1 = Galvanic, 2 = Faradic, 3 = 60 cycle, or 4 = 25 cycle.

Provide a complete analysis of these data, and summarize your findings. Once you obtain a target model, be sure to analyze the data for outliers and influential points.

15.7 As in Problem 10.11, an experiment was conducted to investigate the amount of a particular drug that would be absorbed in the liver of a rat. The investigator knew that larger livers would absorb more of the drug than smaller livers, and that liver weight is strongly related to body weight. Further, the investigator hypothesized that if each rat was given about 40 mg of the drug per kilogram of body weight, then each rat would absorb about the same percentage of the dose. Nineteen rats were weighed, placed under light ether anesthesia, and given an oral dose of the drug according to this design. After a fixed length of time, each rat was sacrificed, the liver was weighed (*LiverWt*), and the percent of the dose in the liver (y) was determined.

The investigator was disappointed by the results of the experiment because the linear regression of y on *Dose*, *LiverWt*, and *BodyWt* gave significant results, contradicting his hypothesis that the mean function is constant.

Analyze the results of the experiment with a view towards resolving the dilemma. The data are in file rat.lsp.

15.8   Cloud Seeding. This problem continues Problem 12.8 on the first Florida Area Cumulus Experiment in the data file `cloud.lsp`.

   15.8.1   Starting with the model suggested in Problem 12.8.5, with all the data included, examine the data for influential cases and for outliers. Summarize your conclusions.

   15.8.2   Refit the model without the most influential case and the two most likely outliers. Are the conclusions reached different than the conclusions of Problem 12.8.5? Present in a paragraph a summary of the analysis, and in one plot provide a summary of the effect of seeding on rainfall.

CHAPTER 16

Predictor Transformations

Using transformations is a continuing theme of this book. Power transformation were introduced in Section 5.2, where we found that there is a simple linear relationship between log(*BrainWt*) and log(*BodyWt*). In Section 13.2.1 we introduced general guidelines for transformations including the important empirical rule that positive predictors that have the ratio between their largest and smallest values equal to 10 and preferably 100 or more should very likely be transformed to logarithms. Response and predictor transformations to multivariate normality were discussed in Chapter 13. We now consider a more detailed study of predictor transformations.

16.1 REGRESSION THROUGH TRANSFORMATION

In Problem 9.4 we reasoned that a haystack might be approximated with a hemisphere having circumference $C_1 = (C + 2Over)/2$. The mean function for haystack volume that follows from this approximation is

$$E(Vol \mid C_1) = \frac{C_1^3}{12\pi^2} = \tau(C_1) \tag{16.1}$$

which is the volume of a hemisphere with circumference C_1. The mean function in (16.1) is a linear function of a transformation τ of C_1, specifically a cubic transformation, where τ is the Greek letter tau. Mean function (16.1) has little flexibility because it includes no unknown parameters.

16.1.1 Power Curves and Polynomial Fits

We could modify mean function (16.1) to get more flexibility and obtain a better approximation of the volume of a haystack. For example, we might use

$$E(Vol \mid C_1) = \eta_0 + \eta_1 C_1^{(\lambda)} \tag{16.2}$$

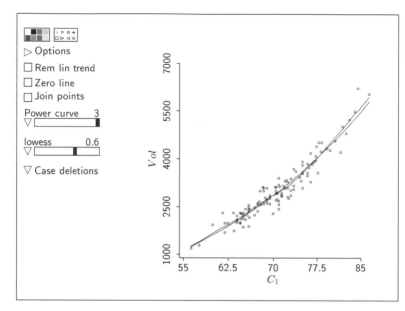

FIGURE 16.1   Plot of *Vol* versus $C_1 = (C + 2Over)/2$ for the haystack data with a cubic power curve and a *lowess* smooth.

Here η_0 and η_1 are unknown regression coefficients and we have allowed for the possibility of raising C_1 to a power other than 3. The superscript "(λ)" indicates the scaled power transformation defined in Section 5.2. This mean function is a transformation of C_1 given by $\tau(C_1) = \eta_0 + \eta_1 C_1^{(\lambda)}$ that includes parameters which may be estimated to increase flexibility of the fitted function.

Using the data in the file `haystack.lsp`, construct a plot of *Vol* versus C_1 and select the item "Power curve" from the pop-up menu for the parametric smoother slidebar. As you move the slider to the right, various power curves are estimated, as described on page 319, and superimposed on the plot. The power curve for $\lambda = 3$ is shown on the plot in Figure 16.1 along with a *lowess* smooth. To construct this plot, you will need to add the value 3 to the power curve slidebar using a pop-up menu item.

As the transformation parameter is varied in the plot of Figure 16.1, the fitted power curve can be judged against the data to select a transformation. For the cubic transformation, the power curve and *lowess* smooth nearly match, suggesting the cubic transformation matches the mean function. Analysis could now proceed based on the model

$$Vol \mid C_1 = \eta_0 + \eta_1 C_1^3 + e$$

Of course, checking residual plots and watching for outliers and influential cases is necessary.

When choosing power transformations in this way, it is always appropriate to prefer meaningful powers, particularly $\lambda = \pm 2, \pm 1, \pm 2/3, \pm 1/3, \pm 1/2$, or 0, provided the choice is sustained by the data. Polynomial fits could be used in the same way as power curves, selecting the OLS option from the parametric smoother slidebar.

16.1.2 Transformations via Smoothing

Generalizing even further, we could adopt the view that

$$Vol \mid C_1 = \tau(C_1) + e \tag{16.3}$$

where $\tau(C_1)$ is now an *unknown* transformation of C_1. The transformation (16.1) contains no unknown parameters, while (16.2) contains up to three unknown parameters that would need to be estimated. In (16.3) the entire transformation τ is unknown; in effect, the transformation function $\tau(C_1)$ becomes a *superparameter* that must be estimated.

Starting with (16.3), the transformation τ could be estimated from a plot of *Vol* versus C_1 as shown in Figure 16.1. That plot can also be described as a plot of $\tau(C_1) + e$ versus C_1, and thus it can be used to estimate τ by smoothing. A *lowess* estimate $\hat{\tau}(C_1)$ of $\tau(C_1)$ is shown in Figure 16.1. We don't have a functional form for the estimate, but we can extract its values $\hat{\tau}(C_{i1})$ at the data points C_{i1}, $i = 1, \ldots, 120$, and add them to the data set. This is done by using the item "Extract mean" from the pop-up menu for the *lowess* slidebar. After selecting this item, you will be presented with a dialog for naming the transformation. The default name is h, but this can be changed. However, do not use just the letter "t" since this is reserved as a system constant in *Arc*. The transformed values will be added to the data set as a new variable with the name you selected. The fitted values from power curves and polynomial fits can be extracted similarly.

16.1.3 General Formulation

Consider a regression with response y and k u-terms, including a constant for an intercept. Let u_2 be a selected nonconstant u-term and collect all remaining u-terms including the constant into a $(k-1) \times 1$ vector $\mathbf{u}_1$. In many applications, u_2 will be one of the predictors. We want to decide if the initial linear model

$$y \mid \mathbf{x} = \eta_1^T \mathbf{u}_1 + \eta_2 u_2 + e$$

can be improved by transforming u_2. To this end, we modify the model by including an unknown transformation $\tau(u_2)$ of u_2,

$$y \mid \mathbf{x} = \eta_1^T \mathbf{u}_1 + \tau(u_2) + e \tag{16.4}$$

The goal is to estimate the unknown transformation function $\tau(u_2)$. Generally, we can estimate $\tau(u_2)$ only up to an unimportant additive constant. As long as the mean function includes an intercept, adding a constant c to any transformation results in a new transformation $c + \tau(u_2)$ that is equivalent to the original transformation $\tau(u_2)$ with a different value for the intercept. Since constant shifts are unimportant for judging the form of the transformation and for subsequent data analysis, they will be neglected.

As suggested by the previous examples, $\tau(u_2)$ will be estimated by smoothing or by fitting a power curve or a polynomial to an appropriate plot to be discussed in Section 16.2. Once selected, the transformation can be substituted for $\tau(u_2)$ in (16.4).

We can distinguish between two general outcomes:

- If $\tau(u_2)$ is judged to be linear in u_2, then we can set $\tau(u_2) = \eta_2 u_2$ and the model reduces to the usual multiple linear regression model. In this way the methods presented in this chapter can be used as diagnostics for the mean function.

- If $\tau(u_2)$ is judged to be nonlinear in u_2, then we have evidence that a transformation of u_2 may result in an improved model.

We describe in the next section graphical displays that can be used to infer about the unknown transformation τ. These displays are called *Ceres plots*; *Ceres* is an acronym for **C**ombining conditional **E**xpectations and **RES**iduals. *Ceres* plots extend the basic ideas just introduced.

16.2   *Ceres* PLOTS

Suppose we know the value of η_1 in (16.4). Rearranging terms, we obtain

$$(y - \eta_1^T \mathbf{u}_1) \mid \mathbf{x} = \tau(u_2) + e \qquad (16.5)$$

A plot of the *adjusted response* $y - \eta_1^T \mathbf{u}_1$ versus u_2 would be the same as a plot of $\tau(u_2) + e$ versus u_2, and we would be able to visualize $\tau(u_2)$ and use the plot to assess the evidence for transforming u_2. If we know the adjusted response, we can proceed by using the methods discussed in Section 16.1.

Since η_1 is not usually known, we need to estimate it using the data. Assuming we have a good estimate $\hat{\eta}_1$ of η_1, we construct the adjusted responses using $\hat{\eta}_1$ and plot

$$y - \hat{\eta}_1^T \mathbf{u}_1 \quad \text{versus} \quad u_2 \qquad (16.6)$$

This is the general version of a *Ceres* plot. The only issue left is how to estimate η_1 without knowing $\tau(u_2)$ in the first place. Exactly how η_1 can be estimated without knowing τ depends on the $k - 2$ *term mean functions* $E(u_{1j} \mid u_2)$, $j = 1, \ldots, (k - 2)$, where u_{1j} is the jth u-term in $\mathbf{u}_1$ excluding the constant term $u_{10} = 1$. We can infer about these mean functions from the $k - 2$ scatterplots of u_{1j} versus u_2.

16.2.1   Constant $E(u_{1j} \mid u_2)$, No Augmentation

When the mean functions $E(u_{1j} \mid u_2)$, $j = 2, \ldots, k - 2$, are *all* constant, η_1 can be estimated by using OLS to fit the working model

$$y = \eta_1^T \mathbf{u}_1 + \text{error} \tag{16.7}$$

We have written out "error" explicitly to remind us that this is a working model fit only with the goal of estimating η_1. It need not provide a useful description of a regression. The *Ceres* plot in this case is just a plot of the residuals from the OLS fit of (16.7) versus u_2. We will call this a *Ceres plot with no augmentation*.

If the term mean functions are not constant, then the estimate $\hat{\eta}_1$ obtained from the fit of (16.7) will not estimate η_1 in (16.4), and the *Ceres* plot with no augmentation may fail to give a reasonable visualization of τ.

16.2.2   Linear $E(u_{1j} \mid u_2)$, Linear Augmentation

When the term mean functions $E(u_{1j} \mid u_2)$ are all linear functions of u_2, including the possibility that some may be constant, η_1 is estimated from the OLS fit of the working model

$$y = \eta_1^T \mathbf{u}_1 + \eta_2 u_2 + \text{error} \tag{16.8}$$

Linear augmentation adjusts for the linear form of the term mean functions so we can still get a good estimate of η_1 for use in *Ceres* plots. The only thing that has changed from the previous case is the way in which η_1 is estimated. The *Ceres* plot is still given by (16.6), but with the estimate of η_1 taken from the OLS fit of (16.8).

The *Ceres* plot in this case is called a *Ceres plot with linear augmentation*.

16.2.3   Quadratic $E(u_{1j} \mid u_2)$, Quadratic Augmentation

Suppose that $E(u_{1j} \mid u_2)$ is a quadratic function of u_2 for at least one j, while the mean function is either linear or constant otherwise. We now augment with both linear and quadratic terms in u_2, and use OLS to fit the working model

$$y = \eta_1^T \mathbf{u}_1 + \eta_2 u_2 + \eta_3 u_2^2 + \text{error} \tag{16.9}$$

The *Ceres plot with quadratic augmentation* is a *Ceres* plot (16.6) with $\hat{\eta}_1$ estimated from this fit. This *Ceres* plot will again allow visualization and estimation of $\tau(u_2)$.

16.2.4   General $E(u_{1j} \mid u_2)$, Smooth Augmentation

In this case, we make no special assumption about the term mean functions $E(u_{1j} \mid u_2)$. Rather, suppose that the function $m_j(u_2)$ is an estimate of $E(u_{1j} \mid u_2)$ based on a *lowess* smooth of the scatterplot of u_{1j} versus u_2. Excluding the constant term, there are $k - 2$ of these m_j, one for each nonconstant component of $\mathbf{u}_1$. We then use OLS to fit the augmented mean function

$$y = \eta_1^T \mathbf{u}_1 + \eta_{2,1} m_1 + \cdots + \eta_{2,k-2} m_{k-2} + \text{error} \qquad (16.10)$$

Taking $\hat{\eta}_1$ from this fit, the *Ceres plot with smooth augmentation* is again given by (16.6). This method can be applied in any regression provided that good estimates m_j are available. Smooth augmentation requires more work to construct the *Ceres* plot, but it may be necessary if the other augmentation types do not apply.

16.3   BERKELEY GUIDANCE STUDY

In the Berkeley guidance study for girls in the file BGSgirls.lsp, consider the regression of *HT18*, the height at age 18, on three of the age 9 predictors *HT9*, *WT9*, and *ST9*. It is possible that we will need to transform one, two or all three of the predictors. It is also possible that none of the predictors will require transformation. To explore these possibilities, we will first consider transforming each predictor in turn. When considering transformation of *HT9*, we require knowledge of the mean functions $E(WT9 \mid HT9)$ and $E(ST9 \mid HT9)$. When considering transformation of *WT9*, we require knowledge of the mean functions $E(HT9 \mid WT9)$ and $E(ST9 \mid WT9)$. Similarly, we need information about two mean functions when considering *ST9*. All of these mean functions can be visualized easily in a scatterplot matrix, extracting the individual plots that may require more detailed analysis with smoothers. In this example we used a scatterplot matrix (not shown) to infer that all the mean functions are strongly linear, but a few present a weak indication of a quadratic trend. To be on the safe side, we decided to use *Ceres* plots with quadratic augmentation, although linear augmentation leads to the same qualitative conclusions.

To construct the *Ceres* plot, use OLS to fit the linear regression of *HT18* on the three predictors and then select the item "Ceres plots–2D" from the resulting model menu. This will produce the dialog shown in Figure 16.2 that is used to specify the type of augmentation, and the terms to be considered for transformation. The "Selection" box must include at least one of the term names; the default is to consider transforming all terms as shown. Choose an augmentation method using one of the four methods outlined in the last section.

We used quadratic augmentation and selected all three terms. This produced a 2D plot with an extra slidebar and pop-up menu. The initial view in this plot

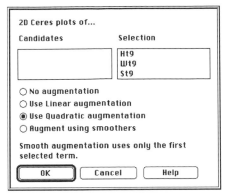

FIGURE 16.2 The *Ceres* dialog.

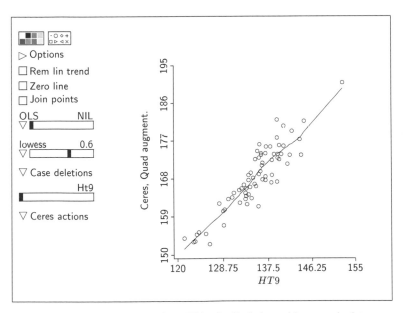

FIGURE 16.3 *Ceres* plot for *HT9* in the Berkeley guidance study data.

is the *Ceres* plot for the first term selected. The slidebar can be used to cycle through all the terms. The order of the terms on the slidebar is the same as the order in the "Selection" box of the *Ceres* dialog.

The *Ceres* plot for *HT9* is shown in Figure 16.3. It has an approximately linear mean function. The *Ceres* plots for the other two predictors also show no clear evidence of curvature. Thus no transformations are indicated and the *Ceres* analysis supports the possibility that the linear regression mean function may be appropriate for these data.

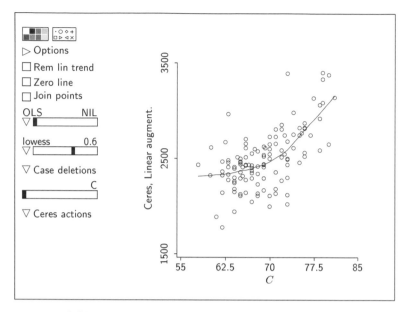

FIGURE 16.4   Initial *Ceres* plot for *C* in the haystack data.

16.4   HAYSTACK DATA

We start the analysis of the haystack data by using OLS to fit the tentative mean function

$$E(Vol \mid C, Over) = \eta_0 + \eta_1 C + \eta_2 Over \tag{16.11}$$

There is clear curvature in a 3D plot of the residuals versus $(C, Over)$, indicating that this mean function is deficient. It seems possible that transformation of one or both of the predictors could result in considerable improvement. When transforming C, we require knowledge of the mean function $E(Over \mid C)$ and when transforming *Over* we need to know about $E(C \mid Over)$. Plots of *Over* versus C and C versus *Over* (not shown) indicate that both of these mean functions are strongly linear. Thus we next investigate *Ceres* plots for C and *Over* using linear augmentation.

The *Ceres* plot for C is shown in Figure 16.4 and the *Ceres* plot for *Over* is shown in Figure 16.5. Judging from the smooths, both plots show curvature and thus some transformation may indeed improve the mean function. Which predictor should we transform? In general, *if one or more predictors require transformation, the Ceres plot for a predictor that does not require transformation may still show curvature*, because the nonlinear effects of one predictor can leak through to the *Ceres* plots for other predictors. The potential confusion caused by this *leakage effect* can be avoided by transforming the predictor

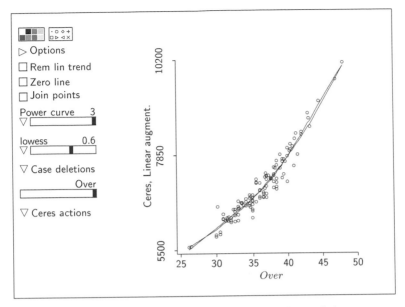

FIGURE 16.5 Initial *Ceres* plot for *Over* in the haystack data.

that shows the strongest visual fit in its *Ceres* plot. In the haystack data we choose to transform *Over*.

The next step is to estimate $\tau(Over)$ in

$$E(Vol \mid C, Over) = \eta_0 + \eta_1 C + \tau(Over)$$

in an attempt to improve the mean function. There are two fits superimposed on the *Ceres* plot for *Over* in Figure 16.5. One is a *lowess* smooth and the other is the cubic power curve. The smooths agree quite well and either may be adequate, but we prefer to proceed using the cubic power curve because of its relative simplicity and the connection with the volume of a hemisphere. The next step is to extract the cubic power curve from the *Ceres* plot for *Over*. Starting with the plot in Figure 16.5,

- remove the *lowess* smooth,
- select "Extract mean" from the pop-up menu from the power curve slide-bar, and
- name the transformation *t[Over]* in the resulting dialog.

The data set now contains the new variable *t[Over]* which is our estimate of $\tau(Over)$. *Arc* provides a shortcut for these last two steps: Select the item "Add smooth to data set" from the pop-up menu for "Ceres actions" to ex-

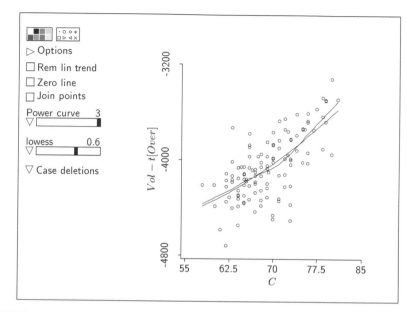

FIGURE 16.6   *Ceres* plot for C in the haystack data using the adjusted response $Vol - t[Over]$ after transforming *Over* to *t[Over]*.

tract the smooth and add the transformed variable, which will be named $t[$"name on x axis"$]$, to the data set.

Now that we have transformed *Over*, our current model is

$$Vol \mid (C, Over) = \eta_0 + \eta_1 C + t[Over] + e$$

Moving *t[Over]* to the left side, we obtain

$$(Vol - t[Over]) \mid (C, Over) = \eta_0 + \eta_1 C + e$$

We now have a working model with one predictor, and we can begin again, using this last model as a starting point for investigating transformations of C, just as we started with model (16.11).

To see if C should be transformed, we plot the adjusted response $Vol - t[Over]$ versus C and estimate the transformation in the usual way. This plot is shown in Figure 16.6. Again we see that the cubic power curve fits quite well, suggesting a cubic transformation of C. In *Arc* the adjusted responses can be created and added to the data set by selecting the item "Add adjusted response to data set" from the pop-up menu for "Ceres Actions."

The *Ceres* plots we have used in this analysis indicate that both C and *Over* should be cubed, so we next consider the model

$$Vol \mid (C, Over) = \eta_0 + \eta_1 C^3 + \eta_2 Over^3 + e \qquad (16.12)$$

This model is the first firm alternative to the starting model (16.11). The fitted mean function is

$$\hat{E}(Vol \mid (C, Over)) = -141.252 + 0.00239258C^3 + 0.0483743Over^3$$

with $\hat{\sigma} = 195.552$. An inspection of a 3D plot of Vol versus $(C^3, Over^3)$ and various residual plots supports this model. We therefore adopt it as the final model for this example and turn to interpretation.

We can gain further understanding of the fitted model by changing the units of the predictors to correspond to the volume of a hemisphere. Accordingly, define the predictors $VC = C^3/(12\pi^2)$ and $VOver = Over^3/(1.5\pi^2)$ to be the volumes of hemispheres based on C and $Over$. The fitted mean function can now be written in terms of the volume predictors by using the results in Section 10.1:

$$\hat{E}(Vol \mid C, Over) = -141.252 + (0.00239258)(12\pi^2)VC$$

$$+ (0.0483743)(1.5\pi^2)VOver$$

$$= -141.252 + 0.2834VC + 0.7162VOver$$

Remarkably, the coefficients of VC and $VOver$ are both positive and essentially add to 1. This means that, apart from the intercept term, the estimated mean values can be interpreted as a weighted average of two different estimates of the volume of a haystack, one based on C and one based on $Over$.

16.5 TRANSFORMING MULTIPLE TERMS

When transforming two or more terms, we need to distinguish between two general types of transformation: *additive* and *nonadditive*. When transforming two terms, u_2 and u_3, an additive transformation is of the form $\tau(u_2, u_3) = \tau_2(u_2) + \tau_3(u_3)$, so the terms are transformed individually and then the transformed values are added to get the joint transformation τ. A nonadditive transformation is any transformation that is not additive. For example, $\tau(u_2, u_3) = u_2 u_3$ is a nonadditive transformation while $\tau(u_2, u_3) = u_2^2 + \log(u_3)$ is an additive transformation. In the next section we discuss a method based on *Ceres* plots for estimating additive transformations of several terms. The method is a generalization of that used for the haystack data in the last section. Nonadditive transformations are discussed in Section 16.7.

16.5.1 Estimating Additive Transformations of Several Terms

After fitting the multiple linear regression model, inspect the 2D *Ceres* plots for convincing nonlinear trends using an appropriate form of augmentation.

If none are found, there is no evidence that any terms should be transformed, and the *Ceres* plots sustain the model. Otherwise, transform the term with the strongest visual fit in its *Ceres* plot, say u_{21}. The estimated transformation $\hat{\tau}(u_{21})$ could be a polynomial, a power, or a smooth.

Next, subtract $\hat{\tau}(u_{21})$ from the response and fit the working model

$$y - \hat{\tau}(u_{21}) = \eta_1^T \mathbf{u}_1 + \text{error} \qquad (16.13)$$

where $\mathbf{u}_1$ is the $(k-1) \times 1$ vector of terms excluding u_{21}. This is a linear model with adjusted response $y - \hat{\tau}(u_{21})$ and $k-1$ u-terms, so we can begin again and use the *Ceres* plots to explore the need to transform components of $\mathbf{u}_1$.

The third item "Fit new model with adjusted response" in the "Ceres actions" pop-up menu executes the first two items in the menu—adding the smooth and adjusted response to the data set—and then fits model (16.13), producing an associated model menu.

After transforming q terms, $u_{21}, \ldots, u_{2q}$, the working model is of the form

$$y_{adj} = \eta_q^T \mathbf{u}_q + \text{error}$$

where $\mathbf{u}_q$ is the $(k-q) \times 1$ vector of terms excluding $u_{21}, \ldots, u_{2q}$ and

$$y_{adj} = y - \hat{\tau}_1(u_{21}) - \cdots - \hat{\tau}_q(u_{2q})$$

The adjusted response y_{adj} is the original response minus all of the extracted transformation curves. The transformation of a component of $\mathbf{u}_q$ is determined as usual and the estimated transformation curve is then subtracted from y_{adj} to produce the next working model. The process continues until all relevant terms have been considered.

16.5.2   Assessing the Transformations

As a final diagnostic check on the transformations, inspect the *Ceres* plots for the working model

$$y_{adj} = \eta_0 + \eta_1 u_1 + \cdots + \eta_{k-1} u_{k-1} + \text{error}$$

where the adjusted response y_{adj} is now the original response minus all of the estimated transformation curves. If all the *Ceres* plots have linear mean functions, then no further transformations are required. If some have nonlinear mean functions, then further transformations may be useful.

16.6   *Ceres* PLOTS WITH SMOOTH AUGMENTATION

We have concentrated on *Ceres* plots with linear and quadratic augmentation since these augmentation types will be adequate in most applications. However, smooth augmentation is necessary when the mean functions for the regression of one term on another cannot be described adequately by a linear or a quadratic curve. In this section we use constructed data to show how to use smooth augmentation in *Arc* and to illustrate that using the wrong type of augmentation can produce misleading results.

The file `ceresdat.lsp` contains data on a response y and three u-terms, u_{11}, u_{12}, and u_2. As suggested by the notation, we will consider transforming only u_2 in this example, so only u_2 should be placed in the "Selection" box of the *Ceres* dialog for this discussion. The response was constructed as

$$y = u_{11} + u_{12} + \frac{1}{1 + \exp(-u_2)}$$

without any error. By not including error, we will be able to illustrate conclusions more clearly. A plot of the true transformation

$$\tau(u_2) = \frac{1}{1 + \exp(-u_2)}$$

versus u_2 is shown in Figure 16.7a. This is the transformation curve we would like to recover by using *Ceres* plots.

These data were generated so that $E(u_{11} \mid u_2) = u_2^{-1}$ and $E(u_{12} \mid u_2) = \log(u_2)$. If these two mean functions were known, a *Ceres* plot for u_2 should be developed by first using OLS to fit the working model

$$y = \eta_0 + \eta_{11} u_{11} + \eta_{12} u_{12} + \eta_2 u_2^{-1} + \eta_3 \log(u_2) + \text{error}$$

and then constructing the plot from (16.6) with the estimate of $\eta_1 = (\eta_0, \eta_{11}, \eta_{12})^T$ taken from this fit. The resulting *Ceres* plot, which does a very good job of capturing the transformation, is shown in Figure 16.7b.

The *Ceres* plot with linear augmentation is shown in Figure 16.7c. Apart from three outliers in the lower left corner, the plot has a strong linear trend; this would not normally be seen as providing evidence for a transformation. This *Ceres* plot fails because linear augmentation does not capture the term mean functions. The *Ceres* plot with quadratic augmentation is shown in Figure 16.7d. This plot does a better job of indicating the need for a transformation, but the results are still not very good. Neither linear nor quadratic augmentation provides a sufficient approximation to the term mean functions, and consequently they both fail to give a reasonable visualization of $\tau(u_2)$. For progress in this example, we must turn to smooth augmentation.

Select the option "Augment using smoothers" in the *Ceres* dialog. With smooth augmentation, only one u-term can be considered for transformation at a time; if more than one term is placed in the "Selection" box, only the first one will be used. Two plots are produced, one plot for estimating the

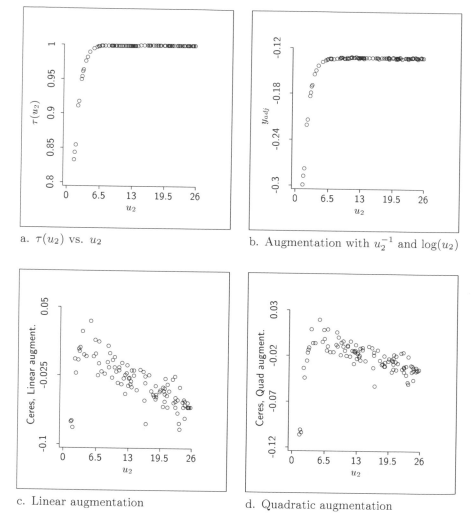

a. $\tau(u_2)$ vs. u_2

b. Augmentation with u_2^{-1} and $\log(u_2)$

c. Linear augmentation

d. Quadratic augmentation

FIGURE 16.7 Four *Ceres* plots for the constructed data discussed in Section 16.6.

term mean functions and a *Ceres* plot. The term plot has an extra slidebar for moving between the terms in $\mathbf{u}_1$. For the constructed data, there are two term plots, u_{11} versus u_2 and u_{12} versus u_2. The term mean functions can be estimated from these plots using *lowess*, a power curve or a polynomial. The default is to estimate all term mean functions using *lowess* with tuning parameter 0.5.

Once the term mean functions have been estimated on the plots, click the new box "Update *Ceres*" on the *Ceres* plot. This will extract the smooths from the term plots, fit the working model with the corresponding augmentation using (16.10), and redraw the *Ceres* plot using the new estimate of η_1.

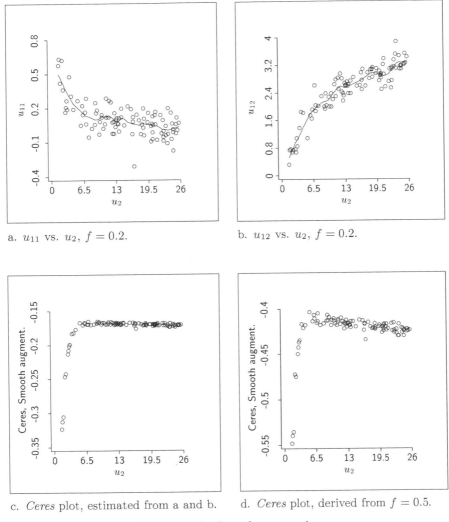

a. u_{11} vs. u_2, $f = 0.2$. b. u_{12} vs. u_2, $f = 0.2$.

c. *Ceres* plot, estimated from a and b. d. *Ceres* plot, derived from $f = 0.5$.

FIGURE 16.8 *Ceres* plot construction.

Two plots versus terms for the example are shown in Figures 16.8a and 16.8b with *lowess* at 0.2 and the plot controls removed. The corresponding *Ceres* plot obtained by clicking the "Update *Ceres*" box is shown in Figure 16.8c. This *Ceres* plot gives a good representation of the true transformation in Figure 16.7a. For contrast, the *Ceres* plot with *lowess* at 0.5 is shown in Figure 16.8d. The estimate of the transformation has clearly deteriorated. *Ceres* plots are influenced more by bias in the estimates of the term mean functions than by variation, so undersmoothed estimates as in Figures 16.8a and 16.8b are desirable.

TABLE 16.1   Augmenting Functions for 2D and 3D *Ceres* Plots[a]

Augmentation	Conditions
None	Each mean function $E(u_{1j} \mid u_2, u_3)$ is constant. No augmentation needed.
Linear	Each mean function $E(u_{1j} \mid u_2, u_3)$ is linear or constant. The augmenting function is $\eta_2 u_2 + \eta_3 u_3$.
Quadratic	Each mean function $E(u_{1j} \mid u_2, u_3)$ is quadratic or linear or constant. The augmenting function is $\eta_2 u_2 + \eta_3 u_3 + \eta_{22} u_2^2 + \eta_{33} u_3^2 + \eta_{23} u_2 u_3$.
Smooth	For arbitrary relationships between the terms. In principle, smooth each u_{1j} on (u_2, u_3), and use these smooths to augment the mean function. This is implemented in *Arc* only for 2D *Ceres* plots for transforming one term at a time.

[a]In this table, $\mathbf{u}_1$ is the vector of $k-2$ terms not considered for transformation, and u_2 and u_3 are the two terms being considered. For 2D *Ceres* plots, delete reference to u_3.

16.7 TRANSFORMING TWO TERMS SIMULTANEOUSLY

The same ideas can be used to transform two terms simultaneously. Let u_2 and u_3 denote the terms that may require transformation and collect the remaining terms into the $(k-2)$-dimensional vector $\mathbf{u}_1$. The required transformation will be denoted by $\tau(u_2, u_3)$.

16.7.1 Models for Transforming Two Terms

Consider the possibility of transforming two terms u_2 and u_3 simultaneously:

$$E(y \mid \mathbf{x}) = \eta_1^T \mathbf{u}_1 + \tau(u_2, u_3)$$

Regardless of whether $\tau(u_2, u_3)$ is additive or nonadditive, the procedure for getting a *Ceres* plot is analogous to the method for transforming terms one at a time. We obtain the OLS fit of an augmented mean function, where the augmenting function depends on the relationship between the terms, and then use the 3D plot

$$\hat{\eta}_1^T \mathbf{u}_1 \quad \text{versus} \quad (u_2, u_3) \tag{16.14}$$

to visualize $\tau(u_2, u_3)$. The estimate of η_1 comes from the augmented mean function as before. The augmenting functions needed are described in Table 16.1. As in the case of transforming a single term, the *Ceres* plot is used to estimate $\tau(u_2, u_3)$, and the procedure should be effective if the mean functions $E(u_{1j} \mid u_2, u_3)$ satisfy the augmentation conditions described in Table 16.1. The distinction between additive and nonadditive transformations is not important for constructing the plot, but may be useful for interpretation and for estimating transformations.

16.7.2   Example: Plant Height

One use for examining two terms simultaneously is to check for *spatial trends*. Agricultural field trials are often done on a rectangular field that is divided into a number of plots. Even if all the plots are treated identically, the expected responses may differ systematically because of fertility gradients. For example, the expected response may decrease from north to south. The 3D *Ceres* plot provides a method of studying this sort of environmental variation after the experiment is done.

Consider an experiment done in the Spring of 1951 to investigate the effects of varying doses (treatments) of cathode rays on the growth of tobacco seeds. The data are in the file `plant-ht.lsp`. Seven doses were used, one of which is a control dose of zero. The experimental area was divided into 56 plots laid out in a grid with eight rows and seven columns. As is common in experiments of this type, the experimenters believed that spatial trends might be possible, so they laid out the experiment as a *randomized complete block design*, using the rows as blocks. This means that within each row each of the seven treatments appears exactly once, allocated to plots within the row at random. Because of the blocking, any systematic affects due to rows should affect each treatment equally, and they should therefore be eliminated from comparisons between treatments. Any spatial trends due to *columns*, however, have not been eliminated and may influence a comparison of the treatments.

For each of the field plots we know the row number *Row*, the column number *Col*, the treatment number *Trt*, and the response *Ht*, the total height in centimeters of twenty plants.

Imagine a *response surface* over the experimental area that reflects the spatial trends. We investigate this surface using a 3D *Ceres* plot with linear augmentation. The *Row* and *Col* indices of a plot are coordinates of points on the surface. We do not have the quantitative values for the treatment levels, so we will regard *Trt* as a factor. Let the indicator term v_j equal 1 if the *j*th treatment level is applied on the plot and 0 otherwise: we get the following general model for the experiment:

$$Ht = \eta_0 + \eta_2 v_2 + \cdots + \eta_7 v_7 + \tau(Row, Col) + e$$

where $\tau(Row, Col)$ is the value of the response surface for a particular value of *Row* and *Col*. To construct a 3D *Ceres* plot we need to fit the initial model with response *Ht* and terms $\{F\}Trt, Row$ and *Col*. This is not the usual model for a randomized complete block design, which would use $\{F\}Trt$ and $\{F\}Row$ as terms, where $\{F\}Row$ is a factor for rows to represent the blocking effects. *Col* would not appear in the usual randomized complete block analysis. The analysis here could be used in any regression in which the units have coordinates in the plane, regardless of the experimental design.

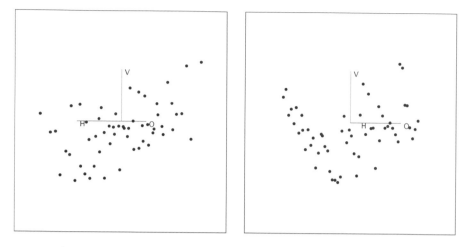

a. OLS view.   b. Quadratic trend view.

FIGURE 16.9   Two views of the 3D *Ceres* plot for the plant-height data.

From the menu for the OLS fit of the working model

$$Ht = \eta_0 + \eta_2 v_2 + \cdots + \eta_7 v_7 + \eta_8 Row + \eta_9 Col + \text{error}$$

select the item "Ceres plot–3D," move *Row* and *Col* to the right "Selection" box and select "Linear augmentation." The resulting plot is the 3D *Ceres* plot with linear augmentation for *Row* and *Col*. It should give us information on the underlying surface $\tau(Row, Col)$.

Use the "Pitch" control to rotate the 3D *Ceres* plot to the 2D view *Col* versus *Row*. The points fall on a regular 8×7 grid corresponding to the field layout. Now return the plot to the "Home" position and rotate the plot by using one of the "Yaw" buttons to gain a feeling for the function $\tau(Row, Col)$. Two 2D views of the plot are shown in Figure 16.9. Substantial spatial differences across the field are evident. The plot seems to be composed of a linear trend, visible in the OLS view in Figure 16.9a, and a quadratic trend shown in Figure 16.9b. This is useful information, particularly for future experimental designs in the same area, but it would be a help in the analysis of this particular experiment if we could characterize the surface more specifically.

Recall the OLS view and then display the screen coordinates:

```
Linear combinations on screen axes in 3D plot.
Horizontal: + 0.4820 - 0.2502 H + 0.1610 O
Vertical: + .01331 + .00252 V
```

Since both *Row* and *Col* clearly contribute to the horizontal screen term, the strongest linear trend runs diagonally across the field and thus does not align with blocks, which were rows. Remove the linear trend and view the resulting

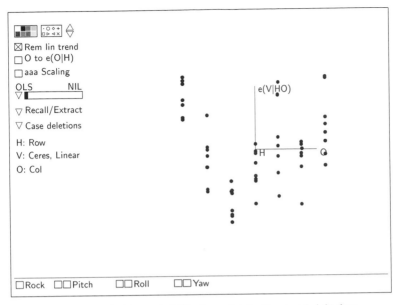

FIGURE 16.10   Detrended 3D *Ceres* plot for the plant-height data.

plot while rotating. One view of the detrended plot with a relatively strong visual trend is shown in Figure 16.10. Here, the residual variation seems to be largely a function of *Col*, so we might be able to simplify the transformation to the additive form

$$\tau(Row, Col) = \eta_8 Row + \eta_9 Col + \tau_1(Col)$$

$$= \eta_8 Row + \tau_2(Col)$$

which leads to the simplified model

$$Ht = \eta_0 + \eta_2 v_2 + \cdots + \eta_7 v_7 + \eta_8 Row + \tau_2(Col) + e \qquad (16.15)$$

Here, $\tau_2(Col)$ incorporates the linear trend from $\eta_9 Col$ plus the nonlinear trend from $\tau_1(Col)$.

To determine the transformation to use in (16.15) we can use the 2D *Ceres* plot for *Col*, as shown with linear augmentation in Figure 16.11. Neither the quadratic polynomial fit nor the *lowess* smooth shown on the plot provide a fully satisfactory approximation to τ_2. The quadratic fails to capture the points at *Col* = 3. Because of the few discrete values of *Col*, the *lowess* smooth just connects the averages of each group of points. If the primary goal is reducing variation to allow more powerful treatment comparisons, then using the quadratic fit may be sufficient. At the other extreme, *Col* could be included as a factor, which is essentially the solution suggested by the *lowess* curve.

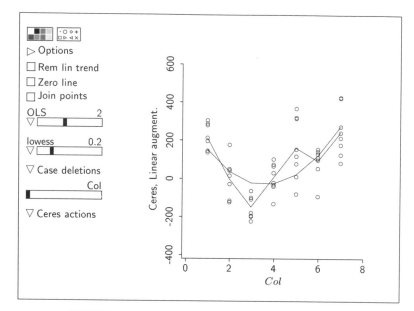

FIGURE 16.11   *Ceres* plot for *Col* in the plant-height data.

The form of τ_2 in this case is

$$\tau_2(Col) = \gamma_2 c_2 + \cdots + \gamma_7 c_7 \tag{16.16}$$

where the c_k are the indicator terms for columns.

The analysis of the plant-height data continues in Exercise 16.7.

16.8   COMPLEMENTS

16.8.1   Mixed forms of $E(u_{1j} \mid u_2)$

The various approaches discussed in Sections 16.2.1 to 16.2.4 can be combined to allow other appropriate forms of augmentation. For example, suppose we have the following regression with three terms:

$$y \mid \mathbf{x} = \eta_0 + \eta_{11} u_{11} + \eta_{12} u_{12} + \tau(u_2) + e$$

To construct a *Ceres* plot for u_2 we would first need to inspect scatterplots of u_{11} versus u_2 and u_{12} versus u_2 to infer about the term mean functions $E(u_{11} \mid u_2)$ and $E(u_{12} \mid u_2)$. Suppose it was inferred that $E(u_{11} \mid u_2) = \alpha_0 + \alpha_1 u_2$ and $E(u_{12} \mid u_2) = \beta_0 + \beta_1 \log(u_2)$. Since one of the mean functions is linear and one is logarithmic, we would construct the working model for estimation of

$\eta_1 = (\eta_0, \eta_{11}, \eta_{12})^T$ by augmenting with linear and logarithmic terms in u_2,

$$y = \eta_0 + \eta_{11}u_{11} + \eta_{12}u_{12} + \eta_2 u_2 + \eta_3 \log(u_2) + \text{error}$$

The estimate of η_1 is taken from the OLS fit of this working model, and the *Ceres* plot for u_2 is once again constructed according to (16.6).

16.8.2 References

Ceres plots were developed by Cook (1993), who provides the justification that the mean function in a *Ceres* plot will estimate the needed transformation. *Ceres* plots consolidate and extend two other methods.

The basic idea with linear augmentation was given by Ezekiel (1924), and the linear augmentation method was fairly well known through the mid-sixties. For example, it is given in Bliss (1970, p. 309). *Ceres* plots with linear augmentation were rediscovered in the early seventies in two papers: Larsen and McCleary (1972), who called them *partial residual plots*, and Wood (1973), who called them *component plus residual plots*. This latter name comes from an alternative way of computing the vertical axis of the plot. Starting with (16.8), we see that the vertical axis can be computed as $\hat{\eta}_2 u_2 + \hat{e}$, where $\hat{e}$ are the residuals from the fit of (16.8). *Ceres* plots with quadratic augmentation were suggested by Mallows (1986), although his justification for them was quite different from the justification used here. The role of the term mean functions in controlling the behavior of these early plots was first discovered by Cook (1993). The performance of *Ceres* plots relative to other methods was studied by Berk and Booth (1995) and Berk (1998).

A formal iterative process of model building that uses linear augmentation is called *generalized additive modeling*; see Hastie and Tibshirani (1990) or Green and Silverman (1994). The general idea, which originated with Ezekiel (1924), is to cycle through all relevant terms until the transformations no longer change. One cycle through the relevant terms is often sufficient.

The plant-height data are from Federer and Schlottfeldt (1954). The data for Exercise 16.8 are from Federer (1955). The constructed data used in Section 16.6 are from Cook (1993). The sniffer data in Problem 16.4 were provided by John Rice.

PROBLEMS

16.1 Reconstruct the *Ceres* plot of Figure 16.3.

16.2 Reconstruct the *Ceres* plots of Figures 16.7 and 16.8. Also construct *Ceres* plots equivalent to those in Figure 16.8, but using the power curve option in the parametric smoother slidebar instead of *lowess* smooths.

16.3   Load the file `cs-prob1.lsp` which contains 100 cases on a response y and three terms, u_{11}, u_{12}, and u_2, generated according to the model

$$y \mid \mathbf{u} = \eta_0 + \eta_{11} u_{11} + \eta_{12} u_{12} + \tau(u_2) + e$$

where the e are independent $N(0, .25)$. Using *Ceres* plots with an appropriate form of augmentation, estimate the unknown transformation.

16.4   The file `sniffer.lsp` contains data from an experiment to study the quantity of hydrocarbons y emitted when gasoline is pumped into a tank. There are 125 observations and four predictors: initial tank temperature, temperature of dispensed gasoline, initial vapor pressure in the tank, and vapor pressure of the dispensed gasoline.

     Use $\sqrt{y}$ as the response and, starting with a linear mean function in the four predictors, investigate transformations of initial tank temperature. Give the final transformation and briefly summarize your conclusions about the value of transforming in this analysis.

16.5

     16.5.1   Show that a *Ceres* plot with linear augmentation has vertical axis given by $\hat{\tau}(u_2) = \hat{\eta}_2 u_2 + \hat{e}$, where $\hat{\eta}_2$ and the residuals $\hat{e}$ are from the fit of (16.8).

     16.5.2   Suppose we use OLS to fit the equation

$$\hat{\tau}(u_2) = \gamma_0 + \gamma u_2 + \text{error}$$

        Verify that $\hat{\gamma}_0 = 0$ and that $\hat{\gamma} = \hat{\eta}_2$, where $\hat{\eta}_2$ is the OLS estimate of η_2 in the working model

$$y = \eta_1^T \mathbf{u}_1 + \eta_2 u_2 + \text{error}$$

        that is used as the basis for constructing *Ceres* plots with linear augmentation. Do this by algebraically manipulating the usual formula for the slope in a simple linear regression and by using a numerical example to illustrate that $\hat{\gamma} = \hat{\eta}_2$ apart from rounding error.

16.6   In the Big Mac data in the file `big-mac.lsp`, investigate the need to transform the predictors in the initial model

$$\log(BigMac) \mid \mathbf{x} = \eta_0 + \eta_1 \, TeachTax + \eta_2 \, TeachSal + \eta_3 \, Service + e$$

Which predictor is the most likely to need a transformation? Determine an appropriate transformation by superimposing a curve on the plot.

The curve could be based on fitting a polynomial using the OLS slide-bar, or it could be a power curve or a smoother. Follow the steps in Section 16.5 to see if another predictor needs to be transformed. If so, repeat the entire procedure to see if the remaining predictor requires transformation. State your final transformations.

16.7 Complete the analysis of the plant-height data in Section 16.7.2 by comparing conclusions on treatment effects under three versions of model (16.15). In the first, set $\tau_2(Col) = \eta_9 Col$. In the second, set $\tau_2(Col) = \eta_9 Col + \eta_{10} Col^2$. Finally, set $\tau_2(Col)$ to be as given in equation (16.16).

16.8 The file `rubber.lsp` contains the results of an experiment to compare the rubber yield of seven varieties of guayule. The experimental area consisted of 35 plots arranged in a 5×7 grid, the rows of the grid forming 5 randomized complete blocks. The response for this regression is the total grams $P_1 + P_2$ of rubber for the two selected plants on each plot.

   Aside from needing to add the terms P_1 and P_2 to obtain the response, the structure of these data is the same as that in Section 16.7.2. Conduct a graphical analysis of these data following the rationale and general steps of Section 16.7.2. Were the blocks selected to be in the best direction?

16.9 The data in the data file `allomet.lsp` and used previously in Problem 6.9 contain information on brain weight, body weight, gestation period, and litter size for 96 placental mammal species.

   Use the methodology of this and the last few chapters to study the regression with response litter size and predictors brain weight, body weight, and gestation period.

CHAPTER 17

Model Assessment

For the linear regression model, the residuals are the weighted differences between the observed values of the response and the fitted values. As described in Section 14.1.2, plots of residuals against predictors, terms, or linear combinations of predictors or terms provide a basis for discovering lack of fit of a model. A nonconstant mean function in a residual plot indicates a problem with the corresponding regression model. Residual plots look for lack of fit by focusing on deviations from the fitted model.

A closely associated problem is assessing how *well* a model matches the data. For example, return to Figure 8.6, page 195, which is a 3D residual plot for an artificial data set with a response y and predictors x_1 and x_2. The curved pattern in any view of this residual plot indicates that the mean function fit to these data is incorrect, and so the plot correctly indicates that the linear regression model may not be appropriate for these data. Figure 17.1 shows a plot of y versus the fitted values from the OLS fit to these data; this figure can be reproduced by loading the file `assess.lsp`. If we removed the linear trend from this plot, we would get a standard plot of residuals versus fitted values, and this plot would display a curved pattern. The curvature can be seen in Figure 17.1, although with less resolution, by observing that the points do not cluster about a straight line, but rather they appear to follow some other smooth curve.

While the linear regression model does not match the data perfectly, it is apparent from Figure 17.1 that most of the variation in y is explained by the incorrect linear regression model because the residuals are all very small relative to the range of the response. An incorrect linear regression model can give useful results, but determining if this is so in any given regression depends on context. For the artificial data of Figure 17.1 there is no context, so the usefulness of the linear regression model as a summary of fit cannot be judged.

Figure 17.1 is a simple example of a *model checking plot*, in which the responses rather than the residuals are plotted against a function of the predictors to judge adequacy of a model. These plots are closely related to residual

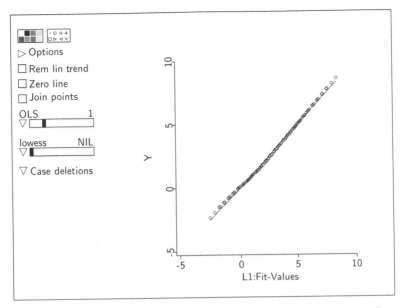

FIGURE 17.1   A plot of y versus $\hat{y}$ for the artificial data in the file `assess.lsp`.

plots, since removing the linear trend from Figure 17.1 would give a plot of residuals versus fitted value. As we will see, model checking plot can have several advantages over residual plots.

17.1   MODEL CHECKING PLOTS

Let's first think about a regression with one predictor in which we have fitted a simple linear regression model. The goal is to decide if a particular model provides an adequate summary of a set of data, as previously discussed in Section 3.6.1, in the development of lack-of-fit tests in Section 9.3, and finally in the development of residuals in Section 14.1. In particular, we saw in Section 9.3.1 that a visual lack-of-fit test, at least for the mean function, can be obtained by comparing the estimated mean function computed from the model to the estimated mean function computed from a smooth. If these are very different, then the model is not reproducing the data. The graphical procedure therefore is to compare the estimated mean function obtained by smoothing the points on the plot to the estimated mean function obtained from fitting the model. Support for the model is available if these two estimated mean functions are similar, while we have evidence against the model if the mean functions do not agree.

For example, Figure 17.2 is a plot of *Vol* versus *C* for the haystack data in the file `haystack.lsp`. We use this plot to explore the simple regression

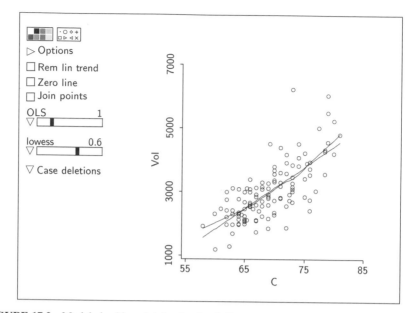

FIGURE 17.2   Model checking plot for the simple linear regression of *Vol* on *C* in the haystack data.

model

$$\mathrm{E}(\mathit{Vol} \mid C) = \eta_0 + \eta_1 C \qquad \text{and} \qquad \mathrm{Var}(\mathit{Vol} \mid C) = \sigma^2$$

From our previous work with these data and from simple geometric considerations, we should not expect this mean function to match the data very well. Two smooths are given on the plot: the OLS fit, which is a straight line that estimates $\mathrm{E}(\mathit{Vol} \mid C)$ only if the simple linear regression model is correct, and the *lowess* fit, which estimates the mean function regardless of the fit of the simple linear regression model. If we judge these two fits to be different, then we have visual evidence against the simple linear regression mean function. We see in the figure that the differences are small but systematic: The straight line slightly underestimates when C is small and large, and it overestimates for middle values of C. This is similar to the artificial data in Figure 17.1, where the mean function is wrong, but relative to the variation in the data the magnitude of the differences may be small enough to ignore. If we chose not to ignore the systematic failure of the model, we could improve the mean function by adding a quadratic or possibly cubic terms in C.

Although Figure 17.2 is a plot of data, the primary focus in this plot is comparing the two curves, using the data as background mostly to help choose the smoothing parameter for the *lowess* smooth, to help visualize variation, and to locate any extreme or unusual points.

17.1.1   Checking Mean Functions

Now suppose we have a regression with $p > 1$ predictors. We could in principle use the same method of comparing a fitted curve based on a model to a fitted curve based on a smoother. If the number of predictors exceeds one, we can't draw the curves, so another method is needed. The approach we take uses marginal models.

Suppose that the model we have fitted has mean function $E(y \mid \mathbf{x}) = \eta^T\mathbf{u}$ for some set of terms $\mathbf{u}$. We will draw a plot with y on the vertical axis and a quantity h on the horizontal axis, where h is any function of the original predictors $\mathbf{x}$ used to define the terms $\mathbf{u}$. Possible candidates for h include any of the original predictors, any linear combination of them, or any linear combination of the terms in the mean function such as the fitted values $\hat{\eta}^T\mathbf{u}$. If we smooth the plot of y versus h, we can get an estimate of $E(y \mid h)$ without any assumptions. Can we get an estimate of $E(y \mid h)$ from the model? If the answer is "yes," then we have an analogy to simple linear regression: If the smooth to the data and the estimate determined by the model fail to agree adequately within the context of the problem, then the model itself is not adequate.

Getting an estimate of $E(y \mid h)$ from the model requires an application of equation (11.2). Under the model, we have

$$E(y \mid h) = E[E(y \mid \mathbf{u}) \mid h]$$

$$= E(\eta^T\mathbf{u} \mid h) \qquad (17.1)$$

Suppose we substitute $\hat{y} = \hat{\eta}^T\mathbf{u}$ for $\eta^T\mathbf{u}$ in (17.1). Then, *for any h, we can estimate $E(y \mid h)$ from the model that produced $\hat{y}$ by smoothing the scatterplot of $\hat{y}$ versus h.* If the model is correct, then the smooth of y versus h and the smooth of $\hat{y}$ versus h should agree; if the model is not correct, then these smooths may not agree.

Let's consider again the transactions data in the file `transact.lsp`, and suppose $h = T_1$, the first of the two predictors. Figure 17.3 shows plots of $y = Time$ versus h and of $\hat{y}$ versus h. The smooth in Figure 17.3a estimates $E(y \mid T_1)$ whether the model is right or not, but the smooth in Figure 17.3b may not give a useful estimate of $E(y \mid T_1)$ if the linear regression model is wrong. Comparison of these two estimated mean functions provides a visual assessment of the adequacy of the mean function for the model. Superimposing the smooth in Figure 17.3b on Figure 17.3a gives a *model checking plot*.

Arc includes a menu item in the regression menu called "Model checking plots" to draw these plots. When this item is selected, a dialog appears like that in Figure 17.4. The dialog allows the user to specify choices for h. This is necessary because *the model has the correct mean function if and only if the model checking plots agree for all h,* not just for one particular choice of h.

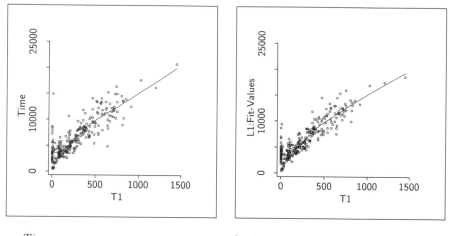

a. *Time.*   b. $\hat{y}$.

FIGURE 17.3   Plots for *Time* versus T_1 and $\hat{y}$ versus T_1. In both plots the curves are lowess smooths with smoothing parameters equal to 0.4. If the model has the correct mean function, then these two smooths estimate the same quantity.

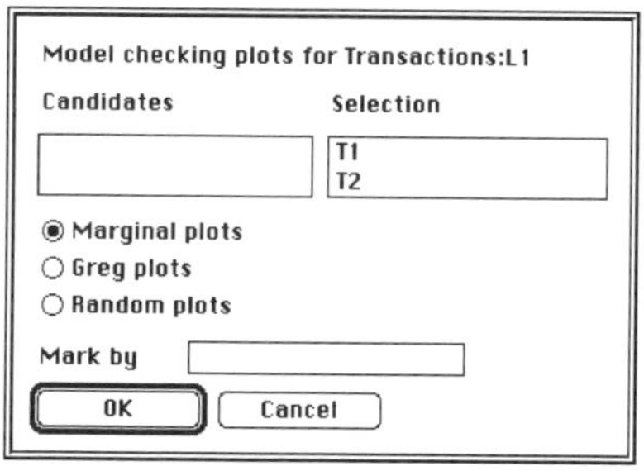

FIGURE 17.4   The model checking plot dialog.

Radio buttons give three choices:

Marginal Plots. This will give a multipanel plot allowing the user to see the model checking plots for h = fitted values, which are called eta'u on the plot, and for h equal to each of the selected predictors whose names are in the list at the right of the dialog.

Greg Plots. GREG is an acronym for *graphical regression.* Both GREG and the GREG predictors are discussed in Chapter 20.

The first GREG plot selects a linear combination of the selected terms that is most likely to show curvature in the plot of residuals versus that linear combination. Thus, the first GREG plot is an approximation to the worst case. The remaining GREG linear combinations are uncorrelated with the first GREG linear combination, and each is less likely than the one before it to show curvature.

Random Plots. If this item is selected, the program chooses h to be a random linear combination of the selection. More random linear combinations can be chosen by pushing the mouse in a slidebar on the plot. According to theory, the smooths on model checking plots should match for all linear combinations, so checking a number of random directions can provide useful confirmatory information.

Model checking plots for the transactions data are shown in Figure 17.5 for four choices of h. In Figure 17.5a, the horizontal axis is $h = \hat{\eta}^T \mathbf{u} = \hat{y}$; this linear combination is almost always of interest. The smooth of the response versus h is shown on the computer screen as a blue solid line, while the smooth of $\hat{y}$ versus h is shown as a red dashed line. A slidebar controls the smoothing parameter for the two *lowess* smooths that are added to the plot. In Figure 17.5a, the two smooths overlap almost exactly, indicating that at least for this h the model is reproducing the information available from the data. Figure 17.5b is for $h = T_1$, as also shown in Figure 17.3. Once again the two smooths are virtually identical. It is interesting perhaps that the slight curvature in the plot of *Time* versus T_1 is reproduced by the linear mean function fit in this regression. This is even more evident in the remaining two plots which have mean functions that are not linear. Even so, the mean functions are matched by the model, although Figure 17.5c does show some disagreement at the extreme left of the plot. On the basis of these four plots and every other model checking plot we tried, there is no compelling evidence that the mean function estimated by (14.7) is inadequate for the data.

17.1.2 Checking Variance Functions

Model checking plots can also be used to check for model inadequacy in the variance function. The basic idea is similar to checking for the mean function. The plot of y versus h can be used to estimate the variance function $\text{Var}(y \mid h)$, as discussed in Section 3.6.3, page 51; this estimate of the variance function does not depend on a model. We will write $\text{sd}_{data}(y \mid h)$ to be the square root of this estimated variance function.

Next, we need an estimate of $\text{Var}(y \mid h)$ that is implied by the model, and this requires a slight generalization of (11.5) in Section 11.1.2, page 265. Given the model and again substituting $\hat{y} \approx \text{E}(y \mid \mathbf{u})$, we obtain

$$\text{Var}(y \mid h) = \text{E}[\text{Var}(y \mid \mathbf{u}) \mid h] + \text{Var}[\text{E}(y \mid \mathbf{u}) \mid h] \qquad (17.2)$$

$$\approx \text{E}[\sigma^2 \mid h] + \text{Var}[\hat{y} \mid h]$$

$$= \sigma^2 + \text{Var}[\hat{y} \mid h] \qquad (17.3)$$

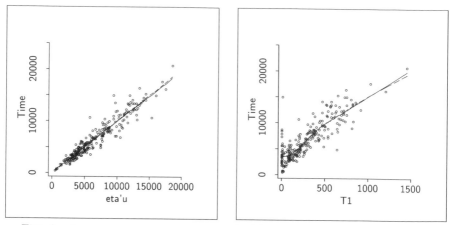

a. Fitted values.   b. T_1.

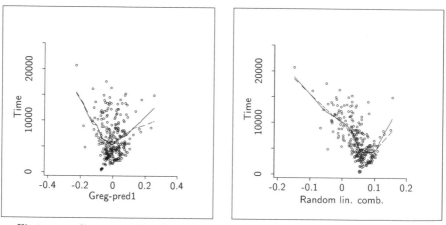

c. First GREG linear combination.   d. Random linear combination.

FIGURE 17.5   Model checking plots for the transactions data. In each, the smoothing parameter is 0.4, the solid line is the smooth from the data, and the dashed line is the smooth from the model.

Equation (17.2) is the general result that holds for any model. Equation (17.3) holds for the linear regression model in which the variance function $\text{Var}(y \mid \mathbf{u})$ $= \sigma^2$ is constant. According to this result, we can estimate $\text{Var}(y \mid h)$ under the model by getting a variance smooth of $\hat{y}$ versus h and then adding to this an estimate of σ^2. We will call the square root of this estimated variance function $\text{sd}_{model}(y \mid h)$. If the model is appropriate for the data, then apart from sampling error, $\text{sd}_{data}(y \mid h) = \text{sd}_{model}(y \mid h)$, but if the model is wrong, these two functions need not be equal.

For visual display, we show the mean function estimated from the plot $\pm \text{sd}_{data}(y \mid h)$ using blue solid lines and the mean function estimated from

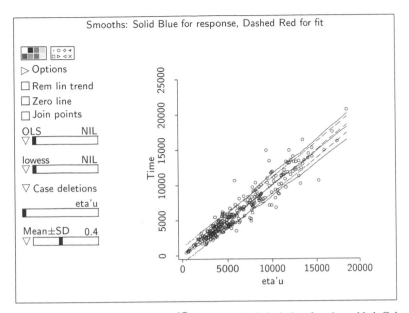

FIGURE 17.6   Model checking plot for $\hat{\eta}^T\mathbf{u}$ with standard deviation function added. Colors are visible on the computer screen.

the model $\pm\mathrm{sd}_{model}(y\mid x)$ using red dashed lines. These can be obtained in the model checking plots by selecting the item "Mean+−SD" from the pop-up menu at the lower left of the plot and then using the slidebar to choose a smoothing parameter. The same smoothing parameter is used for all the smooths.

Figure 17.6 reproduces the model checking plot in Figure 17.5a, except with the standard deviation functions added. The mean smooths are the same as in Figure 17.5a, so they continue to match, but the variance smooths do not match as well. The smooth under the model, given on the computer screen by the red dashed lines, are constant, reflecting the constant variance assumption, while the smooth to the data reflects the increasing variance. We have further confirmation that the constant variance assumption may not be appropriate for these data, depending on the context. Figure 17.7 shows the model checking plot for T_1 with the standard deviations included. For this choice of horizontal axis, the standard deviation functions from the plot and from the model agree. This illustrates that deficiencies may not show up in all model checking plots.

17.2   RELATION TO RESIDUAL PLOTS

In the linear regression model, the model checking plot with fitted values on the horizontal axis contains all the information in a plot of residuals versus fitted values, since the latter can be obtained from the former by simply re-

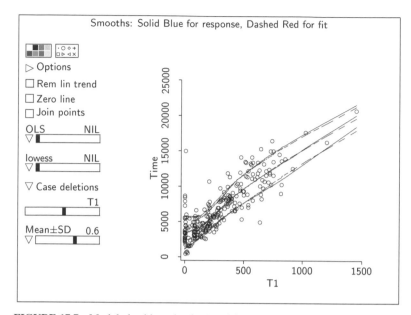

FIGURE 17.7   Model checking plot for T_1 with standard deviation function added.

moving the linear trend. With any other quantity on the horizontal axis, the model checking plots and residual plots are not identical, as detrending the model checking plot will not give the residuals. Residual plots emphasize disagreement between the data and the fitted model, while model checking plots emphasize agreement. Both residual plots and model checking plots in principle require examining plots for a variety of values of h. The same strategy for choosing directions can be used for both type of plots.

One possible systematic pattern in a residual plot is curvature, and a test for curvature is provided in Section 14.2. While this test does not correspond to a comparison of the two smooths on a model checking plot, it can be used to calibrate these plots as well, when the difference between the two smooths appears to be due to curvature. Other types of departure can be detected visually from either residual plots or from model checking plots.

17.3   SLEEP DATA

As an example of the use of model checking plots, we return to the sleep data in the file `sleep.lsp`, introduced in Section 10.4. Suppose we fit the model suggested by (10.12),

$$E(TS \mid \mathbf{x}) = \eta_0 + \eta_1 \log_2(BW) + \eta_2 \log_2(BrW) + \eta_3 \log_2(Life)$$

$$+ \eta_4 \log_2(GP) + \eta_5 D_1 \tag{17.4}$$

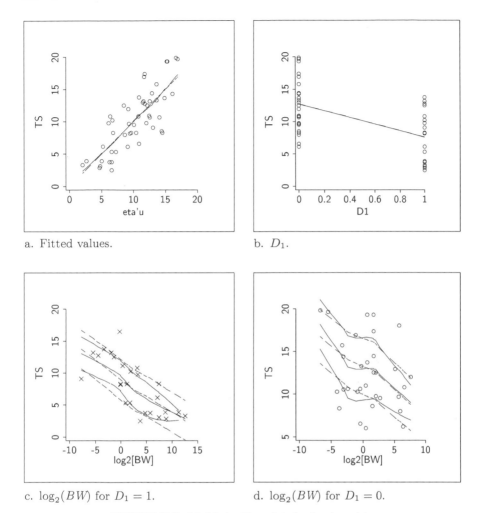

a. Fitted values.

b. D_1.

c. $\log_2(BW)$ for $D_1 = 1$.

d. $\log_2(BW)$ for $D_1 = 0$.

FIGURE 17.8 Model checking plots for the sleep data.

where the variables are defined in Table 10.2, page 239. This model has four continuous terms and one indicator term, D_1.

Figure 17.8a is the model checking plot with fitted values on the horizontal axis. This plot, which is typical of the plots of any of the continuous terms, suggests close agreement between the fitted model and the data. The plot in Figure 17.8b is the model checking plot for the indicator term D_1. Because D_1 has only two values, this plot is not very useful, and in general plotting against indicator terms won't help much. An alternative is to use D_1 as a marking variable. For example, in Figure 17.8c, we have a model checking plot for $\log_2(BW)$, but using only the points for which $D_1 = 1$, and in Figure 17.8d for $D_1 = 0$ (these plots are obtained by clicking the mouse on

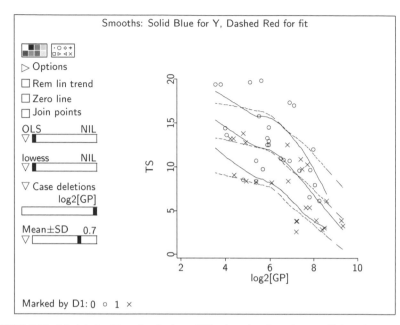

FIGURE 17.9   Model checking plot for $\log_2(GP)$ when the sleep data are fit ignoring $\log_2(GP)$. Colors are visible on the computer screen.

the category label in the marking legend). According to the theory, if the model is correct the smooths should match for subsets of the data. If the smooths were to match for $D_1 = 1$ but not for $D_1 = 0$, we would conclude that the model is matching the data for high-danger animals but not for low-danger animals, which might suggest the need for interactions with D_1 in the mean function. This is not the case here: The smooths match the groups as well as for all the data at once.

Model checking plots can also be used with variables on the horizontal axis that are not included in the model. Suppose we fit a smaller model to the sleep data, obtained from (17.4) by deleting the term $\log_2(GP)$. Figure 17.9 is the model checking plot with $\log_2(GP)$ on the horizontal axis; the agreement between the smooths may be sufficient, depending on the context. The model does not use information about $\log_2(GP)$ to determine estimates, so the other variables must be supplying the information contained in $\log_2(GP)$.

17.4   COMPLEMENTS

Model checking plots were proposed by Cook and Weisberg (1997). They can be used as outlined in just about any regression, not just in linear regression as described here. Model checking plots emphasize mean and variance functions, which are the most important properties of conditional distributions that can be seen in plots.

The GREG predictors are based on principal Hessian directions (Li, 1992) and are discussed in the Complements to Chapter 20.

PROBLEMS

17.1 Use the haystack data, in the file `haystack.lsp`.

17.1.1 Fit the model

$$\text{E}(Vol \mid C, Over) = \eta_0 + \eta_1 C + \eta_2 Over \quad \text{and}$$

$$\text{Var}(Vol \mid C, Over) = \sigma^2$$

Draw model checking plots with fitted values on the horizontal axis and with each of the two terms on the horizontal axis, and summarize results. Is there evidence that this mean function fails to match the data?

17.1.2 Continuing with Problem 17.1.1, use model checking plots to decide if constant variance is reasonable.

17.1.3 Fit the model developed in Section 16.4,

$$\text{E}(Vol \mid C, Over) = \eta_0 + \eta_1 C^3 + \eta_2 Over^3 \quad \text{and}$$

$$\text{Var}(Vol \mid C, Over) = \sigma^2$$

Draw the model checking plots with C on the horizontal axis and with C^3 on the horizontal axis. Do these plots contain different information? Why or why not? Summarize the results.

17.1.4 Continuing with 17.1.3, examine the model checking plot with fitted values on the horizontal axis for constant variance, and summarize results.

17.1.5 Continuing with 17.1.3, draw the model checking plot with the first GREG predictor on the horizontal axis. If the model were "true," then the smooths on any model checking plot of a function of any predictors should agree. Do they agree here? If not, explain a possible cause.

PART III

Regression Graphics

Part II of this book centered on tools for using multiple linear regression models to summarize a regression. Multiple linear regression methodology is powerful and widely used.

Graphical analysis of a regression with one predictor can be based on a scatterplot of the response versus the predictor. We can tell if a simple linear regression model is appropriate for the data, and if it isn't we can use the plot to judge the success of remedial action like transformations or fitting polynomials. By adding fitted curves to the scatterplot, we can tell if any fitted model matches the data.

With many predictors, there is no direct generalization because we can't draw simple plots in many dimensions. Consequently, multiple linear regression seems to be more mysterious than simple linear regression, because we can't actually *see* the fitted regression model against the background of the data.

In this part of the book we present graphical methods that can help us visualize regression in many dimensions. In Part II we assumed that the linear model holds or can be made to hold, and we worked within that framework. In this part of the book, we assume no models at the outset of an analysis and show how plots can be used to find structure and to select a first model.

In Chapter 18 we discuss regressions with two predictors. In Chapter 19 we extend these ideas to many predictors. In Chapter 20 we introduce graphical regression, which can be used to understand regressions through plots based on minimal assumptions.

Visualizing Regression

This is the first of three chapters on graphical regression. A main change from Part II of this book is that at the outset *no models are required*. Our goal is to obtain regression information using plots. This approach may lead to a purely graphical solution in which a plot shows all the information that is available about the regression. It may also lead to using a particular model.

Multiple linear regression models start with p predictors $\mathbf{x} = (x_1, \ldots, x_p)^T$, and from these the terms $\mathbf{u} = (u_0, \ldots, u_{k-1})^T$ are determined so that the mean function is given by $E(y \mid \mathbf{x}) = \eta^T \mathbf{u}$. Terms are not needed in the graphical approach; instead, we use the predictors directly.

One goal in regression through graphics is to find a *sufficient summary plot*, which is a plot that contains all the sample regression information about the conditional distribution of the response given the predictors. For example, with one predictor the scatterplot of the response versus the predictor is always a sufficient summary plot because all the sample information about the regression is contained in this one plot. Similarly, with two predictors the 3D plot of the response versus the predictors is always a sufficient summary plot, and in principle all the regression information can be obtained from this plot. 3D plots are more complex than 2D plots, and understanding the information in a 3D plot can occasionally be quite hard, but useful descriptions are possible.

18.1   PINE TREES

The file `pines.lsp` contains data on 70 short-leaf pine trees. The response is wood volume *Vol* in cubic feet. There are two predictors: the height H of the tree in feet, and the diameter D of the tree at breast height in inches. Let $\mathbf{x} = (H, D)^T$ be the vector of predictors. The goal of regression is to understand, as far as possible with the available data, how the conditional distribution of $Vol \mid (H, D)$ varies with the height and diameter. At the outset, no model is assumed.

Construct a 3D plot of *Vol* versus (H, D). This is a sufficient summary plot for this regression. Rotate the plot about its vertical axis to get a feeling for

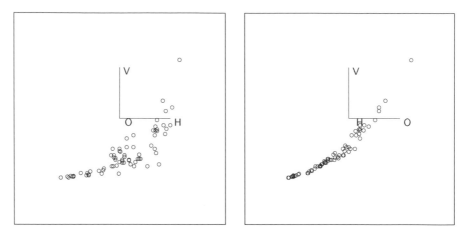

FIGURE 18.1   Two views of the 3D plot of *Vol* versus (H,D) for the pines data.

any patterns in the data; two views of the plot are shown in Figure 18.1. Even in this relatively simple 3D plot, useful verbal descriptions may be difficult to formulate. For example, the two views in Figure 18.1 suggest that the mean function $E(Vol \mid D,H)$ is curved and that the variance function is nonconstant. These statements characterize the 2D views, but may not have much to do with the full 3D plot; describing features that can be viewed only while rotating is more difficult.

A general goal of statistical analysis is to reduce complex summaries to simpler ones without loss of important information. Applying this idea to plots, we could replace the 3D plot by a 2D plot if all the regression information were contained in the 2D plot. Recall from Section 8.4 that when the 3D plot of *Vol* versus (H,D) is rotated around its vertical axis by an angle θ, the 2D plot of the response versus a linear combination $h(\theta) = b(\cos\theta)H + c(\sin\theta)D$ of the two predictors is displayed on the computer screen. We will call the plot of *y* versus $h(\theta)$ a *2D view* of the 3D plot. If *Vol* depends on (H,D) only through the linear combination $h(\theta)$, then the 2D view of *Vol* versus $h(\theta)$ will be a sufficient summary plot: All the sample information about the dependence of *Vol* on the predictors will be contained in this plot. Further analysis can be based on this 2D view rather than the more complex 3D plot.

We might ask, Is there a sufficient 2D summary plot for the pines data, and if so how can we estimate it?

18.2   THE ESTIMATED 2D SUMMARY PLOT

Every 2D view of the 3D plot of *Vol* versus (H,D) has a mean function $E(Vol \mid h(\theta))$ and a variance function $Var(Vol \mid h(\theta))$. Rotate the 3D plot and stop

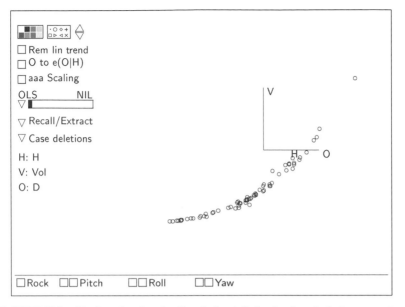

FIGURE 18.2   2D view showing the "best" visual fit for the short-leaf pine regression.

at the 2D view that visually has the smallest variation about the mean function, or equivalently where the average variance $E(\text{Var}(Vol \mid h(\theta)))$ is minimized. You may need to use the item "Slower" in the plot's menu to slow the rate of rotation to get what you think is the best view. The plot we chose is shown in Figure 18.2. We can find the linear combination h of the predictors on the horizontal screen axis by using the "Display screen coordinates" item in the "Recall/Extract" pop-up menu:

```
Linear combinations on screen axes in 3D plot.
Horizontal: - 2.106 + .01112 H + .09672 O
Vertical:   - 1.025 + .01238 V
```

Your output will be somewhat different if you stopped rotating at a nearby view. The 2D view in Figure 18.2 is a plot of *Vol* versus

$$h = \mathbf{b}^T \mathbf{x} = 0.01112H + 0.09672D$$

where $\mathbf{b} = (0.01112, 0.09672)^T$.

Rotating to the 2D view that has the smallest average variance about the mean function in the plot is a visual method for estimating a summary plot. The mean function can have any shape at all; in Figure 18.2 it is curved. Next, we need a method to decide if the estimated summary plot loses regression information. Before doing this we need to define a few new concepts.

18.3   STRUCTURAL DIMENSION

We suppose that we have a regression with response y and p predictors $\mathbf{x} = (x_1, \ldots, x_p)^T$. The *structural dimension* of the regression is the smallest number of linear combinations of $\mathbf{x}$ needed to characterize the regression without loss of information. The structural dimension is always an integer between 0 and p. We will speak of regressions having 0D, 1D, $\ldots$, pD structure.

The structural dimension of a regression with two predictors and of the corresponding 3D plot is either zero, one, or two. We will say the plot or the regression has either 0D structure, 1D structure, or 2D structure.

We develop and explain the idea of structural dimension in this section. The general discussion is in terms of p predictors, but all illustrations and examples are based on two predictors.

18.3.1   Zero-Dimensional Structure

If $y \mid \mathbf{x}$ does not depend on $\mathbf{x}$, we have 0D structure because no linear combination of $\mathbf{x}$ provides any information about y. With 0D structure, a histogram of y is a sufficient summary plot since any plot involving $\mathbf{x}$ contains no more information about y. We have encountered 0D structure in model checking, where residual plots from a correct model will exhibit 0D structure. Given 0D structure, both the mean function $\mathrm{E}(y \mid \mathbf{x})$ and the variance function $\mathrm{Var}(y \mid \mathbf{x})$ are constant for every 2D view.

Return to the plot of *Vol* versus (H, D) for the pines data. We will have 0D structure if and only if y is independent of $h(\theta)$ for all 2D views of the 3D plot. This does not hold, since dependence is apparent in both views in Figure 18.1.

18.3.2   One-Dimensional Structure

A regression has 1D structure if y depends on $\mathbf{x}$ through only one linear combination $\beta^T \mathbf{x}$ of the predictors. This idea is very important, so we will restate it in several ways.

In the pines data, suppose there is 1D structure and we know the value of β. Then there is no more information in the separate values of H and D than that contained in the linear combination $\beta^T \mathbf{x}$. To be explicit, suppose that $\beta = (0.1, 1.7)^T$, and suppose for one tree we have $\mathbf{x}_1^T = (H_1, D_1) = (80, 5)$ so $\beta^T \mathbf{x}_1 = 16.5$ and for a second tree we have $\mathbf{x}_2^T = (H_2, D_2) = (46, 7)$ and $\beta^T \mathbf{x}_2 = 16.5$. If 1D structure is appropriate, then for both trees the distribution of $Vol \mid (H, D)$ is the same because $\beta^T \mathbf{x}_1 = \beta^T \mathbf{x}_2$. The additional knowledge that the first tree is narrow and tall and the second tree is wider and short has no regression information. A formal way of indicating this is to say that y is independent of $\mathbf{x}$ given $\beta^T \mathbf{x}$. If the regression has 1D structure then a 2D plot of y versus $\beta^T \mathbf{x}$ is a sufficient summary plot.

Many of the models commonly used in regression have 1D structure. One model is

$$E(y \mid \mathbf{x}) = \text{M}(\beta^T \mathbf{x}) \quad \text{and} \quad \text{Var}(y \mid \mathbf{x}) = \sigma^2 \quad (18.1)$$

where M is a function that may be known or unknown. We will call M the *kernel mean function*. The word *kernel* means "a central or essential part," and M is an essential part of the mean function, since M determines its shape. The plot y versus $\beta^T \mathbf{x}$ is a sufficient summary plot for this model because it contains all the information about y that is available from $\mathbf{x}$. The multiple linear regression mean function with terms equal to the predictors is a special case of (18.1) with $\text{M}(\beta^T \mathbf{x}) = \eta_0 + \eta_1^T \mathbf{x} = \eta_0 + \gamma(\beta^T \mathbf{x})$, and γ is a parameter that converts β to the right scale, $\gamma \beta = \eta_1$.

Suppose that the pines data has 1D structure as represented in model (18.1) and that the 2D view in Figure 18.2 is a good summary, so $h \approx \beta^T \mathbf{x}$. Then the mean function for the summary view should be close to the kernel mean function M, and M can be estimated from the summary view. A nonparametric estimate of M could be obtained using a *lowess* smooth. Alternatively, one might use a 1D quadratic model with mean function

$$E(y \mid \mathbf{x}) = \eta_0 + \eta_1 (\beta^T \mathbf{x}) + \eta_2 (\beta^T \mathbf{x})^2$$

If β is known, this is the mean function for a usual quadratic polynomial in the single predictor $\beta^T \mathbf{x}$. With β unknown, this is actually a nonlinear model.

In most regression models the dependence of the distribution of $y \mid \mathbf{x}$ on $\mathbf{x}$ is through the mean function but the dependence on $\mathbf{x}$ can be more general. Another model with 1D structure is

$$E(y \mid \mathbf{x}) = \eta_0 \quad \text{and} \quad \text{Var}(y \mid \mathbf{x}) = \text{V}(\beta^T \mathbf{x}) \quad (18.2)$$

The function $\text{V}(\beta^T \mathbf{x})$ is called the *kernel variance function* and is a nonnegative function that may have a different value for each value of $\beta^T \mathbf{x}$. For example, one possibility for the kernel variance function is given by (14.14), page 346, by setting $\text{V}(\beta^T \mathbf{x}) = \sigma^2 \exp(\beta^T \mathbf{x})$. In (18.2) the mean function is constant, $E(y \mid \mathbf{x}) = \eta_0$ for all $\mathbf{x}$, but the variance function changes with $\mathbf{x}$.

The most general model with 1D structure that we consider in this book is

$$E(y \mid \mathbf{x}) = \text{M}(\beta^T \mathbf{x}) \quad \text{and} \quad \text{Var}(y \mid \mathbf{x}) = \text{V}(\beta^T \mathbf{x}) \quad (18.3)$$

We will call (18.3) the *1D model*. It includes models (18.1) and (18.2) as special cases. For the 1D model, both the mean function and the variance function depend on the same linear combination $\beta^T \mathbf{x}$.

Suppose now that the pines data has 1D structure as represented in model (18.3) and that the 2D view in Figure 18.2 is a good summary. Then the kernel mean function M and kernel variance function V can be both visualized and estimated from the 2D summary plot.

In summary, with 1D structure the 2D plot y versus $\beta^T\mathbf{x}$ is a sufficient summary plot because this one plot contains all the information about $y \mid \mathbf{x}$ available from $\mathbf{x}$. If we knew $\beta^T\mathbf{x}$, we could replace $\mathbf{x}$ by $h^* = \beta^T\mathbf{x}$ without loss of regression information. This is a very powerful result because it allows us to use the methodology from simple regression to understand a multiple regression.

For example, with two predictors, we can estimate $\beta^T\mathbf{x}$ visually using the method of Section 18.2. Given the estimate, we show in Section 18.4 how to check for 1D structure in a 3D plot. With more predictors, we will need conditions on the predictors to estimate $\beta^T\mathbf{x}$, as outlined in Section 19.1.

18.3.3   Two-Dimensional Structure

A regression has 2D structure if two linear combinations $\beta_1^T\mathbf{x}$ and $\beta_2^T\mathbf{x}$ are needed to characterize the regression, so that y is independent of $\mathbf{x}$ given $(\beta_1^T\mathbf{x}, \beta_2^T\mathbf{x})$. With 2D structure, every 2D plot of the response versus one linear combination of predictors loses information, and a 3D plot is required to summarize regression information. 2D structure is the most general form of dependence in regressions with two predictors.

One model with 2D structure is

$$\mathrm{E}(y \mid \mathbf{x}) = \mathrm{M}(\beta_1^T\mathbf{x}) \qquad \text{and} \qquad \mathrm{Var}(y \mid \mathbf{x}) = \mathrm{V}(\beta_2^T\mathbf{x}) \qquad (18.4)$$

provided β_1 and β_2 are not exactly collinear. One linear combination of the predictors $\beta_1^T\mathbf{x}$ is required as the argument to the kernel mean function, while a different linear combination $\beta_2^T\mathbf{x}$ is required in the kernel variance function; both are needed to understand fully the distribution of $y \mid \mathbf{x}$.

Another model with 2D structure is

$$\mathrm{E}(y \mid \mathbf{x}) = \mathrm{M}(\beta_1^T\mathbf{x}, \beta_2^T\mathbf{x}) \qquad \text{and} \qquad \mathrm{Var}(y \mid \mathbf{x}) = \sigma^2 \qquad (18.5)$$

The mean function $\mathrm{E}(y \mid \mathbf{x}) = \mathrm{M}(\beta_1^T\mathbf{x}, \beta_2^T\mathbf{x})$ depends on the two linear combinations $\beta_1^T\mathbf{x}$ and $\beta_2^T\mathbf{x}$, while the variance function is constant. A simple example of a model of this type with two predictors has kernel mean function

$$\mathrm{M}(\beta_1^T\mathbf{x}, \beta_2^T\mathbf{x}) = [(x_1 + x_2)^2 - (x_1 - x_2)^2]/4$$
$$= [(\beta_1^T\mathbf{x})^2 - (\beta_2^T\mathbf{x})^2]/4$$
$$= x_1 x_2$$

where $\beta_1^T\mathbf{x} = x_1 + x_2$ and $\beta_2^T\mathbf{x} = x_1 - x_2$. A product of two predictors can be obtained from two linear combinations of them, and so this mean function

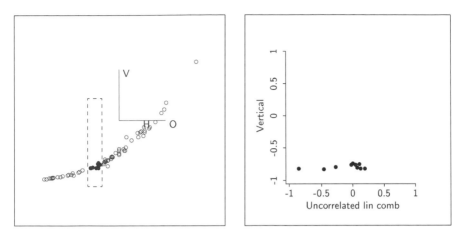

a. The summary plot.   b. The uncorrelated view, y versus h_{unc}.

FIGURE 18.3 Assessing the summary plot for Figure 18.2.

gives 2D structure. Most polynomial models in two or more predictors have dimension two or higher.

18.4 CHECKING AN ESTIMATED SUMMARY PLOT

Choosing between 0D structure and greater than 0D structure in a 3D plot is often easy: If any 2D view has either a nonconstant mean or variance function, then 0D structure must be abandoned in favor of dimension one or higher. The choice between 1D structure and 2D structure can be harder. Let's suppose we have an estimate for a sufficient summary plot, say y versus $h = \mathbf{b}^T\mathbf{x}$. We know that if 1D structure holds, then y is independent of $\mathbf{x}$ given the linear combination $h^* = \beta^T\mathbf{x}$, and this provides the basis for checking to see if a summary plot misses important information. We examine the plot to find information that contradicts the possibility that y is independent of $\mathbf{x}$ given h.

Imagine conditioning approximately on h by selecting points in a vertical slice of the 2D plot of y versus h. The summary plot shown in Figure 18.2 is repeated in Figure 18.3a, with a slice selected. If h is all we need to know about the predictors, then y should be independent of $\mathbf{x}$ within the slice and the selected points should form a horizontal band with constant mean and variance functions as the plot is rotated. If any *within-slice* dependence of y on $\mathbf{x}$ is apparent, then the estimated summary plot does not contain all the information about the distribution of $y \mid \mathbf{x}$; either h is a poor estimate of h^*, or 1D structure does not hold. For full confidence that the estimated summary plot y versus h is adequate, y should be independent of $\mathbf{x}$ within a series of slices that covers the range of h.

This basic checking procedure is too time-consuming to be of much practical value, so we present a simpler procedure based on one 2D view. The horizontal axis of this 2D view is the linear combination of the predictors that is uncorrelated with h. Call this linear combination h_{unc}, and call the plot y versus h_{unc} the *uncorrelated 2D view*. If the predictors are uncorrelated, this is the 2D plot you would see if your computer screen were actually a 3D solid, and you could look at the plot from the side. Figure 18.3b is the uncorrelated 2D view for the 3D plot in Figure 18.3a. The points corresponding to the slice in Figure 18.3a are shown in Figure 18.3b. If no within-slice dependence is apparent in any slice, then y versus h is a good summary plot for these data, and we have support for 1D structure.

Here is how to use *Arc* to obtain the uncorrelated 2D view for checking a summary plot:

1. Obtain an estimated summary plot as in Section 18.2. This view can be remembered for later use by selecting the item "Remember view" from the "Recall/Extract" menu.

2. Select "Extract uncorrelated 2D plot" from the "Recall/Extract" menu to get the uncorrelated 2D view y versus h_{unc} in a separate window. The 2D view of the 3D plot y versus (x_1, x_2) is left unchanged.

3. You now have two plots, y versus h, which is the 2D view visible in the 3D plot, and y versus h_{unc}. The plot y versus h_{unc} contains an extra slider with a pop-up menu. The slider allows you to slice on h. As you move the slider, the corresponding points will be selected in both plots. The pop-up menu for the slider allows you to change the number of slices; the default is usually adequate.

    If in the uncorrelated 2D view, y appears to be independent of h_{unc} within each slice, then there is no evidence that the summary plot is insufficient. The points in the uncorrelated 2D view will usually move up or down as the slider is moved, but this does not contradict the possibility that y is independent of h_{unc} *within* each slice.

4. The plot y versus h misses information if within-slice dependence is visible in the uncorrelated 2D view. It may be that there is no sufficient 2D plot, implying that the plot has 2D structure, and only the full 3D plot can summarize the regression of y on **x**. Alternatively, it may be that there is a sufficient 2D plot, but the present estimated summary plot is far from it. In this case try rotating the 3D plot to obtain another estimated summary plot, and return to Step 1.

Draw the plots equivalent to Figure 18.3, and examine all the slices. To increase visual resolution in the plot of y versus h_{unc}, rescale it to fill the plotting area by selecting the item "Rescale plot for each slice" from the slicer slidebar's pop-up menu. Repeating step 3 above, the plot for one of the slices is shown in Figure 18.4. Although the range on the vertical axis is

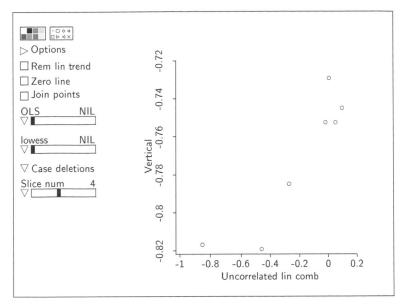

FIGURE 18.4 An uncorrelated 2D view for the pines data with the plot rescaled to fill the plotting area.

very narrow, a within-slice increasing trend is visible in this slice and also in neighboring slices. This provides evidence against 1D structure, at least for trees with values of h in the middle of its range. We explore the implications of this further in Problem 18.1.

18.5 MORE EXAMPLES AND REFINEMENTS

18.5.1 Visualizing Linear Regression in 3D Plots

What will a 3D plot look like when the multiple linear regression model holds? Let's suppose that

$$E(y \mid \mathbf{x}) = \eta_0 + \eta_1 x_1 + \eta_2 x_2 \tag{18.6}$$

This is a 1D model, so a sufficient summary plot is y versus $h^* = c(\eta_1 x_1 + \eta_2 x_2)$ for any nonzero constant c. We examine the question using simulated data. The predictors x_1 and x_2 were generated as independent $N(0, 1)$ random variables, and the response was then constructed as

$$y \mid \mathbf{x} = 1 + 2x_1 + 3x_2 + e$$

$$= 1 + h^* + N(0, 1)$$

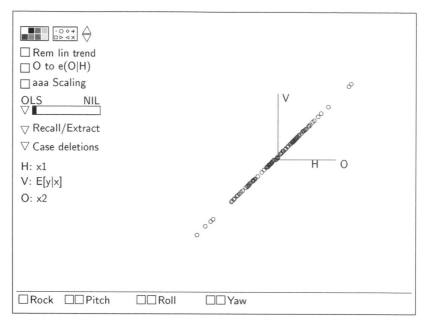

FIGURE 18.5   A 2D view of the "$E[y \mid x]$ linear, predictors uncorrelated" demonstration, with $E(y \mid \mathbf{x})$ on the vertical axis.

where $h^* = 2x_1 + 3x_2$. The errors are independent, normal random variables with mean zero and variance one. The mean function is $E(y \mid \mathbf{x}) = 1 + 2x_1 + 3x_2$ with $\eta = (1,2,3)^T$, and the variance function is $\text{Var}(y \mid \mathbf{x}) = 1$. The data can be obtained by loading the file `demo-3d.lsp` and then selecting the item "$E[y \mid x]$ linear, predictors uncorrelated" from the Demos: 3D menu.

First, construct the plot of $E(y \mid \mathbf{x})$ versus (x_1, x_2). What do you see as the important characteristics of the plot? Rotate the plot until the view on the computer screen gives a single straight line, as in Figure 18.5. This is the sufficient summary plot. The horizontal screen axis is equal to $c(2x_1 + 3x_2)$ for some nonzero c, as can be verified using the "Display screen coordinates" plot control. You can verify that for this example $c = 0.09955$.

Next, construct a plot of y versus (x_1, x_2). The points in the 3D plot y versus (x_1, x_2) scatter about a plane, as is evident during rotation. They do not fall exactly on a plane because of the errors.

Rotate the 3D plot to estimate a summary plot, the view that minimizes the average variance about the mean function in the 2D view. Your plot should look like Figure 18.6 or its mirror image. The linear combination of the predictors on the horizontal axis of the 2D plot should be nearly a constant times $h^* = 2x_1 + 3x_2$, or equivalently the ratio of the multipliers for x_1 and x_2 should be close to $2/3$. Because there is error, we should not expect these results to be exact. Selecting "Display screen coordinates" from the "Recall/Extract"

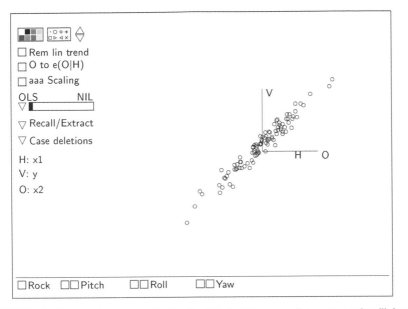

FIGURE 18.6 Estimated summary plot for the "E[$y \mid x$] linear, predictors uncorrelated" demonstration, with y on the vertical axis.

pop-up menu, we get:

```
Linear combinations on screen axes in 3D plot.
Horizontal: - .04333 + 0.1997 H + 0.2984 O
Vertical:   - .03352 + .08713 V
```

The value of the horizontal screen variable h is almost equal to $0.1\times$ $(2x_1 + 3x_2)$, and the ratio $0.1997/0.2984 = 0.6692$ is not far from 2/3. We have nearly recovered the true linear combination of the predictors that determines the response.

Fitting the mean function E($y \mid x_1, x_2$) = $\eta_0 + \eta_1 x_1 + \eta_2 x_2$ by OLS will determine a linear combination $h_{\text{OLS}} = \hat{\eta}_1 x_1 + \hat{\eta}_2 x_2$. The values of this linear combination differ from the OLS fitted values $\hat{y}$ only by the addition of the constant $\hat{\eta}_0$, so $\hat{y} = \hat{\eta}_0 + h_{\text{OLS}}$. Imagine summarizing the 3D plot by using the 2D plot of y versus $\hat{y}$. What would this plot look like? If $\hat{\eta}$ is a good estimate of η, then a 2D plot of y versus $\hat{y}$ should be a good summary of the regression. Select the item "Recall OLS" from the "Recall/Extract" pop-up menu in the plot y versus (x_1, x_2). The view on the computer screen is now y versus h_{OLS}. How does your visually determined summary plot compare to the OLS summary plot? Most people are pretty good at finding the OLS estimate by eye.

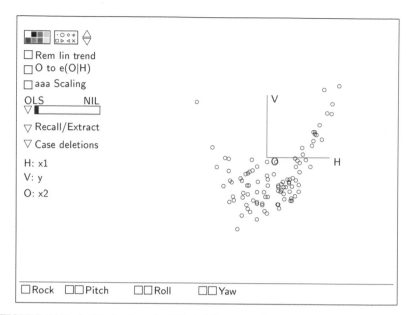

FIGURE 18.7   A 3D plot from the "E[y | x] linear, predictors nonlinear" demonstration.

18.5.2   Linear Regression Without Linearly Related Predictors

In the last example every 2D view of the 3D plot has a linear mean function and a constant variance function. As the plot is rotated, only the slope and the amount of variability change. In general, 2D views of a 3D plot can have nonlinear mean functions or nonconstant variance functions, even when the multiple linear regression model holds. The relationship between the predictors x_1 and x_2 is the key. If the predictors are linearly related, then the mean functions for all 2D views of y versus (x_1, x_2) will be linear. If the predictors are not linearly related, some 2D views will exhibit a nonlinear trend, even when the multiple linear regression model (18.6) holds. Linearly related predictors were introduced in Section 13.1.3.

Select the item "E[y | x] linear, predictors nonlinear" from the Demos: 3D menu. This example again uses artificial data, with x_1 standard normal, but with $x_2 = x_1^2 + N(0, 0.25)$. The response y is again computed as $y | x = 1 + 2x_1 + 3x_2 + N(0, 1)$. The linear regression model holds so all the data scatter about a plane in 3D. The only change between this and the last example is that $E(x_2 | x_1)$ is now nonlinear. The initial 2D view of the plot y versus (x_1, x_2) is given in Figure 18.7. This initial view can be used to estimate the function $E(y | x_1)$, which is nonlinear. Nevertheless, while rotating the 3D plot about its vertical axis, the points are seen to lie near a plane as required by the regression model; this can be seen perhaps a bit more clearly if you add the OLS plane to the plot. Using the methodology of Section 18.4, an estimated summary plot should be close to the OLS view, with horizontal axis close to $h^* = 2x_1 + 3x_2$. After returning the plot to the "Home" position, use a "Pitch" button to display the plot of x_2 versus x_1, which has a curved mean function.

Thus $E(y \mid \mathbf{x}) = E(y \mid x_1, x_2)$ is a *linear* function of $\mathbf{x}$ while $E(y \mid x_1)$ is a *nonlinear* function of x_1. This could have been predicted using Section 11.1. Repeating results from (11.1)

$$E(y \mid x_1) = \eta_0 + \eta_1 x_1 + \eta_2 E(x_2 \mid x_1) \tag{18.7}$$

This equation shows that $E(y \mid x_1)$ depends on x_1 and on $E(x_2 \mid x_1)$, the mean function for the regression of x_2 on x_1. This should clarify the example: We constructed the example to have $E(x_2 \mid x_1) = x_1^2$, and so

$$E(y \mid x_1) = \eta_0 + \eta_1 x_1 + \eta_2 E(x_2 \mid x_1)$$
$$= 1 + 2x_1 + 3x_1^2$$

The mean function $E(y \mid x_1)$ is therefore quadratic, as can be observed in Figure 18.7.

Rotate the 3D plot to display the marginal response plot y versus x_2. This plot is approximately linear. Is this expected? Interchanging the roles of x_1 and x_2 in (18.7), we find

$$E(y \mid x_2) = \eta_0 + \eta_1 E(x_1 \mid x_2) + \eta_2 x_2$$

But $E(x_1 \mid x_2)$ is zero in this example since for each value of x_2, x_1 has a distribution that is symmetric about zero. Thus, $E(y \mid x_2) = \eta_0 + \eta_2 x_2$, and a linear plot should be expected.

The effects of nonlinearly related predictors on variance functions are described in Problem 18.8.

18.5.3 More on Ordinary Least Squares Summary Views

After loading the file demo-3d.lsp, select the item "$E[y \mid x]$ nonlinear, predictors linear" from the Demos: 3D menu. These simulated data comprise 100 observations on two predictors x_1 and x_2, along with an error e, all generated as independent $N(0, 1)$ random variables. The response y was computed using

$$y \mid \mathbf{x} = 3e^{(2 + 2x_1)/2.5} + e$$

so $y \mid \mathbf{x}$ does not depend on x_2, and its dependence on x_1 is nonlinear. This problem has 1D structure with kernel mean function $M(\beta^T \mathbf{x}) = 3 \exp[(2 + \beta^T \mathbf{x}) / 2.5]$, $\beta = (2, 0)^T$, and $\beta^T \mathbf{x} = 2x_1$. The sufficient summary plot of y versus x_1 will recover the kernel mean function.

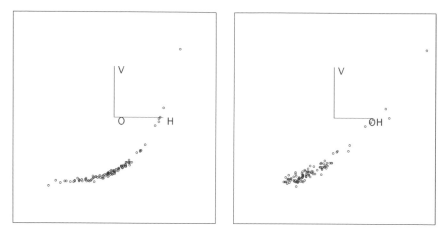

a. Ideal 2D view.   b. OLS view.

FIGURE 18.8 Two 2D views of the 3D plot of y versus (x_1, x_2) from the "$E[y \mid x]$ nonlinear, predictors nonlinear" demonstration.

Use the 3D plot of y versus (x_1, x_2) to verify that 1D structure is indeed appropriate for this regression. Rotate the plot to the estimated 1D summary plot. Next, use the "Rotate to OLS" control from the "Recall/Extract" pop-up menu to rotate to the OLS linear combination. Assuming your visual fit was reasonable, the OLS view and your visual fit should agree very closely. Perhaps surprisingly, fitting by OLS gives a good estimate of the sufficient summary plot *even through the regression of y on* **x** *is clearly not linear.* Does this happen in general? The answer is *yes*, with linearly related predictors, but *no* otherwise. In particular, with 1D structure and linearly related predictors, then the plot of y versus $\hat{y}$ can be a good summary plot even if the kernel mean function $M(\beta^T \mathbf{x})$ is not linear.

Consider another version of the example by selecting the item "$E[y \mid x]$ nonlinear, predictors nonlinear" from the Demos: 3D menu. The data were generated as in the last example, except now $x_2 = x_1^2 + N(0, 1)$. Once again, use the methodology of Section 18.2 to choose an estimated summary plot from the plot of y versus (x_1, x_2). Your solution should be similar to Figure 18.8a. Compare this plot to the OLS summary view, shown in Figure 18.8b. The linear combinations on the horizontal axis of these two plots are very different. Whereas by eye we can easily pick out the 2D view with minimal average variation, the OLS estimator gets fooled because it looks for the best view with a linear mean function, and the nonlinear relationship between the predictors translates through (18.7) into a linear combination where the mean appears to be fairly linear. Examination of uncorrelated 2D views as in Section 18.4 will confirm that the visual fitting provides a useful 2D summary view, but the OLS view misses relevant information.

The only difference between the two examples of this section is the distribution of the predictors: In the first example the predictors were linearly related (actually they are independent, which is a special case of linearly related), while in the second example the predictors were not linearly related. We will see that using linearly related predictors is essential for regression graphics; without them, plots can miss or distort information.

Here are the essential points of the discussion of using the OLS view as a summary view for a 3D plot.

- If a multiple linear regression model is appropriate, then the distribution of the predictors doesn't matter and the OLS view, which is essentially a plot of y versus $\hat{y}$, is a good summary plot.
- With 1D structure and linearly related predictors, the plot of y versus $\hat{y}$ is a good summary plot even if the kernel mean function $M(\beta^T \mathbf{x})$ is not linear.
- If the kernel mean function is not linear and the predictors have a nonlinear relationship, then y versus $\hat{y}$ should not be trusted as a good summary, even if the true model is 1D.
- If the regression has structural dimension greater than one, then the plot y versus $\hat{y}$ must necessarily miss information that may be relevant.

18.6 COMPLEMENTS

The material in this chapter is drawn from Cook and Weisberg (1994b), which was based on Cook (1994). Procedures for checking summary plots, including the method in Section 18.4, were investigated by Cook and Wetzel (1993). A comprehensive discussion of methods for checking summary plots was given by Cook (1998b, Chapter 4).

The data for Problem 18.7 is from Weisberg (1985), originally provided by Doug Tiffany. The data on short-leaf pines is from Atkinson (1994).

PROBLEMS

18.1 The following problems relate to the data on short-leaf pines in the file pines.lsp. Suppose that the view in Figure 18.2 *is* a sufficient summary plot so all the information about *Vol* is contained in the one linear combination $\mathbf{b}^T \mathbf{x}$. We can then study the 2D view shown in Figure 18.2 using all of the tools we have for investigating the simple regression of *Vol* on the single predictor $h = \mathbf{b}^T \mathbf{x}$. For example, using the graphical response transformation method discussed in Section 13.1.2, we conclude that the cube root transformation will likely linearize the mean function of the plot in Figure 18.2. As shown in Figure 18.9, there is a strong linear relation between $Vol^{1/3}$ and $\mathbf{b}^T \mathbf{x}$. Our graphical analysis

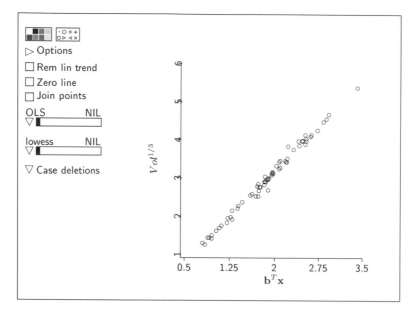

FIGURE 18.9   Scatterplot of $Vol^{1/3}$ versus $\mathbf{b}^T\mathbf{x}$ for the short-leaf pine regression.

then leads us to the model

$$Vol^{1/3} \mid (H,D) = \eta_0 + \gamma\mathbf{b}^T\mathbf{x} + e \tag{18.8}$$

$$Vol^{1/3} \mid (H,D) = \eta_0 + \eta_1 H + \eta_2 D + e \tag{18.9}$$

In (18.8), $\mathbf{b}$ was estimated visually from Figure 18.2, and the parameters are an intercept η_0 and a slope γ. In (18.9), $(\eta_1,\eta_2)^T$ was substituted for $\gamma\mathbf{b}$ so that the coefficients of H and D can be re-estimated using OLS.

18.1.1   Use the uncorrelated 2D view to check the adequacy of the estimated summary plot of Figure 18.9. What do you conclude? Since the distribution of $Vol^{1/3} \mid D,H$ must have the same structural dimension as the distribution of $Vol \mid D,H$, you should get the same answer given in Section 18.4.

18.1.2   We didn't use any prior information on the nature of trees in the graphical analysis of Section 18.1. If we model the shape of a tree as a cone, we get the following mean function for tree volume,

$$\mathrm{E}(Vol \mid H,D) = \alpha D^2 H$$

where α is an unknown parameter. If trees were exact cones, then $\alpha = \pi/12$. This suggests that taking logarithms might yield a useful linear model,

$$\log(Vol) \mid (H,D) = \eta_0 + \eta_1 \log(D) + \eta_2 \log(H) + e$$

(18.10)

If trees were cones, then $\eta_0 = \log(\pi/12)$, $\eta_1 = 2$, and $\eta_2 = 1$.

a. Rotate the 3D plot of $\log(Vol)$ versus $(\log H, \log D)$ to the 2D view with $2\log(D) + \log(H)$ on the horizontal axis. This can be done in *Arc* using the "Move to horizontal" plot control from the "Recall/Extract" menu. Is there evidence in the data to suggest that this 2D view is insufficient? If so, can you find a sufficient 2D view of the 3D plot of $\log(Vol)$ versus $(\log H, \log D)$?

b. Using model (18.10), construct an F test of NH: $\eta_1 = 2$ and $\eta_2 = 1$ versus AH: either $\eta_1 \neq 2$ or $\eta_2 \neq 1$. Do the results of this F test agree with your visual impressions in part **a**?

18.1.3 By examining the coefficients of the OLS fit using mean function (18.10), comment on the applicability of the assumption that a tree can be approximated by a cone.

18.2 Select the item "$E[y \mid x]$ linear, correlated predictors, x_2 not needed" from the Demos: 3D menu. These data were generated with two highly correlated normal predictors, $\rho(x_1, x_2) = 0.95$, and response $y = 1 + 2x_1 + N(0,1)$, so x_2 is not needed.

18.2.1 Construct a scatterplot matrix of the response and the two predictors. That the marginal response plot y versus x_1 shows a linear trend is unsurprising, since x_1 is the important predictor. Explain why the marginal response plot y versus x_2 shows a strong linear trend as well. Describe how you might construct an example so that the same model holds but the marginal response plot y versus x_2 shows a strong *nonlinear* trend.

18.2.2 Construct the 3D plot y versus (x_1, x_2) and then push the "O to e(O|H)" button. The resulting plot shows that only x_1 is relevant, which is the correct conclusion in this case. Why does this work? Would it work if you plotted y versus (x_2, x_1)?

18.3 From the Demos: 3D menu, select the item "$E[y \mid x]$ linear, predictors nonlinear." Construct the 3D plot y versus (x_1, x_2) and then push the "Rem lin trend" button. Describe the contents of this plot and what it shows in this example.

18.4 Suppose model (18.6) holds. Under what conditions on (x_1, x_2) is the marginal mean function $E(y \mid x_2)$ linear? Which plot can help decide if the required condition is satisfied?

18.5   This problem uses the data file BGSgirls.lsp from the Berkeley Guidance Study for Girls born in Berkeley, California in 1928–1929.

   18.5.1   Draw the 3D plot *HT18* versus (*HT9*, *WT9*). After observing the rotating plot, write a short description of the plot.

   18.5.2   Obtain a visual fit to this plot by rotating to the 2D view that minimizes the average variation. Use the "Display screen coordinates" item on the "Recall/Extract" menu to find the linear combination of the age-nine variables that correspond to the horizontal axis. Compare the ratio b_1/b_2 from the screen coordinates to the ratio $\hat{\eta}_1/\hat{\eta}_2$ from the regression output obtained from the OLS regression of *HT18* on the two age-nine variables.

   18.5.3   Without rotating the plot from the view selected in Problem 18.5.2, select the "Extract axis" item from the "Recall/Extract" menu to save the horizontal screen variable with the name *h1*. Now compute the regression with *h1* as the response and *HT9* and *WT9* as predictors using the "Fit Linear LS" item in the Graph&Fit menu. Verify that the regression coefficient estimates for *HT9* and *WT9* are the same as given by the "Display screen coordinates" item.

   18.5.4   Next, create a regression model with *HT18* as the response and *h1* as the predictor. How do you think the fitted values from the regression of *HT18* on *HT9* and *WT9* will compare to the fitted values from the regression of *HT18* on *h1*? After thinking about this, draw the plot of one set of fitted values versus the other.

   18.5.5   For the estimated summary plot obtained in Problem 18.5.2, follow the steps in Section 18.4 to decide if your 2D view misses relevant information about the relationship between the response and the predictors.

18.6   Select the "E[y | x] linear, correlated predictors" item from the "Demo: 3D" menu. This example is similar to the demonstration in Section 18.5.1, except that the correlation between the predictors is 0.99 rather than zero.

   Verify the correlation between x_1 and x_2 by drawing the plot of x_1 versus x_2, and observing that these points fall nearly on a line. Before drawing the 3D plot *y* versus (x_1, x_2), how do you think it will look? Now draw and describe the plot.

   Try fitting by eye to this plot. Most static 2D views of the 3D plot show a linear relationship of about equal strength; all that changes is the slope. Why is this? Fitting by eye is highly variable. To improve resolution, use the "O to e(O|H)" plot control. Explain what this control does to the plot, and try to fit by eye again. What do you conclude?

18.7 The data for this problem are from an economic study of the variation in rent paid for agricultural land planted with alfalfa in 1977. Alfalfa is a high-protein crop used to feed dairy cows. The unit of analysis is a county in Minnesota, and the data include: Y = average rent per acre planted with alfalfa; X_1 = average rent paid for all tillable land; X_2 = density of dairy cows, number per square mile; X_3 = proportion of farmland in the county used as pasture; and X_4 = an indicator with value 1 if liming is required to grow alfalfa, and 0 otherwise.

Load the file `landrent.lsp`. View Y as the response variable and X_1 and X_2 as the predictors. We will not use X_3 or X_4 in this problem.

18.7.1 Examine the scatterplot matrix of Y, X_1, and X_2. Does the assumption of linearly related predictors seem plausible for these data? Why or why not? If the assumption of linearly related predictors does not seem reasonable, use the transformation controls on the plot to transform the predictors to a set of more nearly linear predictors. We then view the transformed predictors as the base predictors.

18.7.2 Examine the 3D plot of the response and the two transformed predictors from Problem 18.7.1 for the assumption of linearly related predictors. What do you conclude?

18.7.3 Assuming that the predictors are nearly linear in the scales you have chosen, draw the OLS summary plot y versus $\hat{y}$. What would the summary plot look like if the linear model is true? What does it look like here? What do you conclude?

18.8 Load the file `demo-3d.lsp` and from the resulting Demos: 3D menu select the item "E[y | x] linear, predictors nonlinear." This example is discussed in Section 18.5.2, where we see that nonlinear relationships among the predictors can lead to a nonlinear mean function for some 2D views, even when the multiple linear regression model holds.

18.8.1 Use the results of Section 11.1.2 to show that nonlinearly related predictors can effect the variance function of a 2D view in a 3D plot. For the data in the demonstration, using the fact that $x_2 = x_1^2 + N(0,0.25)$, show that some 2D views will show nonconstant variance.

18.8.2 Find a 2D view of the 3D plot of y versus (x_1, x_2) that clearly demonstrates nonconstant variance in the 2D view. You will probably want to remove the linear trend from this plot to show the result more clearly.

CHAPTER 19

Visualizing Regression with Many Predictors

With two predictors, a summary plot can be either a histogram if the regression has 0D structure, a 2D scatterplot of the response versus a linear combination of the predictors with 1D structure, or a 3D plot if the regression has 2D structure. With $p \geq 2$ predictors, the structural dimension can be any integer between 0 and p: With k-dimensional structure, for example, the dependence of y on the predictors is through k linear combinations of them that would require a $(k + 1)$-dimensional scatterplot to be viewed fully.

The methods described in the last chapter show how a 3D plot can be used to estimate structural dimension visually when $p = 2$. With more than two predictors, we must continue to rely on 2D and 3D plots to give us information about the regression because we can't draw a plot of y versus all p predictors. To be useful the low-dimensional plots must not distort high-dimensional relationships. Whether or not distortion occurs depends on the relationships between the predictors.

In this chapter we discuss the required conditions on the predictors for estimating structural dimension, and we show how the conditions can be checked and often induced. The central graphical object is the scatterplot matrix. It provides useful though necessarily incomplete information on the distribution of the predictors and on the dimension of a regression. In the next chapter we will take a different approach that will allow us to use just a few 3D plots to understand a high-dimensional regression.

19.1 LINEARLY RELATED PREDICTORS

We need conditions on the predictors to guarantee that lower-dimensional plots do not distort higher-dimensional relationships. The conditions depend on the structural dimension of the regression.

Linearly Related Predictors with 1D Structure. Suppose that the regression has 1D structure so y depends on $\mathbf{x}$ through only $\beta^T\mathbf{x}$, and let x_j be one of the predictors. Linearly related predictors requires that $\mathrm{E}(x_j \mid \beta^T\mathbf{x})$ be a linear function of $\beta^T\mathbf{x}$,

$$\mathrm{E}(x_j \mid \beta^T\mathbf{x}) = a_j + b_j\beta^T\mathbf{x} \tag{19.1}$$

for all predictors, $j = 1,\dots,p$.

In Section 13.1.3 we required *linearly related terms* in the context of response transformations. Condition (19.1) is the same, except it is applied to the predictors rather than to the terms.

Linearly Related Predictors with 2D Structure. Suppose that the regression has 2D structure so y depends on $\mathbf{x}$ through only $\beta_1^T\mathbf{x}$ and $\beta_2^T\mathbf{x}$, and again let x_j be one of the predictors. Then we will require that $\mathrm{E}(x_j \mid \beta_1^T\mathbf{x}, \beta_2^T\mathbf{x})$ be a linear function of $\beta_1^T\mathbf{x}$ and $\beta_2^T\mathbf{x}$,

$$\mathrm{E}(x_j \mid \beta^T\mathbf{x}) = a_j + b_j\beta_1^T\mathbf{x} + c_j\beta_2^T\mathbf{x} \tag{19.2}$$

for all predictors, $j = 1,\dots,p$.

Linearly Related Predictors with kD Structure. If the regression has kD structure, so y depends on $\mathbf{x}$ only through the k linear combinations $\beta_1^T\mathbf{x},\dots,\beta_k^T\mathbf{x}$, we will require that the mean function $\mathrm{E}(x_j \mid \beta_1^T\mathbf{x},\dots,\beta_k^T\mathbf{x})$ be linear for each predictor x_j. We will use the phrase *linearly related predictors* to indicate these conditions, depending on the structural dimension of the regression.

19.2   CHECKING LINEARLY RELATED PREDICTORS

The condition of linearly related predictors cannot be checked directly because the β's needed in the mean functions are unknown. However, there are good indirect checks available. The condition should hold to a good approximation if every frame in a scatterplot matrix of the predictors has a mean function that is either linear, or at least not noticeably curved. If any frame has a clearly curved mean function, then the condition may fail. In that case we suggest inducing linearly related predictors by transforming to multivariate normality as far as possible. If the predictors follow a multivariate normal distribution, then they will be linearly related regardless of the structural dimension of the regression.

The results in the rest of this chapter and in the next chapter will still apply if the predictors are only *approximately* linearly related, but gross nonlinearity should not be present.

19.3   LINEARLY RELATED PREDICTORS AND THE 1D MODEL

Suppose we have the 1D model specified in (18.3),

$$E(y \mid \mathbf{x}) = \mathrm{M}(\beta^T \mathbf{x}) \qquad \text{and} \qquad \mathrm{Var}(y \mid \mathbf{x}) = \mathrm{V}(\beta^T \mathbf{x})$$

so both the mean function and variance function can depend on the linear combination $\beta^T \mathbf{x}$. A plot of y versus $\beta^T \mathbf{x}$ is a sufficient summary plot, since from this one plot we can visualize both M and V. This plot requires knowing $\beta^T \mathbf{x}$. Can we estimate β without making any assumptions about M or V? If we have linearly related predictors, it turns out that we can get useful information on β. Let

$$\hat{y} = \hat{\mathbf{b}}_0 + \hat{\mathbf{b}}^T \mathbf{x}$$

denote the fitted values from the OLS regression of y on $\mathbf{x}$. *In performing this regression we are* not *assuming that a multiple linear regression model holds or even that it yields a sensible fit to the data.* Rather, OLS is just a convenient method of computing an estimate.

The 1D Estimation Result. Here is a remarkable result that enables us to construct a good summary plot. Assuming linearly related predictors, $\hat{\mathbf{b}}$ is an estimate of $c\beta$ for some constant c. Since the magnitude of β doesn't matter in a plot, the 2D summary plot of y versus $\hat{\mathbf{b}}^T \mathbf{x}$ is an estimate of the sufficient summary plot of y versus $\beta^T \mathbf{x}$. Equivalently, we can take the plot of y versus the OLS fitted values $\hat{y}$ as the summary plot. The summary plot enables us to visualize the kernel mean function M and the kernel variance function V. We call this the *1D estimation result.*

The 1D estimation result is still useful when the structural dimension is higher than one. The linear combination $\hat{\mathbf{b}}^T \mathbf{x}$ will now be one of the linear combinations needed, so the plot of y versus the OLS fitted values $\hat{y}$ may miss information if the dimension exceeds one, but it is still relevant.

With many predictors we require an assumption—linearly related predictors—that was not required in the two-predictor case. This assumption allows us to infer about regression from 2D and 3D plots and is the price paid for not being able to see in high dimensions.

19.4   TRANSFORMING TO GET LINEARLY RELATED PREDICTORS

Even if predictors are not linearly related, they can often be replaced by transformations that are more nearly linearly related. Both the visual and numeric methods for transforming to multivariate normality discussed in Section 13.2, page 324, can be used, but transform *only* the predictors, *not* the response.

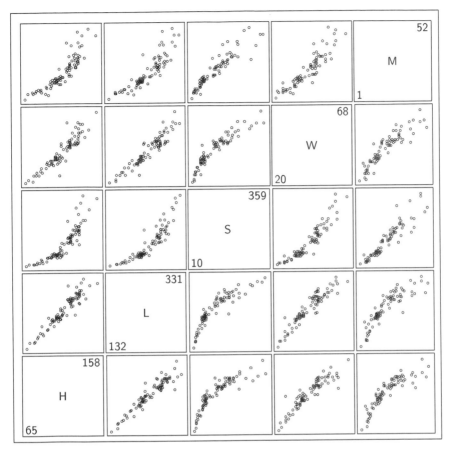

FIGURE 19.1 Scatterplot matrix for the mussels data.

As an example, consider the mussels data described in Problem 14.3, page 351. The goal of the analysis is to understand how muscle mass M depends on the length L, the width W, and height H of the mussel's shell and on the shell mass S. First examine a scatterplot matrix, as shown in Figure 19.1.

Many of the predictor plots have approximately linear mean functions, but the mean functions for the plots including S are clearly curved. Consequently, we may not have linearly related predictors. Transforming at least S, and possibly other predictors, may induce linearly related predictors. Use the Box–Cox method of Section 13.2 to determine suitable transformations: Select the item "Find normalizing transformations" from the "Transformations" pop-up menu. If you just select the four predictors and then push the OK button, *Arc* gives an error message and stops computing: Evidently the default starting values are sufficiently bad that the algorithm fails. Try again, but use "Marginal

Box–Cox starting values." This method will succeed, and it gives the following output:

```
Summary of Transformations to Normality
Variable Lambda-hat   SE    Wald test    Wald test
                             lambda =0    lambda =1
H              0.639   0.314    2.04        -1.15
L              0.439   0.289    1.52        -1.94
S              0.022   0.068    0.32       -14.47
W              0.104   0.212    0.49        -4.23

Likelihood ratio tests...
    of all lambda = 0:    6.925 df = 4 p = .140
    of all lambda = 1:  132.665 df = 4 p = .000
```

The likelihood ratio test that all four predictors should be transformed to logarithms has p-value 0.140. The scatterplot matrix in log-scale is shown in Figure 19.2. Linearity is supported by each 2D plot of the predictors, and so transformation to log scale seems to work. We will consider two further refinements. First, one point, identified in Figure 19.2 with an $\times$, is separated from the remaining points. This is case number 77, and since it seems to be different from the other mussels, all further analysis will be done without this point. We leave as an exercise to verify that deleting case 77 has little influence on the choice of transformation. Second, we could examine the transformed predictors in 3D plots to check for deviations from linearity. If no 2D view shows curvature, perhaps after using the "Rem lin trend" and "O to e(O | H)" buttons to improve visual resolution, then we have further support for linearly related predictors. In this regression, examining the four possible 3D plots further confirms linearly related predictors.

19.5   FINDING DIMENSION GRAPHICALLY

Given the linearly related predictors for the mussels data shown in Figure 19.2, can we say anything about the structural dimension of the regression? If we conclude that we have 1D structure, then the 1D estimation result can be used to get a summary plot. Since the response M is related to at least one of the predictors, we can eliminate 0D structure. How can we decide if we have 1D structure?

19.5.1   The Inverse Regression Curve

Suppose that 1D structure is appropriate so the distribution of $y \mid \mathbf{x}$ depends on $\mathbf{x}$ only through a single linear combination $\beta^T\mathbf{x}$. We can turn the regression around and, rather than study $y \mid \mathbf{x}$, we study the *inverse regression* of $\mathbf{x} \mid y$. The inverse regression is simpler, since it is a collection of p simple regressions,

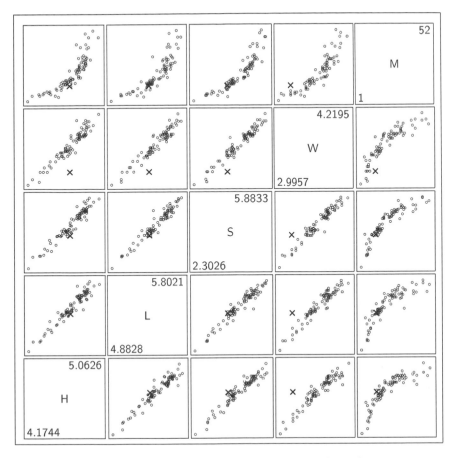

FIGURE 19.2 Mussel data with all predictors in log scale.

$x_1 \mid y$, $x_2 \mid y, \ldots,$ $x_p \mid y$, and each of these p regressions can be studied with a 2D scatterplot.

1D Checking Conditions. We are again aided by a remarkable result. Assume linearly related predictors and 1D structure. Then the mean and variance functions for each of the simple inverse regressions have the form, for $j = 1, \ldots, p$,

$$E(x_j \mid y) = E(x_j) + \alpha_j m(y) \qquad (19.3)$$

$$\text{Var}(x_j \mid y) \approx \text{Var}(x_j) - \alpha_j^2 s(y) \qquad (19.4)$$

where α_j has the same value in (19.3) and (19.4). Each α_j can be positive, negative, or zero. These equations tell us about the mean and variance functions of the *p inverse marginal response plots* of x_j versus y when there is 1D structure in the regression of y on **x**.

Equation (19.3) says that for each predictor x_j the inverse mean function $E(x_j \mid y)$ equals the overall mean $E(x_j)$ plus an unknown function $m(y)$ of y multiplied by the scale factor α_j. The important point is that, assuming 1D structure, each plot of x_j versus y is expected to have one of three forms:

- $m(y)$ versus y if $\alpha_j > 0$,
- $-m(y)$ versus y if $\alpha_j < 0$, and
- a constant versus y if $\alpha_j = 0$.

Thus, if $E(x_j \mid y)$ is linear for one plot, then it must be linear for all other plots to be consistent with 1D structure. If $E(x_j \mid y)$ is ∩-shaped for one plot, then it must be ∩-shaped or ∪-shaped or constant for all other plots. If this is not the case, then 1D structure must be abandoned.

Condition (19.4) requires that the inverse variance function $\mathrm{Var}(x_j \mid y)$ be approximately the overall variance $\mathrm{Var}(x_j)$ minus α_j^2 times an unknown function $s(y)$ that can depend on y, but not on j. When looking at all p plots, the variability must change in the same way in each of the plots. If the variability does not change in the same way, then 1D structure must be abandoned.

For some plots, we may have that $\alpha_j = 0$ in (19.3) and (19.4). To be consistent with 1D structure, plots that show no dependence on y in the inverse mean function ($\alpha_j = 0$) must show no dependence in the inverse variance function, even if the variance is not constant in other plots.

We will call (19.3) and (19.4) *1D checking conditions* since they must be satisfied for each predictor if 1D structure is to hold. They provide a basis for deciding between 1D structure and greater than 1D structure.

19.5.2 Inverse Marginal Response Plots

The $p = 4$ inverse marginal response plots for the mussels data are given in the last column of Figure 19.2. All these plots have the same shape, a finding that is consistent with 1D structure. By the 1D estimation result, this suggests that a summary plot of M versus the OLS fitted values will provide a complete summary for these data, as shown in Figure 19.3. To obtain this figure yourself, fit the OLS regression of M on the four predictors, each on log scale. Case 77, which was not used in fitting, is shown on the plot. It conforms nicely to the distribution of points in the plot, so it can be restored if further analysis is required.

The next step in the analysis depends on its goals. For example, if we would like a simple model, we might want to transform M to achieve linearity.

The inverse response plot of $\hat{y}$ versus M is shown in Figure 19.4, with case 77 restored, and with the logarithmic power curve superimposed. According to the discussion in Section 13.1, taking the logarithm of the response will

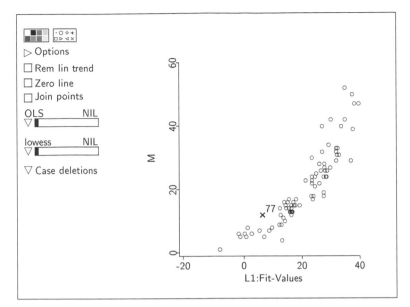

FIGURE 19.3 Summary plot for the mussels data. Case 77 is shown on the plot, but it was not used to compute the fitted values.

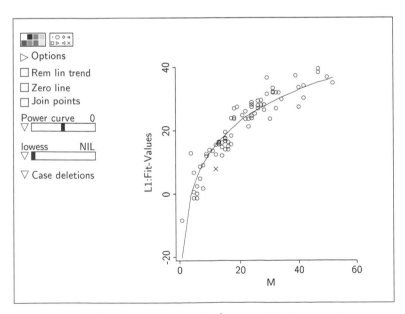

FIGURE 19.4 Inverse response plot $\hat{y}$ versus M for the mussels data.

TABLE 19.1   The Australian Athletes Data

Variable	Description
LBM	Lean body mass, kg.
Ht	Height, cm.
Wt	Weight, kg.
BMI	Body mass index, $Wt/(Ht)^2$.
%Bfat	Percent body fat.
Ferr	Plasma ferritin concentration.
Hc	Hematocrit.
Hg	Hemoglobin level.
RCC	Red cell count.
Sex	0 = male or 1 = female.
SSF	Sum of skin folds.
WCC	White cell count.
Label	Case labels.
Sport	Sport.

likely result in a linear regression model. Thus, we are led to the first model

$$\log M = \eta_0 + \eta_1 \log H + \eta_2 \log L + \eta_3 \log S + \eta_4 \log W + e \qquad (19.5)$$

which we could now analyze using all the methods discussed in Part II of this book. Because of the graphics, we can see that (19.5) must provide a reasonable starting point for modeling. Diagnostic methods for model checking may occasionally lead to further refinements of the graphical starting point.

19.6   AUSTRALIAN ATHLETES DATA

The data in the file `ais.lsp` give several measurements taken on 202 elite Australian athletes who trained at the Australian Institute of Sport. The variables are defined in Table 19.1. The members of the sample participate in different sports, and are about equally split between men and women. For this example, we will consider three predictors *Ht*, *Wt*, and the red blood cell count *RCC*, and we use lean body mass *LBM* as the response.

Our first task is to examine the assumption of linearly related predictors, using the scatterplot matrix shown in Figure 19.5. Apart from a few straggling points the assumption of linearly related predictors seems plausible; viewing the three predictors in a 3D plot with trends removed leads to the same conclusions. Removal of the straggling points, as was illustrated in the last example, might improve matters, but we will continue with all the data included.

With linearly related predictors, we can use the 1D checking conditions to examine the plausibility of 1D structure. This requires examining the inverse marginal response plots in the last column of Figure 19.5. To examine them

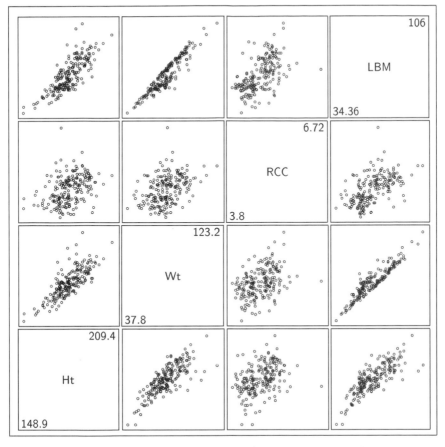

FIGURE 19.5 Scatterplot matrix for the Australian athletes data.

more carefully, select the item "Multipanel plot of" from the Graph&Fit menu, and then select *Ht*, *Wt*, and *RCC* to go on the "Changing axis" and select the response *LBM* to go on the "Fixed axis." Using the radio buttons, set the fixed axis to be horizontal, rather than the default of vertical. This will produce a 2D scatterplot, initially showing the 2D plot with the response *LBM* on the horizontal axis, and one of the predictors on the vertical axis. The extra slidebar on this plot allows cycling through choices for the changing axis.

Fit a smoother to each of the p inverse marginal response plots. The smoother in the jth plot is an estimate of the inverse regression function $E(x_j \mid y)$, which, according to checking condition (19.3), should approximate $E(x_j) + \alpha_j m(y)$ if 1D structure is appropriate. A different smoother can be used for each of the plots, with the goal of obtaining a useful estimate of the inverse regression functions. Any of the methods for simple regression can be used to

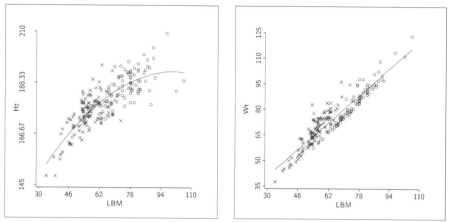

a. *Ht.*   b. *Wt.*

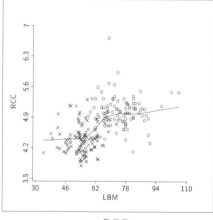

c. *RCC.*

FIGURE 19.6   Inverse marginal response plots for the Australian athletes data.

help determine a suitable fit in each of the inverse marginal response plots. The three inverse marginal response plots are shown in Figure 19.6 with smoothers added. The smoother used for *Ht* in Figure 19.6a is a quadratic fit with OLS. This plot shows both curvature and variance increasing to the right, but these characteristics are influenced by the straggling points mentioned earlier. The inverse marginal response plot for *Wt* in Figure 19.6b appears reasonably well-matched by a linear fit, and the smoother shown is the OLS line. The third inverse marginal response plot for *RCC* is in Figure 19.6c. The *lowess* smooth with parameter 0.6 on this plot suggests a nonlinear relationship, with different linear phases for low and high values of *LBM*. The three inverse regression functions are evidently different, so 1D structure is doubtful. Further evidence for this conclusion comes from consideration of variances. The variance of *Ht* increases with *LBM*, but increasing variance is not evident in the other plots. We are again led to the conclusion that 1D structure cannot be supported.

We must now face the difficult question of how to proceed. There is no fixed prescription since the regression is evidently quite complicated with greater than 1D structure. However, there are guidelines that can help. One possibility is to use the methods discussed in the next chapter to estimate the structural dimension and produce a 3D summary plot. This line of inquiry is continued in Problem 20.3.

Another possibility is to seek additional variables that might remove the necessity for 2D or greater structure. For this particular regression, the point marking in Figure 19.6 provides a clue on how to proceed. The points marked ○ correspond to males, while those marked × correspond to females. In all three, the males and females seem to fall into separate populations. Perhaps, therefore, the greater than 1D structure is caused by the need for a separate analysis for each gender. If this were so, we would expect 1D structure when analyzing each gender separately.

From the Graph&Fit menu, select the item "Set marks," and then select *Sex* to be a marking variable. This will mark the points in the scatterplot matrix using *Sex*. Push the mouse button over the "0" in the mark legend to display only the points corresponding to males in the scatterplot matrix. For males only, linearly related predictors still seems plausible. Examining the 1D checking conditions, the inverse marginal response plot for *RCC* displays both constant variance and constant mean function, while the remaining two plots are approximately linear with increasing variance. In the plot for *Ht*, two relatively short athletes with high *LBM* don't follow the general linear trend. These athletes cause a bit of curvature in this plot; otherwise, 1D structure is plausible for understanding the dependence of *LBM* on *Ht*, *Wt*, and *RCC* for males. We can repeat this analysis for females, and we come to essentially the same conclusions.

To continue the analysis, we would use the 1D estimation result and produce separate summary plot for males and females. The methods in Section 12.4 could then be used to combine the two summary plots into a single analysis.

19.7 COMPLEMENTS

The checking condition (19.4) becomes an equality if the predictors are normally distributed. If they are not normally distributed, this checking condition is not exact, but the approximation is generally quite good.

19.7.1 Sliced Inverse Regression

The 1D checking conditions in Section 19.5.1 provide a purely graphical approach to determining dimension based on the inverse marginal response plots.

Duan and Li (1991) and Li (1991) suggested that the inverse regression $\mathbf{x} \mid y$ can be informative about $y \mid \mathbf{x}$, and they developed a numerical procedure called *sliced inverse regression* or *SIR* in which the mean function is estimated in each of the inverse marginal response plots via smoothing. Li (1991) showed that the estimated smooths can then be compared to give a test of dimension.

Similar to SIR, Cook and Weisberg (1991) suggested a method called *sliced average variance estimation*, or *SAVE*, that uses both the mean checking condition (19.3) and the variance checking condition (19.4) to find inverse structure; *SIR* uses only the mean checking condition. Both methods are implemented in *Arc* using the "Inverse regression" item in the Graph&Fit menu. A full discussion of their use is given at the Internet site for this book.

19.7.2   References

Hall and Li (1993) have shown that for regressions with a large number of predictors the condition of linearly related predictors will generally be satisfied. In addition, the condition will generally be satisfied for data from experiments that use standard experimental designs. Derivations of the 1D checking conditions are available from Cook (1998b, Chapter 10).

The condition (19.1) required for linearly related predictors is generally not checkable because it depends on unknown parameters. However, with 1D structure, if $\mathbf{x}$ has an *elliptically contoured distribution*, then (19.1) holds for any β (Eaton, 1986). This condition is much more restrictive than (19.1), but it can be checked in practice by checking if $\mathbf{x}$ has an elliptically coutoured distribution. The most important member of this class of distributions is the multivariate normal.

The result in Section 19.2 that OLS estimates can give a consistent estimate of $c\beta$ even when the kernel mean function is not linear is based on Li and Duan (1989) using earlier work by Brillinger (1983). The results in Section 19.2 depend on linearly related predictors for all the cases in the data. As we have seen in Chapter 15, a few unusual points can strongly influence the OLS estimates. When a few such points are observed in the data, the result $\hat{\mathbf{b}}^T\mathbf{x} \approx c\beta^T\mathbf{x}$ need not hold.

The OLS summary plot can also fail to provide a useful summary plot if the mean function is symmetric. In this case, the OLS estimate $\hat{\beta}$ will estimate zero. For further discussion of such occurrences, see Cook and Weisberg (1991), Cook, Hawkins, and Weisberg (1992), Cook (1998b), and Exercise 19.5. The evaporation data in Exercise 19.4 is taken from Freund (1979).

PROBLEMS

19.1   Suppose we have a regression with p linearly related predictors and a structure that is at most 1D. Let $\hat{y}$ denote the fitted values from the

OLS regression of y on $\mathbf{x}$. If the plot y versus $\hat{y}$ appears as a random scattering of points with no clear systematic features, what would you choose as the structural dimension of the regression? Why?

19.2 This problem uses the Big Mac data in the file `bigmac.lsp`.

    19.2.1 Draw the scatterplot matrix of (*BusFare*, *Bread*, *TeachTax*, *TeachSal*, *BigMac*). Identify the point for Lagos; explain why it is an extreme point, and delete it from further consideration. Use the automatic procedure described in Section 13.2.2 to suggest normalizing transformations of the predictors (excluding the response *BigMac*). Test the hypothesis that the transformation parameters are all equal to the log transformation against a general alternative.

    19.2.2 Using log-transformed predictors, graphically explore the adequacy of the transformations to normality by checking for a linear mean function in each frame of a scatterplot matrix of the transformed predictors.

    19.2.3 Use inverse marginal response plots to check for 1D structure versus greater than 1D structure. What do you conclude?

    19.2.4 Find a transformation of the response *BigMac* so that the regression of the transformed response on the log-transformed predictors is approximately linear.

    19.2.5 Use model checking plots and residual plots to explore the adequacy of the model obtained in the last section. Summarize the results for this data set.

19.3 Consider a regression with a response y and three predictors $\mathbf{x} = (x_1, x_2, x_3)^T$. Write mean functions that correspond to the following situations. For each, give the structural dimension of the regression, assuming that the dependence of y on $\mathbf{x}$ is only through the mean function.

    a. A linear regression mean function with four terms, $u_0 = 1$, $u_1 = x_1$, $u_2 = x_2$, and $u_3 = x_3$.

    b. A mean function that cannot be written as a linear combination of the predictors, but is still a linear regression mean function.

    c. A mean function that can be written as a linear combination of the predictors, but is a nonlinear function of the parameters and hence not a linear regression mean function.

    d. A mean function that can neither be written as a linear combination of the predictors nor as a linear function of the parameters.

19.4 The file `evaporat.lsp` contains data on daily soil evaporation *Evap* for a period of 46 days. There are ten possible predictors that characterize the air temperature, soil temperature, humidity, and wind speed during a day; use the minimum and maximum of the daily air temperature, soil temperature, and humidity, for a total of six predictors. Use the response *Evap* and the predictors *Maxat*, *Minat*, *Maxst*, *Minst*, *Maxh*, and *Minh*.

19.4.1 Check the assumption of linearly related predictors. Explain how you do this, and explain your conclusions for these data.

19.4.2 Use inverse marginal response plots to explore the dimension of this regression. What do you conclude? What is the evidence?

19.4.3 Repeat the first two parts of this problem after removing the predictor *Minat*. Compare the results of your analysis to those in the first two parts of this problem for the full data.

19.5 In this problem we will construct an example that is almost identical to the example given in Section 18.5.3, except that the mean function is slightly modified. In the text window, type the following three statements:

```
(def x1 (normal-rand 100))
(def x2 (normal-rand 100))
(def e (normal-rand 100))
```

This has created three lists of standard normal random numbers, each of length one hundred. For this example, we want the mean function to be $E(y \mid \mathbf{x}) = x_1^2$, and we want the variance function to be constant with $\sigma = 1$. We can compute y and create a data set as follows:

```
(def y (+ (^ x1 2) e))
(make-dataset :data (list x1 x2 e y))
```

This will give you a series of dialog boxes to name the data set and the variables (use *x1*, *x2*, *e*, and *y*). Every time you do this problem, you will get slightly different answers because the data are generated at random each time. By contrast, in the demonstrations used throughout the book, the same data values are used each time. You can create a file from the data you have created using the "Save to file" item in the data set menu.

19.5.1 Compare the data generated here to the data used in Section 18.5.3. How do the mean functions differ? Are the variance functions the same or different? Is the assumption of linearly related predictors satisfied by the data you generated?

19.5.2 Examine the 3D plot y versus (x_1, x_2). What is the structural dimension of this plot? Rotate to the strongest 2D view in the plot. Does it correspond to the way you generated the data? How do you know? Mark this view by selecting the item "Remember view" from the "Recall/Extract" pop-up menu.

19.5.3 Modify the problem so it has mean function $E(y_1 \mid \mathbf{x}) = (1 + x_1)^2$. Do this by creating a new response y_1 using the "Add a variate" item in the data set menu and typing y1=(x1+1)^2+e.

Draw the plot y_1 versus (x_1, x_2), and observe that the best 2D view is very similar to the view chosen by OLS.

Explain why OLS gave a useful answer with y_1 as the response, but did not give a useful answer with y as the response.

CHAPTER 20

Graphical Regression

In Chapter 18 we saw how 3D plots can be used to construct graphical summaries in regressions with two predictors. In this chapter we show how to construct summary plots in regressions with more than two predictors. We assume throughout this chapter that the predictors are linearly related, but we do not require a model for the distribution of $y \mid \mathbf{x}$.

20.1 OVERVIEW OF GRAPHICAL REGRESSION

Suppose we have a regression with response y and p linearly related predictors $\mathbf{x} = (x_1, \ldots, x_p)^T$. Since we can't draw simple plots of more than three variables at a time, we will build up to the full regression by looking at the variables in pairs. Select two of the predictors, say x_1 and x_2, and for notational convenience collect the remaining $(p - 2)$ predictors in the vector $\mathbf{x}_3$.

The following question is at the heart of graphical regression: Can we replace x_1 and x_2 with a linear combination of them

$$x_{12} = b_1 x_1 + b_2 x_2 \tag{20.1}$$

without loss of sample information on the regression? As we have seen before, this is the same as asking, Can we find b_1 and b_2 so that y is independent of $\mathbf{x}$ given both x_{12} and $\mathbf{x}_3$?

- If so, and $b_1 = b_2 = 0$, then both x_1 and x_2 can be deleted from the regression without loss of sample information.
- If so, and either $b_1 \neq 0$ or $b_2 \neq 0$, then we can replace the pair of predictors (x_1, x_2) with the single predictor x_{12} without loss of sample information, effectively reducing the number of predictors by one. We could now select two other predictors to combine in the new regression with $p - 1$ predictors.

- If not, then the chosen predictors cannot be combined and the structural dimension of the regression of y on $\mathbf{x}$ must be at least two. Combining other predictors may still be possible.

Graphical regression is thus a sequential procedure for combining pairs of predictors using (20.1). The process stops when no further combining is possible. If the structural dimension of the regression is at most two, then the original p predictors can always be reduced to at most two linear combinations without loss of sample information on the regression.

A 3D added-variable plot for x_1 and x_2 after $\mathbf{x}_3$ can be used to tell if x_1 and x_2 can be combined, and to combine them into x_{12} if appropriate. Recall from Section 10.7 that the 3D added-variable plot for (x_1, x_2) is the 3D plot of $\hat{e}(y \mid \mathbf{x}_3)$ versus $(\hat{e}(x_1 \mid \mathbf{x}_3), \hat{e}(x_2 \mid \mathbf{x}_3))$. The following cases show how the structural dimension of this 3D added-variable plot is related to the study of (x_1, x_2):

- If the 3D added-variable plot has 0D structure, then $b_1 = b_2 = 0$ and (x_1, x_2) can be deleted from the regression.
- If the 3D added-variable plot has 1D structure, then (x_1, x_2) can be combined into x_{12}. The coefficients b_1 and b_2 are the same as the coefficients of $\hat{e}(x_1 \mid \mathbf{x}_3)$ and $\hat{e}(x_2 \mid \mathbf{x}_3)$ in the sufficient 2D view of the 3D added-variable plot.
- If the 3D added-variable plot has 2D structure, then x_1 and x_2 cannot be combined, and two other predictors should be selected.

All of the tools discussed in Chapter 18 for assessing the structural dimension of 3D plots can be used for the analysis of 3D added-variable plots in graphical regression.

20.2 MUSSELS' MUSCLES

In this section we use graphical regression to analyze the mussels data discussed in Section 19.4 and given in the datafile mussels.lsp. The four predictors are the length L, width W, height H and mass S of a mussel's shell; L, W and H are in mm and S is in grams. The response M is the mass in grams of a mussel's muscle. We are interested in studying the regression of M on (H, L, W, S).

20.2.1 The GREG Predictors

The first step in graphical regression is to insure that the condition of linearly related predictors is not seriously violated. As we found in Section 19.4, deleting case 77 and replacing the predictors by their logarithms gives a new set of transformed predictors that reasonably satisfies the assumption of linearly related predictors. We therefore begin analysis with response M and predictors

$$\mathbf{x} = (\log H, \log L, \log W, \log S)^T$$

At this point we could start graphical regression as suggested in Section 20.1, but useful answers can be obtained more quickly by using the 1D estimation result discussed in Section 19.2, page 431. We do this by replacing the predictors $\mathbf{x}$ with p new predictors which are *uncorrelated linear combinations* of $\mathbf{x}$. The new predictors have the following properties.

- The first linear combination is called *Fit*. It differs from the OLS fitted values from the regression for y on $\mathbf{x}$ only by subtraction of the intercept, so $Fit = \mathbf{b}_{OLS}^T \mathbf{x} = \hat{y} - \hat{\eta}_0$. If the regression has 1D structure and the predictors are linearly related, then we know from Section 19.2 that the plot of the response versus *Fit* is most likely all that we need to continue the analysis. If the dimension is bigger than one, this linear combination is likely to be one of the linear combinations needed.
- The second linear combination is $gr_1 = \mathbf{b}_1^T \mathbf{x}$. It is defined to be a linear combination that is uncorrelated with the first linear combination and, if the regression has 2D structure, it is a good candidate for the second linear combination we will need in addition to *Fit*.
- The third linear combination $gr_2 = \mathbf{b}_2^T \mathbf{x}$ is uncorrelated with the first two linear combinations, *Fit* and gr_1. If the regression has 3D structure, then this is a good candidate for the third linear combination we will need in addition to *Fit* and gr_1.
- The rest of the linear combinations are defined in the same way. In general, the linear combinations are uncorrelated and are ordered according to a measure of their likely importance in the analysis. The ordering is not certain, however, and we still need graphics to complete the analysis. We will refer to these and all linear combinations constructed during graphical regression as GREG predictors, which is an acronym for *gr*aphical *reg*ression.

20.2.2   Graphical Regression

We are now ready to apply graphical regression using *Arc*. The data set contains the original four predictors, the four transformed predictors $\mathbf{x}$, and the response. Case 77 should be deleted.

Select "Graphical regression" from the Graph&Fit menu to get the dialog shown in Figure 20.1. The predictors $\mathbf{x}$ and the response are specified in the usual way. The two buttons on the bottom of the dialog allow you to use the predictors specified in the dialog, or use the GREG predictors; we select the latter choice, which is the default. After clicking the "OK" button, a graphical regression menu called "G1" will be added to the menu bar, and the

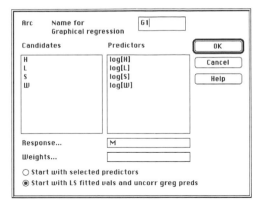

FIGURE 20.1 The graphical regression dialog.

TABLE 20.1 *Arc* **Output Showing the** GREG
Predictors as Functions of the Original Predictors
for the Mussels Data

	log[H]	log[L]	log[S]	log[W]
G1.Fit	−0.252	−0.595	0.472	0.599
G1.gr1	0.906	0.322	−0.275	−0.017
G1.gr2	0.625	−0.743	0.111	−0.212
G1.gr3	0.092	−0.592	−0.139	0.789

information shown in Table 20.1 will be displayed in the text window. The names of the GREG predictors in the table have a prefix, "G1" in this case, that corresponds to the name of the graphical regression menu. The table lists multipliers that define the GREG predictors as combinations of the original predictors. For example, $G1.gr1 = \mathbf{b}_1^T\mathbf{x}$ with

$$\mathbf{b}_1 = (0.906, 0.322, -0.275, -0.017)^T$$

or $gr_1 = 0.906\log(H) + 0.322\log(L) - 0.275\log(S) - 0.017\log(W)$.

We are now ready to use 3D added-variable plots to conduct a graphical regression analysis using the GREG predictors. From the "G1" menu select the item "3D AVP." In the resulting dialog shown in Figure 20.2, select two of the predictors to combine using the ideas outlined in Section 20.1. While in theory you can choose any pair, starting either with the first two predictors (*Fit* and gr_1), or with the last two predictors is easiest. We chose to combine the last two gr_2 and gr_3.

The dialog of Figure 20.2 produces a 3D added-variable plot for the selected predictors after the rest, as shown in Figure 20.3.

Select two terms to combine

Candidates   Selection

| G1.Fit | G1.gr2 |
| G1.gr1 | G1.gr3 |

OK   Cancel

FIGURE 20.2   The dialog for selecting GREG predictors to combine in graphical regression.

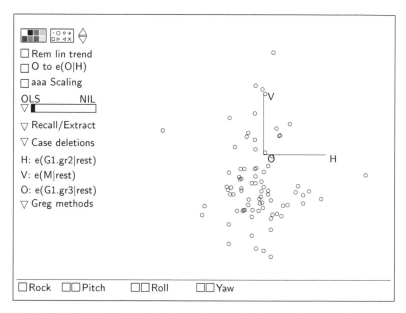

FIGURE 20.3   3D added-variable plot from the graphical regression analysis of the mussels data.

- If the 3D added-variable plot has 0D structure, then we can delete the GREG predictors gr_2 and gr_3. This is done by selecting the item "Dimension 0" in the "Greg methods" pop-up menu that appears on the bottom left of the 3D added-variable plot in Figure 20.3.

- If the plot has 1D structure, then we can combine the two GREG predictors by first rotating to the sufficient summary view and then selecting "Dimension 1" in the "Greg methods" pop-up menu. *Arc* evaluates the horizontal screen variable $b_2 gr_2 + b_3 gr_3$ and adds it to the data set as a new GREG predictor gr_k, incrementing the index k by one. The new GREG predictor gr_4 corresponds to x_{12} in the general discussion of Section 20.1.

- If the plot has 2D structure, we cannot combine the predictors, and two new predictors could be selected. Two-dimensional structure is specified

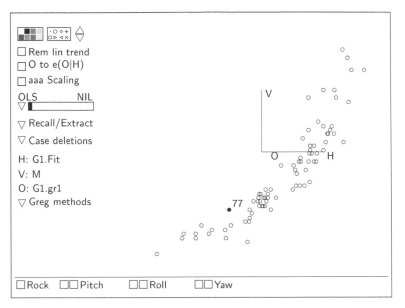

FIGURE 20.4   Final summary plot for the mussels data.

by selecting "Dimension 2" in the "Greg methods" pop-up menu. In this case, *Arc* returns you to the stage just prior to the construction of the 3D added-variable plot.

- After taking one of these three actions, *Arc* displays either the final 3D plot if only two GREG predictors remain or the dialog for selecting two of the remaining predictors to combine in another 3D added-variable plot.

At any stage you can find how the present GREG predictors are constructed from the original linear predictors **x** by selecting the item "Display active predictors" in the graphical regression menu. A table like Table 20.1 will then be displayed in the text window.

We judged Figure 20.3, the 3D added-variable plot for (gr_2, gr_3) after (Fit, gr_1), to have 0D structure and selected "Dimension 0" in the "Greg Methods" pop-up menu.

We now have a regression with response M and the two remaining GREG predictors, *Fit* and gr_1. The plot produced by *Arc* is just the 3D plot of M versus (Fit, gr_1). We judged this plot to have 1D structure. After rotating to the summary view shown in Figure 20.4 and selecting "Dimension 1" in the "Greg methods" pop-up menu, the horizontal screen variable

$$gr_4 = b_0 Fit + b_1 gr_1$$
$$= 0.31 \log H - 0.50 \log L + 0.38 \log S + 0.71 \log W \qquad (20.2)$$

TABLE 20.2 Reaction Yield Data in the File `ryield.lsp`

Variable	Desicrption
T_1	$(Temp_1 - 122.5)/7.5$, where $Temp_1$ = temperature °C during first stage.
T_2	$(Temp_2 - 32.5)/7.5$, where $Temp_2$ = temperature °C during second stage.
Lt_1	$1 + 2[\log(Time_1) - \log(5)]/\log(2)$, where $Time_1$ = run time during first stage in hours.
Lt_2	$1 + 2[\log(Time_2) - \log(1.25)]/\log(5)$, where $Time_1$ = run time during second stage in hours.
C	$(Con - 70)/10$, where Con = concentration in percent.
y	Yield from the two-stage chemical reaction.

is computed and added to the data set. We found the coefficients shown in (20.2) by selecting the item "Display active predictors" in the graphical regression menu.

We have a regression with response M and single GREG predictor gr_4. This brings us to the end of the graphical regression analysis. The summary plot shown in Figure 20.4 is now used to guide our study of the regression of M on **x**.

As might have been expected, we have reached the same point in the analysis of these data that was reached in Section 19.5.2. In that section, we concluded 1D structure, with summary plot of M versus the OLS fitted values, whereas here the 1D structure is summarized by M versus gr_4. The OLS fitted values and gr_4 are almost the same (see Problem 20.4), so both analyses give essentially the same result. Depending on the goals of the analysis, we would continue as outlined in Section 19.5.2.

20.3 REACTION YIELD

In this section we consider a different regression in which the response y is the yield from a two-stage chemical process. There are five coded predictors defined in Table 20.2 in terms of five measured predictors. Since the times varied over a wide range, they were replaced by their logarithms.

Prior experiments had indicated that the optimum settings that maximize the mean function would probably be found within the ranges of the predictor values used in this experiment. The mean function therefore is unlikely to be a monotone function of the coded predictors. Accordingly, the investigators used a full quadratic model, Section 7.3.4, in the five coded predictors to analyze the experiment. This model contains 21 terms—an intercept, five linear terms, five quadratic terms and ten interaction terms—leading to a very complex summary with a model that automatically forces a high structural dimension. We consider how graphical regression can be used to gain insights into the results of this experiment. The data are in the file `ryield.lsp`.

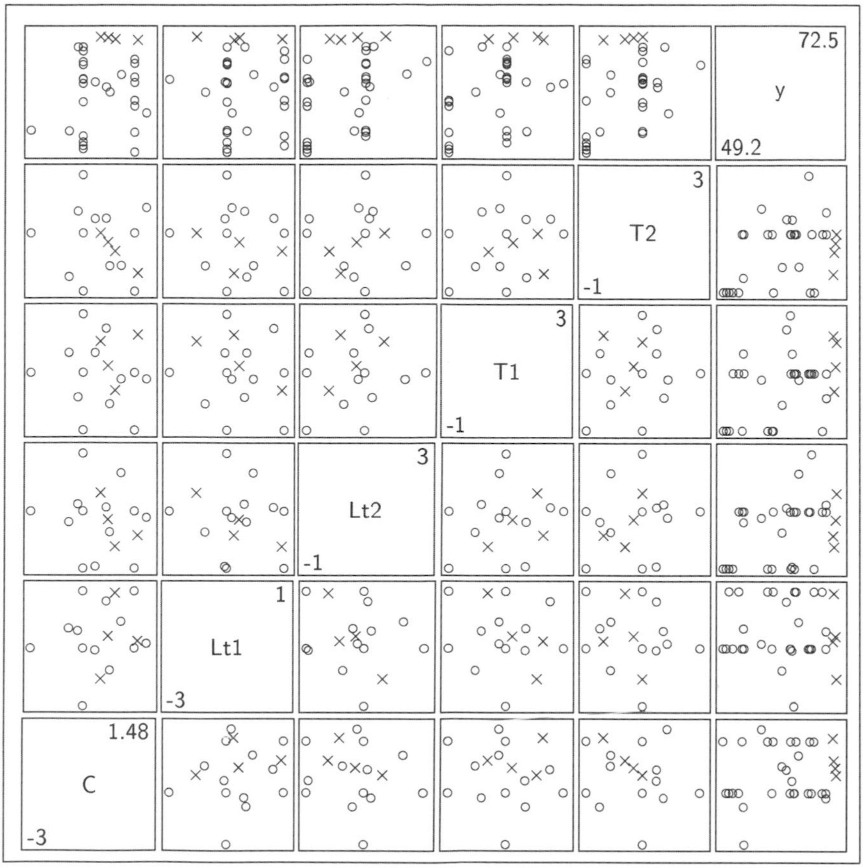

FIGURE 20.5   Scatterplot matrix for the reaction yield data.

20.3.1 Linearly Related Predictors

The first step in graphical regression is always to insure that the condition of linearly related predictors is not seriously violated. A scatterplot matrix of the five coded predictors and the response is shown in Figure 20.5; ignore the two plotting symbols for now. There are no clear nonlinear relationships between the predictors, so we can proceed by assuming that the coded predictors do not seriously violate the condition of linearly related predictors, and set

$$\mathbf{x} = (T_1, T_2, Lt_1, Lt_2, C)^T$$

The marginal response plots in the top row of Figure 20.5 show only weak relationships between y and each of the predictors individually, suggesting the possibility of large unexplainable variation in the response. We can expect that finding structure graphically will be harder in this example than it was in the mussels data.

20.3.2   Graphical Regression

Start graphical regression as in Section 20.2.2 by select "Graphical regression" from the Graph&Fit menu and specify the response and predictors in the resulting dialog. We again use the default GREG predictors *Fit* and gr_k, $k =$ 1,2,3,4, and begin by combining gr_3 and gr_4.

The interpretation of the 3D added-variable plot for gr_3 and gr_4 after (Fit, gr_1, gr_2) is not as clear as it was in the previous example. Weak dependence is apparent in the plot as can be seen a bit more clearly by adding a full quadratic surface to the plot while rotating. The fitted surface is only slightly curved. Our goal is to obtain a parsimonious summary that includes most of the regression information. We chose to characterize the plot as having 0D structure, ignoring the relatively small quadratic trend to get to a relatively simple solution. After selecting "Dimension 0" from the "Greg methods" pop-up menu, we now have a regression with response y and three remaining GREG predictors (Fit, gr_1, gr_2).

There is clear dependence in the 3D added-variable plot for (gr_1, gr_2) after *Fit*. We characterized the plot as having 1D structure with summary plot shown in Figure 20.6. The horizontal axis of this summary view is just the variable gr_1. The dependence in this plot can be further visualized by adding the curve pln+H^2, which fits a quadratic only in the H-variable gr_1, to the plot. We judge that y is independent of gr_2 given gr_1. According to the rules for graphical regression, we can therefore simply delete gr_2.

After selecting "Dimension 1" from the "Greg methods" pop-up menu, the final 3D plot of y versus (Fit, gr_1) is shown by *Arc*. Two views of this 3D plot are given in Figure 20.7. We judged this plot to have 2D structure, with a generally linear trend in one direction, and a curved trend in a second direction. This 3D plot is the summary plot for the experiment. As in the previous example, we can now use it to guide the analysis.

If we had concluded that the 3D added-variable plot for (gr_1, gr_2) after *Fit* and the 3D added-variable plot for (Fit, gr_1) after gr_2 both had 2D structure, then the full regression would have at least 3D structure, and we would not have been able to combine any of the predictors further.

20.3.3   Continuing the Analysis

Two views of the 3D summary plot are shown in Figure 20.7, but the full structure can be seen only while rotating. The view in Figure 20.7a shows generally a linear increasing trend. Contrary to expectations of the investigators, the maximum yield may occur at the edge or even outside the experimental range. The view in Figure 20.7b shows a maximum more clearly in

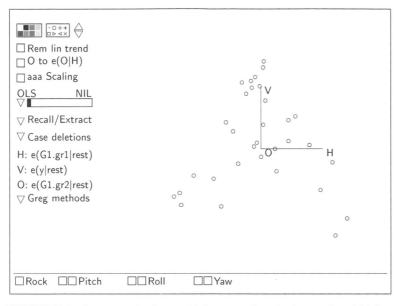

FIGURE 20.6 Summary plot for combining gr_1 and gr_2 in the reaction yield data.

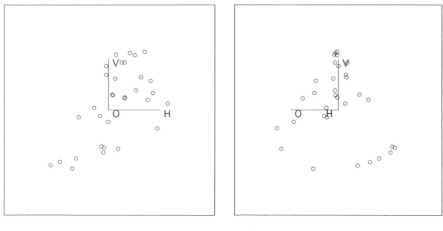

a. y versus *Fit* b. View 2

FIGURE 20.7 Two 2D views of the 3D plot y versus (Fit, gr_1) from the reaction yield data.

the interior of the range. Combining these two views, the general shape is of a cone cut off on one side.

Two additional graphical aids may help. First, add the full quadratic fit to the plot to give our eyes something to follow when trying to understand the

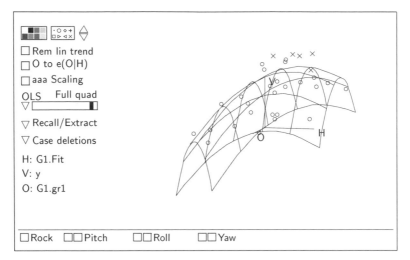

FIGURE 20.8 The 3D summary plot for the reaction yield data with a full quadratic polynomial added.

summary plot. A view of the summary plot with the quadratic fit added is shown in Figure 20.8. The maximum of the quadratic is near the edge of the experimental range. As a consequence, the point of maximum yield is very poorly estimated.

The second tool is to link Figure 20.8 to the scatterplot matrix of the predictors in Figure 20.5. Using plot linking, the four points where the predicted yield is highest are marked by $\times$, while the remaining points are marked by $\circ$. We see in the scatterplot matrix that for Lt_1 at least one of the high-yield points is at an extreme of the variable's range, another indicator that the optimum is not well determined.

There are several ways to proceed from here, depending on the specific goals of the investigators. Here are three possibilities:

- The summary plot itself might give sufficient information on how to achieve near-maximum yields by controlling the two linear combinations *Fit* and gr_1 to keep the yield near the observed peak.

- We might attempt to formulate a statistical model of the conical shape to refine the graphical results.

- Finally, we might reason that additional experimental runs are required because the maximum yield is not nearly as well-determined in Figure 20.7a as it is in Figure 20.7b. Additional data can be used to investigate the missing part of the cone in the summary plot, perhaps by expanding the range of Lt_1 and Lt_2.

20.4 VARIATIONS

20.4.1 Standardizing the Linear Predictors

Our graphical analysis of the mussels data in Section 20.2 ended with a summary plot of M versus the GREG predictor (20.2),

$$gr_4 = 0.31 \log H - 0.50 \log L + 0.38 \log S + 0.71 \log W$$

As in a linear regression model, the coefficients of the linear predictors depend on the units of those predictors. This dependence on the scale of measurement can make interpreting the coefficients difficult. Rescaling each of the linear predictors by dividing by their sample standard deviation can simplify interpretation, as in Section 10.1.6. One unit in the standardized scale is then one standard deviation in the original scale.

For the mussels data, reexpressing gr_4 as a linear combination of the standardized predictors gives

$$gr_4 = 0.15 \frac{\log H}{\text{sd}(\log H)} - 0.27 \frac{\log L}{\text{sd}(\log L)} + 0.81 \frac{\log S}{\text{sd}(\log S)} + 0.50 \frac{\log W}{\text{sd}(\log W)}$$

$$(20.3)$$

Coefficient ratios for the standardized predictors often give a better idea about the relative importance of the predictors than the coefficients in the original scale. For example, this representation indicates that the rate of change in the mean function $E(M \mid gr_4)$ produced by a one standard deviation change in $\log S$ is about $(0.81/0.15) = 5.4$ times as large as that produced by a one standard deviation change in $\log H$. This interpretation is subject to the cautions given in Section 10.1.6.

At any point in a graphical regression analysis, you can use the item "Display std. active predictors" in the graphical regression menu to display the GREG predictors in terms of the standardized linear predictors. This is how we obtained the coefficients shown in (20.3).

20.4.2 Improving Resolution in 3D Added-Variable Plots

As with the analysis of the 3D added-variable plot for (gr_3, gr_4) in the reaction yield data, the structural dimension of a 3D added-variable plot may not be certain. While we decided to take a parsimonious view and choose 0D structure, there were hints of dependence. To increase the power of the 3D added-variable plot to show structural dimension, the residuals on the vertical axis can be replaced with the residuals from the *full quadratic regression* of y on x_3, without changing the horizontal axes of the plot. This is accomplished in *Arc* by selecting the item "Quadratic in 'rest' residuals" from the "Greg methods" pop-up menu; select "e(y|rest)" to return to the standard 3D added-variable plot. The vertical axis of the plot can also be changed to the response

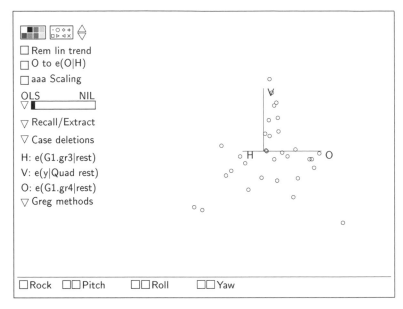

FIGURE 20.9   3D added-variable plot with the quadratic option for the residuals on the vertical axis.

y by selecting "Response" in the "Greg methods" pop-up menu. In theory, these 3D plots have the same structural dimension, and they have the same horizontal axes for a summary plot.

The residuals used on the vertical axis are intended to remove gross variation in the marginal mean function $E(y \mid \mathbf{x}_3)$, and so the choice "e(y | rest)" will be adequate as long as this mean function has a dominant linear trend. When $E(y \mid \mathbf{x}_3)$ is nonlinear without much linear trend, using the residuals from a full quadratic fit in $\mathbf{x}_3$ can result in considerable improvement. When $E(y \mid \mathbf{x}_3)$ is linear, using the full quadratic residuals results in overfitting and increased variability. Of course, we can try both options as needed.

Returning to the reaction yield data, Figure 20.9 shows one view of the initial 3D added-variable plot for (gr_3, gr_4) after (Fit, gr_1, gr_2) with the quadratic option for the vertical axis. It is now clear that this plot has at least 1D structure, confirming the hints from the previous analysis. Continuation of this analysis is given in Problem 20.7.

20.4.3   Model Checking

Graphical regression provides a means of model checking by using residuals from a fit of the target model in place of the response. Except for this change, all of the ideas discussed in this chapter apply. In particular, if the model is correct, then all 3D added-variable plots encountered during graphical regression will have 0D structure. Evidence of 1D or 2D structure in any plot is evidence that the model is not correct and may therefore require remedial action.

For example, suppose we have developed a linear model for our regression, say

$$y \mid \mathbf{x} = \eta^T \mathbf{u}(\mathbf{x}) + e \qquad (20.4)$$

where the $\mathbf{u}$ terms in the model are written explicitly as functions of the linearly related predictors $\mathbf{x}$. Letting $\hat{e}$ denote the residuals from an OLS fit of this model, we would then apply graphical regression with response $\hat{e}$ and predictors $\mathbf{x}$.

When using OLS residuals as the response in graphical regression, the GREG predictors may not include *Fit*. As discussed in Section 14.1.2, the residuals $\hat{e}$ are uncorrelated with the u-terms in the model and with *Fit* from (20.4), so *Fit* contains no useful information. *Arc* automatically excludes *Fit* from the GREG predictors; the other GREG predictors are still ordered on their likely importance in finding model deficiencies, with gr_1 being the most important.

In addition, *Arc* computes a diagnostic test for the structural dimension of the regression of the residuals on $\mathbf{x}$. The test is of NH: 0D structure versus AH: greater than 0D structure. If the null hypothesis is rejected then we have information that the model is deficient. The test supplements, but does not replace, the graphical procedure. No test is available when using the response y in graphical regression.

With these ideas we can use graphical regression to check model (19.5) for the mussels data. The first step is to fit model (19.5) using OLS with case 77 deleted. Next, we need to add the residuals from this fit to the data set so we can use them as the response in graphical regression. This is most easily done with the item "Add to data set" in the model menu for the fit of (19.5). Once the residuals are added to the data set we again start graphical regression using the Graph&Fit menu. In the resulting dialog, the linearly related predictors are selected in the usual way and the residuals are used as the response.

Figure 20.10 shows the plot for (gr_3, gr_4) after (gr_1, gr_2) in the mussels data. Aside from a few outlying points that may require attention, there is little evidence to suggest that its structural dimension is not zero. After selecting "Dimension 0" from the "Greg Methods" pop-up menu, the analysis should be continued by inspecting the plot of $\hat{e}$ versus (gr_1, gr_2); all this will lead to 0D structure for the residual regression, suggesting that the model matches the data quite well.

The diagnostic test for structural dimension is displayed in the text window when the graphical regression is created in *Arc*:

```
Test for 0D structure versus 1D or higher
value = 8.89109 df = 10 p-value = 0.54247
```

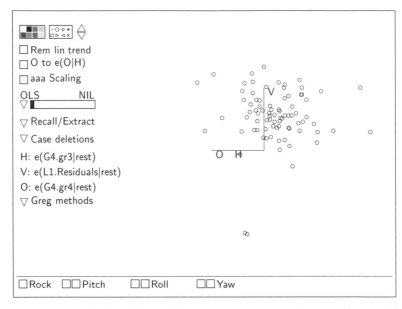

FIGURE 20.10   3D added-variable plot for (gr_3, gr_4) after (gr_1, gr_2) for checking model (19.5) for the mussels data.

The test statistic 8.89 is compared to a chi-squared distribution with 10 degrees of freedom to get the p-value near 0.54. The visual assessment of 0D structure in the residual regression agrees with the large p-value given by the test.

20.4.4   Using the Linearly Related Predictors

In the examples here, we have started graphical regression with the GREG predictors. We could use the linearly related predictors **x** instead simply by clicking the "Start with selected predictors" button on the bottom of the graphical regression dialog box. Using the original linearly related predictors **x** may be an advantage when application-specific information is available that can help combine variables. There are a few changes when using **x** instead of the GREG predictors.

   Unlike the GREG predictors, the linearly related predictors are not naturally ordered on likely importance. Although not required by the theory, ordering the predictors in a useful way can lead to useful answers more quickly. A useful idea is to view the predictors in 2D added-variable plots, using the "All 2D AVP of active predictors" item in the graphical regression menu. The two predictors with the strongest dependence in their 2D added-variable plots can be combined first. The process is repeated: The item "All 2D AVP of active predictors" is used to construct all 2D added-variable plots for the *remaining* predictors, and the predictors exhibiting the strongest dependence are again selected for combining in a 3D added-variable plot. After each dimension decision, *Arc* automatically gives a dialog for choosing the next two predictors to combine. 2D added variable plots can be used at this point by canceling the dialog and returning to the main GREG menu, effectively beginning again with the current active predictors.

When studying the regression of y on linearly related predictors, most of the 3D added-variable plots will probably exhibit 1D or 2D structure. A useful starting point for examining each of these plots is to rotate to the OLS linear combination, since by the 1D estimation result we know that this is a good summary if the plot has 1D structure.

Finally, there is a direct connection between graphical regression using the linearly related predictors and fitting via OLS: *If every 3D added-variable plot is judged to have 1D structure and the OLS view is taken as the summary then, apart from unimportant scale factors, the final 2D summary plot will be of y versus $\hat{y}$, the fitted values from the OLS regression of y on all predictors.*

20.5 COMPLEMENTS

20.5.1 GREG **Predictors and Principal Hessian Directions**

Apart from a constant, the first GREG predictor *Fit* is the fitted values from the OLS regression of y on the linearly related predictors **x**. If there is 1D structure and linearly related predictors, then the plot of y versus *Fit* may be a good summary of the regression.

The other GREG predictors $gr_1, \ldots, gr_{(p-1)}$ can be based either on the *method of principal Hessian directions* which was suggested by Li (1992), or on *sliced average variance estimation* (SAVE) which was suggested by Cook and Weisberg (1991) and discussed in Section 19.17.1. The method of principal Hessian directions is the default. In that case, gr_1 is formed from the part of the first principal Hessian direction (linear combination) that is orthogonal to the OLS direction. Similarly, the remaining GREG predictors are formed from part of the sequence of principal Hessian direction that are orthogonal to the OLS direction and the previous principal Hessian directions. Background on principal Hessian directions, including computing formulas, is available from Li (1992) or from Cook (1998a), who gave the justification for combining OLS with principal Hessian directions as used in this chapter. The GREG predictors based on SAVE are formed similarly. *Arc* can be instructed to use SAVE by using the settings menu.

Arc can be used to obtain the principal Hessian directions and a few related statistics by selecting "Inverse regression" from the Graph&Fit menu. In the resulting dialog, specify the response and the linearly related predictors in the usual way and click the option "pHd(OLS residuals)" on the dialog's right. The output in the text window contains a variety of information.

- First, a brief description of the regression is displayed, including any deleted cases.
- Second, a table is displayed giving the coefficients of the raw and standardized predictors used in computing the first two GREG predictors.
- Next, a table of test statistics is given for testing the structural dimension of the regression of the OLS residuals on the linear predictors. The kth row of the table is for testing the hypothesis NH: $(k - 1)$D structure versus AH: greater than $(k - 1)$D structure. The table is used by starting with the test in the first row and continuing until the p-value is no longer significant. If the p-value for the kth row is significantly small, while the p-value for the $(k + 1)$st row is not, then the structural dimension is inferred to be k. Equivalently, the structural dimension is just the number of significant p-values. The two different columns of p-values correspond to two different tests. Since they are based on different assumptions, the two tests need not agree.

More information on this type of analysis is available on the Internet site for this book.

20.5.2   References

Graphical regression was developed by Cook (1994) and by Cook and Wetzel (1993). A comprehensive treatment including theoretical development was given by Cook (1998a,b), who discussed the possibility of using principal Hessian directions to form the GREG predictors. The reaction yield data are from Box and Draper (1987).

PROBLEMS

20.1   What is the goal of graphical regression, and how might it help in a regression analysis?

20.2   Give two reasons why using GREG predictors instead of the linearly related predictors may facilitate a graphical regression.

20.3   We left the analysis of the Australian Athlete Data in Section 19.6 with the finding that the structural dimension is greater than 1.

    20.3.1   Conduct a graphical analysis of the same regression using the GREG predictors. Do you again reach the conclusion that the structural dimension is greater than 1?

    20.3.2   Assuming the regression has 2D structure, describe the final summary plot. Does the plot suggest that there might be two different regressions involved? Now mark the points in your summary plot according to *Sex*. What do you conclude?

20.4 Use the mussels data, starting as in Section 20.2.1 with predictors in log scale, and with case 77 deleted.

    20.4.1 Use the graphical regression methodology, but unlike the text start by combining *Fit* and gr_1, then adding gr_2 and finally gr_3. Does this lead to essentially the solution obtained as the one in the text?

    20.4.2 Compute the OLS regression of M on the predictors in log scale, and then draw the plot of the fitted values from this regression versus the final linear combination you obtained in Problem 20.4.1. Summarize the information in this plot. Should you have anticipated the outcome, given the 1D estimation result discussed in the last chapter?

    20.4.3 Repeat Problem 20.4.1, with two exceptions. Use the original linearly related predictors, not the GREG predictors, and in each 3D plot always use the OLS view as a summary plot. Now repeat Problem 20.4.2, and summarize the information in the plot.

20.5 This problem concerns the regression of $\log M$ on

$$\mathbf{x} = (\log H, \log L, \log S, \log W)^T$$

in the full mussels data after restoring case 77.

    20.5.1 Suppose that the following model holds:

$$\log M \mid \mathbf{x} = \eta_0 + \eta^T \mathbf{x} + e$$

      a. Describe how you expect the process of graphical regression to unfold using the GREG predictors and starting with a 3D added-variable plot for (gr_2, gr_3).

      b. Describe how you expect the process of graphical regression to unfold using the linearly related predictors $\mathbf{x}$ and starting with a 3D added-variable plot for $(\log S, \log W)$.

      c. Describe how you expect the process of graphical regression to unfold using $\hat{e}(\log M \mid \mathbf{x})$ as the response and the GREG predictors.

      d. Describe how you expect the process of graphical regression to unfold using $\hat{e}(\log M \mid \mathbf{x})$ as the response and the linearly related predictors $\mathbf{x}$.

    20.5.2 Conduct a graphical regression of $\log M$ on $\mathbf{x}$ using the GREG predictors.

Outliers generally stand out in at least one of the 3D added-variable plots during graphical regression. One way to deal with outliers is to delete them as they are encountered, and then begin graphical regression again based on the reduced data. The outlier should be restored in the final summary plot so they can be assessed against the reduced data. These ideas may be helpful in this and the next two problems.

20.5.3   Conduct a graphical regression of $\hat{e}(\log M \mid \mathbf{x})$ on $\mathbf{x}$ using the GREG predictors.

20.5.4   Conduct a graphical regression of $\log M$ on $\mathbf{x}$ using the predictors $\mathbf{x}$.

20.6   Repeat the analysis of the mussels data given in the text leading to the summary plot in Figure 20.4. Extract the 2D summary view and estimate its mean function using a *lowess* smooth. Next, extract the estimated mean function, naming it h. Now form the residuals $\hat{e} = M - h$, and carry out a graphical regression of $\hat{e}$ on $\mathbf{x}$ using the GREG predictors. Write a brief summary of your findings.

This is one way to check a summary plot without transforming to a linear model. If a 1D model holds with constant variance, say $E(M \mid \mathbf{x}) = M(\beta^T\mathbf{x})$ and $\text{Var}(y \mid \mathbf{x}) = \sigma^2$, then we would expect the regression of $\hat{e}$ on $\mathbf{x}$ to have 0D structure. Otherwise we expect it to have structural dimension greater than 0. For example, if the 1D model has nonconstant variance $\text{Var}(y \mid \mathbf{x}) = V(\beta^T\mathbf{x})$, then the regression of $\hat{e}$ on $\mathbf{x}$ will have 1D structure.

20.7   Repeat the graphical regression analysis of the data set on reaction yield, but this time change the vertical axis of the 3D added-variable plots to the residuals from the quadratic regression, as described in Section 20.4.2. Do you still find that the regression has 2D structure? Write a brief summary of your findings, including a description of your final summary plot.

20.8   Conduct a graphical regression for the Big Mac data in file `big-mac.lsp`, using *BigMac* as the response and the four predictors *Service*, *TeachSal*, *EngSal*, and *EngTax*. Give a brief summary of your analysis along with a description of your final summary plot.

PART IV

Logistic Regression And Generalized Linear Models

In this part of the book we discuss more fitting methods and models that might be used in regression. We first discuss logistic regression models, in which the response is a count of the number of successes in a known number of trials, giving the use and interpretation of this model in Chapter 21 and then discussing graphical methods and diagnostics in Chapter 22. Logistic regression is an important special case of *generalized linear models*, and these are presented in Chapter 23. These models are based on additional information about the relationship between the mean function and the variance function, but they all assume that the dependence of y on $\mathbf{x}$ is through a linear combination $\eta^T \mathbf{u}$, where $\mathbf{u}$ is a set of terms derived from $\mathbf{x}$.

CHAPTER 21

Binomial Regression

In most of the regressions we have considered, the response has been continuous, taking any value in a particular interval. So, for example, *Vol* in the haystacks data could take any value between around 1,200 and 6,230 cubic feet. In this chapter we consider regressions where the response can take only a few values, in particular when the response is a count of the number of successes observed in a known number of trials. Often, each response is binary, either success or failure of a single trial. We present the standard logistic regression model and make connections to earlier chapters on the multiple linear regression models. The next chapter covers graphical methods for binary response data.

21.1 RECUMBENT COWS

Either just before or just after calving, some dairy cows become unable to support their own weight, and they become recumbent—they lie down. Some cows with this condition will recover, but many will not, and it is of interest to understand how survival probability varies with characteristics of the cow. The data in the file downer.lsp are from a study of 435 recumbent cows, collected at the Ruakura Animal Health Laboratory, New Zealand, during 1983–1984. A variety of blood tests and physical measurements were taken shortly after the condition was diagnosed, and the eventual outcome of survival or death was later determined. The goal of the study is to determine if any of the measurements are related to survival probabilities for the cows, as understanding these relationships may lead to a treatment for this condition. The variables we consider are defined in Table 21.1.

The analysis of these data is likely to be complicated because many records are incomplete. The inflammation variable, for example, was measured only during the second year of the study and is available for only 136 of the 435 cows; apart from the response, none of the other variables is observed for every cow. Recall the strategy that *Arc* uses for missing values: Each compu-

TABLE 21.1 **Data on Recumbent Cow Survival, in the File** `downer.lsp`

Name	n	Description of Variable
AST	429	Serum asparate amino transferase, IU/l at 30°C.
Calving	431	0 if condition first occured before calving, 1 if post-calving.
CK	413	Serum creatine phosphokinase, IU/l at 30°C.
Daysrec	432	Days recumbent when measurements were taken, rounded down to the nearest day.
Inflamat	136	Inflammation: 1 if present, 0 if absent.
Myopathy	222	Muscle disorder: 1 if present, 0 if absent.
Outcome	435	Outcome: 1 if survived, 0 if died or killed.
PCV	175	Packed cell volume (hematocrit), %.
Urea	266	Serum urea, mmol/l.

tation uses as many fully observed cases as possible, so, for example, a regression including the inflammation variable can be based on no more than 136 cases.

The second important feature of this regression is that the response is categorical: The cow either survived or it died. By convention, we code the outcome as an indicator variable, with value 1 for survival and 0 for death. The association of values with categories is arbitrary; for example, actuaries often code a death as 1 and a survival as 0.

The overall goal of the analysis is a familiar one: We want to understand how the probability of survival changes as the predictors are varied, so we are interested in the conditional distribution of *Outcome* given the predictors.

21.1.1 Categorical Predictors

We start by looking at a few summary statistics. For example, what fraction of the cows in the study survived? The fraction surviving is the number of cows with *Outcome* = 1 divided by the total number. This is the sample mean of *Outcome*, which can be viewed using the "Display summaries" item in the data set menu. The relevant output is given in the first part of Table 21.2. The fraction of cows that survived is about 0.38. We can think of this number as follows: If we select a cow at random from this sample, the probability that it survived is about 0.38. If the 435 cows were a random sample of all recumbent cows, then the survival probability for a cow chosen at random from the general population of cows with this condition is estimated to be about 0.38. These data might be a random sample of cows with this condition only in the area of New Zealand served by the Ruakura Animal Health Laboratory. We have no direct information as to whether the 38% survival can be applied to all of New Zealand or to other countries.

Is the survival probability the same for cows with the muscle disorder *Myopathy* as for cows without *Myopathy*? This question concerns the condi-

TABLE 21.2 Marginal Survival, and Survival as a Function of *Myopathy*

```
Data set = Downer, Summary Statistics
373 cases are missing at least one value.
Variable    N    Average   Std. Dev   Minimum   Median   Maximum
Outcome    435   0.38161   0.48634       0         0        1.
```

```
Data set = Downer, Table of included cases
373 cases are missing at least one value.
Table columns are levels of Myopathy

Table of Outcome: Count
Column variable: Myopathy
    0     1
|  ———————
| 127    95

Table of Outcome: Mean
Column variable: Myopathy
         0               1
| ————————————————————————
| .38582677      .06315789
```

tional distributions of *Outcome* given *Myopathy*. We can summarize these distributions by the probability of survival given the value of *Myopathy*, and we estimate these survival probabilities using the observed fraction that survive. Select "Table data" from the data set menu, make *Myopathy* a conditioning variable, make *Outcome* a variate, and get the mean of *Outcome* given *Myopathy*. The result is shown in Table 21.2. Because of missing data, we see that these fractions are based on relatively few observations— 127 with *Myopathy* = 0 and 95 with *Myopathy* = 1. However, the survival fractions, about 0.39 with no myopathy and only 0.06 with myopathy, are quite different. Relatively few cows in the sample with myopathy survive. An important question, which we will address shortly, is whether or not we can believe that this observed difference in the survival rates is due to a real difference in population survival rates or simply due to chance variation.

We can study the conditional distributions of the response given several categorical predictors by computing the survival fraction given the combination of the categorical predictors. In these data, for example, we could use the predictors *Calving*, *Inflamat*, and *Myopathy* as conditioning variables, giving eight possible combinations of conditions (since each variable has two levels and $2^3 = 8$) and with eight sample proportions surviving. The counts and proportions for each combination are shown in Table 21.3. We see that the sample sizes are now greatly reduced—only seven cows that became recumbent before calving were measured on the three predictors.

TABLE 21.3   Survival Fraction as a Function of
Calving, *Inflamat*, **and** *Myopathy*

```
Table of Outcome: Count
Column variable: Calving
         0      1
 _|_____

Myopathy = 0, Inflamat:
0|      2      9
1|      3     28
Myopathy = 1, Inflamat:
0|      1      3
1|      1     18

Table of Outcome: Mean
Column variable: Calving
         0      1
 _|_____

Myopathy = 0, Inflamat:
0|            0.5   .55555556
1|            0.0   .32142857
Myopathy = 1, Inflamat:
0|            0.0          0.0
1|            0.0   .05555556
```

21.1.2   Continuous Predictors

Computing observed proportions of survivors works well as a method of summarizing the data as long as sample sizes per group remain large, and as long as none of the predictors are continuous. With a continuous predictor like *CK*, using tables to present survival fractions will require turning *CK* into a categorical variable by slicing. While this can be an effective method of summarizing data in some problems, it can lose important information if the categories are too coarse, or it may not be effective if the categories are too fine. As an alternative we can adapt regression methodology developed in the earlier parts of this book.

Let's begin with a graph of the response $y = $ *Outcome* versus the predictor $\log_2(CK)$, as shown in Figure 21.1 using base-two logarithms. We used $\log_2(CK)$ instead of *CK* because the sample distributions of $CK \mid Outcome$ are quite skewed, the ratio of the largest to the smallest value of *CK* within each outcome exceeding 1,000. Consequently, using logarithms will likely be a better place to start the analysis. The mean function $\mathrm{E}(y \mid \log_2(CK))$ can be estimated to be the fraction of ones in a narrow vertical slice of this plot constructed by using the "slicing" mouse mode described in Section 2.3.2. As the slice is moved across the plot from left to right, the fraction of points with *Outcome* = 1 can be seen to decrease, suggesting that the probability of survival decreases as $\log_2(CK)$ increases.

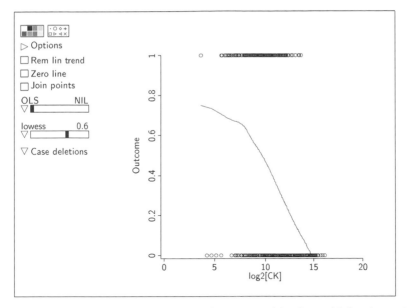

FIGURE 21.1 *Outcome* versus $\log_2(CK)$. A *lowess* smooth has been added to the graph.

A smoother may be necessary to visualize the mean function when the response is binary. Shown on Figure 21.1 is a *lowess* smooth with smoothing parameter 0.6. The smoother indicates that the mean function has maximum value of about 0.75 at the smallest values of $\log_2(CK)$, and the survival probability decreases as $\log_2(CK)$ increases, with a faster decrease for $\log_2(CK)$ exceeding about eight. Consequently, $\log_2(CK)$ alone is an indicator of survival, with large values of $\log_2(CK)$ suggesting that survival is much less likely.

Binomial regression is regression in which the response is a binary variable or the sum of binary variables. It permits the use of both continuous predictors and factors. Before turning to methods and models for binomial regression, we review briefly the Bernoulli distribution and its generalization to the binomial distribution. These provide the random element in considering binomial responses, much as the normal distribution provides the random element in many regressions with a continuous response.

21.2 PROBABILITY MODELS FOR COUNTED DATA

21.2.1 The Bernoulli Distribution

Suppose we observe a binary random variable. Examples of binary variables include whether or not a student passes a test, whether or not a patient given a treatment survives, whether or not an athlete wins a race, whether or not a project is completed on time, whether or not a manufactured part chosen at random from production meets specifications, and so on. Such a variable has a Bernoulli distribution.

Let y have a Bernoulli distribution. We use the symbol $\theta = \Pr(y = 1)$ to be the probability that $y = 1$. $\Pr(y = 0) = 1 - \theta$ is the probability that $y = 0$. We can easily compute the mean of y,

$$E(y) = (0 \times \Pr(y = 0)) + (1 \times \Pr(y = 1))$$
$$= 0 + \theta = \theta$$

so the mean is the probability θ that y is equal to 1. Although y can equal only 0 or 1, its mean value is between 0 and 1. A similar calculation gives the variance:

$$Var(y) = E(y - E(y))^2$$
$$= (0 - \theta)^2 \Pr(y = 0) + (1 - \theta)^2 \Pr(y = 1)$$
$$= \theta^2(1 - \theta) + (1 - \theta)^2\theta$$
$$= \theta(1 - \theta)(\theta + 1 - \theta)$$
$$= \theta(1 - \theta)$$

With normally distributed data the mean and variance are two separate parameters, but the variance of a Bernoulli random variable is determined by the mean θ. The variance attains its maximum value of $1/4$ when $\theta = 1/2$, and it is close to its minimum value of 0 when θ is close to either 0 or 1.

If θ were known, then we would know everything there is to know about the distribution of y. When θ is unknown, we can use data to estimate it. With only a single observation equal to either zero or one, we cannot expect to do a very good job of estimating θ, but if we can take many observations we might be able to get a good estimate of θ. This leads to the *binomial distribution*.

21.2.2   Binomial Random Variables

Suppose, as in the recumbent cows example, we have a sample of m cows, and let y be the number of cows out of m that survive. When y is the number of successes in m *independent* trials, each with the same probability of success θ, we say that y has a *binomial distribution* and write $y \sim \text{Bin}(m, \theta)$. If $m = 1$, then y has a Bernoulli distribution, and so the Bernoulli is the simplest case of the binomial.

The value of a binomial variable y can be any integer between zero and the number m of trials. The probability that y equals a specific integer $j = 0, 1, \ldots, m$, is given by

$$\Pr(y = j) = \binom{m}{j} \theta^j (1 - \theta)^{(m-j)} \tag{21.1}$$

Here $\binom{m}{j}$ is the number of different orderings of j successes in m trials, computed from the formula

$$\binom{m}{j} = \frac{m!}{j!(m-j)!}$$

where $m!$ is read as m factorial and is equal to the product of all the integers from 1 up to and including m (and $0! = 1$). We will not have much need for computing these values. Equation (21.1) is called the *probability mass function* for the binomial.

The probabilities (21.1) can be computed using the function `binomial-pmf`. For example, to get the probability of 5 successes in 9 trials when $\theta = 0.5$, use

```
>(binomial-pmf 5 9 .5)
0.246094
```

The function `binomial-cdf` can be used to compute the probability that y is less than or equal to some value; for example, to compute the probability that $y \leq 5$ when $m = 9$ and $\theta = 0.5$, use

```
>(binomial-cdf 5 9 .5)
0.746094
```

The mean and variance of a binomial random variable are easy to compute because a binomial is a sum of m independent Bernoulli random variables, each with the same value of θ. Using the fact that the expected value of a sum of random variables is the sum of the expectations and equation (4.7), we obtain

$$E(y) = m\theta \tag{21.2}$$

$$\text{Var}(y) = m\theta(1-\theta) \tag{21.3}$$

Both the mean and variance depend on θ and m.

If $y_1 \sim \text{Bin}(m_1, \theta_1)$ and $y_2 \sim \text{Bin}(m_2, \theta_2)$, then the sum $y_1 + y_2$ is binomially distributed only if $\theta_1 = \theta_2$; we can then write $y_1 + y_2 \sim \text{Bin}(m_1 + m_2, \theta_1)$.

Data can be used to estimate θ. Our estimation method of choice is called *maximum likelihood estimation*, and the resulting estimate is called the maximum likelihood estimate, which is usually abbreviated as MLE. The basic idea is simple: Suppose we have observed y successes in m trials. The MLE of θ is then the value of θ that makes the probability of observing y successes in m trials as large as possible. This amounts to rewriting (21.1) as a function of θ with y held fixed at its observed value,

$$L(\theta) = \binom{m}{y} \theta^y (1-\theta)^{(m-y)} \tag{21.4}$$

$L(\theta)$ is called the *likelihood function* for θ. The MLE $\hat{\theta}$ is the value of θ that maximizes $L(\theta)$. Since the same value maximizes both $L(\theta)$ and $\log(L(\theta))$, we work with the more convenient log-likelihood, given by

$$\log(L(\theta)) = \log \binom{m}{y} + y \log(\theta) + (m - y)\log(1 - \theta) \qquad (21.5)$$

Maximization of (21.5) is an elementary calculus problem. Differentiating (21.5) with respect to θ and setting the result to zero gives

$$\frac{d \log(L(\theta))}{d\theta} = \frac{y}{\theta} - \frac{m - y}{1 - \theta} = 0$$

Solving for θ gives the MLE,

$$\hat{\theta} = \frac{y}{m} = \frac{\text{Observed number of successes}}{\text{Observed fixed number of trials}}$$

which is the observed fraction of successes.

Computing the variance of the MLE is easy, since

$$\text{Var}(\hat{\theta}) = \text{Var}\left(\frac{y}{m}\right) = \frac{1}{m^2}\text{Var}(y) = \frac{\theta(1 - \theta)}{m} \qquad (21.6)$$

This variance is estimated by substituting $\hat{\theta}$ for θ.

21.2.3 Inference

In the recumbent cow data, we have $m = 435$ cows, of which $y = 166$ survived. If all cows have the same probability of survival θ, and survival is independent between cows, then the number who survive will be binomially distributed. The maximum likelihood estimate of θ is $\hat{\theta} = 166/435 = 0.38$. The standard error of $\hat{\theta}$ is obtained by substituting $\hat{\theta}$ for θ in (21.6) and then taking square roots, $(\hat{\theta}(1 - \hat{\theta})/m)^{1/2} = (0.38(0.62)/435)^{1/2} = 0.02$.

As long as θ is not too close to 0 or 1, the normal distribution can be used as a basis for tests and confidence statements concerning θ. In large samples

$$\hat{\theta} \sim N(\theta, \theta(1 - \theta)/m) \qquad (21.7)$$

The large sample variance is estimated by substituting $\hat{\theta}$ for θ. For example, an approximate 95% confidence interval for θ is the set of all points

$$\hat{\theta} - Q(z, 0.975)\sqrt{\hat{\theta}(1 - \hat{\theta})/m} \le \theta \le \hat{\theta} + Q(z, 0.975)\sqrt{\hat{\theta}(1 - \hat{\theta})/m}$$

$$0.38 - 1.96(0.02) \le \theta \le 0.38 + 1.96(0.02)$$

$$0.34 \le \theta \le 0.42$$

Suppose that the probability of survival depends on whether or not the cow has myopathy, and let $\Pr(y = 1 \mid Myopathy = j) = \theta_j$, for $j = 0, 1$. We can use large-sample normality to test the hypothesis that $\theta_0 = \theta_1$, assuming that cows within a myopathy group have the same probability of survival and are independent of each other. The relevant data are given in Table 21.2, where the MLE for θ_0 is 0.39 based on $m_0 = 127$ trials, and the MLE for θ_1 is 0.06, based on $m_1 = 95$ trials. Each of these estimates is approximately normally distributed and they are independent, and so their difference is also approximately normal. The statistic

$$
z = \frac{\hat{\theta}_0 - \hat{\theta}_1}{\sqrt{\hat{\theta}_0(1 - \hat{\theta}_0)/m_0 + \hat{\theta}_1(1 - \hat{\theta}_1)/m_1}}
$$

$$
= \frac{0.38 - 0.06}{\sqrt{0.38(0.62)/127 + 0.06(0.94)/95}} = 6.47
$$

can be compared to the standard normal distribution to get a p-value. Whether the alternative hypothesis is two-sided, $\theta_0 \neq \theta_1$, or one sided, $\theta_0 > \theta_1$, the resulting p-value is very small, suggesting that the rate of survival is not the same for the two levels of myopathy and that cows with myopathy absent have a higher rate of survival.

21.3 BINOMIAL REGRESSION

The binomial distribution for a fixed number of trials is determined by the probability θ of success. Both the mean and the variance depend only on θ and the known number m of trials.

Suppose that we have n binomial random variables y_i, $i = 1, \ldots, n$. For each y_i we know the number of trials m_i, and in addition there is an associated vector of p predictors $\mathbf{x}_i$. We suppose further that the probability of success $\theta(\mathbf{x}_i)$ depends on $\mathbf{x}_i$. We can write this compactly as $y_i \mid \mathbf{x}_i \sim \text{Bin}(m_i, \theta(\mathbf{x}_i))$, $i = 1, \ldots, n$. It is convenient to study the random variable y_i/m_i, which is the observed fraction of successes at each i, because the range of y_i/m_i is always between 0 and 1, whereas the range of y_i is between 0 and m_i and can be different for each i. Using (21.2) and (21.3), the mean and variance functions are

$$
\text{E}(y_i/m_i \mid \mathbf{x}_i) = \theta(\mathbf{x}_i) \tag{21.8}
$$

$$
\text{Var}(y_i/m_i \mid \mathbf{x}_i) = \theta(\mathbf{x}_i)(1 - \theta(\mathbf{x}_i))/m_i \tag{21.9}
$$

The value of $\theta(\mathbf{x}_i)$ determines both the mean function and the variance function, so we need to estimate $\theta(\mathbf{x}_i)$. If the m_i are all large, we could simply estimate $\theta(\mathbf{x}_i)$ by y_i/m_i, the observed proportion of successes at $\mathbf{x}_i$. In many applications the m_i are small—often $m_i = 1$ for all i—so this method won't always work.

21.3.1 Mean Functions for Binomial Regression

As with the linear regression models in Section 7.2, we assume that we can find a set of k u-terms such that y_i depends on x_i only through a linear combination of the u-terms. Since the u-terms may be any functions of the elements of x_i, we are allowing for regressions of dimension one or higher. Given the u-terms u_i, we can write the mean function as

$$\theta(x_i) = M(\eta^T u_i)$$

$\theta(x_i)$ depends on two quantities: (1) the vector of regression parameters η that will be estimated from the data and (2) the kernel mean function M, which we will select according to the regression at hand. The kernel mean function for binomial regression must be bounded between 0 and 1, and for generality it should cover the whole range from 0 to 1. The most frequently used kernel mean function for binomial regression is the *logistic function*. The logistic regression model specifies that $M(\eta^T u_i)$ is given by

$$\theta(x_i) = M(\eta^T u_i) = \frac{\exp(\eta^T u_i)}{1 + \exp(\eta^T u_i)} = \frac{1}{1 + \exp(-\eta^T u_i)} \qquad (21.10)$$

A graph of this equation is shown in Figure 21.2. The logistic mean function is always between 0 and 1, and has no additional parameters. For the data in Figure 21.1, we have added the logistic function in Figure 21.3. This is done by using the parametric slidebar's pop-up menu, selecting "Logistic," and then moving the slidebar to 1. Here, the number of trials, m_i, is equal to 1 for all i; if the number of trials is greater than 1, the logistic fit can be obtained by plotting the observed fraction of successes against a term and then specifying the number of trials as weights in the "Plot of" dialog from the Graph&Fit menu. The logistic function shown in this figure is obtained from the one in Figure 21.2 by reflection, so the coefficient for $\log_2(CK)$ must be negative. Also, the range of probabilities in the figure is smaller than the maximum range from zero to one, so only a portion of the logistic function is needed. The smooth and the logistic fit don't seem to match very well for very large and small values of $\log_2(CK)$, but over most of the range the two curves are similar.

21.3.2 Summary

The logistic regression model has three components, as follows:

The Data. The data consist of n independent observations. The ith observation is the number of successes y_i observed in a known number m_i of trials.

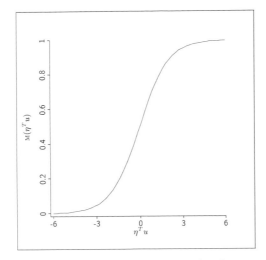

FIGURE 21.2 The logistic mean function.

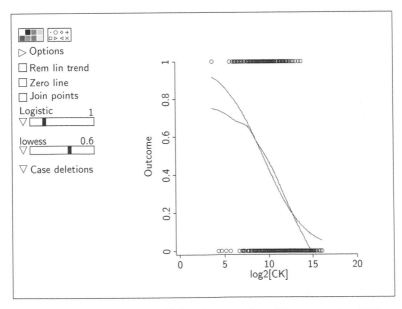

FIGURE 21.3 Logistic fit added to the regression of *Outcome* on $\log_2(CK)$.

In many regressions, $m_i = 1$ for all i, but in general the m_i can be any positive integers. Associated with the ith observation are the values of p predictors in $\mathbf{x}_i$, and from these we compute k terms in $\mathbf{u}_i$.

  The Random Part. For observation i, we assume that the m_i trials are independent, each with the same probability of success. Given these two assumptions, y_i has a binomial distribution. Consequently, both the mean and variance of y_i depend only on m_i and the probability $\theta(\mathbf{x}_i)$ that a trial is a success.

The Fixed Part. We need to specify $\theta(\mathbf{x}_i)$. First, assume that $\theta(\mathbf{x}_i) = \text{M}(\eta^T\mathbf{u}_i)$ for some vector of parameters η and kernel mean function M. As with the multiple linear regression model, $\eta^T\mathbf{u}_i$ may include functions of $\mathbf{x}_i$. Next, we specify the kernel mean function, and for this we use the logistic function (21.10),

$$\text{M}(\eta^T\mathbf{u}_i) = \frac{\exp(\eta^T\mathbf{u}_i)}{1 + \exp(\eta^T\mathbf{u}_i)}$$

In the multiple linear regression model, the kernel mean function used is the identity function, $\text{M}(\eta^T\mathbf{u}_i) = \eta^T\mathbf{u}_i$.

21.4 FITTING LOGISTIC REGRESSION

The maximum likelihood estimates for η in logistic regression can be computed using *Arc*. The computational method is outlined in Section 21.6.4; for now we go directly to estimates and their general properties. As we discuss the methodology, we will draw parallels with the multiple linear regression model.

Consider first a single predictor $x = CK$, and fit with two u-terms $u_0 = 1$ for the intercept and $u_1 = \log_2(CK)$.

From the Graph&Fit menu, select the item "Fit binomial response." This gives a dialog like the one shown in Figure 21.4. This dialog differs slightly from the dialog for the multiple linear regression model. First, the name of the model will start with the letter "B" for binomial rather than "L" for linear; of course you can choose any name you want. Second, the choice of kernel mean functions is different. The default is the logistic function discussed so far; the other two choices are discussed in the complements. For the example, the response variable is *Outcome*. You must also specify an additional quantity called *Trials*. For a Bernoulli response, the number of trials is always 1; and since this occurs so often, an additional variable called *Ones* is provided by the program. If the number of trials is not always 1, you will need to have a variable in the data set that gives the number of trials. The resulting output is shown in Table 21.4.

The computational method maximizes the likelihood function iteratively, and a history of the iterations is displayed in Table 21.4. We are reminded of the name of the data set, the missing values, the response, and the terms. The kernel mean function is the logistic, and the variable *Ones* specifies the number of trials. The coefficient estimates are $\hat{\eta}_0 = 4.00065$ and $\hat{\eta}_1 = -0.424022$. The estimate of $\eta^T\mathbf{u}$ is approximately

$$\hat{\eta}^T\mathbf{u} = 4.00 - 0.42\log_2(CK)$$

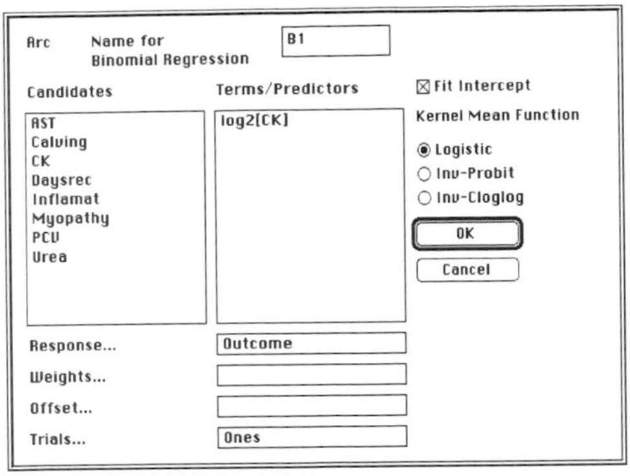

FIGURE 21.4   The binomial regression dialog.

TABLE 21.4   Logistic Regression of *Outcome* **on** $\log_2(CK)$ **for the Recumbent Cows Data**

```
Iteration 1: deviance = 477.902
Iteration 2: deviance = 475.193
Iteration 3: deviance = 475.175

Data set = Downer, Name of Fit = B1
372 cases are missing at least one value.
Binomial Regression
Kernel mean function = Logistic
Response    = Outcome
Terms       = (log2[CK])
Trials      = Ones
Coefficient Estimates
Label           Estimate    Std. Error    Est/SE
Constant         4.00065     0.580883      6.887
log2[CK]        -0.424022    0.0549681    -7.714

Scale factor:                    1.
Number of cases:               435
Number of cases used:          413
Degrees of freedom:            411
Pearson X2:                402.960
Deviance:                  475.175
```

From this, we can compute both the fitted probabilities $\hat{\theta}(\mathbf{x}_i) = \mathrm{M}(\hat{\eta}^T\mathbf{u}_i)$, which are called :fit-prob in *Arc*, and the fitted values $\hat{y}_i = m_i\hat{\theta}(\mathbf{x}_i)$, called :fit-values. The logistic curve drawn on Figure 21.3 is of $\mathrm{M}(\hat{\eta}^T\mathbf{u})$ versus $\log_2(CK)$, with the points joined to give the smooth curve. The standard errors of the coefficient estimates are given in the column marked Std. Error, and the column marked Est/SE gives the ratios of the estimates to their standard errors.

Since the variance of a binomial is determined by the mean, there is not a variance parameter that can be estimated separately. The output reports a Scale factor that is of use in more complicated settings, but is not relevant here. The number of cases is the number of rows in the data set. The number of cases used is computed by taking the number of cases and subtracting (a) the number of cases missing at least one value on the response or predictors in the model and (b) the number of cases deleted using plot controls. This is followed by the degrees of freedom, which is the number of cases used minus the number of η-coefficients estimated.

The remaining two values are goodness-of-fit statistics called Pearson X2 and Deviance. These are discussed in Section 21.4.3.

21.4.1   Understanding Coefficients

As with multiple linear regression, the estimated covariance matrix of the coefficient estimates can be obtained using the "Display variances" item in the model's menu. In multiple linear regression with normal errors the estimates are exactly normally distributed, but for logistic regression the estimates are only approximately normal, with the approximation improving as the sample size increases. The ratio of estimates to standard errors in Table 21.4 can be used to provide tests of the hypothesis that the corresponding coefficients are equal to zero, and the normal distribution can be used to get approximate p-values. As usual, a test based on an estimate divided by its standard error is called a Wald test.

Coefficients have a useful interpretation in logistic regression. We start with equation (21.10) for the logistic function,

$$\theta(\mathbf{x}) = \frac{\exp(\eta^T \mathbf{u})}{1 + \exp(\eta^T \mathbf{u})}$$

and solve this equation for $\eta^T \mathbf{u}$. We find

$$\log\left(\frac{\theta(\mathbf{x})}{1 - \theta(\mathbf{x})}\right) = \eta^T \mathbf{u} \qquad (21.11)$$

The left side of (21.11) is a function of $\theta(\mathbf{x})$ and the right side is $\eta^T \mathbf{u}$. For every kernel mean function used in practice, we can always obtain a function of θ that is equal to $\eta^T \mathbf{u}$ and is called a *link function*. The link function $\log(\theta(\mathbf{x})/(1 - \theta(\mathbf{x})))$ for logistic regression is called a *logit*.

The ratio $\theta(\mathbf{x})/(1 - \theta(\mathbf{x}))$ is the *odds of success*. For example, if the probability of success is 0.25, the odds of success are $0.25/(1 - 0.25) = 1/3$, one success to each three failures. If the probability of success is 0.8, then the odds of success are $0.8/0.2 = 4$, or four successes to one failure. Whereas probabil-

TABLE 21.5 Logistic Regression Summary with Response *Outcome* **and the Single Binary Predictor** *Myopathy*

```
Response          = Outcome
Terms             = (Myopathy)
Trials            = Ones
Coefficient Estimates
Label          Estimate     Std. Error     Est/SE
Constant       -0.464889    0.182287       -2.550
Myopathy       -2.23199     0.459425       -4.858

Scale factor:                   1.
Number of cases:              435
Number of cases used:         222
Degrees of freedom:           220
Pearson X2:               221.968
Deviance:                 214.137
```

ities are bounded between 0 and 1, odds can be any nonnegative number. The logit is the logarithm of the odds; natural logs are generally used in defining the logit. According to equation (21.11), the logit is equal to a linear combination of the terms.

It is (21.11) that provides the basis for understanding coefficients in logistic regression. For example, the estimate of the coefficient for $\log_2(CK)$ in Table 21.4 is -0.424022. This means that if $\log_2(CK)$ were increased by one unit (since this is a base-two logarithm, increasing $\log_2(CK)$ by one unit means that the value of *CK doubles*), then the (natural) logarithm of the odds will decrease by 0.424022 and the odds will be *multiplied* by $\exp(-0.424022) = 0.65$. For example, a cow with $CK = 1000$ has odds of survival that are 0.65 times the odds of survival of a cow with $CK = 500$. In any logistic regression, if $\hat{\eta}_j$ is an estimated regression coefficient, and assuming that the corresponding term u_j can be changed without affecting the other terms, then increasing u_j by one unit will multiply the estimated odds of success by $\exp(\hat{\eta}_j)$.

As a second example, suppose we fit a logistic regression model with response *Outcome* and the single term *Myopathy*. The fit is shown in Table 21.5. The effect of observing *Myopathy* = 1 is to multiply the odds of survival by $\exp(-2.23199) = 0.107$. The odds of survival when *Myopathy* = 0 is estimated to be $\exp(-0.464889) = 0.628$, while the odds of survival when *Myopathy* = 1 is $0.628(0.107) = 0.067$. When the predictor is an indicator variable, it is common to summarize the result by looking at the *odds ratio*, $0.628/0.067 = 9.32$: The odds of survival are estimated to be over nine times higher when *Myopathy* is not present.

There is a close connection between (a) the logistic regression model with *Myopathy* as the single predictor and (b) the analysis of the conditional distribution of *Outcome* given *Myopathy* presented in Table 21.2 and Section 21.2.3. We have previously seen that the fractions surviving with *Myopathy* = 0 and

TABLE 21.6   Logistic Regression with Six Terms

```
Data set = Downer, Name of Fit = B3
372 cases are missing at least one value.
Binomial Regression
Kernel mean function = Logistic
Response       = Outcome
Terms          = (Calving Myopathy Daysrec log2[AST] log2[CK])
Trials         = Ones
Coefficient Estimates
```

Label	Estimate	Std. Error	Est/SE
Constant	1.23651	1.61115	0.767
Calving	-0.303328	0.407659	-0.744
Myopathy	-1.84066	0.611761	-3.009
Daysrec	-0.0545999	0.112338	-0.486
log2[AST]	0.0388272	0.263829	0.147
log2[CK]	-0.152586	0.144471	-1.056

Scale factor:	1.
Number of cases:	435
Number of cases used:	216
Degrees of freedom:	210
Pearson X2:	207.681
Deviance:	211.268

Myopathy = 1 were, respectively, 0.386 and 0.0631. If we convert these to odds, we get $0.386/(1 - 0.386) = 0.628$ and $0.0631/(1 - 0.0631) = 0.067$, which are the same as the odds estimated by logistic regression. The logistic regression is formally identical to the earlier analysis, as it produces the same estimates and inferences. In Section 21.2.3 we derived a test based on the large-sample normality of sample proportions of the hypothesis that the probability of surviving is the same for both levels of *Myopathy*. From Table 21.5 we have another test, given by the ratio of the estimate for *Myopathy* to its standard error; the test statistic is equal to -4.858, and it also can be compared to the normal distribution to get *p*-values.

21.4.2   Many Terms

Table 21.6 summarizes the fit of a logistic regression with six terms, an intercept, indicator variables for *Calving* and *Myopathy*, the variable *Daysrec*, and base-two logarithms of *AST* and *CK*. Choosing terms is a topic in the next chapter, and for now we will simply assume that this is a reasonable model for the conditional distribution of survival given the predictors.

From Table 21.6, the number of cases is now only 216; this is the number of cows for which all the terms and the response were observed. Interpretation of point estimates is the same as in logistic regression with one predictor: The effect of observing *Myopathy* = 1 while holding the others terms fixed is to multiply the estimated odds of survival by exp(−1.84) = 0.16, and the effect of doubling *CK*, equivalent to adding one to $\log_2(CK)$, is to multiply the odds by exp(−0.15) = 0.86.

As with logistic regression with a single term, the estimates are approximately normally distributed, with estimated covariance matrix that can be displayed using the "Display variances" item in the model's menu. Apart from *Myopathy*, the ratios of the estimates to their standard errors are all less than 2.0 in absolute value; upon comparing them to a standard normal distribution, only the coefficient for *Myopathy* appears to be nonzero. We have found that $\log_2(CK)$ is an important predictor of *Outcome* ignoring the other predictors, but that it is probably not very important after adjusting for the other predictors. The cause of this parallels the multiple linear regression case, as discussed in Chapter 11: The predictors are probably collinear. We will examine this graphically in the next chapter. For now, we turn to the testing of submodels obtained by deleting terms, getting results that parallel Section 11.2.

21.4.3 Deviance

In multiple linear regression, the residual sum of squares provides the basis for tests for comparing mean functions. In logistic regression the residual sum of squares is replaced by the *deviance*, which is often called G^2. The deviance is defined to be twice the difference between (a) the log-likelihood evaluated at the MLE $\hat{\theta}(\mathbf{x}_i) = \hat{\eta}^T\mathbf{u}_i$ and (b) the log-likelihood evaluated by setting $\theta(\mathbf{x}_i) = y_i/m_i$, effectively fitting one parameter for every observation. For logistic regression, recalling that the fitted values are $\hat{y}_i = m_i\hat{\theta}(\mathbf{x}_i)$, the formula for the deviance is

$$G^2 = 2\sum_{i=1}^{n}\left[y_i\log\left(\frac{y_i}{\hat{y}_i}\right) + (m_i - y_i)\log\left(\frac{m_i - y_i}{m_i - \hat{y}_i}\right)\right] \qquad (21.12)$$

The degrees of freedom associated with the deviance is equal to the number of cases used in the calculation minus the number of elements of η that were estimated; in the example, 216 − 6 = 210.

Pearson's X^2 is an approximation to the deviance defined for logistic regression by

$$X^2 = \sum_{i=1}^{n}\left[(y_i - \hat{y}_i)^2\left(\frac{1}{\hat{y}_i} + \frac{1}{m_i - \hat{y}_i}\right)\right]$$

$$= \sum_{i=1}^{n}\frac{m_i(y_i/m_i - \hat{\theta}(\mathbf{x}_i))^2}{\hat{\theta}(\mathbf{x}_i)(1 - \hat{\theta}(\mathbf{x}_i))} \qquad (21.13)$$

X^2 and G^2 have the same large sample distribution, and for large samples they will give the same inferences. In small samples there may be differences, and sometimes X^2 may be preferred to G^2. In this book we will use G^2 exclusively because it is more appropriate for comparing models.

Methodology for comparing models parallels the results in Section 11.2. Write $\eta^T\mathbf{u} = \eta_1^T\mathbf{u}_1 + \eta_2^T\mathbf{u}_2$, and consider testing

$$\text{NH}: \quad \theta(\mathbf{x}) = \eta_1^T\mathbf{u}_1$$

$$\text{AH}: \quad \theta(\mathbf{x}) = \eta_1^T\mathbf{u}_1 + \eta_2^T\mathbf{u}_2$$

to see if the terms in $\mathbf{u}_2$ have zero coefficients. Obtain the deviance G^2_{NH} and degrees of freedom df_{NH} under the null hypothesis, and then obtain G^2_{AH} and df_{AH} under the alternative hypothesis. As with linear models, we will have evidence against the null hypothesis if $G^2_{NH} - G^2_{AH}$ is too large. To get p-values, we compare the difference $G^2_{NH} - G^2_{AH}$ to the χ^2 distribution with $\mathrm{df}_{NH} - \mathrm{df}_{AH}$ df, not to an F-distribution as was done for linear models.

Suppose we set $\mathbf{u}_1 = (Ones, Myopathy)^T$, where $Ones$ is the vector of ones to fit the intercept, and $\mathbf{u}_2 = (Calving, Daysrec, \log_2(AST), \log_2(CK))^T$ to test the hypothesis that all the coefficients except those for the constant term and *Myopathy* are zero against the alternative that they are not all zero. Fitting under the alternative hypothesis is summarized in Table 21.6, where we see that $G^2_{AH} = 211.27$ with $\mathrm{df}_{AH} = 210$. We now need to fit under the null hypothesis. While we would normally do this by fitting from the Graph&Fit menu, missing data make the computation a bit more complicated because fewer observations may have missing data under the null hypothesis, and so more cases may be used in the fitting. For the test, the same cases must be used to fit both the null and alternative models. This can be guaranteed by using the "Examine submodels" item from the alternative hypothesis model's menu. After selecting this item, double-click on the name for *Myopathy* to make this part of the base model. The constant term is always part of the base model if an intercept has been fit for the alternative hypothesis model. The remaining predictors are those in $\mathbf{u}_2$. Select "Fit in specified order" in the dialog; the resulting output is shown in Table 21.7.

The interpretation of this table parallels closely the results for multiple linear regression models in Section 11.4. The first line of the table, with the label "Base Model," gives the deviance when a model is fitted that includes the constant term and *Myopathy*, so it is the fit of the mean function specified by the null hypothesis. The deviance is 213.33, with 214 df. We can now perform the test of the hypothesis of interest: The change in deviance is $213.33 - 211.27 = 2.06$ with $214 - 210 = 4$ df. The p-value from the χ^2_4 distribution is larger than 0.7, giving no evidence that any of the variables in $\mathbf{u}_2$ are needed after $\mathbf{u}_1 = (Ones, Myopathy)$.

As with the Sequential Analysis of Variance table discussed in Section 11.4, the *Sequential Analysis of Deviance table* in Table 21.7 provides all the in-

TABLE 21.7 Analysis of Deviance for Recumbent Cow Data

```
Sequential Analysis of Deviance
All fits include an intercept.
Base model = (Ones Myopathy)
```

		Total		Change	
Predictor	df	Deviance		df	Deviance
Base Model	214	213.327			
Calving	213	212.961		1	0.366156
Daysrec	212	212.940		1	0.0209456
log2[AST]	211	212.378		1	0.562029
log2[CK]	210	211.268		1	1.10996

formation that is needed to examine a sequence of hypothesis tests. For example, the line in Table 21.7 marked *Calving* gives the deviance for fitting the base model plus *Calving*, along with its degrees of freedom. In the columns marked "Change", the difference in deviance and df between this mean function and the one excluding *Calving* are given, and so the "Change" columns provide the information needed to test the null hypothesis that only the terms in $\mathbf{u}_1$ are needed against the alternative that both $\mathbf{u}_1$ and *Calving* are needed. Upon comparing the change in deviance 0.366 to the χ_1^2 distribution, we get a *p*-value larger than 0.5, providing no evidence that we need to adjust for whether the cow becomes recumbent before or after calving in modeling survival.

There are three remaining items in the "Examine submodels" dialog that allow using stepwise methods to examine subsets further. The use of these items is similar to their use in the multiple linear regression model in Section 11.5.2, except that the criterion ordering mean functions is different. An example with Poisson regression is given in Section 23.3.

21.4.4 Goodness-of-Fit Tests

When the number of trials m_i is greater than 1, the deviance G^2 and Pearson's statistic X^2 can be used to provide a goodness-of-fit test for a logistic regression model, essentially comparing the null hypothesis that the mean function used is adequate versus the alternative that a separate parameter needs to be fit for each value of i (this latter case is called the *saturated model*). When all the m_i are large enough, either G^2 or X^2 can be compared to the χ_{n-k}^2 distribution to get an approximate *p*-value.

If the m_i are small, then the χ^2 approximation to the distribution of G^2 or X^2 is likely to be quite poor (the χ^2 approximation when comparing two non-saturated models as in the last section is generally better than for comparing a model to the saturated model). If all the m_i equal 1, then the value of G^2 depends on the fitted values but not on the particular y_i observed (Problem 21.4), and so G^2 cannot be used for a goodness-of-fit test.

21.5   WEEVIL PREFERENCES

Many regressions with a categorical response will have only categorical predictors, and we now consider an example of this type.

Insect herbivores often specialize to prefer just a few host plant species. The mechanism that determines specialization is not completely understood, with theories suggesting both genetic and environmental determinants. We discuss part of an experiment using the milfoil weevil, *Euhrychiopsis lecontei*, that is designed to explore these theories (the full experiment is discussed in Problem 21.9). Eight male weevils, called sires, were obtained from a lake where an exotic host plant has been present for at least 90 generations of weevils (about 30 years). Progeny of each sire were assigned to each of four rearing environments or treatments: (1) complete development on an exotic host plant; (2) complete development on a native host plant; (3) early development on the exotic and later development on the native host plant; or (4) early development on the native and later development on the exotic host plant.

After the progeny matured to adults, each was moved to a new environment that included both the exotic and native plant species. The observed response was the type of host selected by the progeny, either the exotic or native. The data consist of the number of adult insects in each sire/treatment combination that preferred the exotic; the number of trials is the number of adults tested in each sire/treatment combination. The data are in the file weevila.lsp.

The experiment included eight sires to represent the genetic variability in the population of weevils. If the progeny of some sires preferred the exotic species more than others, then we will have evidence of a genetic component to host plant preference. The four treatments explore the environmental effects on host plant preference. If progeny with treatments one and two differ with respect to host plant preference, then an environmental component will be apparent. Treatments three and four explore the environmental effects further: By comparing the results of these treatments to the results for treatments one and two, we can see if any environmental effects happen earlier or later in the development of the weevil. Finally, a sire by treatment interaction would mean that the environmental effect is potentially different for each sire, a very complicated outcome. However, a mean function including factors for sire, for treatment, and for treatment by sire interaction will have as many parameters as observations; seven for sire, three for treatment, $7 \times 3 = 21$ for the interaction, and one for the overall mean, giving 32 in total. As a result, this mean function will estimate one parameter for each observation and will therefore fit the data exactly. Without some simplification, we cannot assess this interaction.

Table 21.8 shows the observed proportion of adults preferring the exotic plant for each combination of *Trt* and *Sire*. Examining this table without further summaries is not easy, as there are no clear trends. A reasonable step is to look at logistic regression models. Define factors corresponding to *Trt* and *Sire*, and fit logistic regression with response *Exotic*, trials *Total*, and terms

TABLE 21.8   Table of Observed Fraction of Weevils Preferring the Exotic Host Plant, by Sire and Treatment[a]

Trt:	1	2	3	4
Sire:				
1|	1.0	0.625	0.625	0.75
2|	.33333333	0.0	.55555556	0.625
3|	0.625	0.625	.66666667	.83333333
4|	1.0	0.875	0.5	0.5
5|	.88888889	0.5	.57142857	.66666667
6|	.55555556	.55555556	.44444444	.28571429
7|	.88888889	.57142857	.71428571	.88888889
8|	.77777778	.44444444	.55555556	.88888889

[a]This table was created by computing a new variable given by *frac* = *Exotic*/*Total* and then getting a table of the mean of *frac*.

{*F*}*Trt* and {*F*}*Sire*, as summarized in Table 21.9. The coefficient estimates in Table 21.9 are not of immediate interest, since our first goal might be to test the overall hypothesis of genetic effects (coefficients for {*F*}*Sire* are not all zero) and environmental effects (coefficients for {*F*}*Trt* are not all zero) using differences in deviance as in Section 21.4.3. *Arc* includes a shortcut for computing these tests by selecting the item "Examine submodels" from the model's menu and then pushing the button for "Change in deviance for fitting each term last." The results are shown in Table 21.10. For both factors, the change in deviance is large, and the corresponding *p*-values are small, so the probability of selecting the exotic appears to depend on both factors. Preference for the exotic host has both a genetic and an environmental component.

Next, examine the coefficient estimates for {*F*}*Trt* in Table 21.9. The estimate for treatment one is 0 by the parameterization, and the estimate for treatment four, which is really the difference between treatment four and treatment one, is not too far from zero: The ratio of the estimate to its standard error is −0.97. Similarly, the estimates for treatments two and three, which again are really the differences between these treatments and treatment one, are each nearly equal to about −1. We might be able to get some model simplification by dividing the treatment effect into two parts. The first part compares the average of treatment one and four to the average of treatments two and three, while the second part summarizes all other comparisons of the four treatments.

This can be accomplished by defining a new variable that has the value 0 for treatments one and four and has the value 1 for treatments two and three. Select the item "Add a variate" from the data set menu, and then in the text area, type

```
Trt1 = (recode-values Trt '(1 2 3 4) '(0 1 1 0))
```

TABLE 21.9   Logistic Regression Fit to the Weevil Data, with Main Effects

```
Data set = weevils, Name of Fit = B1
Binomial Regression
Kernel mean function = Logistic
Response        = Exotic
Terms           = ({F}Trt {F}Sire)
Trials          = Total
Coefficient Estimates
```

Label	Estimate	Std. Error	Est/SE
Constant	1.75996	0.491763	3.579
{F}Trt[2]	-1.12771	0.390000	-2.892
{F}Trt[3]	-0.837973	0.394181	-2.126
{F}Trt[4]	-0.394187	0.405360	-0.972
{F}Sire[2]	-1.71376	0.547554	-3.130
{F}Sire[3]	-0.372908	0.569378	-0.655
{F}Sire[4]	-0.159740	0.591349	-0.270
{F}Sire[5]	-0.477048	0.559369	-0.853
{F}Sire[6]	-1.28051	0.543074	-2.358
{F}Sire[7]	0.104899	0.599354	0.175
{F}Sire[8]	-0.445690	0.548221	-0.813

```
Scale factor:            1.
Number of cases:         32
Degrees of freedom:      21
Pearson X2:          26.977
Deviance:            31.333
```

TABLE 21.10   Likelihood Ratio Tests for the Weevil Data

```
Data set = weevils, Name of Fit = B1
Binomial Regression
Kernel mean function = Logistic
Response        = Exotic
Terms           = ({F}Trt {F}Sire)
Trials          = Total
Change in deviance for fitting each term last.
All fits include an intercept.
```

Term	df	LR-test	p-value
{F}Trt	3	10.082	0.0179
{F}Sire	7	20.860	0.0040

This variate will be used for the first part of the treatment effect, comparing treatments one and four to treatments two and three. Fitting $\{F\}Trt$ after $Trt1$ will take care of the second part of the treatment effect. Next, fit the logistic regression model with the same response and trials as before, but with terms

TABLE 21.11 Sequential Deviance with Treatments Divided into Two Parts, with the First Part Comparing Treatments One and Four to Two and Three

```
Binomial Regression
Kernel mean function = Logistic
Response      = Exotic
Terms         = ({F}Sire Trt1 {F}Trt)
Trials        = Total
Sequential Analysis of Deviance
All fits include an intercept.
```

		Total		Change	
Predictor	df	Deviance	|	df	Deviance
Ones	31	62.2451	|		
{F}Sire	24	41.4148	|	7	20.8303
Trt1	23	32.8972	|	1	8.51761
{F}Trt	21	31.3327	|	2	1.56452

$\{F\}Sire$, $Trt1$, and $\{F\}Trt$. From this model's menu, select the item "Examine submodels," and then get a sequential deviance table, in the order $\{F\}Sire$, $Trt1$, $\{F\}Trt$. The results are shown in Table 21.11. The change in deviance for $Trt1$ is 8.52 with 1 df, with a corresponding p-value from the χ_1^2 distribution of about 0.004, suggesting that treatments one and four differ from two and three. The change in deviance for adding $\{F\}Trt$ after $Trt1$ is only 1.56 with 2 df, with a p-value near 0.46. Combining these two results, progeny reared under treatments one and four showed the same preference for host plant. Treatment one progeny grew with the exotic host plant, while treatment four progeny grew at first with a native host plant but were later switched to the exotic. In treatment two, progeny grew on the native host plant; and in treatment three, progeny were switched to the native host plant after initial growth on the exotic. The conclusion is that the host plant during growth does affect host plant preference, but preference is determined late in the development of the progeny, as the early host plant does not have an effect. The analysis of this experiment is continued in Problem 21.8.

21.6 COMPLEMENTS

21.6.1 Normal Approximation to the Binomial

If $y \sim \text{Bin}(m, \theta)$, then the fraction of successes $\hat{\theta} = y/m$ will be approximately normally distributed as long as the true value of θ is not too close to 0 or 1. Exactly how close is too close depends on the number of trials m. Here is an empirical rule that can be used in practice: The normal approximation is acceptable if $m\theta(1-\theta) \geq 2$ (McCullagh and Nelder, 1989). For very small m, or for θ near zero or one, other methods of inference are required; see Collett (1991, Chap. 2).

21.6.2 Smoothing a Binary Response

The *lowess* estimate of the mean function in Figure 21.1 would be negative for larger values of $\log_2(CK)$, but of course the probability of survival must be in the interval $[0, 1]$. The smoother does not use this information about the survival probability, so it could give results outside the permissible range. For our purpose of providing a general description of the dependence of the mean on the predictor, this smoother is adequate, and the smooth is truncated if it goes outside the interval between zero and one. When fitting models, we will require that methods produce probabilities in the range of zero to one.

21.6.3 Probit and Clog-Log Kernel Mean Functions

Arc includes two other kernel mean functions that can be used in place of the logistic mean function for binomial regression.

The first of these is called the *inverse probit* mean function. It was suggested by Finney (1947, 1971), who derived it as follows. Suppose we administer a stimulus, such as the dose or log-dose of a toxic chemical to a sample of subjects, perhaps insects. At dosage x, some subjects die and some survive. We assume that each subject has a *tolerance* v for the stimulus. If for that subject $v \leq x$ the subject dies, but if $v > x$ the subject survives. Each subject has a different value of v drawn from a *tolerance distribution* $H(v)$ which gives the probability that a randomly chosen subject has tolerance less than or equal to v. We can then write

$$\text{Pr(survival} \mid x) = \theta(x)$$

$$= \text{Pr(Random subject has tolerance} > x)$$

$$= 1 - H(x)$$

Under the tolerance approach, the mean function $\theta(x)$ is equivalent to $1 - H(x)$. For example, if the tolerances are assumed to be normally distributed $N(0, 1)$, then the kernel mean function is $\theta(x) = 1 - \Phi(x)$, where Φ is the cumulative distribution function of the standard normal distribution. In general, with a linear predictor $\eta^T \mathbf{u}$, the kernel mean function for probit regression is

$$\theta(\mathbf{x}) = 1 - \Phi(\eta^T \mathbf{u})$$

which uses the cumulative distribution function for the standard normal distribution as a kernel mean function. If we invert this function to get the link function, we have

$$\Phi^{-1}(1 - \theta(\mathbf{x})) = \eta^T \mathbf{u}$$

The quantity $\Phi^{-1}(1 - \theta(\mathbf{x}))$ is called a *probit*, hence the name inverse probit for this kernel mean function.

Whether this tolerance approach makes sense in any specific binomial regression is open to debate. The stochastic element is in the subject, not in the observation: Given the subject, the subject's tolerance, and the stimulus, the outcome can be known perfectly. In any case, there is little practical difference between probit and logistic regression because the shape of the inverse probit and the logistic kernel mean functions are very similar.

The remaining kernel mean function that is occasionally used in practice is most easily defined in terms of the link function, defined by

$$\log(-\log(1 - \theta(\mathbf{x}))) = \eta^T \mathbf{u}$$

This is called the *complementary log–log link*. The corresponding kernel mean function, found by inverting this equation, is called the inverse complementary log–log kernel mean function.

21.6.4 The Log-Likelihood for Logistic Regression

Suppose that we have the structure laid out in Section 21.3.2. We proceed to obtain the MLE for η. We will then get the MLE for $\theta(\mathbf{x}_i)$ through the equation $\hat{\theta}(\mathbf{x}_i) = \text{M}(\hat{\eta}^T \mathbf{u}_i)$. Equation (21.4) gives the likelihood function for a single observation y_i based on m_i trials. The likelihood for the complete data is obtained by multiplying the likelihood for each observation,

$$L = \prod_{i=1}^{n} \binom{m_i}{y_i} (\theta(\mathbf{x}_i))^{y_i} (1 - \theta(\mathbf{x}_i))^{m_i - y_i}$$

$$\propto \prod_{i=1}^{n} (\theta(\mathbf{x}_i))^{y_i} (1 - \theta(\mathbf{x}_i))^{m_i - y_i}$$

In the last expression we have dropped the binomial coefficients $\binom{m_i}{y_i}$ because they do not depend on parameters. After minor rearranging, the log-likelihood is

$$\log(L) \propto \sum_{i=1}^{n} \left[y_i \log\left(\frac{\theta(\mathbf{x}_i)}{1 - \theta(\mathbf{x}_i)} \right) + m_i \log(1 - \theta(\mathbf{x}_i)) \right]$$

Next, we substitute for $\theta(\mathbf{x}_i)$ using equation (21.11) to get

$$\log(L(\eta)) = \sum_{i=1}^{n} [(\eta^T \mathbf{u}_i) y_i - m_i \log(1 + \exp(\eta^T \mathbf{u}_i))] \qquad (21.14)$$

The log-likelihood is now explicitly a function of η that we can maximize. Unlike the multiple linear regression model, there is no formula that will give the value of η that maximizes (21.14), but rather an iterative procedure is needed. Most software for logistic regression including *Arc* uses a simple but effective algorithm for this computation that has the following steps:

1. Obtain an initial guess for $\hat{\eta}$, say $\mathbf{h}_0$. The choice $\mathbf{h}_0 = 0$ is often used in practice.
2. Set the iteration counter $j = 1$.
3. Approximate the log-likelihood close to $\mathbf{h}_{j-1}$ by a quadratic curve.
4. Use the known formula for maximizing the quadratic curve to produce a new guess $\mathbf{h}_j$.
5. Check a stopping criterion. A common criterion is to stop iteration if $\log(L(\mathbf{h}_j)) - \log(L(\mathbf{h}_{j-1}))$ is smaller than a preselected tolerance value. If the criterion is satisfied, report $\mathbf{h}_j$ as the estimator, otherwise increment j by one and go to step 3.

In logistic regression this algorithm usually attains convergence in just a few iterations, although problems can arise with unusual data sets (for example, if one or more of the predictors can determine the value of the response exactly); see Collett (1991), Sec. 3.12.

Readers familiar with numerical methods for maximizing a function will recognize that this algorithm is related to the *Newton–Raphson* method. If the approximation used at step 3 is based on a Taylor series expansion, then the algorithm is exactly Newton–Raphson. Most software for logistic regression including *Arc* uses a slightly modified method called *Fisher scoring* that estimates the quadratic equation in a slightly different way; details are provided by McCullagh and Nelder (1989, Sec. 2.5), Collett (1991), and Agresti (1990, 1996), among others.

Using the Fisher scoring method, the update in Step 4 of the algorithm has the form

$$\mathbf{h}_j = (\mathbf{U}^T \mathbf{W}_{j-1} \mathbf{U})^{-1} \mathbf{U}^T \mathbf{W}_{j-1} \mathbf{z}_{j-1} \tag{21.15}$$

where both the diagonal matrix $\mathbf{W}_{j-1}$ of estimated weights and the "working response" $\mathbf{z}_{j-1}$ depend on $\mathbf{h}_{j-1}$ and the kernel mean function. For logistic regression, if $\theta_{j-1}(\mathbf{x}_i)$ is the estimated probability of success assuming that $\eta = \mathbf{h}_{j-1}$, then the ith diagonal element of $\mathbf{W}_{j-1}$ is

$$w_i = m_i \theta_{j-1}(\mathbf{x}_i)(1 - \theta_{j-1}(\mathbf{x}_i))$$

and the ith element of $\mathbf{z}_{j-1}$ is

$$z_i = \mathbf{h}_{j-1}^T \mathbf{u}_i + (y_i - m_i \theta_{j-1}(\mathbf{x}_i))/[m_i \theta_{j-1}(\mathbf{x}_i)(1 - \theta_{j-1}(\mathbf{x}_i))]$$

We can recognize (21.15) as a weighted least squares estimate, with weights and working response that change from iteration to iteration. The algorithm is therefore sometimes called *iteratively reweighted least squares*.

At convergence, the estimated covariance matrix of the estimates is given by the usual weighted least squares formula,

$$\text{Var}(\hat{\boldsymbol{\eta}}) = (\mathbf{U}^T \widehat{\mathbf{W}} \mathbf{U})^{-1}$$

where $\widehat{\mathbf{W}}$ is a diagonal matrix with entries $m_i \hat{\theta}(\mathbf{x}_i)(1 - \hat{\theta}(\mathbf{x}_i))$.

21.6.5 References

Book-length treatments of binomial regression are given by Collett (1991) and by Hosmer and Lemeshow (1989). McCullagh and Nelder (1989) study these models at a higher mathematical level. The Bernoulli distribution was named for James Bernoulli who first described it in 1713.

The recumbent cow data were provided by Harold Henderson and are discussed in Clark, Henderson, Hoggard, Ellison, and Young (1987). Data for the weevil experiment in Section 21.5 were provided by Susan L. Solarz, Ray Newman, Diane L. Byers, and Ruth G. Shaw. The Challenger data in Problem 21.6 are given by Dalal, Fowlkes, and Hoadley (1989). The Donner party data were collected by Johnson (1996). The Titanic data in Problem 21.7 were discussed by Dawson (1995).

PROBLEMS

21.1 In the recumbent cow data, why might the variable *Daysrec* be included among the predictors? After all, it is a characteristic of the study, not of the cows.

21.2 Starting with (21.10), prove (21.11).

21.3 The usual formula for the sample variance of a sample $y_1, \ldots, y_n$ is $\text{sd}_y^2 = \sum (y_i - \bar{y})^2 / (n - 1)$. Show that if the y_i can only have values zero and one, then

$$\text{sd}_y^2 = \frac{n}{n-1} \hat{\theta}(1 - \hat{\theta})$$

where $\hat{\theta}$ is the fraction of ones in the data. Hence, apart from the multiplier $n/(n-1)$ that is nearly equal to one in large samples, the usual formula for the sample variance gives the same answer as the formula for a Bernoulli variance evaluated at $\theta = \hat{\theta}$.

21.4 For the special case of binary regression where $m_i = 1$ for all i, show that the formula for the deviance (21.12) reduces to

$$G^2(y, \hat{\boldsymbol{\eta}}) = -2 \sum_{i=1}^{n} \left\{ \hat{\theta}(\mathbf{x}_i) \log \left[\frac{\hat{\theta}(\mathbf{x}_i)}{1 - \hat{\theta}(\mathbf{x}_i)} \right] + \log \left[1 - \hat{\theta}(\mathbf{x}_i) \right] \right\}$$

The curious feature of this deviance is that it depends only on the fitted probabilities but not on the observed patterns of zeroes and ones. Any two data sets that give the same values of the estimates will therefore have the same deviance, regardless of how well they fit. Consequently, in binary regression the deviance cannot be used as a summary measure of the overall fit of a model. (*Hint*: First, since y_i can only equal zero or one, we must have that $y \log(y) = (1 - y) \log(1 - y) = 0$. To complete the proof, you must show that $\sum y_i \log\{\hat{\theta}(\mathbf{x}_i)/[1 - \hat{\theta}(\mathbf{x}_i)]\} = \sum \hat{\theta}(\mathbf{x}_i) \log\{\hat{\theta}(\mathbf{x}_i)/[1 - \hat{\theta}(\mathbf{x}_i)]\}$. This latter result is proved by examining the likelihood function in Section 21.6.4 evaluated at the maximum.)

21.5 In the winter of 1846–1847, about ninety wagon train emigrants in the Donner party were unable to cross the Sierra Nevada Mountains of California before winter, and almost one-half of them starved to death. Perhaps because they were ordinary people—farmers, merchants, parents, children—their story captures the imagination. The data in file donner.lsp include some information about each of the members of the party. The variables include *Age*, the age of the person; *Sex*, whether male or female; *Status*, whether the person was a member of a family group, a hired worker for one of the family groups, or a single individual who did not appear to be a hired worker or a member of any of the larger family groups, and *Outcome*, coded one if the person survived and zero if the person died.

    21.5.1 How many men and women were in the Donner Party? What was the survival rate for each sex? Obtain a test that the survival rates were the same against the alternative that they were different. What do you conclude?

    21.5.2 Fit the logistic regression model with response *Outcome* and predictor *Age*, and provide an interpretation for the fitted coefficient for *Age*.

    21.5.3 Draw the graph of *Outcome* versus *Age*, and add both a *lowess* smooth and a fitted logistic curve to the graph. The logistic regression curve apparently does not match the data; explain what the differences are, and how this failure might be relevant to understanding who survived this tragedy. On your graph click on the slidebar for the logistic curve to fit a quadratic (so the terms in the mean function are then a constant, *Age*, and Age^2). Does this mean function match the *lowess* curve more accurately? You can get an even better match if you draw a graph of *Outcome* versus Age^5 and then fit the logistic with terms $(1, Age^5, Age)$.

    21.5.4 Fit the logistic regression model with terms for an intercept, Age^5, *Age*, *Sex*, and {*F*}*Status*. Provide an interpretation for the parameter estimates for *Sex* and for each of the parameter estimates for $\{F\}Status$. Use the "Examine submodels" item on the model's menu to obtain tests based on the deviance for adding each of the terms to a mean function that already includes the other terms, and summarize the results of each of the tests via a p-value and a one-sentence summary of the results.

21.5.5 The standard error of the coefficient for $\{F\}Status[Single]$ appears to be very large. Why do you think this happened? (*Hint*: Examine the number of people in each of the *Status* groups and examine the proportion surviving in each group.)

21.5.6 Assuming that the logistic regression model provides an adequate summary of the data, give a one-paragraph written summary on the survival of members of the Donner Party.

21.6 The file `challeng.lsp` contains data on O-rings on 23 U.S. space shuttle missions prior to the Challenger disaster of January 20, 1986. For each of the previous missions, the temperature at take-off and the pressure of a pre-launch test were recorded, along with the number of O-rings that failed (out of six).

Use these data to try to understand the probability of failure as a function of temperature and also as a function of temperature and pressure. Use your fitted model to estimate the odds of failure of an O-ring when the expected temperature was 31°F, the launch temperature on January 20, 1986.

21.7 The Titanic was a British luxury passenger liner that sank when it struck an iceberg about 640 km south of Newfoundland on April 14–15, 1912, on its maiden voyage to New York City from Southampton, England. Of 2201 known passengers and crew, only 711 are reported to have survived. The data in the file `titanic.lsp` classifies the people on board the ship according to their *Sex*, *Age*, either child or adult, and *Class*, either first, second, third, or crew. For each age/sex/class combination, the number of people N and the number surviving *Surv* is also reported.

21.7.1 Obtain a table that gives the fraction surviving for each of the combinations of the conditioning variables, and give a descriptive summary of the results.

21.7.2 Fit a logistic regression model with predictors $\{F\}Sex$, $\{F\}Age$, and $\{F\}Class$. Based on the tables you reviewed in the first part of this problem, explain why you expect that this mean function will be adequate to explain these data.

21.7.3 Fit a logistic regression model that includes all the terms of the last part, plus all the two-factor interactions. Use appropriate testing procedures to decide if any of the two-factor interactions can be eliminated. Assuming that the mean function you have obtained matches the data well, summarize the results you have obtained by interpreting the parameters to describe different survival rates for various factor combinations. (*Hint*: How does the survival of the crew differ from the passengers? First class from third class? Males from females? Children versus adults? Did children in first class survive more often than children in third class?)

21.8   This problem continues the analysis of the weevil data in file `weevila.lsp` in Section 21.5.

   21.8.1   Obtain a quantitative estimate and standard error for the difference in exotic host plant preference between the average of treatments one and four and the average of treatments two and three (this will require fitting the logistic regression model with terms *Trt1* and {*F*}*Sire* and examining the coefficient estimates).

   21.8.2   Given that treatment differences can be summarized by only one degree of freedom, the difference between the average of treatments one and four, and the average of treatments two and three, we can examine the possibility of an environmental by treatment interaction by fitting a mean function with terms *Trt1*, {*F*}*Sire* and a *Trt1*×{*F*}*Sire* interaction. Fit the model, perform the relevant test, and summarize your results.

21.9   The data in the file `weevils.lsp` gives the full experiment on which the results in Section 21.5 are based. In the full experiment, eight sires were taken from each of two lakes. The first lake, used in Section 21.5, has had the exotic host plant species present for at least 90 generations of weevils, about 30 years. In the second lake, the exotic host plant has never been present. The full experiment therefore has an additional factor, the source of the sires. Examine these data for genetic and environmental effects, and summarize your results. (*Hint*: In fitting logistic regression models that include *Sire* effects, you must include a {*F*}*Sire* × *Source* interaction to account for the arbitrary numbering of the sires: Since sire one at lake one has nothing in common with sire one at lake two, we need to fit a separate parameter for each of these, and this is done by fitting this interaction.)

21.10   Suppose y is binary, and let $z = 1 - y$ and consider two logistic regression models. (a) The regression of y on terms **u** and (b) the regression of z on terms **u**. How are the two sets of coefficient estimates related (see (21.11))? How are the two deviances and df related?

CHAPTER 22

Graphical and Diagnostic Methods for Logistic Regression

In this chapter we consider graphical and diagnostic methods for logistic regression, in which the response is the number of successes in a fixed number of trials. For most of the chapter, we discuss the case in which the number of trials $m_i = 1$ for all cases, so the response variable y_i can only have values 0 or 1.

22.1 ONE-PREDICTOR METHODS

With one continuous predictor x and a continuous response y, the 2D plot of y versus x is a sufficient summary plot: All the sample information about the regression of y on x is available in this one plot. When the response is binary, with $y = 0$ or $y = 1$, a 2D plot of y versus x is still a sufficient summary plot, but it may not be very useful for visualizing changes in the conditional distribution of $y \mid x$. Consider again Figure 21.3, page 477, which is a plot of *Outcome* versus $\log_2(CK)$ for the recumbent cow data in file downer.lsp. Concentrating on the points and not the smoothers, all we learn visually from the plot is that y is either zero or one, and that the range of points with $y = 1$ differs somewhat from the range of points with $y = 0$.

Information about the dependence of y on x is conveyed through the fraction or *relative density* of ones at each value of x. The mean function at a particular value of x can be estimated as the fraction of ones in a narrow vertical slice centered at that value. In Figure 21.3, visualizing how the mean changes with x is very difficult because of both (a) the discrete response and (b) overplotting of predictor values. We need plot enhancements to see the dependence. The *lowess* smoother, which estimates the relative density by local averaging, is one enhancement that can help. Point jittering can also be useful.

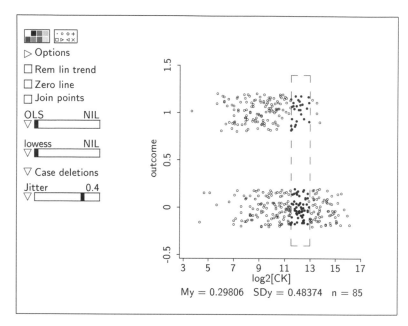

FIGURE 22.1   Scatterplot of *Outcome* versus $\log_2(CK)$ using jittering and the slicing mouse mode.

22.1.1   Jittering to See Relative Density

Jittering spreads out overplotted points to making seeing individual points easier. Click the mouse on the plot control "Options" on the plot, and then check the button "Jitter slidebar." This will add a slidebar to the plot. From this slidebar's pop-up menu, select "Jitter vertical only" and then move the slidebar to the right. This will add a random number, with standard deviation equal to the value above the slidebar times the range of the data on the vertical axis, to the y-coordinate of each of the plotted points, making the density of the points easier to see. This is illustrated in Figure 22.1 along with results from one slice. We see from the figure that there are 85 cases within the slice and that the fraction of ones is about 0.3. While we can't tell visually that the fraction of ones in the slice is 0.3, we should be able to see that it is roughly of that magnitude.

22.1.2   Using the Conditional Density of $x \mid y$

We can get a more complete look at the relative density of ones by estimating the conditional density function $f(x \mid y = j)$, $j = 0, 1$, of the predictor given the value of the response. To construct estimates of the two conditional densities $f(\log_2(CK) \mid Outcome = j)$, $j = 0, 1$ for the recumbent cow data, draw a

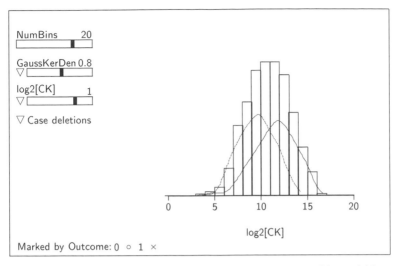

FIGURE 22.2   Histogram for $\log_2(CK)$ using *Outcome* as a marking variable.

histogram of $\log_2(CK)$ using the response *Outcome* as a marking variable, as shown in Figure 22.2. Estimates of the two conditional densities can now be obtained by first selecting the item "Fit by marks" from the density estimate pop-up menu and then moving the density estimate slider to about 0.8. We see that the densities have the same roughly normal-looking shape for both *Outcome* = 0 and *Outcome* = 1.

How do the conditional densities $f(x \mid y)$ provide information about the regression of y on x? Using the notation of Chapter 21, the next equation provides a connection between the mean function $E(y \mid x) = \theta(x)$ and the conditional densities:

$$\frac{\theta(\mathbf{x})}{1 - \theta(\mathbf{x})} = \frac{\Pr(y = 1)}{1 - \Pr(y = 1)} \times \frac{f(\mathbf{x} \mid y = 1)}{f(\mathbf{x} \mid y = 0)}$$

$$= c \times \frac{f(\mathbf{x} \mid y = 1)}{f(\mathbf{x} \mid y = 0)} \tag{22.1}$$

To emphasize that this equation holds for multiple predictors, we have written it in terms of the vector $\mathbf{x}$, although in the present discussion we are still concerned with just a single predictor. The factor $c = \Pr(y = 1)/(1 - \Pr(y = 1))$ is the marginal odds of success ignoring the predictors.

Equation (22.1) tells us how to use information from the density estimates in Figure 22.2 to gain information on the odds ratio. First, in the previous chapter we estimated $\Pr(y = 1)$ to be about 0.382, so we can estimate c to be $0.382/(1 - 0.382) \approx 0.62$. Now from Figure 22.2 we see that the conditional density for surviving cows at $\log_2(CK) = 9$ is about twice that for the other cows, $f(9 \mid y = 1) \approx 2f(9 \mid y = 0)$. Thus,

$$\frac{\theta(9)}{1 - \theta(9)} \approx 0.62 \times 2 = 1.24$$

The odds of a randomly chosen cow with $\log_2(CK) = 9$ surviving are about 1.24. Apart from the constant c the ratios of the conditional densities at x in Figure 22.2 estimate the odds of survival given that value of the predictor.

22.1.3   Logistic Regression from Conditional Densities

Taking logarithms of both sides of (22.1), we have

$$\log \left(\frac{\theta(x)}{1 - \theta(x)} \right) = \log(c) + \log \left(\frac{f(x \mid y = 1)}{f(x \mid y = 0)} \right) \tag{22.2}$$

and so the log-odds equals the sum of $\log(c)$, which does not depend on x, and the *log density ratio*, which generally will depend on x. This is just a function of x, and, as was shown in (21.11), page 480, (22.2) is the equation for the logit link function in binary regression. Inverting this equation will give the logistic kernel mean function (21.10). Thus, the logistic regression model for y on x is always appropriate, with u-terms that depend on the log density ratio.

22.1.4   Specific Conditional Densities

When the densities $f(\mathbf{x} \mid y)$ are from a known parametric family, the u-terms that are needed in the mean function can be determined.

Normal Densities. Suppose that $f(x \mid y = j)$ is a normal density, with mean μ_j and variance σ_j^2, $j = 0, 1$. These normal densities can be written as

$$f(x \mid y = j) = \frac{1}{\sqrt{2\pi}\sigma_j} e^{[-(x-\mu_j)^2 / 2\sigma_j^2]}, \qquad j = 0, 1, \quad -\infty < x < \infty$$

Substitute this for $f(x \mid y = j)$ in (22.2) and simplify to get

$$\log \left(\frac{\theta(x)}{1 - \theta(x)} \right) = c_0 + \left(\frac{\mu_1}{\sigma_1^2} - \frac{\mu_0}{\sigma_0^2} \right) x + \frac{1}{2} \left(\frac{1}{\sigma_0^2} - \frac{1}{\sigma_1^2} \right) x^2 \tag{22.3}$$

where

$$c_0 = \log(c) + \log(\sigma_0/\sigma_1) + (\mu_0^2/2\sigma_0^2 - \mu_1^2/2\sigma_1^2)$$

is a constant that does not depend on x. Let's look first at the important special case of $\sigma_1^2 = \sigma_0^2 = \sigma^2$, so the variance is the same in each of the two

TABLE 22.1   Appropriate Terms for Various Densities or Probability Mass Functions[a]

Density or Probability Mass Function of $x \mid y$	$\mathbf{u}^T =$
Normal, common variance	$(1, x)$
Normal, different variances	$(1, x, x^2)$
Skew, (exact for gamma densities)	$(1, x, \log(x))$
$x \in [0, 1]$	$(1, \log(x), \log(1 - x))$
Bernoulli (that is, an indicator variable)	$(1, x)$
Poisson count	$(1, x)$

[a]Logistic regression should be fit with linear predictor given by $\eta^T \mathbf{u}$.

populations. Substituting into (22.3), the term multiplying x^2 is zero and we are left with the result

$$\log \left(\frac{\theta(x)}{1 - \theta(x)} \right) = \eta_0 + \eta_1 x = \eta^T \mathbf{u} \qquad (22.4)$$

with u-terms $\mathbf{u} = (1, x)^T$, $\eta_0 = c_0$, and $\eta_1 = (\mu_1 - \mu_0)/\sigma^2$. So, for the case of normal densities with the same variance, the log-odds depend linearly on x, the intercept is a function of c and of the parameters of the normal distributions, and the slope parameter η_1 depends on the difference of the means scaled by the common variance.

If the two normal populations have different variances, then (22.3) can be used directly, and $\mathbf{u}$ has three terms, an intercept, x, and x^2. The values of the corresponding η's as functions of the parameters of the normal distributions can be read off from (22.3). Of course, since the parameters of the normal distributions are unknown, we will need to estimate the regression coefficients.

Let's summarize the results when the predictor $x \mid y$ is normally distributed for $y = 0, 1$. We know that logistic regression is appropriate, and the only question is the terms needed in the mean function. If the variances of the two normal populations are equal, then only x itself is needed, but if they are unequal, then x^2 is also required. If the conditional distributions are identical for $y = 0$ and $y = 1$, then the log-odds are constant, and functions of x are not needed in the mean function.

Other Densities. Table 22.1 summarizes similar results for non-normal conditional densities. If the densities $f(x \mid y = j)$ are skewed, then the mean function should include terms for the intercept and either x or possibly $\log(x)$. Both x and $\log(x)$ may be required. When conducting a binary regression with a skewed predictor, it is often easiest to assess the need for x and $\log(x)$ by including them both in the model so that their relative contributions can be assessed directly.

TABLE 22.2   Logistic Regression of *Outcome* **on** *CK* **and** $\log_2(CK)$ **for the Recumbent Cow Data**

```
Data set = Downer, Name of Fit = Bl
Binomial Regression
Kernel mean function = Logistic
Response      = Outcome
Terms         = (CK log2[CK])
Trials        = Ones
Coefficient Estimates
Label           Estimate       Std. Error       Est/SE
Constant        2.09917        0.845382          2.483
CK             -0.000144639    0.0000594375     -2.433
log2[CK]       -0.200180       0.092866         -2.155

Scale factor:                      1.
Number of cases:                 413
Degrees of freedom:              410
Pearson X2:                  388.718
Deviance:                    465.415
```

Discrete predictors like indicator variables or Poisson counts don't have density functions, but rather they have probability mass functions. Poisson counts, as used for events that occur randomly in time or space, are discussed in Section 23.3. The same results hold, however, when we substitute the probability mass function for densities in (22.2). When the math is done, a predictor x that is an indicator variable or a Poisson count requires terms for an intercept and x itself.

22.1.5   Implications for the Recumbent Cow Data

We apply these single predictor results for the regression of *Outcome* on three of the predictors in the recumbent cow data.

Outcome on CK. Returning to the regression *Outcome* on *CK* rather than on $\log_2(CK)$, we see from histograms constructed in the manner of Figure 22.2 that the conditional densities $f(CK \mid outcome = j)$, $j = 0, 1$, are skewed to the right. According to Table 22.1, this means that *CK* and $\log_2(CK)$ are likely predictors. Our previous use of $\log_2(CK)$ is consistent with this result. It remains to determine if we need *CK* also. If the densities estimated in Figure 22.2 are normal with the same variance then we don't need *CK*. But if they are sufficiently non-normal, then *CK* probably will be needed. The easiest way to decide the issue is to fit a logistic regression model including both *CK* and $\log_2(CK)$, as shown in Table 22.2. Both predictors appear to be needed in the regression.

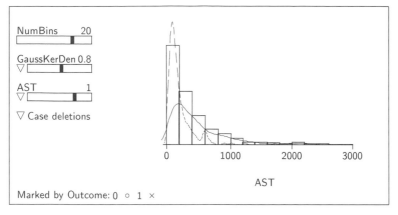

FIGURE 22.3 Histogram of *AST* in the recumbent cow data.

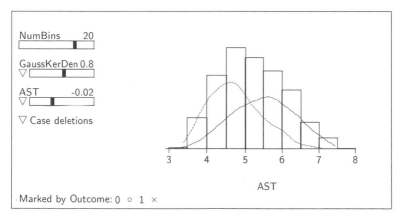

FIGURE 22.4 Histogram of *AST*, transformed nearly to logarithms, with separate density estimates for the two values of *Outcome*.

Outcome on AST. Figure 22.3 gives conditional density estimates for *AST*. These estimates are clearly not normal, but perhaps approximate normality can be obtained via transformation; the transformation plot controls can be used to explore this possibility. Select the item "Find normalizing transformation" from the transformation slidebar's pop-up menu. You will then get a dialog that allows a choice between conditioning on the marking variable or ignoring the marking variable. By conditioning on the marking variable, we will select a transformation that will make $f(AST^{(\lambda)} \mid y)$ as normal as possible for $y = 0$ and $y = 1$, which is our goal here. The result of this automatic procedure is shown in Figure 22.4, with *AST* raised to the power -0.02, essentially replacing *AST* by its logarithm. The densities in this scale appear to be reasonably symmetric,

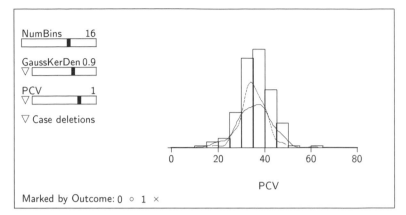

FIGURE 22.5   Histogram of *PCV* with separate density estimates for the two values of *Outcome*.

and the variances are approximately the same; this can be verified by selecting the item "Display summary statistics" from the density estimate slidebar's pop-up menu. The within-group standard deviations are about 0.8 for *Outcome* = 0 and 0.7 for *Outcome* = 1. Logistic regression with terms for an intercept and $\log_2(AST)$ is likely appropriate here. The results from including both *AST* and $\log_2(AST)$ in a logistic model supports this possibility.

Outcome on PCV. Finally, consider the regression of *Outcome* on *PCV*. Conditional density estimates for *PCV* are shown in Figure 22.5. The densities appear roughly symmetric with very similar means but noticeably different variances. Using the normal case in Table 22.1 for guidance, the logistic regression of *Outcome* on *PCV* may require *u*-terms *PCV* and PCV^2.

On the other hand, recall from Table 21.1 that *PCV* is a percentage and so must be between 0% and 100%. Equivalently $PCV/100$ must be between 0 and 1. Again using Table 22.1 for guidance, we may require the terms $\log_2(PCV)$ and $\log_2(100 - PCV)$. These log transformations for a variable with a restricted range are most likely needed when there are many values near the limits of the range. The minimum and maximum values of *PCV* are 13% and 61% so it may be difficult to distinguish between a logistic regression model with terms (PCV, PCV^2) and one with terms $(\log_2(PCV), \log_2(100 - PCV))$. A good way to tell if one set of terms is clearly better than the other is to try them both.

22.2   VISUALIZING LOGISTIC REGRESSION WITH TWO OR MORE PREDICTORS

In Section 22.1.5 we concluded using separate marginal regressions that *Outcome* is dependent on *AST* and *PCV*. To introduce graphical methods for visualizing binary response regressions with two predictors, we next investigate the regression of *Outcome* on *AST* and *PCV* simultaneously.

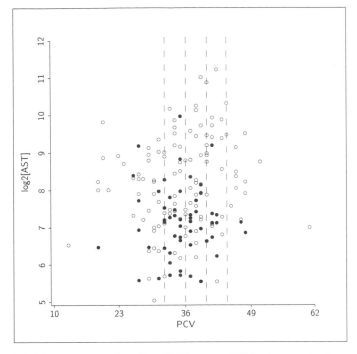

FIGURE 22.6 Binary response plot of $\log_2(AST)$ versus PCV for the recumbent cow data. Filled circles represent survivors. The vertical slices were added to facilitate discussion.

22.2.1 Assessing the Predictors

We begin by asking if there is graphical information to indicate that AST furnishes information about $Outcome$ beyond that available from PCV alone, without a specific model. Equivalently, we ask if there is information to contradict the possibility that $Outcome$ is independent of AST given PCV.

Figure 22.6 shows a *binary response plot* of $\log_2(AST)$ versus PCV. The figure was constructed by plotting $\log_2(AST)$ versus PCV and marking the points to reflect survival, with filled circles representing surviving cows and open circles representing nonsurvivors. Color is usually helpful in assessing plots of this type. The vertical slices in the plot were added to aid discussion; they represent subpopulations in which the value of PCV is relatively constant.

We can use this plot to see if $\log_2(AST)$ contains information about $Outcome$ after adjusting for PCV, or, equivalently, if $Outcome$ and $\log_2(AST)$ are independent given PCV. If $Outcome$ is independent of AST given PCV, then the relative density of survivors should be constant throughout any narrow vertical slice that approximately conditions on PCV. This means that the fraction of survivors should be constant as we move from the bottom of a slice to the top, recognizing that there will be random variation along the way. The density of survivors will usually change from slice to slice, but must be constant throughout any slice. Viewing the three slices in Figure 22.6, the relative density of survivors is higher at the bottom of the slices than at the top. This indicates that *Outcome* is *dependent* on *AST* given *PCV*. Thus, any logistic model that includes terms in *PCV* should include terms in *AST* also.

We can turn the question around and ask if there is visual information to contradict the possibility that *Outcome* is independent of *PCV* given *AST*. To answer this question, imagine a narrow *horizontal* slice that intersects the vertical axis of Figure 22.6 between 8 and 9. The relative density of survivors is higher in the middle of the slice than at either end, so *Outcome* is dependent on *PCV* given *AST*. Thus, any logistic model that includes terms in *AST* should include terms in *PCV* also. Combining results, we conclude further that any logistic model should include terms in both *AST* and *PCV*.

22.2.2 Assessing a Logistic Model with Two Predictors

We know from our visual analysis of Figure 22.6 that both *AST* and *PCV* will likely be needed in any logistic model for the regression of *Outcome* on (AST, PCV). We now consider a specific logistic model with

$$\eta^T \mathbf{u} = \eta_0 + \eta_1 \log_2(AST) + \eta_2 PCV \tag{22.5}$$

If this is the correct model, then *Outcome* is independent of (AST, PCV) given $\eta^T \mathbf{u}$. If *Outcome* depends on (AST, PCV) given $\eta^T \mathbf{u}$, then the model does not provide a complete characterization of the regression and additional terms may be required.

To assess model (22.5) graphically, begin by constructing a 3D plot with $\log_2(AST)$ on the horizontal axis, *PCV* on the vertical, and case numbers on the out-of-page axis. The variable on the out-of-page axis is only a place holder selected because it contains no missing values, and it plays no role in this discussion. Mark the points in the plot according to the value of *Outcome* to obtain a *binary response plot*. Finally, select the item "Recall logistic (H,V)" from the plot's "Recall/Extract" menu. This causes *Arc* to perform two operations. First, it fits the logistic model with the binary marking variable as the response and the horizontal and vertical axis variables as the predictors, and then it determines $\hat{\eta}^T \mathbf{u}$. Next, the 3D plot is rotated about its out-of-page axis so that the horizontal screen axis corresponds to $\hat{\eta}^T \mathbf{u}$. The results of this operation are shown in Figure 22.7 after extracting the 2D view to facilitate discussion.

The process of interpreting Figure 22.7 is the same as we used for Figure 22.6. If (22.5) is a good model, then the plot should appear as if *Outcome* is independent of the quantity on the vertical axis given $\hat{\eta}^T \mathbf{u}$ which is on the horizontal axis. Because the density of survivors seems higher in the middle of the slice shown in Figure 22.7 than at the top or the bottom, we have visual evidence to indicate that the model can be improved. The mean function

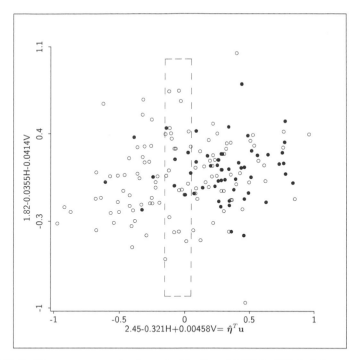

FIGURE 22.7   Binary response plot from the fit of model (22.5). Filled circles represent survivors. The vertical slice was added to facilitate discussion.

with u-terms 1, PCV, and $\log_2(AST)$ is a 1D model in the base predictors $(PCV, \log_2(AST))$. We therefore have evidence *against* 1D structure for this regression.

The same process can be used to fit logistic models by eye. Start by constructing a 3D binary response plot with the two predictors or terms on the H and V axes. Next, rotate the plot about the out-of-page axis by hand until the relative density of successes seems constant in slices perpendicular to the horizontal screen axis. The linear combination of H and V on the horizontal screen axis is then the visual estimate of $\eta^T u$. With a little practice, many people can become quite good at reproducing logistic fits in this way.

22.2.3   Assessing a Logistic Model with Three Predictors

We can visually assess the fit of a logistic model with three predictors or terms by using a process similar to that described in the last section for two predictors. Our discussion is in the context of the logistic model

$$\eta^T u = \eta_0 + \eta_1 \log_2(AST) + \eta_2 PCV + \eta_3 \log_2(CK) \qquad (22.6)$$

for the recumbent cow data.

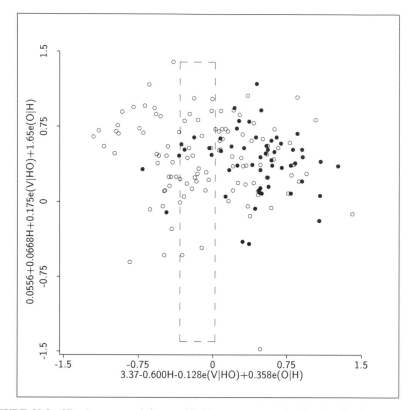

FIGURE 22.8 2D plot extracted from a 3D binary response plot for the fit of model (22.6). Filled circles represent survivors. The vertical slice was added to facilitate discussion.

Construct a *3D binary response plot*, assigning the terms $\log_2(AST)$, *PCV*, and $\log_2(CK)$ to the axes and marking with the binary response *Outcome*. If desired, visual resolution in the plot can be improved by clicking the "Rem lin trend" and "O to e(O | H)" buttons. Next, select the item "Recall logistic (H,V,O)" from the plot's "Recall/Extract" menu. *Arc* will then fit the logistic model (22.6) and rotate the plot so that $\hat{\eta}^T\mathbf{u}$ is on the horizontal screen axis. Assessing this fit is a bit more complicated than with two predictors: For the model to be sustained, the relative density of survivors must be constant within any slice perpendicular to the horizontal screen axis ($\hat{\eta}^T\mathbf{u}$) *in all 2D views obtained by rotating about the horizontal screen axis* using the "Pitch" control. Adding a predictor has in effect added a dimension to our visual assessment.

Shown in Figure 22.8 is one slice of a view we encountered while rotating about the fixed horizontal screen axis $\hat{\eta}^T\mathbf{u}$. The survivors are localized in the middle of the slice so again we find visual evidence that the model is deficient, and against 1D structure for the base predictors $(\log_2(AST), PCV, \log_2(CK))$.

TABLE 22.3 **Appropriate Choices of Terms u for Logistic Regression with Many Continuous Predictors**[a]

Density $f(\mathbf{x} \mid y)$	$\mathbf{u}^T =$	Dimension
Independent Predictors		
Σ = diagonal matrix	Univariate results	$\geq$ 1D
Multivariate Normal		
Common Σ	$(1, \mathbf{x})$	1D
Different Σ	$(1, \mathbf{x}, \text{quadratics, possibly interactions})$	$\geq$ 1D
Other Multivariate		
General case	?	?

[a]In this table, Σ is the covariance matrix of $\mathbf{x} \mid y$.

22.3 TRANSFORMING PREDICTORS

22.3.1 Guidelines

In examining Figure 22.6 on page 505, the points for the survivors and nonsurvivors seem to form approximately elliptical point clouds that are characteristic of bivariate normal distributions. If the density of the predictors within each group is bivariate normal, the results in Section 22.1 concerning log-density ratios generalize. Results for many predictors are summarized in Table 22.3. Here are a few highlights:

- If any predictor x_j is independent of the rest of the predictors, then the univariate methods can be applied to x_j to find u-terms that depend on x_j. If all the predictors in $\mathbf{x}$ are independent, then the univariate methods can be applied to each predictor separately.
- If the conditional distributions of $\mathbf{x} \mid y$ are multivariate normal distributions with common covariance matrix but different means, then use $\mathbf{x}$ for the terms.
- If the conditional distributions of $\mathbf{x} \mid y$ are multivariate normal distributions with different means and different covariance matrices, both quadratics and interactions may be required in addition to $\mathbf{x}$.
- In the case of $p = 2$ predictors $\mathbf{x} = (x_1, x_2)^T$, an interaction is required only if the regression of x_1 on x_2 has a different slope in each of the two groups ($y = 0$ and $y = 1$). A quadratic in x_i, $i = 1, 2$, is needed if the variance of x_i is different in the two groups.

Returning to Figure 22.6, there seems no clear evidence to indicate that $\log_2(AST)$ is dependent on PCV within either the survivors or nonsurvivors. Thus, we can use the single-predictor results of Section 22.1.5 to guide

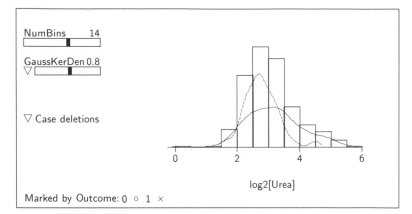

FIGURE 22.9 Conditional density estimates for $\log_2(Urea)$. The dashed line is for *Outcome* = 1, and the solid line for *Outcome* = 0.

the construction of a logistic model for the regression of *Outcome* on $(\log_2(AST), PCV)$.

22.3.2 Transforming x | y to Multivariate Normality

Without normal distributions or independence, there are no generally useful results, so transforming toward the normal distribution provides a good start for an analysis. Let's pursue this idea using the recumbent cow data with continuous predictors *AST*, *CK*, *PCV*, and *Urea*, which are all computed from a blood sample, and also the predictor *Daysrec*. Viewing these five predictors in a scatterplot matrix using *Outcome* as a marking variable, the joint conditional distributions are not multivariate normal because the mean functions are not all linear. However, this same scatterplot matrix can be used to find power transformations that may normalize their joint conditional distributions. We concluded that *AST*, *CK*, and *Urea* should be replaced by their logarithms because the estimated normalizing powers are all close to zero, and *PCV* should be left untransformed. We didn't attempt to transform *Daysrec* for two reasons. First, there doesn't seem to be much dependence between *Daysrec* and the other predictors so we can consider *Daysrec* alone. Second, *Daysrec* is a count, suggesting that it might follow a Poisson distribution and, according to Table 22.1, Poisson predictors should not be transformed.

As a next step, we examined conditional density estimates for each of the transformed predictors. Estimates for $\log_2(CK)$, $\log_2(AST)$, and *PCV* are shown in Figures 22.2, 22.4, and 22.5. The discussion of those figures applies here. Density estimates for $\log_2(Urea)$ are shown in Figure 22.9.

If normality is reasonable for $\log_2(Urea)$ in Figure 22.9, then we should include $\log_2(Urea)$ and $[\log_2(Urea)]^2$ as terms. If normality is not reason-

TABLE 22.4 Initial Logistic Regression Summaries for the Recumbent Cow Data

```
Data set = Downer, Name of Fit = B1
372 cases are missing at least one value.
Binomial Regression
Kernel mean function = Logistic
Response        = Outcome
Terms           = (log2[AST] log2[CK] log2[Urea] log2[Urea]^2
                   PCV Daysrec)
Trials          = Ones
Coefficient Estimates
Label           Estimate    Std. Error    Est/SE
Constant        -1.01734    4.73578       -0.215
log2[AST]       -0.505871   0.281431      -1.797
log2[CK]        -0.147766   0.155148      -0.952
PCV             0.0674652   0.0329217      2.049
log2[Urea]      4.49064     3.00449        1.495
log2[Urea]^2    -1.04146    0.526232      -1.979
Daysrec         -0.320013   0.147248      -2.173

Sigma hat:                     1.
Number of cases:             435
Number of cases used:        165
Degrees of freedom:          158
Pearson X2:              141.416
Deviance:                151.262
```

able, we should include *Urea* as well. The possibility of interactions between $\log_2(Urea)$ and the other predictors can be explored by looking at a multi-panel plot of each of the other transformed predictors versus $\log_2(Urea)$. If the within-group OLS regression lines in these plots are not parallel, then interactions may be needed; examination of these plots provides some evidence of the need for interactions, so we cannot eliminate this possibility.

Based on this discussion, a reasonable first logistic model for the recumbent cow data would include the nine terms suggested by the separate density estimates: $\log_2(AST)$, CK, $\log_2(CK)$, PCV, PCV^2, $Urea$, $\log_2(Urea)$, $[\log_2(Urea)]^2$ and *Daysrec*. It is quite likely that several of these terms will not be needed in the model, and there is still the possibility that interactions may be needed.

To facilitate and enrich the discussion in the rest of this chapter, we fit a simpler model with terms

$$\mathbf{u} = (\log_2(AST), \log_2(CK), PCV, \log_2(Urea), [\log_2(Urea)]^2, Daysrec)^T$$

The summary of the fit of this regression is shown in Table 22.4. This regression is based on 165 fully observed cases.

As with any regression, we must check to see if the fit of this model matches the data, and perform other diagnostic checks. Model checking plots and residual plots can be of use here.

22.4 DIAGNOSTIC METHODS

22.4.1 Residual Plots

Let $\hat{y}_i = m_i \hat{\theta}(\mathbf{x}_i)$ denote the fitted values from a logistic regression of y_i on $\mathbf{x}_i$, where m_i denotes the number of trials for case $i = 1, \ldots, n$. In Chapter 14, residuals $\hat{e}_i$ were defined to be

$$\hat{e}_i = \sqrt{w_i}(y_i - \hat{y}_i) \tag{22.7}$$

where the weight w_i comes from the variance function, $\text{Var}(y_i \mid \mathbf{x}_i) = \sigma^2 / w_i$. In logistic regression the variance function $\text{Var}(y_i \mid \mathbf{x}_i) = m_i \theta(\mathbf{x}_i)(1 - \theta(\mathbf{x}_i))$ can be put in the earlier notation by setting $\sigma^2 = 1$, and

$$w_i = 1/(m_i \theta(\mathbf{x}_i)(1 - \theta(\mathbf{x}_i))) \tag{22.8}$$

Unlike the weights with the linear regression model, these weights depend on unknown parameters, and therefore the weights must be estimated along with the fitted values. Substituting $\hat{\theta}(\mathbf{x}_i)$ for the unknown $\theta(\mathbf{x}_i)$, we obtain

$$\hat{e}_{Xi} = \frac{y_i - \hat{y}_i}{\sqrt{m_i \hat{\theta}(\mathbf{x}_i)(1 - \hat{\theta}(\mathbf{x}_i))}} \tag{22.9}$$

These are called *chi-residuals* because the sum of their squares add to Pearson's X^2 statistic, as is easily verified by examining equation (21.13). In *Arc*, the chi-residuals are called :Chi-Residuals.

The chi-residuals can be plotted as described in Section 14.1, but interpreting chi-residual plots requires more care than interpreting plots with a continuous response. For example, shown in Figure 22.10 is a plot of $\hat{e}_{Xi}$ versus $\hat{\eta}^T \mathbf{u}_i$ for the fitted model summarized in Table 22.4. Since each fitted value $\hat{y}_i$ is just a nonlinear transformation of $\hat{\theta}(\mathbf{x}_i)$, the residuals fall on two smooth curves: one for cases with $y = 0$, for which all residuals are negative, and one for cases with $y = 1$, for which all residuals are positive. The essential information in a plot of the chi-residuals $\hat{e}_{Xi}$ versus a linear combination $\mathbf{b}^T \mathbf{x}$ of the predictors is contained in the mean function $\text{E}(\hat{e}_{Xi} \mid \mathbf{b}^T \mathbf{x})$. If this mean function is constant, then there is no information to contradict the model. But if the mean function is not constant, then we do have information to contradict the model. The *lowess* smooth on Figure 22.10 is nearly constant, suggesting no evidence against the fitted mean function in Table 22.4. In plots against

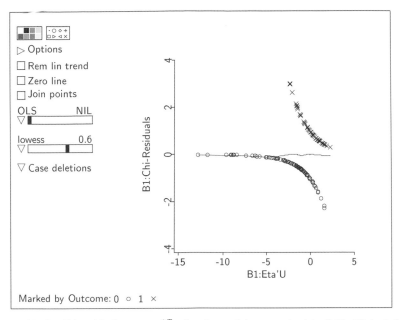

FIGURE 22.10 Chi-residuals versus $\hat{\eta}^T\mathbf{u}$ for the model summarized in Table 22.4. A *lowess* smooth is shown on the plot.

other linear combinations of the predictors—for example, in Figure 22.11 for $\log_2(Urea)$—the *lowess* smooth also seems fairly constant.

As the number m_i of trials increases, chi-residual plots begin to look more like residual plots for a continuous response, and all the methods described in Section 14.1 apply. In general, we find that the model checking plots to be described in Section 22.4.3 are better at examining a fit of a model than residual plots, particularly when the m_i are small.

22.4.2 Influence

Cook's distance can be applied directly to binomial regression or to most other regression models, with exactly the same methodology that was used for multiple linear regression. Recall that Cook's distance for case i provides a summary of the difference between an estimator $\hat{\eta}$ based on all the data and $\hat{\eta}_{(i)}$ obtained without using case i. In multiple linear regression, $\hat{\eta}_{(i)}$ can be obtained without actually recomputing the regression, but in logistic regression, recomputing is needed to get $\hat{\eta}_{(i)}$ exactly. However, a good approximation can be obtained using the formula for multiple linear regression. The formula for D_i for logistic regression is

$$D_i = \frac{1}{k}\left(\frac{\hat{e}_{Xi}^2}{1-h_i}\right)\frac{h_i}{1-h_i} \qquad (22.10)$$

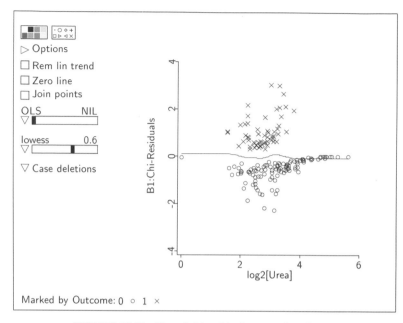

FIGURE 22.11   Plot of chi-residuals versus $\log_2(Urea)$.

which is the same as (15.9), but with the chi-residual divided by $(1 - h_i)^{1/2}$ replacing the Studentized residual, and with the ith leverage h_i computed with estimated weights w_i given by (22.8).

22.4.3   Model Checking Plots

Model checking plots were introduced in Section 17.1 for linear models. The general idea is the same for a binomial response: For any linear combination $\mathbf{a}^T\mathbf{u}_i$ of the predictors or terms, we imagine drawing two plots: one of the observed fraction of successes y_i/m_i versus $\mathbf{a}^T\mathbf{u}_i$, and the other of the estimated probabilities of success $\hat{\theta}(\mathbf{x}_i)$ versus $\mathbf{a}^T\mathbf{u}_i$. We then smooth each of these two plots using an appropriate mean smoother. If the smoothers agree for this choice of $\mathbf{a}^T\mathbf{u}_i$, and in principle for every other choice of $\mathbf{a}^T\mathbf{u}_i$, then we say that the model is reproducing the data; if we find an $\mathbf{a}^T\mathbf{u}_i$ for which the two smooths do not agree, then the model does not reproduce the data.

We begin with the model summarized in Table 22.4 and select the item "Model checking plots" from the model's menu. For the selected variables, choose $\log_2(AST)$, $\log_2(CK)$, PCV, $\log_2(Urea)$, and $Daysrec$. Plots versus indicator variables are not informative and should be skipped. Using the option for marginal plots, the initial view shown in Figure 22.12 is of y_i/m_i on the vertical axis ($m_i = 1$ in the recumbent cow data) versus $\hat{\eta}^T\mathbf{u}_i$ estimated from the model. The solid line shows the *lowess* smooth fit to this plot. The dashed

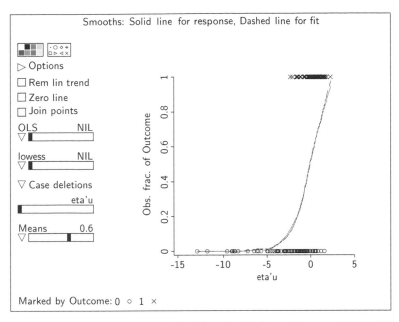

FIGURE 22.12   Initial view of the model checking plot for the model summarized in Table 22.4 for the recumbent cow data. The horizontal axis is $\hat{\eta}^T\mathbf{u}$. The smooth fit to the data on the plot is shown as a solid (and blue on the computer screen) line, while the smooth fit from the model is shown as a dashed (and red on the computer screen) line. When plotting against $\hat{\eta}^T\mathbf{u}$, the smooth from the model is just the kernel mean function.

line is obtained by smoothing the fitted probabilities $\hat{\theta}(\mathbf{x}_i)$ versus $\hat{\eta}^T\mathbf{u}_i$ and then adding this smooth to the plot. Since by (21.10), page 476, the fitted probabilities are determined by $\hat{\eta}^T\mathbf{u}$ through the kernel mean function, the dashed curve will reproduce the mean function. If the horizontal axis is anything other than $\hat{\eta}^T\mathbf{u}_i$, the dashed line from the model will not be the kernel mean function.

In examining Figure 22.12, we see most of the survivors ($y = 1$) have values of $\hat{\eta}^T\mathbf{u}$ that are large, while the nonsurvivors have a much wider range of values for $\hat{\eta}^T\mathbf{u}$. Since the two curves are essentially identical, the model is reproducing the data in this view. We show in Figure 22.13 four of the five additional model checking plots for this illustration. In all four cases the two smooths agree reasonably well, except possibly at the extremes of the range where apparent differences may be due to small sample sizes.

In Figure 22.13a the smooth fit to the data is strongly influenced by four cases with small values of $\log_2(AST)$ that died. The estimated probabilities of survival for three of these cases are around 0.8; you can discover this by drawing a plot of the fitted probabilities, called :Fit-Fraction in *Arc*, versus $\log_2(AST)$ and then selecting in the model checking plot the four points

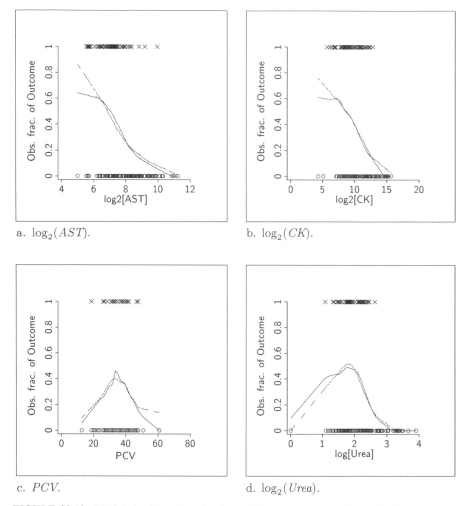

a. $\log_2(AST)$.   b. $\log_2(CK)$.

c. PCV.   d. $\log_2(Urea)$.

FIGURE 22.13   Model checking plots for four of the continuous predictors in the recumbent cow data. Solid lines are fit to the data, dashed lines are fit from the model. In each case, a *lowess* smooth with $f = 0.6$ is shown.

for the animals with low $\log_2(AST)$ that died. Observing a few animals with moderately high estimated survival probabilities that died is to be expected now and then.

For Figure 22.13b, the agreement between the two smooths appears acceptable throughout the range, in spite of two cases with very low $\log_2(CK)$ that died. (Are the low $\log_2(CK)$ cases also low $\log_2(AST)$ cases?)

We next turn to PCV in Figure 22.13c. The model reproduces a roughly quadratic trend, even though a quadratic term for PCV is not in the mean function. The two smooths again are reasonably close, with largest disagreement at the upper end of the range for *PCV*, where we have only one data point.

The final plot shown in Figure 22.13d is for $\log_2(Urea)$. We see that the model successfully reproduces the roughly quadratic shape of the dependence of the probability of survival on $\log_2(Urea)$: Survival is low if $\log_2(Urea)$ is far from the middle of its range. The two curves do appear to disagree for very small values of $\log_2(Urea)$, but this may be due to one case with an exceptionally low value of $\log_2(Urea)$. When this case is deleted, the smooths for the remaining data look very similar.

In Section 17.1.2, page 401, we showed how to use model checking plots to check for assumptions about the variance function. In models with a binary response, this check is not applicable because the variance is completely determined by the probability of success, regardless of the model. The checks on the variance function discussed in Section 17.1.2 can be used in regressions with a binomial response with $m_i > 1$ because correlations between the responses, or heterogeneity of the observations, can lead to *overdispersion*. References for appropriate methodology are given in the complements section.

22.5 ADDING FACTORS

Many binary regressions include both continuous variables and factors. There are at least two issues to consider: (1) How should the factors be included? Are factor-by-factor or factor by continuous-term interactions required? (2) How should the continuous predictors be transformed? The study of interactions parallels the discussion in Chapter 12 for the multiple linear regression model, and so we will not repeat the discussion here. The choice of transformations is a bit trickier.

There seem to be two strategies for transforming continuous predictors with factors in the regression. First, we could try to select transformations after conditioning not only on the response *y* but also on the factors. For example, if the only factor was a single indicator variable with two categories, then we would transform toward normality in each of four groups, all combinations of *y* and the indicator variable. This can be done in *Arc* by creating a marking variable with four values, one for each combination of *y* and the indicator variable. With two indicator variables, we would have eight groups for conditioning. It is clear that this approach will quickly get out of hand, although, in theory, this approach will obtain the correct transformations of predictors into terms.

The second approach is more practical, but it is not as well-supported by theory: Transform the continuous predictors first, and then add the indicator variables for the factors. This method will not always give the best transformations of the predictors, but it will generally provide a reasonable starting point for further analysis.

TABLE 22.5   Logistic Regression Summaries for the Recumbent Cow Data with Two Categorical Predictors Added

```
Data set = Downer, Name of Fit = B2
372 cases are missing at least one value.
Binomial Regression
Kernel mean function = Logistic
Response       = Outcome
Terms          = (log2[AST] log2[CK] PCV log2[Urea] log2[Urea]^2
                 Daysrec Calving Inflamat)
Trials         = Ones
Coefficient Estimates
Label             Estimate     Std. Error     Est/SE
Constant          -1.62345      5.80178       -0.280
log2[AST]         -0.421156     0.334006      -1.261
log2[CK]          -0.137520     0.187697      -0.733
PCV                0.0435961    0.0426326      1.023
log2[Urea]         4.39140      3.63961        1.207
log2[Urea]^2      -1.04630      0.650270      -1.609
Daysrec           -0.341009     0.172844      -1.973
Calving            1.77105      0.761026       2.327
Inflamat          -0.675952     0.511157      -1.322

Sigma hat:                        1.
Number of cases:                435
Number of cases used:           124
Degrees of freedom:             115
Pearson X2:                 101.504
Deviance:                   113.589
```

In the recumbent cow data, for example, there are three indicator variables, or factors with two levels: *Inflamat*, *Calving*, and *Myopathy*, and these can be added to the mean function. However, because *Myopathy* was not observed on the majority of animals, using this term in the mean function will reduce the sample size from 165 to only 71; we leave consideration of this predictor to Problem 22.5 and add only the other two indicator variables to **u**. This assumes that the factors enter additively, so there are no interactions, and that the transformations of the continuous predictors obtained previously do not need to be modified. Both of these assumptions can be checked, using tests, residual plots, and model checking plots. The summary of this regression is given in Table 22.5.

Compare Tables 22.4 and 22.5: some of the terms that had relatively large coefficient estimates in Table 22.4 have smaller coefficient estimates in Table 22.5. As usual, this is caused by close relationships between the terms. We leave further analysis of this example to Problem 22.5.

22.6 EXTENDING PREDICTOR TRANSFORMATIONS

22.6.1 Power Transformations with a Binomial Response

The methods described in this chapter for choosing transformations of predictors are easily extended to a binomial, rather than binary, response. Suppose that y_i is the number of successes in m_i trials. The automatic procedures for finding transformations can be computed by allocating y_i copies of x_i to the population of successes and $m_i - y_i$ copies of x_i to the population of failures.

This can be done for the Box–Cox method for selecting transformations from a scatterplot matrix by pushing the button to "Condition using binomial weights" in the "Find normalizing transformations" dialog obtained from the "Transformations" pop-up menu. This will bring up a dialog box to specify the variable to use for the number of successes and the variable to use for the number of trials.

22.6.2 *Ceres* Plots

The transformation model

$$y \mid \mathbf{x} = \eta_1^T \mathbf{u}_1 + \tau(u_2) + e$$

used to motivate *Ceres* plots in Chapter 16, equation (16.4), can be adapted straightforwardly to logistic regression:

$$\log\left(\frac{\theta(\mathbf{x})}{1 - \theta(\mathbf{x})}\right) = \eta_1^T \mathbf{u}_1 + \tau(u_2) \tag{22.11}$$

where τ is the unknown transformation of the term u_2. *Ceres* plots for logistic regression can be used to visualize τ in the same way that *Ceres* were used for multiple linear regression in Chapter 16 provided that one additional condition holds: $\theta(\mathbf{x})$ should stay away from its extremes, say $0.1 < \theta(\mathbf{x}) < 0.9$. If $\theta(\mathbf{x})$ can be near 0 or 1, then a *Ceres* plots may provide a biased representation of τ. This possibility was not relevant for the discussion of Chapter 16.

Ceres plots for logistic regression can be constructed in the *Arc* by selecting the item "Ceres plots – 2D" from the logistic regression menu and then selecting the augmentation form, as discussed in Chapter 16. The form of τ can be judged by smoothing the resulting *Ceres* plots.

22.7 COMPLEMENTS

22.7.1 Marginal Odds Ratio

The value and interpretation of $c = \Pr(y = 1)/(1 - \Pr(y = 1))$ as used in equation (22.1) depend on the sampling plan that generated the data. If the data were a random sample from a population, then c is the odds of success in the

population, as we used for the recumbent cow data. If the data were selected
in some other way, then c need not reflect a population quantity. For example,
if the sample were selected to have 100 successes and 100 failures, then $c = \frac{1}{2}/\frac{1}{2} = 1$ is fixed by the sampling design. If the number of successes and
failures is fixed by the sampling plan, we have a *case-control study*.

22.7.2   Relative Density

Equation (22.1), which provides a connection between the mean function and
the conditional densities of the predictor, can be derived with the aid of Bayes'
theorem.

Let $f(x)$ be the *marginal density* of x. Using Bayes' theorem we write

$$E(y \mid x) = \theta(x) = \Pr(y = 1 \mid x) = \frac{f(x \mid y = 1)\Pr(y = 1)}{f(x)} \qquad (22.12)$$

We can write a similar formula relating $\theta(x)$ to $f(x \mid y = 0)$. Recalling that
$\Pr(y = 0) = 1 - \Pr(y = 1)$, we get

$$1 - \theta(x) = \Pr(y = 0 \mid x) = \frac{f(x \mid y = 0)(1 - \Pr(y = 1))}{f(x)} \qquad (22.13)$$

Taking the ratio of (22.12) to (22.13), the marginal density $f(x)$ cancels, leaving (22.1).

22.7.3   Deviance Residuals

Several alternative definitions of residuals to replace the $\hat{e}_X$ have been proposed in the literature to overcome at least in part the systematic problems
discussed above or to meet other specific needs. The most commonly used of
these are the *deviance residuals*, which are defined so their squares add up to
the deviance G^2. For logistic regression, these are defined by

$$\hat{e}_{Di} = \sqrt{2} \, \text{sign}(y_i - \hat{y}_i) \left[y_i \log\left(\frac{y_i}{\hat{y}_i}\right) + (m_i - y_i)\log\left(\frac{m_i - y_i}{m_i - \hat{y}_i}\right) \right]^{1/2}$$
$$(22.14)$$

where $\text{sign}(z)$ is equal to $+1$ if $z \geq 0$ and -1 if $z < 0$. That $G^2 = \sum \hat{e}_{Di}^2$ can
be verified by inspection of (21.12). In *Arc*, the deviance residuals are called
:Dev-Residuals.

22.7.4   Outliers

Outliers in logistic regression are harder to define and identify than outliers in multiple linear regression. Informally, an outlier in a binary response
plot will appear as an isolated success (failure) that is surrounded by fail-

ures (successes). Response outliers in logistic data were discussed by Copas (1988), who proposed a misclassification model, Williams (1987), Collett (1991, Chapter 5), and Bedrick and Hill (1990).

22.7.5 Overdispersion

When all the $m_i = 1$, the distribution of $y \mid \mathbf{x}$ can depend on $\mathbf{x}$ only through the mean function because the only distribution on the values 0 and 1 is the Bernoulli distribution, which depends only on the probability that the response is equal to one. If $m_i > 1$ and the y_i are binomially distributed, then once again the variance is determined by the mean only. However, the binomial distribution requires two assumptions that need not hold in practice: for an observation with m_i trials, each trial must have the same probability of success, and all trials must be independent of each other. If, for example, the m_i trials correspond to the m_i pigs in a litter, then the pigs might compete for resources, invalidating the independence assumption. In the literature, this is called *overdispersion*. Methodology for overdispersed binomial data is presented by Williams (1982), Collett (1991, Chapter 6), and McCullagh and Nelder (1989, Section 4.5).

22.7.6 Graphical Regression

The graphical regression methods discussed in Chapter 20 can be used with a binary response. In *Arc* select "Graphical regression" from the Graph&Fit menu, specify the predictors, and specify the binary response in the resulting dialog. The buttons at the bottom of the dialog allow you to work in terms of the original predictors or the GREG predictors, as discussed in Chapter 20. Next, select "Binary response plot" from the graphical regression menu. This will bring up a dialog box in which you can choose two or three predictors to combine. This discussion is for combining two predictors. It is possible to combine three predictors simultaneously, but the visual analysis takes more practice than that for two predictors.

After selecting two predictors to combine, you will be presented with a binary response plot with two GREG predictors on the horizontal and vertical axes, and a place holder on the out-of-page axis. The points in the plot are marked according to the states of y and rotation is allowed only about the out-of-page axis. The structural dimension of this plot can be assessed visually as follows:

- If the density of successes is constant throughout the plot, then the structural dimension is zero and the two GREG predictors can be deleted from the analysis. In this case, select "Dimension 0" from the "Greg methods" pop-up menu.
- If the structural dimension is greater than 0, try to rotate the plot about the out-of-page axis so that the density of successes is constant in slices that

are perpendicular to the horizontal screen axis. If this is possible, then there is 1D structure which should be indicated by selecting "Dimension 1" from the "Greg methods" pop-up menu. *Arc* will then compute the linear combination of the original predictors on the horizontal screen axis and add the new GREG predictor to the data set.

- Otherwise, the plot must have 2D structure and the selected predictors cannot be combined. It may still be possible to combine two other predictors, however.

- The linear combination of the predictors from the fit of a logistic regression model is often a good place to start when assessing 1D or 2D structure. The binary response plot can be rotated to this linear combination by selecting "Recall logistic (H,V)" from the "Recall/Extract" menu.

This iterative procedure is continued until no further combining is possible, in which case the final binary response plot is the summary of the regression.

Apart from the use of binary response plots and the possibility of combining three predictors simultaneously, graphical regression with a binary response is the same as that discussed in Chapter 20. In particular, it is still necessary to start with linearly related predictors.

22.7.7   References

The results in Section 22.1 for transforming one predictor were adapted from Kay and Little (1987). The results in Section 22.2 are from Cook (1996). *Ceres* plots for logistic regression were developed by Cook and Croos-Dabrera (1998). Additional discussion of graphics for regressions with a binary response at a more advanced level was given by Cook (1998b). The collision data in Problem 22.7 is from Härdle and Stoker (1989). The bank note data are from Flury and Riedwyl (1988).

PROBLEMS

22.1   Describe how a binary response plot for two predictors (x_1,x_2) would look if the binary response were independent of (x_1,x_2). Your answer should be in terms of the relative density.

22.2   Assume that the binary response y is independent of the two predictors (x_1,x_2) given the linear combination $ax_1 + bx_2$.

    22.2.1   Describe the essential characteristics of a binary response plot for (x_1,x_2). Your answer should again be in terms of the relative density.

22.2.2 Suppose now that you fit the logistic model

$$\eta^T \mathbf{u} = \eta_0 + \eta_1 x_1 + \eta_2 x_2$$

How would you expect the estimates of the η's to be related to a and b in the linear combination $ax_1 + bx_2$?

22.3 In the recumbent cow data, assume that

$$(\log_2(CK), \log_2(AST)) \mid (Outcome = j)$$

follows a bivariate normal distribution. Based on a binary response plot of $\log_2(CK)$ versus $\log_2(AST)$, what terms do you think will be required in a logistic model for the regression of *Outcome* on $(\log_2(CK), \log_2(AST))$. Give the rationale you used for your choices.

22.4 Consider the regression of *Outcome* on *Inflamat* and *Daysrec* in the recumbent cow data.

  22.4.1 Assuming that *Inflamat* and *Daysrec* are independent and that *Daysrec* is a Poisson random variable, give the appropriate logistic model for the regression.

  22.4.2 Give an appropriate logistic model for the regression assuming that *Inflamat* and *Daysrec* are *dependent* and that *Daysrec* is a Poisson random variable.

22.5 We left the recumbent cow data Section 22.3.2 with the initial logistic model consisting of an intercept and the nine terms $\log_2(AST)$, CK, $\log_2(CK)$, PCV, PCV^2, *Urea*, $\log_2(Urea)$, $[\log_2(Urea)]^2$, and *Daysrec*. The following problems are based on this *full model*.

  22.5.1 Using Cook's distance and the full model, are there any clearly influential cases in the data? If so, describe how they influence the fit and then delete them for all subsequent problems.

  22.5.2 Investigate plots of the chi-residuals from the fit of the full model and describe any deficiencies you see. Do the plots sustain the model?

  22.5.3 Investigate model checking plots using $\log_2(AST)$, $\log_2(CK)$, PCV, *Urea*, *Daysrec*, $\hat{\eta}^T \mathbf{u}$, and several randomly chosen plots. Are there any notable deficiencies?

  22.5.4 Using the full model, perform backward elimination, sequentially deleting terms where the coefficient estimate divided by its standard error is smaller than a cutoff you select. For the reduced model, investigate model checking plots using $\log_2(AST)$, $\log_2(CK)$, PCV, *Urea*, *Daysrec*, $\hat{\eta}^T \mathbf{u}$, and several randomly chosen plots. Does your reduced model seem to fit as well as the full model you investigated in Problem 22.5.3?

TABLE 22.6 The Collision Data

Age	"Age" of the crash dummy used in the test.
Acl	Maximum acceleration on impact measured on the dummy's abdomen.
Vel	Velocity of automobile at impact.
Y	1 if the accident would be fatal, 0 otherwise.

    22.5.5   Finally, restore any influential cases deleted at the start in Problem 22.5.1. Do they agree with the final model?

22.6   The data in the file `banknote.lsp` contain six physical measurements on 100 genuine Swiss bank notes and on 100 counterfeit Swiss bank notes. The goal of the analysis is to be able to predict whether a bill is genuine or counterfeit based on the physical measurements alone.

    22.6.1   Draw a scatterplot matrix of the six predictors, using *Status* as a marking variables, and summarize results.

    22.6.2   Based on examination of the scatterplot matrix, use logistic regression to estimate the probability that a note is counterfeit.

    22.6.3   Apply appropriate diagnostic methods to decide if the mean function you obtained in the last part is reasonable.

    22.6.4   Are any "outliers" (either a counterfeit bill surrounded by genuine ones, or a genuine bill surrounded by counterfeit ones in a plot) apparent in these data?

    22.6.5   If all counterfeit and genuine bills in the future will have the same characteristics as these bills, will the model you fit provide good discrimination between genuine and counterfeit bills?

22.7   The data in the file `collide.lsp` describe the outcomes of 58 simulated side-impact collisions using crash dummies. In each crash, the variables described in Table 22.6 were measured.

    22.7.1   Using a scatterplot matrix and histograms of each of the three predictors, both using *Y* as a marking variable, use the graphical methods of this chapter to support replacing each predictor by its logarithm. If you think the some other terms are more appropriate than logs, give the evidence for a different scaling.

    22.7.2   From the scatterplot matrix, present an argument that would support deleting at least one of the three predictors. Then, fit the logistic models with and without the deleted variable to support what you concluded graphically.

    22.7.3   Examine the model you obtained in the last part of this problem using the diagnostic tools outlined in this chapter.

CHAPTER 23

Generalized Linear Models

In this chapter we introduce *generalized linear models*, or GLMs. There are several families of generalized linear models, including both the normal family that leads to multiple linear regression and the binomial family that leads to logistic regression.

23.1 COMPONENTS OF A GENERALIZED LINEAR MODEL

A generalized linear model has the following structure:

The Data. The data consist of cases $(\mathbf{x}_i, y_i)$, $i = 1,\dots,n$, where $\mathbf{x}_i$ is a $p \times 1$ vector of predictors and y_i is a scalar response. We assume that cases are independent. As usual, the goal is to study the distribution of $y \mid \mathbf{x}$ as $\mathbf{x}$ is varied, with emphasis on the mean function $\mathrm{E}(y \mid \mathbf{x})$ and the variance function $\mathrm{Var}(y \mid \mathbf{x})$.

The Linear Predictor. Let $\mathbf{u}_i$ be a $k \times 1$ vector of u-terms derived from $\mathbf{x}_i$. We assume that the dependence of y_i on $\mathbf{x}_i$ is through the linear combination $\eta^T \mathbf{u}_i$, for some unknown $k \times 1$ vector η of regression coefficients. By allowing the terms to be nonlinear functions of the predictors or interactions between them, we are implicitly allowing for models of dimension one or higher. Virtually all of the methodology discussed previously for determining dimension and for graphical analysis applies to GLMs.

The dependence of $y_i \mid \mathbf{x}_i$ on $\mathbf{x}_i$ is assumed to be through the mean function, given by $\mathrm{E}(y_i \mid \mathbf{x}_i) = \mathrm{M}(\eta^T \mathbf{u}_i)$, where M is a kernel mean function. While the user has latitude in choosing M, there is a particular kernel mean function for each family—called the *canonical kernel mean function*—that is used most often.

The variance function is $\mathrm{Var}(y_i \mid \mathbf{x}_i) = \mathrm{V}(\eta^T \mathbf{u}_i)$, where V is the kernel variance function, and $\eta^T \mathbf{u}_i$ is the same linear combination used to determine the mean. If we know the mean function and the relationship between the mean and variance, then apart from a possible scale factor we must also know the variance function.

TABLE 23.1   Relationship Between the Mean and Variance for Four Families of Generalized Linear Models[a]

Family Name	Canonical Kernel Mean Function	Variance Function
Normal	$\mathrm{M}(\eta^T\mathbf{u}) = \eta^T\mathbf{u}$	σ^2/w
Binomial	$\mathrm{M}(\eta^T\mathbf{u}) = [1 + \exp(-\eta^T\mathbf{u})]^{-1}$	$\mathrm{M}(\eta^T\mathbf{u})[1 - \mathrm{M}(\eta^T\mathbf{u})]/w$
Poisson	$\mathrm{M}(\eta^T\mathbf{u}) = \exp(\eta^T\mathbf{u})$	$\mathrm{M}(\eta^T\mathbf{u})$
Gamma	$\mathrm{M}(\eta^T\mathbf{u}) = 1/\eta^T\mathbf{u}$	$\sigma^2[\mathrm{M}(\eta^T\mathbf{u})]^2/w$

[a]The canonical kernel mean functions are also given for each family. In each of the variance functions in this table, w is a known weight that may differ from case to case, and σ^2 is a constant.

The Mean–Variance Relationship. The relationship between the mean and the variance determines a *family* of generalized linear models. Estimates are computed differently for each family. *Arc* includes the four most important families of GLMs, and for these the relationship between the mean function and the variance function is given in Table 23.1.

The first family of models is the normal family. For this family, the variance function is constant, and, apart from known weights, it depends only on σ^2. The canonical kernel mean function for the normal family is the identity function. This choice gives the usual multiple linear regression model studied in Part II of this book. If the kernel mean function is not the identity function, we can get other members of the normal family that may be useful in some regressions, as illustrated in Section 23.2.

The second family in Table 23.1 is the binomial family, in which the mean function is bounded between 0 and 1 and the variance function is $\mathrm{M}(\eta^T\mathbf{u})$ $\times(1 - \mathrm{M}(\eta^T\mathbf{u}))/w$. The canonical kernel mean function is the logistic function, which gives logistic regression. Other choices for the kernel mean function were discussed in Section 21.6.3.

In the third family, the response y_i is non-negative and the variance is equal to the mean. This relationship between the mean and the variance holds for Poisson random variables (see Section 23.5.1), and the estimates used are maximum likelihood estimates if $y_i \mid \mathbf{x}_i$ has a Poisson distribution. Even if $y_i \mid \mathbf{x}_i$ is not Poisson but the mean function and the variance function are equal, the estimates have some of the same properties as maximum likelihood estimates. As in the binomial case, there are no additional scale parameters in the kernel variance function; once the mean is determined, so is the variance. The canonical kernel mean function for the Poisson is the exponential function. Weights are generally not used with Poisson models.

For the gamma family the response is positive and the variance is equal to a constant times the square of the mean, and thus variability increases more quickly than it does with the Poisson. The canonical kernel mean function for gamma regression is the inverse function.

TABLE 23.2 Kernel Mean Functions for Various Choices of Fitting Method[a]

Kernel Mean Function	Fitting Method(s)	E($y \mid$ **x**) =
Identity	**Linear LS**, Poisson, Gamma	$\eta^T\mathbf{u}$
Inverse	Linear LS, Poisson, **Gamma**	$1/(\eta^T\mathbf{u})$
Exponential	Linear LS, **Poisson**, Gamma	$\exp(\eta^T\mathbf{u})$
1D-Quad	Linear LS	$\alpha_0 + \alpha_1(\eta^T\mathbf{u}) + \alpha_2(\eta^T\mathbf{u})^2$. This model is described in the context of 3D plots in Section 18.3.2.
Full-Quad	Linear LS	Quadratic linear regression with all possible linear, quadratic, and interaction terms.
Logistic	**Binomial**	$1/(1 + \exp(-\eta^T\mathbf{u}))$
Inv-Probit	Binomial	$\Phi(\eta^T\mathbf{u})$, where Φ is the standard normal cumulative distribution function.
Inv-Cloglog	Binomial	$1 - \exp(-\exp(\eta^T\mathbf{u}))$

[a]Boldface indicates the canonical kernel mean function for that fitting method.

In addition to the canonical kernel mean functions, *Arc* includes alternative choices for each of the four families that cover most practical applications of generalized linear models; these are listed in Table 23.2.

Once we select the terms $\mathbf{u}_i$, a family for the relationship between the mean and variance functions, and a kernel mean function, we can compute estimates of unknown parameters, get standard errors, compute confidence statements and tests, and generally apply all the ideas we have learned for the multiple linear regression model and for the logistic regression model.

23.2 NORMAL MODELS

For the normal generalized linear model, the canonical choice of the identity kernel mean function M leads to the multiple linear regression model. By allowing M to be some other function, we can get nonlinear models. Other choices for M listed in Table 23.2 include the exponential and the inverse functions. Both of these are graphed in Figure 23.1.

The exponential function grows large very quickly as $\eta^T\mathbf{u}$ increases and approaches zero very slowly as $\eta^T\mathbf{u}$ becomes large and negative. This kernel mean function can therefore be useful if the response either increases rapidly or else approaches a limit called an *asymptote*. The inverse kernel mean function is undefined if $\eta^T\mathbf{u} = 0$, and it should only be used if $\eta^T\mathbf{u}$ does not change sign for the range of **x** that might occur in practice. This function approaches

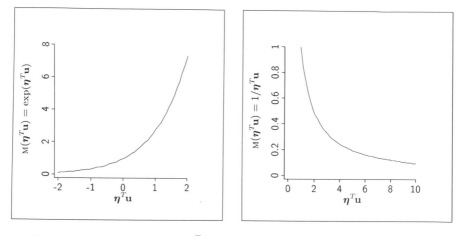

a. The exponential function, $\text{M}(\eta^T\mathbf{u}) = \exp(\eta^T\mathbf{u})$.

b. The inverse function, $\text{M}(\eta^T\mathbf{u}) = 1/\eta^T\mathbf{u}$.

FIGURE 23.1 Two mean functions.

an asymptote for large values of $\eta^T\mathbf{u}$, and grows very large as $\eta^T\mathbf{u}$ becomes small.

As an example of the use of one of these kernel mean functions, we consider the *Michaelis–Menton formula* for enzyme kinetics. This relates the initial speed of a chemical reaction y to the concentration x of an agent via the mean function

$$E(y_i \mid x_i) = \frac{\theta_1 x_i}{x_i + \theta_2} \tag{23.1}$$

Assuming that both of the unknown parameters θ_1 and θ_2 are positive, (23.1) is monotonically increasing; at $x = 0$, $E(y \mid x) = 0$, while as x grows large $E(y \mid x)$ approaches θ_1. The second parameter θ_2 is a rate parameter.

If we transform (23.1) by dividing the top and bottom of the right side of (23.1) by $\theta_1 x_i$, we get

$$E(y_i \mid x_i) = \frac{1}{1/\theta_1 + (\theta_2/\theta_1)(1/x_i)}$$

$$= 1/\eta^T\mathbf{u}_i \tag{23.2}$$

where $\eta = (1/\theta_1, \theta_2/\theta_1)^T$ and $\mathbf{u}_i = (1, 1/x_i)^T$. If the variance function $\text{Var}(y_i \mid x_i)$ is constant, we have a generalized linear model with kernel mean function given by the inverse function, $\text{M}(\eta^T\mathbf{u}_i) = 1/\eta^T\mathbf{u}_i$.

Data from an experiment for which the Michaelis–Menton model is thought to be appropriate are given in the file `pur.lsp`. Load this file, and then draw the graph of the response versus x as shown in Figure 23.2. The shape of the

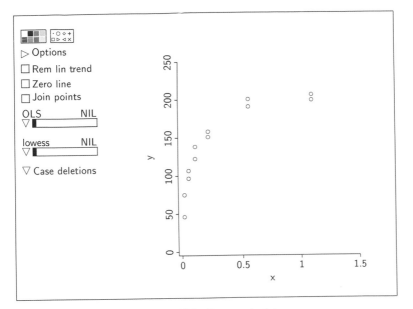

FIGURE 23.2 Puromycin data.

mean function for this graph is similar to Figure 23.1b after multiplying by a negative constant, so use of the inverse kernel mean function seems reasonable. Also, the variability between pairs of points at each value of x appears to be approximately constant, suggesting a constant variance function. We therefore have the three components needed for using a normal GLM: Terms derived from the predictors to define $\eta^T \mathbf{u}$, constant variance, and the inverse function for the kernel mean function.

To fit this model, first transform x to $1/x$ to get the required term. Select the item "Fit Linear LS" from the Graph&Fit menu to get the usual linear regression dialog, as shown in Figure 23.3. Select the terms in the linear predictor as usual, except make use of the buttons at the right of the dialog to select the inverse function to be the kernel mean function. The output is shown in Table 23.3.

As with most generalized linear models, estimates are computed iteratively; four iterations were required before a convergence criterion was met. The summary of the model reminds us that the kernel mean function used was the inverse. The estimates are maximum likelihood assuming $y_i \mid \mathbf{x}_i$ is normally distributed.

Both the deviance and Pearson's X^2 for a normal generalized linear model are computed from the formula

$$G^2 = X^2 = \sum_{i=1}^{n} w_i(y_i - \mathrm{M}(\hat{\eta}^T \mathbf{u}_i))^2 \qquad (23.3)$$

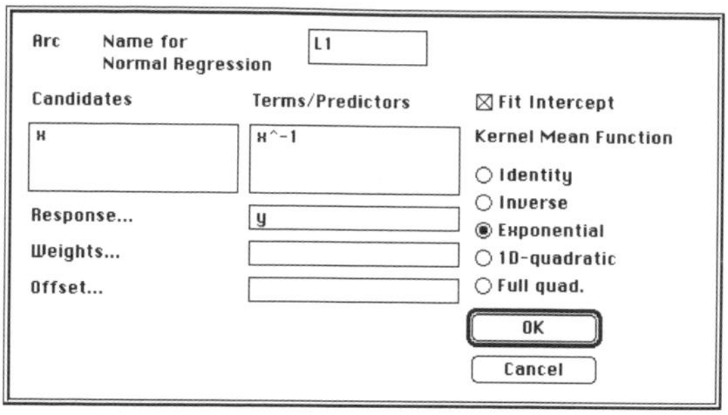

FIGURE 23.3 The LS regression dialog. The inverse mean function is selected.

TABLE 23.3   Fit of the Normal GLM with Inverse Kernel Mean Function for the Michaelis–Menton Model Applied to the Puromycin Data

```
Iteration 1: deviance = 1493.09
Iteration 2: deviance = 1201.27
Iteration 3: deviance = 1195.51
Iteration 4: deviance = 1195.45

Data set = PUR, Name of Fit = L1
Normal Regression
Kernel mean function = Inverse
Response      = y
Terms         = (x^-1)
Coefficient Estimates
```

Label	Estimate	Std. Error	Est/SE
Constant	0.00470184	0.000153584	30.614
x^-1	0.000301480	0.0000320298	9.412

Scale factor:	10.9337
Number of cases:	12
Degrees of freedom:	10
Pearson X2:	1195.449
Deviance:	1195.449

which we recognize as the residual sum of squares (in the current regression, all the weights are equal to one). The estimate $\hat{\sigma}^2$ is the deviance divided by $n - k$. Tests comparing mean functions are based on the F-distribution, as in Section 11.2, with the deviance substituting for the residual sum of squares.

We can judge how well this model matches the data using model checking plots. Select the item "Model checking plots" in the model's menu. Add the original variable x to the "Selection" list, and select the default "Marginal plots." In this regression, there is only one predictor in the model, and so we

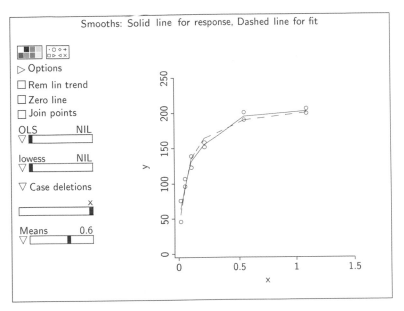

FIGURE 23.4 Model checking plot for x.

only need to consider the model checking plot for x. Use the model checking plot control to get the plot of y versus x, and then add the smoothers as shown on the plot in Figure 23.4. The fitted curve and the smooth fit to the data match very closely, suggesting that this model may be adequate for these data.

23.2.1 Transformation of Parameters

The Michaelis–Menton model was initially specified in terms of the parameters θ_1 and θ_2, as given in (23.1), but we have fitted a model with parameters $\eta_0 = 1/\theta_1$ and $\eta_1 = \theta_2/\theta_1$. We can invert these equations to get $\theta_1 = 1/\eta_0$ and $\theta_2 = \eta_1/\eta_0$. Maximum likelihood estimates of the θs can be obtained by substituting the MLEs of the ηs into these equations:

$$\hat{\theta}_1 = 1/0.004702 = 212.68; \qquad \hat{\theta}_2 = 0.00030148/0.004702 = 0.064$$

23.2.2 Transformation to Simple Linear Regression

Another possible approach to the Michaelis–Menton problem starts with equation (23.2), restated as

$$\frac{1}{\mathrm{E}(y_i \mid x_i)} = 1/\theta_1 + (\theta_2/\theta_1)(1/x_i) \tag{23.4}$$

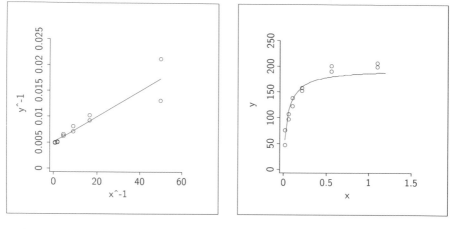

a. Inverse scale, with OLS line.   b. Original scale with fit from (a).

FIGURE 23.5   The Michaelis–Menton model in the inverse scale.

To the extent that $E(1/y_i \mid x_i) \approx 1/E(y_i \mid x_i)$, we might expect that we can turn this regression into a standard simple linear regression by using $1/y_i$ as a response variable and $1/x_i$ as a term; this approach has been advocated previously in discussions of transformations for linearity. The plot of $1/y_i$ versus $1/x_i$ in Figure 23.5 shows why this approach is likely to give worse answers in this particular regression. In the original scale in Figure 23.2, variability seems more or less constant across the plot, as the variation between pairs of observations is reasonably similar everywhere. In Figure 23.5a the variability is much larger for large values of $1/x$, which correspond to small values of x. The fitted OLS line shown on Figure 23.5a may not match the data very well; when plotted on the original plot of y versus x in Figure 23.5b, considerable bias in estimating the asymptote is apparent.

The choice between using a linearizing transformation or a nonidentity kernel mean function depends on the variance function. If linearizing also stabilizes variance, then linearizing is generally preferred. This is often the case with the logarithmic transformation, since in many regressions this transformation of y will both linearize the mean and stabilize variance. On the other hand, if, as in the example, the variability is nearly constant before transformation, fitting the nonidentity mean function using the untransformed response is likely to be preferred.

23.3   POISSON REGRESSION

Poisson random variables are discussed briefly in Section 23.5.1 of the complements to this chapter. Poisson regression is appropriate when the mean

TABLE 23.4   Definitions of Variables in the Possum Data

Name	Description of Variable
y	Number of possum species found on the site.
Acacia	Basal area of acacia +1, m^2.
Bark	Bark quality index.
Habitat	Habitat score for Leadbeater's possum.
Shrubs	Number of shrubs +1.
Stags	Number of hollow trees +1.
Stumps	1 if stumps from past logging are present, 0 if no stumps.
Asp	Primary aspect of plot, either SW-NW, NW-NE, NE-SE or SE-SW.
Dom	Dominant eucalypt species, either *E. regnans, E. delegatensis* or *E. nitens.*

function and the variance function are equal, as will occur if y is a count and $y \mid \mathbf{x}$ has a Poisson distribution.

As an example, we consider a study of possum abundance. In Australia and elsewhere, demand for managing hardwood forests for timber, conservation, and recreation is increasing. Managers need to identify areas with high conservation value, such as areas with ideal habitat for endangered animals. The goal of this study is to find factors that might be associated with good habitat for possums. Data were collected on 151 3ha sites with uniform vegetation in the state of Victoria, Australia. Several factors that describe the site were recorded, as given in Table 23.4. The data are given in the file possums.lsp. The constant 1 has been added to several of the predictors to allow for the use of power transformations. The goal of the analysis is to understand how the count y of the number of possum species found in the plot depends on the other predictors, and to describe sites that are favorable to possums.

Two of the predictors, *Asp* and *Dom*, are categorical, and so they should be converted to factors. Another, *Stumps*, is an indicator variable. The remaining predictors are all more or less continuous, so a first step might be to examine these remaining predictors in a scatterplot matrix to try to get linearly related predictors. Several of the frames in this scatterplot matrix have curved mean functions, so transformations might help. Use the "Find normalizing transformation" item from the "Transformations" pop-up menu to choose a starting point for looking at transformations. After the initial choice of transformation (see Problem 23.1), we deleted two points and used the "Find normalizing transformation" item again. After rounding, we decided on the initial predictors given in Table 23.4, but with $Acacia^{0.5}$, $\log(Bark)$, $Shrubs^{0.33}$, and $\log(Stags)$ replacing the untransformed predictors.

The next step is to examine inverse response plots for 1D structure (a multipanel plot is convenient for this, excluding the indicator predictor and the two factors). All the inverse response plots exhibit high variability, so comparing the shape of mean functions between them is very hard. In any

TABLE 23.5 **Poisson Regression Fit for the Possum Data**

```
Data set = Possums, Name of Fit = P1
Poisson Regression
Kernel mean function = Exponential
Response      = y
Terms         = (Acacia^0.5 log[Bark] Habitat Shrubs^0.33
                 log[Stags] Stumps {F}Asp {F}Dom)
Coefficient Estimates
```

Label	Estimate	Std. Error	Est/SE
Constant	-2.51018	0.591436	-4.244
Acacia^0.5	0.0943778	0.0718576	1.313
log[Bark]	0.422571	0.162026	2.608
Habitat	0.0579924	0.0401229	1.445
Shrubs^0.33	0.0409867	0.224186	0.183
log[Stags]	0.449517	0.124992	3.596
Stumps	-0.165733	0.277153	-0.598
{F}Asp[NW-NE]	0.455143	0.243183	1.872
{F}Asp[NE-SE]	0.484396	0.227447	2.130
{F}Asp[SE-SW]	0.549813	0.222164	2.475
{F}Dom[E.deleg]	-0.131750	0.307621	-0.428
{F}Dom[E.regnans]	-0.0876663	0.267026	-0.328

```
Scale factor:                    1.
Number of cases:                151
Degrees of freedom:             139
Pearson X2:                   95.648
Deviance:                    117.263
```

case, there is little or no evidence against 1D structure, and we will accept this as a useful starting point. However, these plots do not take into account the indicator variable or the factors, so interactions with these, leading to higher-dimensional models, cannot yet be excluded.

Table 23.5 summarizes the fit of a Poisson regression model with **u** derived from all the predictors in Table 23.4, with four of the predictors replaced by their transformations discussed above and *Dom* and *Asp* as factors. This table was obtained by defining the terms as needed using the data set menu and selecting the item "Fit Poisson response" from the Graph&Fit menu. We used the canonical exponential kernel mean function for the fit, as is usual in Poisson regression. The estimated mean function is $\exp(\hat{\eta}^T \mathbf{u})$, with $\hat{\eta}$ given by the column marked "Estimate" in Table 23.5. As with logistic regression, the column marked "Std. Error" provides approximate standard errors of the estimates, and the ratio "Est/SE" provides Wald tests that each of the elements of η is zero, after adjusting for all the other terms in the mean function. These statistics can be compared to the normal distribution to get p-values.

TABLE 23.6 Likelihood Ratio Tests Obtained When One Term is Dropped from the Mean Function in the Possum Data

```
Change in deviance for fitting each term last.
All fits include an intercept.
```

Term	df	LR-test	p-value
Acacia^0.5	1	1.747	0.1863
log[Bark]	1	7.033	0.0080
Habitat	1	2.103	0.1470
Shrubs^0.33	1	0.033	0.8550
log[Stags]	1	13.154	0.0003
Stumps	1	0.373	0.5415
{F}Asp	3	7.106	0.0686
{F}Dom	2	0.182	0.9132

As with logistic regression, there is no scale factor in the variance function, so the value of the scale factor is always reported as one. If $\hat{y}_i = \text{M}(\hat{\eta}^T \mathbf{u}_i)$ is a fitted value, then the deviance and Pearson's X^2 are given by

$$G^2 = \sum_{i=1}^{n} \hat{y}_i \log(y_i/\hat{y}_i) \tag{23.5}$$

$$X^2 = \sum_{i=1}^{n} (y_i - \hat{y}_i)^2/\hat{y}_i \tag{23.6}$$

These two statistics are used in goodness-of-fit tests for comparing mean functions. One set of useful tests can be obtained using the "Examine submodels" item in the model's menu. Select this item and then push the button for "Change in deviance for fitting each term last," and then push "OK." The resulting output is in Table 23.6. For example, the change in deviance between (a) the mean function with all the terms and (b) the mean function with all the terms except *Habitat* is 2.103, with 1 df. This difference can be compared to the χ_1^2 distribution to get a *p*-value of about 0.15. This suggests that, at least after the other terms, *Habitat* provides little additional information. Similarly, the change in deviance for the mean function with all terms and the mean function that does not include {*F*}*Asp* is 7.106, but this has 3 df, because {*F*}*Asp* is a factor with four levels. The *p*-value for this term is about 0.07. We see that at least two of the terms, log(*Bark*) and log(*Stags*), have very small *p*-values, while *Asp* appears to be potentially important after adjusting for the others.

The next step in this analysis is to examine the fit through the use of model checking plots, against $\hat{\eta}^T \mathbf{u}$ and the continuous predictors, in random plots and in GREG plots. We leave it as an exercise (Problem 23.1) to show that this mean function matches the data very closely. Had the fitted model not matched the data, we might next consider that 1D structure does not hold, and we might examine separate fits for values of the factors and the indicator variable.

Before summarizing results, we may try to remove a few of the terms from the mean function to obtain a simpler summary of the regression. The likelihood ratio tests in Table 23.6 indicate that this approach could be useful. We will use backward elimination that deletes a term at each step. Select the item "Examine submodels" from the model's menu, and then select "Delete from full model (Backward Elimination)" from the dialog. An edited version of the resulting output is shown in Table 23.7.

The output is divided into several stages. At the first stage, each line in the table is obtained by re-fitting with one of the terms deleted from the model. For example, at stage one the line marked "Delete: *Stumps*" is for the mean function obtained by removing *Stumps* from the current mean function. This mean function has deviance $G^2 = 117.636$ and Pearson's $X^2 = 95.8036$, both with 140 df. The value $k = 11$ is the number of terms in the subset model, and *AIC*, called *Aikike's information criterion*, is a measure of relative information in this model compared to the model with all possible terms included. *AIC* is defined by the formula

$$AIC = G^2 + 2k \tag{23.7}$$

If all terms in the mean function had 1 df, then the minimum *AIC* mean function would be the same as the minimum deviance mean function; since the factors have several df, *AIC* adds an appropriate adjustment for differences in degrees of freedom. At the first stage, $\{F\}Dom$ is removed. The second stage considers only models that exclude $\{F\}Dom$, examining the mean functions that can be obtained by dropping one additional term. At succeeding stages, $Shrubs^{0.33}$, *Stumps*, and $Acacia^{0.5}$ have minimum *AIC* and are removed. The total increase in deviance for removing these 5 df is only $119.39 - 117.26 = 2.13$. Further deletion causes a large increase in deviance, suggesting tentatively accepting the mean function that includes the remaining terms. Table 23.8 summarizes the fit of this mean function.

As an additional check on this mean function, we can examine model checking plots again, this time using the terms not in the mean function to plot on the horizontal axis. If the information in the terms removed from the mean function can be adequately approximated by the remaining terms in the model, then the two smooths on the model checking plots should agree; showing this agreement is left to Problem 23.1.

Finally, we turn to interpretation of the coefficient estimates in Table 23.8. All the coefficients for the continuous predictors are positive, so higher values for log(*Bark*), *Habitat*, and log(*Stags*) all imply the presence of more possum species. Interpretation of the remaining coefficients is a bit different. Since no indicator variable is included for the category SW-NW, the three coefficients for $\{F\}Asp$ suggest an increase in species abundance in all three aspects other than SW-NW. Since the coefficients for the three categories are nearly equal,

TABLE 23.7 Backward Elimination for the Possum Data, Starting with the Mean Function in Table 23.5[a]

Stage	Action	df	G^2	Pearson X^2	k	AIC
1	Delete: $\{F\}Dom$	141	117.445	95.9924	10	137.445
1	Delete: $Shrubs^{0.33}$	140	117.296	95.7975	11	139.296
1	Delete: $Stumps$	140	117.636	95.8036	11	139.636
1	Delete: $Acacia^{0.5}$	140	119.01	95.4325	11	141.010
1	Delete: $Habitat$	140	119.366	98.2852	11	141.366
1	Delete: $\{F\}Asp$	142	124.369	99.426	9	142.369
1	Delete: $\log(Bark)$	140	124.296	99.1637	11	146.296
1	Delete: $\log(Stags)$	140	130.417	105.794	11	152.417
2	Delete: $Shrubs^{0.33}$	142	117.494	96.1858	9	135.494
2	Delete: $Stumps$	142	117.827	96.1581	9	135.827
2	Delete: $Acacia^{0.5}$	142	119.086	95.5212	9	137.086
2	Delete: $Habitat$	142	120.027	99.1312	9	138.027
2	Delete: $\{F\}Asp$	144	124.578	99.8735	7	138.578
2	Delete: $\log(Bark)$	142	125.04	99.7188	9	143.040
2	Delete: $\log(Stags)$	142	130.887	105.237	9	148.887
3	Delete: $Stumps$	143	117.856	96.2578	8	133.856
3	Delete: $Acacia^{0.5}$	143	119.104	95.4485	8	135.104
3	Delete: $Habitat$	143	120.14	99.5269	8	136.140
3	Delete: $\{F\}Asp$	145	124.65	100.127	6	136.650
3	Delete: $\log(Bark)$	143	127.161	101.748	8	143.161
3	Delete: $\log(Stags)$	143	130.9	105.33	8	146.900
4	Delete: $Acacia^{0.5}$	144	119.39	95.7451	7	133.390
4	Delete: $Habitat$	144	120.385	99.4518	7	134.385
4	Delete: $\{F\}Asp$	146	125.529	99.6707	5	135.529
4	Delete: $\log(Bark)$	144	127.257	101.703	7	141.257
4	Delete: $\log(Stags)$	144	132.371	106.571	7	146.371
5	Delete: $\{F\}Asp$	147	127.093	98.7495	4	135.093
5	Delete: $\log(Bark)$	145	128.78	101.576	6	140.780
5	Delete: $Habitat$	145	128.836	100.491	6	140.836
5	Delete: $\log(Stags)$	145	132.387	106.429	6	144.387
6	Delete: $\log(Bark)$	148	135.097	105.346	3	141.097
6	Delete: $Habitat$	148	135.606	103.546	3	141.606
6	Delete: $\log(Stags)$	148	138.507	109.538	3	144.507
7	Delete: $Habitat$	149	144.087	110.507	2	148.087
7	Delete: $\log(Stags)$	149	149.861	123.14	2	153.861

[a]Each stage is obtained from the one before it by removing the term whose deletion results in the smallest value of AIC.

TABLE 23.8   Subset Mean Function for the Possum Data

```
Data set = Possums, Name of Fit = P2
Poisson Regression
Kernel mean function = Exponential
Response      = y
Terms         = (log[Bark] Habitat log[Stags] {F}Asp)
Coefficient Estimates
Label             Estimate      Std. Error      Est/SE
Constant         -2.35889       0.425536        -5.543
log[Bark]         0.431805      0.140890         3.065
Habitat           0.0910179     0.0302181        3.012
log[Stags]        0.400125      0.110392         3.625
{F}Asp[NW-NE]     0.489444      0.241278         2.029
{F}Asp[NE-SE]     0.511118      0.224037         2.281
{F}Asp[SE-SW]     0.557198      0.220791         2.524

Scale factor:                         1.
Number of cases:                    151
Degrees of freedom:                 144
Pearson X2:                          95.745
Deviance:                           119.390
```

we could replace the three factor levels of *Asp* by a single indicator variable for comparing the SW-NW aspect to any of the others. Fitting this simplified model in *Arc* requires creating an indicator variable, and this is most easily done in two steps. First, use the "Add a variate" item in the data set menu, and enter Asp1 = (recode asp). The new variable *Asp1* will now have values 0, 1, 2, and 3 rather than text strings for values. Then, use the same menu item and type Asp2 = (recode-values asp1 '(0 1 2 3) '(0 1 1 1)), which will recode Asp1 into an indicator variable Asp2. The resulting fit is shown in Table 23.9. The coefficients for the other terms are virtually unchanged, and the coefficient estimate for *Asp2* is 0.52, which suggests that the mean number of species is multiplied by exp(0.52) = 1.68 when the aspect is not SW-NW. The difference in G^2 for the last two models, $119.537 - 119.390 = 0.147$ with $146 - 144 = 2$ df, can be used to test the equality of the three aspect terms, after adjusting for the other terms in the model. Nothing important is lost by using the mean function simplified in this way.

23.3.1   Log-Linear Models

Probably the most common use of Poisson regression is in the analysis of log-linear models for tables of counts, and *Arc* can be used for fitting these models. Log-linear models have much in common with the regression models studied in this book, but there are many important differences in language, notation, forms of models, special cases, and summaries. We provide references to a

TABLE 23.9 Final Fitted Model for the Possum Data

```
Data set = Possums, Name of Fit = P3
Poisson Regression
Kernel mean function = Exponential
Response      = y
Terms         = (log[Bark] Habitat log[Stags] Asp2)
Coefficient Estimates
Label          Estimate    Std. Error    Est/SE
Constant       -2.36938    0.424165      -5.586
log[Bark]       0.430450   0.140483       3.064
Habitat         0.0909858  0.0301258      3.020
log[Stags]      0.405792   0.109113       3.719
Asp2            0.524986   0.202697       2.590

Scale factor:                   1.
Number of cases:              151
Degrees of freedom:           146
Pearson X2:                    96.193
Deviance:                     119.537
```

few of the excellent treatments of log-linear models in the complements section.

23.4 GAMMA REGRESSION

Gamma random variables are introduced in Section 23.5.2. As an example of gamma regression, we return to the transactions data in file transact.lsp, first described in Section 7.3.3. In previous analyses, we have concluded that the mean function $E(Time \mid T_1, T_2) = \eta_0 + \eta_1 T_1 + \eta_2 T_2$ must be appropriate for these data, but that the variance increases with the mean. In gamma regression, we have that $Var(y \mid x) = \sigma^2 E(y \mid x)^2$, which can be rewritten as $\sigma = \sqrt{Var(y \mid x)}/E(y \mid x)$. The quantity $\sqrt{Var(y \mid x)}/E(y \mid x)$, called the *coefficient of variation*, is constant for gamma regression.

Table 23.10 summarizes the fit of a gamma regression model to the transactions data, using the identity kernel mean function. This is not the canonical kernel mean function for the gamma, and it can lead to problems in fitting by giving negative fitted values to a non-negative response variable. We use it here because we have already concluded in Section 7.3.3 that the identity kernel mean function is appropriate for this regression. In this particular application, no such problems with negative fitted values occur. The coefficient estimates for T_1 and T_2 are essentially the same as obtained in the OLS fit given in Table 7.3, page 156. The main difference is the variance function: For the gamma fit, we have estimated the variance function to

TABLE 23.10   Gamma Regression Fit to the Transactions Data

```
Data set = Transactions, Name of Fit = G1
Gamma Regression
Kernel mean function = Identity
Response        = Time
Terms           = (T1 T2)
Coefficient Estimates
Label           Estimate    Std. Error      Est/SE
Constant        152.934     51.4228          2.974
T1                5.70559    0.422661        13.499
T2                2.00713    0.0575820       34.857

Scale factor:                0.170237
Number of cases:                  261
Degrees of freedom:               258
Pearson X2:                     7.582
Deviance:                       7.477
```

be $\widehat{\text{Var}}(Time \mid T_1, T_2) = 0.170^2 \times \hat{E}(Time \mid T_1, T_2)^2$, rather than assuming constant variance.

We can judge the fit of this model to the transactions data using model checking plots. Figure 23.6 shows the model checking plot for $\hat{\eta}^T \mathbf{u}$. Shown on the plot are smooths that represent both the mean function and the variance function. The mean functions match very well. The variance function from the model appears to give variances that are a bit too large for the few branches with large values of T_1 and T_2. This model is an improvement over the OLS fit, since a constant variance assumption cannot be supported with the data, and the variance function for the gamma is reasonably consistent with the data.

23.5   COMPLEMENTS

23.5.1   Poisson Distribution

Normally distributed random variables are continuous: They can take on any real value. Some variables are discrete. The most important example of a discrete distribution is the binomial, discussed in Section 21.2.2, page 472. The second example that is also frequently encountered in practice is the *Poisson distribution*.

Suppose the random variable y can be any non-negative integer. The Poisson distribution assigns probabilities to the integers using the probability mass function:

$$\Pr(y) = e^{-\lambda}\lambda^y/y! \qquad y = 0, 1, \ldots$$

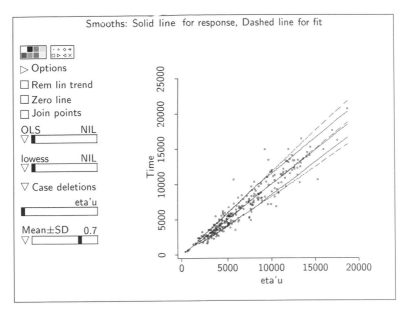

FIGURE 23.6 Model checking plot for the gamma regression fit to the transactions data.

where λ is the parameter of this distribution, $\lambda > 0$, and the factorial notation ($y!$) is defined on page 473. We will write $y \sim \text{Po}(\lambda)$ for this distribution. The mean of y is equal to λ and that the variance of y is also equal to λ. Like the binomial distribution, once the mean of the distribution is fixed, so is the variance of the distribution. Poisson probabilities are computed in *Arc* using the function `poisson-pmf` to get the probability that the variable has a specific value and `poisson-cdf` to get the probability that y is less than or equal to a specific value.

The sum of independent Poisson random variables also has a Poisson distribution, with mean given by the sum of the means of the individual Poisson random variables. In symbols, if $y_i \sim \text{Po}(\lambda_i)$, $i = 1,2$, and if y_1 and y_2 are independent, then $y_1 + y_2 \sim \text{Po}(\lambda_1 + \lambda_2)$.

The Poisson distribution is closely related to the binomial distribution in at least two ways. First, suppose $y \sim \text{Bin}(m, \theta)$. If θ is small and m is large, then y behaves like a $\text{Po}(m\theta)$ random variable. Thus, the Poisson distribution is the limiting distribution of the binomial for small θ. The second connection is through conditioning. Suppose, for example, that doctors in a clinic prescribe a certain type of drug either using a brand-name drug or using a generic equivalent. Let y_1 be the number of brand-name prescriptions in a fixed time period and let y_2 the number of generic prescriptions in that time period. Assuming prescriptions are independent of each other, both y_1 and y_2 are likely to be Poisson random variables, with means λ_1 and λ_2. We can now ask the following question: Given that the drug was prescribed, what is the probability that the prescription was for the brand-name drug? In the data, we have y_1 "successes" in $m_1 = y_1 + y_2$ trials, and the distribution of $y_1 \mid m_1 \sim \text{Bin}(m_1, \lambda_1/(\lambda_1 + \lambda_2))$.

The Poisson distribution was first derived by S. D. Poisson (1837).

23.5.2   Gamma Distribution

Suppose that y is a continuous random variable that can equal any positive value. Let's suppose further that, unlike the normal distribution, the variance of y is a function of the mean of y, and that larger values of $E(y)$ imply larger values of $\text{Var}(y)$.

One family of distributions with this property is the family of *gamma distributions*. For this family, the *coefficient of variation*, defined to be the ratio $\sqrt{\text{Var}(y)}/E(y)$, is constant, which is equivalent to requiring that $\text{Var}(y) = \sigma^2[E(y)]^2$ for some constant σ.

The density function of a gamma distribution depends on two positive parameters, the mean μ and a *shape parameter* $\alpha = 1/\sigma^2$. The variance of y can be shown to equal μ^2/α. The special case of $\alpha = 1$, is called the *exponential distribution*. If $\alpha > 1$, the density is zero at the origin and is unimodal. For values of α not too much larger than one, the density is highly skewed, but for large values of α the density approaches symmetry. The *Arc* function gamma-dens computes values of the gamma density with $\mu = 1$. For example, to obtain a plot of the density for the gamma distribution with $\mu = 1$ and $\alpha = 2$, type

```
(plot-function #'(lambda (y) (gamma-dens y 2)) 0 8)
```

23.5.3   References

Generalized linear models were first suggested by Nelder and Wedderburn (1972). An elegant presentation of them, at a mathematical level higher than that of this book, is given by McCullagh and Nelder (1989). Agresti (1996) provides a book-length introduction to log-linear models at a level similar to this book that includes both logistic and Poisson models; Agresti (1990) covers some of the same ground at a higher mathematical level.

The possum data were provided by Alan Welsh and are discussed in Lindenmayer, Cunningham, Tanton, Nix, and Smith (1991). The puromycin data discussed in Section 23.2 are from Bates and Watts (1988).

PROBLEMS

23.1.

    23.1.1 Find normalizing transformations for the continuous predictors in the possum data, possums.lsp, as suggested in the text. Two points appear to be separated from the others in the transformed scale, and so these may be influential for the transformation choice. Select these points, delete them, and then obtain new normalizing transformations. How do the transformations change?

23.1.2 Use model checking plots to examine the fit of the model in Table 23.5. Check for influence. Are the two points that were influential in the determination of transformations of predictors influential in fitting the model?

23.1.3 Use model checking plots to examine the fit of the model in Table 23.8.

APPENDIX A

Arc

Arc is a computer program written in the *Xlisp-Stat* language. Both *Xlisp-Stat* and *Arc* are included when you down-load the program from the Internet.

Arc is described in this book. The *Xlisp-Stat* language, which is a very powerful environment for statistical programming, is described in Tierney (1990). To use *Arc*, you do not need to know how to program in *Xlisp-Stat* or any other language.

A.1 GETTING THE SOFTWARE

You must obtain the appropriate program for your type of computer. *Arc* and the *Xlisp-Stat* system are nearly identical for Windows OS, Macintosh OS and Unix.

The most recent version of *Arc* can be obtained on the Wiley Mathematics web page at `http://www.wiley.com/mathematics`, by accessing the link for *Applied Regression* and then following the directions given there for your computer.

If you do not have Internet access, you can obtain *Arc* for a nominal fee by writing to *Arc* Software, University of Minnesota, Applied Statistics, 1994 Buford Ave., St. Paul, MN 55108.

A.1.1 Macintosh OS

Arc requires at least a 68020 processor; a Power PC is highly recommended. Double-click on the icon for the *Arc* installer that you down-loaded from the Web or otherwise obtained, and follow the on-screen directions for installing the program. After installation, you will have a folder on your hard disk called `Arc folder`, unless you choose some other name. You can start *Arc* by double-clicking on the icon for *Xlisp-Stat*. The startup screen with Mac OS should resemble Figure A.1.

A *menu bar* appears at the top of the computer screen and the *text window* appears below it. The text window is common to all operating systems, and

545

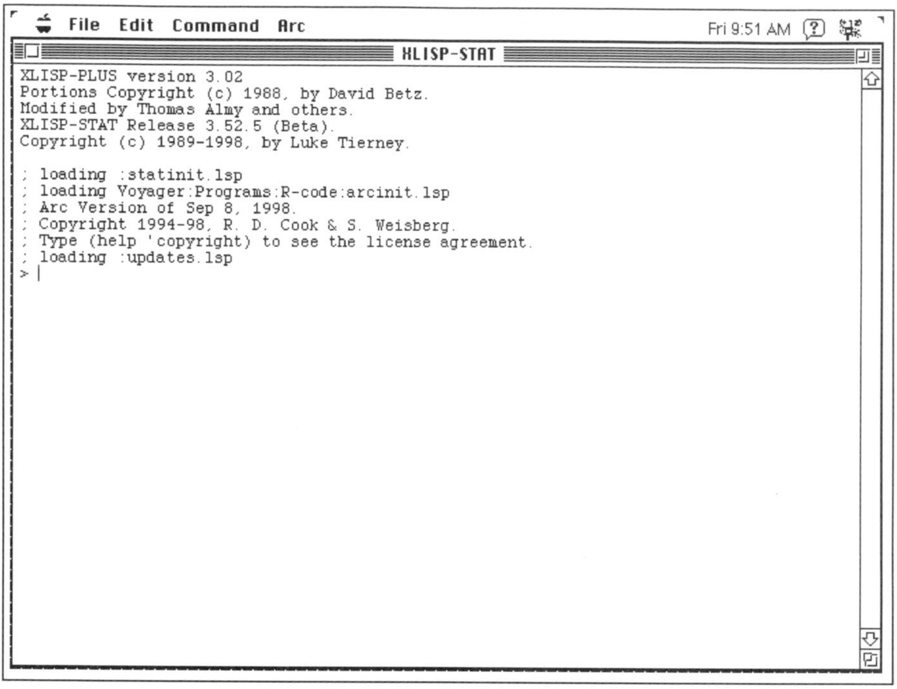

```
  File   Edit   Command   Arc                                    Fri 9:51 AM

                                    HLISP-STAT
 XLISP-PLUS version 3.02
 Portions Copyright (c) 1988, by David Betz.
 Modified by Thomas Almy and others.
 XLISP-STAT Release 3.52.5 (Beta).
 Copyright (c) 1989-1998, by Luke Tierney.

 ; loading :statinit.lsp
 ; loading Voyager:Programs:R-code:arcinit.lsp
 ; Arc Version of Sep 8, 1998.
 ; Copyright 1994-98, R. D. Cook & S. Weisberg.
 ; Type (help 'copyright) to see the license agreement.
 ; loading :updates.lsp
 > |
```

FIGURE A.1   The startup screen with the Mac OS.

its use is described in Section A.2. Menus are a little different on all three operating systems, so they are described separately for each. With Macintosh OS, the menu bar initially displays four menu names plus the Apple menu, but as you use the program more menus will be added to the menu bar.

File Menu. The File menu includes eight items. The item "Load" is used to open data files, and to read files containing *lisp* computer code that can be read by *Xlisp-Stat*. The next six items in the menu, "New edit," "Open edit," "Save edit," "Save edit as," "Save edit copy," and "Revert edit" are used to read and save files with the simple editor that is built into *Xlisp-Stat* for the Mac OS. Use of this editor should be familiar to anyone with Mac OS experience. The final item in this menu is "Quit," which is used to exit *Xlisp-Stat* and *Arc.*

Edit Menu. The Edit menu contains the items "Undo," "Cut," "Copy," "Paste," and "Clear" for use with the built-in editor and with the text window. "Copy-paste" will copy selected text in the text window to the command line, which is the last line of the text window. The last two items in the Edit menu are "Edit selection," which will copy the selection in the text window into a window for the editor, and "Eval Selection," which will evaluate the selected text in an edit window.

Command Menu. The Command menu appears only with Mac OS. The item "Show Xlisp-Stat" will return the text window to the screen if you accidentally make it disappear by pushing the mouse in the go-away box, the small square at the upper-left of the window. The items "Clean Up" and "Top Level" are generally not used with *Arc*; they are helpful when debugging programs in *Xlisp-Stat*. The final item "Dribble" is used for automatically copying all *future* text in the text window to a file for later printing or editing. The same item appears on the Arc menu as described in Section A.6.

Arc Menu. The Arc menu is common to all platforms and is described in Section A.6.

A.1.2   Windows OS

You will need Microsoft Windows 3.11 or newer and at least 8 mb of RAM. About 5 mb of hard disk space is required. Separate installers are available on the Web page for Windows 3.11 (a 16-bit program) and for more recent Windows systems (a 32-bit program). Follow the instructions on the Web page to install the program.

To start *Arc*, double-click on the icon for *Xlisp-Stat*, or else select *Xlisp-Stat* from the Start button in Windows or from the Program Manager in Windows 3.11. The Windows OS startup screen is similar to the Macintosh OS startup screen shown in Figure A.1. At startup, the Windows menu bar contains four menus:

File Menu. The File menu includes only four items: "Load," which is used to read data files and files of *lisp* statements, "Dribble," which is described in Section A.3, "Exit," which is used to leave *Xlisp-Stat* and *Arc*, and "About Xlisp-Stat," which gives information about *Xlisp-Stat*. Unlike the Mac OS version, the Windows OS version does not have a built-in text editor, but it does have a very simple stand-alone text editor called *LspEdit*. *LspEdit* has two useful features. Automatic parenthesis matching can be useful in typing commands. Also, you can highlight text in *LspEdit*, select "Execute" from *LspEdit*'s Edit menu, and the highlighted code will be executed by *Xlisp-Stat*.

Edit Menu. This menu contains the usual items, "Undo," "Cut," "Copy," "Paste," and "Clear." These items are used to copy and paste text in the text window, as well as to copy whole graphics windows to be pasted into other applications (see Section A.3 for information on saving text and plots). An additional item, "Copy-Paste," will copy selected text in the text window and paste it onto the *command line* which is the last line of the text window beginning with a prompt >.

Window. This standard Windows OS menu lists the names of all the open windows and allows selecting one to be the top window. In addition, items appear here to modify the appearance of the screen.

A.1.3 Unix

Unix installation instructions are given on the Web page. On networked systems with many users, setting up this version of *Xlisp-Stat* and *Arc* may require the assistance of a system manager.

When the program is properly installed, *Arc* is usually started by entering the Unix command arc.

Like the Windows OS and Mac OS versions, the Unix version has a menu bar, but this menu bar initially contains only the Arc menu described in Section A.6.

A.1.4 What You Get

With both Windows OS and Mac OS, you will have created a directory including many files. The file Xlisp-Stat is the executable program file needed to run *Arc*. *Arc* is contained in the file xlisp.wks; the files statinit.lsp, arcinit.lsp, and copyrite.lsp must also be present to run the program. The file statinit.lsp is read every time you start *Arc*, and users familiar with *Xlisp-Stat* may want to modify this file to read their own code. The files settings.lsp and updates.lsp will also be read, if they are present. The files copying and copyrite.lsp contain copyright notices for various parts of the code. Most other files and directories are part of the standard *Xlisp-Stat* distribution. You can learn about them in Tierney (1990) or by consulting the Web page for recent Internet references.

A.1.5 Data Files

The directory Data and its subdirectories contain all the data sets described in this book. If you know the name of the data file you want to use, you can load the file without actually knowing its location. For example, the typed command in the text window (load "ais") will search for the file called ais.lsp in several standard *Arc* directories. You can add your own directory to this search list. Suppose that you keep data in the directory MyData, which is the directory from which you start *Arc*. The following command when either typed into the text window or added to the file statinit.lsp will add this directory to your search path:

Mac OS :	(add-data-directory "MyData:")
Windows OS :	(add-data-directory "MyData\\")
Unix :	(add-data-directory "MyData/")

You can add as many directories as you like by repeating the above command with different directory names.

A.2   THE TEXT WINDOW

The text window is used to display numeric output produced by the program, and to enter information for the program to use. Since a full implementation of *Xlisp-Stat* is included with *Arc*, the user can execute functions, get *Arc* to perform tasks, and even write computer programs. However, almost all of the features of *Arc* can be accessed using menus and dialogs, so you do not need to know how to program.

A.2.1   Typing in the Text Window

For a brief tutorial on typing command when using *Arc*, consider two variables: (1) cigarette consumption per person per year in 1930 and (2) 1950 male lung cancer death rate per million males. The variables are measured for the United States, Canada, and nine countries in Europe. One possible analysis goal for these data is to see if lung cancer death rate is associated with cigarette smoking. The smoking data lead the cancer data by twenty years to allow time for cancer to develop.

A.2.2   Typing Data

Most often, data will be entered into *Arc* by reading a data file, but data can be entered directly into the text window. Duplicate the commands on each line following the prompt >. The number of items you put on one line is arbitrary. Each line you type ends by hitting the "Return" key. The computer will respond by providing a new prompt when a command is finished.[1]

```
>(def CancerRate (list 58 90 115 150 165 170 190
   245 250 350 465))
CANCERRATE
>(def CigConsumption (list 220 250 310 510 380
   455 1280 460 530 1115 1145))
CIGCONSUMPTION
>(def Country (list "Iceland" "Norway" "Sweden"
                    "Canada" "Denmark" "Austria"
                    "USA" "Holland" "Switzerland"
                    "Finland" "Great_Britain"))
COUNTRY
```

You have now created three lists of data with the names *CancerRate*, *CigConsumption*, and *Country*. To understand a little about how this happened, it will probably be helpful to know that *lisp* stands for *list* processor and, as the

---

[1] You can reproduce the results without typing data by loading the data file `cancer.lsp`.

name implies, the language works by processing lists. A list is just a series of items separated by spaces and enclosed in parentheses. The items in a list can themselves be lists, or even more complicated constructions. The first item of a list is an instruction that tells the program what to do with the rest of the list. For example, the list (+ 1 2) contains three items—namely +, 1, and 2—and it instructs the program to add 1 and 2.

Let's now return to the first of the three commands you just typed. The main list contains three items: def, CancerRate, and (list 58 90 115 150 165 170 190 245 250 350 465). The first item def is the instruction that tells the program you want to define something. The next item CancerRate is the name to be defined and the final item is the definition. The definition is itself a list consisting of 12 items: list and the numbers 58,...,465. The first item is the instruction for forming a list. In summary, the first command that you just typed instructs the program to define CancerRate to be a list of 11 numbers. After you press the return key, the computer responds with the name of the list, and then the next prompt.

The names of lists or other objects in *Xlisp-Stat* can be of any length, but must not contain spaces. A few names like pi, t and nil are reserved by the system. If you try to use one of these you will get an error message.

Remember these points when typing information into *Xlisp-Stat*:

- The parentheses must match. In Mac OS, Windows OS, and some implementations of Unix, *Xlisp-Stat* flashes matching parentheses to make this easier.

- Quotation marks must match. If they do not match, the program will get very confused, and it will never return you to the prompt or give you any printed output.

- You can have any number of items on a line.

- A command can actually take up several lines on the computer screen.

- Each command ends with a "Return" after the final right parenthesis.

- Items in a list are separated by white space, consisting of either blanks or tabs.

- If the computer does not respond to your typing, you need to escape from the current command to start over. On the Macintosh, you can select the item "Top Level" from the Command menu, or press the command (cloverleaf) and period keys at the same time. With Windows, press Control-Break, and with Unix press Control-C.

- If you have an error, you can use standard cut-and-paste methods to edit the command for reentry. With Windows OS, the standard keys like End and Start can be used to move the cursor around the command line. With Mac OS, the Enter key moves the cursor to the end of the command line.

A.2.3 Working with Lists

You can do many things with lists. The simplest is just to display them. To see the list CigConsumption, type

```
>CigConsumption
(220 250 310 510 380 455 1280 460 530 1115 1145)
```

Don't forget to press enter after CigConsumption. To get the number of items in a list, use the length function:

```
>(length CigConsumption)
11
```

If you do not get 11, then you have incorrectly entered the data: Try to find your error, and then retype or use cut and paste to correct the data before continuing.

To get the mean and standard deviation of *CigConsumption*, you can use the mean and standard-deviation functions:

```
>(def xbar (mean CigConsumption))
XBAR
>xbar
605
>(def sdx (standard-deviation CigConsumption))
SDX
>sdx
384.37
```

In each of these statements, a value is computed and stored as a constant so it can be used in a later calculation. If you do not plan to use the value later, you need not store it. For example, to get the natural logarithm of CigConsumption, type

```
>(log CigConsumption)
(5.39363 5.52146 5.73657 6.23441 5.94017 6.1203 7.15462
6.13123 6.27288 7.01661 7.04316)
```

while the expression

```
>(exp (log CigConsumption))
(220 250 310 510 380 455 1280 460 530 1115 1145)
```

computes exp(log(CigConsumption)), which simply returns the original values. The function log can take two arguments, a value and a base. For example, (log 4 10) will return the logarithm of 4 to the base 10, while (log CigConsumption 2) will return a list of the logarithms to the base 2

of the elements of `CigConsumption`. If the second argument is missing, then
natural logarithms are used.

Arc can also be used as a calculator. For example,

```
>(+ 2 4 5 6)
17
>(- 17 2 3)
12
>(* 3 xbar)
1815
>(/ CigConsumption 12)
(18.3333 20.8333 25.8333 42.5 31.6667 37.9167 106.667
38.3333 44.1667 92.9167 95.4167)
```

These statements illustrate four basic operations that can be applied to lists of
numbers. The general form of the operation may seem a bit unnatural: After the
opening parenthesis comes the instruction, and then the numbers. For addition
and multiplication, this does the obvious thing of adding all the numbers and
multiplying all the numbers, respectively. For subtraction and division, it is
not necessarily obvious what happens with more than two arguments. As
illustrated above, (- 17 2 3) subtracts 2 from 17 and then subtracts 3 from
the result. Similarly, (/ 8 2 4) divides 8 by 2 and then divides the result by
4, giving an answer of 1.

Arithmetic operations are applied elementwise to pairs of lists having the
same length:

```
>(* CancerRate CigConsumption)
(12760 22500 35650 76500 62700 77350 243200 112700
 132500 390250 532425)
```

Multiplication of a list by a scalar is also possible:

```
>(* CancerRate 1000)
(58000 90000 115000 150000 165000 170000 190000
 245000 250000 350000 465000)
```

The statement (^ 4 3) returns $4^3 = 64$, (sqrt 3) returns the square root of 3,
and (log 3) returns the natural logarithm of 3.

A.2.4   Calculating the Slope and Intercept

You can also do more complicated computations. Let x_i refer to the *i*th value
of the predictor *CigConsumption* and let y_i refer to the *i*th value of the response
CancerRate. To get $SXY = \sum(x_i - \bar{x})(y_i - \bar{y})$ and $SXX = \sum(x_i - \bar{x})^2$, you can

type

```
>(def xbar (mean CigConsumption))
XBAR
>(def ybar (mean CancerRate))
YBAR
>(def sxy (sum (* (- CigConsumption xbar)
                  (- CancerRate ybar))))
SXY
>sxy
338495
>(def sxx (sum (^(- CigConsumption xbar) 2)))
SXX
>sxx
1.4774e+06
```

This last result is given in scientific notation and is equal to 1.4774×10^6.

The hard part in calculations like the last one is getting the parentheses to match and remembering that functions like + or * or log go at the start of an expression. *Xlisp-Stat* allows mixing lists and numbers in one expression. In the above example, (- CigConsumption xbar) will subtract xbar, a number, from each element of the list CigConsumption.

To complete the calculations of the slope and intercept obtained by ordinary least squares regression of *CancerRate* on *CigConsumption*, use the usual formulas:

```
>(def b1 (/ sxy sxx))
B1
>(def b0 (- ybar (* b1 xbar)))
B0
>(list b0 b1)
(65.7489 0.229115)
```

A.3   SAVING AND PRINTING

A.3.1   Text

Xlisp-Stat does not have a built-in method for printing. Output from the program must be saved to a file and ultimately printed by a word processor. To save output, select the text you want to save by dragging the mouse across it with the button down; select "Copy" from the Edit menu, switch to the word processor, and select "Paste" from that application's Edit menu. *You should use a fixed-width font like Courier or Monaco, or else columns won't line up properly.*

The second method saves all subsequent input and output in the text window to a file. Saving is started by selecting the item "Dribble" from the Arc menu.

Choose a file name using a dialog. All text will be put in this file until you select "Dribble" a second time. The resulting plain text file can then be read by any word processor or editor. If you prefer, you can type a command:

```
>(dribble "filename")
```

to start saving the text window to the file filename.lsp, or

```
>(dribble (get-file-dialog "Name of file"))
```

to choose the file name using a file dialog. You cannot toggle the "Dribble" file on and off. If you select "Dribble" a third time with the same file name, the file will be overwritten without warning. The file you create in this way can be modified or printed using an editor or word processor.

A.3.2 Graphics

With Mac OS and Windows OS, a plot is saved by copying and pasting it into a word processor document. Make sure the plot you want is the front window by clicking the mouse on it. Select "Copy" from the Edit menu. Switch to the word processor and paste the plot into a document by selecting "Paste" from that program's Edit menu. On Unix workstations, plots can be saved as a PostScript bitmap by selecting the "Save to file" item from the plot's menu.

A.4 QUITTING

To quit from *Arc*, select "Quit" from the Arc menu. You can also type the command (end), followed by enter, in the text window.

A.5 DATA FILES

A.5.1 Plain Data

The simplest format of a data file is a *plain data file*, which is a text file with *cases* given by rows and *variables* given by columns. To read a plain data file, select "Load" from the Arc menu or from the File menu on the Macintosh or Windows. A standard file dialog will then allow you to find the file and read it. With Windows OS, you will initially see file names ending in .lsp, but you can get all files by clicking the mouse in the scrolling list for "Files of type" and then selecting "All Files." After selecting a file, you will be prompted via dialogs to give additional information, such as a name for the data set, an optional description of the data, and names for the variables and descriptions of them. You will be able to save the data set as a formatted data file using the item "Save data set as" in the data set menu.

A.5.2 Plain Data File with Variable Labels

You can put variable labels in the first row of a plain data file. Variable names will be converted to all uppercase letters; if you want lowercase, or your labels include blanks, you must enclose the labels in double quotes.

A.5.3 Importing Data from a Spreadsheet

A spreadsheet can be used to make a plain data file or a plain data file with variable labels. All other information should be removed from the spreadsheet. Missing data should be indicated by a ?, not by a blank. Save the spreadsheet as a plain text file; in Microsoft Excel, save the file as a "Text (Tab delimited) .txt" file.

A.5.4 Special Characters

A few characters have special meaning and should not appear in data files because their use can cause problems. The special characters include ( ) ; " | # . ' '. The period "." should be noted specially, since some statistical packages use this as a code for missing values. Periods surrounded by tabs or blanks in the data file will always cause *Arc* to read the file incorrectly. Of course the symbol "." is acceptable when used as a decimal point.

A.5.5 Getting into Trouble with Plain Data Files

When *Arc* reads a plain data file, it determines the number of columns of data by counting the number of items on the first row of the file. If your data file has many variables, the editor or spreadsheet that created the file may not have the correct line feeds. You may get the following error message:

```
Error: list not divisible by this length
```

This message may also be caused by the presence of special characters characters on the file. Here are a few hints that might help in reading a file.

If you know the data file called `fred.lsp` has, for example, 17 variables, type

```
>(make-dataset :data (read-data-columns "fred.lsp" 17))
```

If you want the program to find the data file for you, you can type

```
>(make-dataset :data
          (read-data-columns (get-file-dialog) 17))
```

If this doesn't work, here are a few *Xlisp-Stat* functions that can help you. The command

```
>(def d (read-data-file "fred.lsp"))
```

will put all the contents of the data file in a list called d, with the contents of the first row first, then the contents of the second row, and so on. Type

```
>(length d)
```

to find out how many data elements were read from the data file. Type

```
>(select d (* 17 (iseq 4)))
```

to display four elements of d, namely elements 0, 17, 34, and 51. If the data file had 17 columns, this would give the first four items in the first column of the file. You can often find problems using these functions, and then use a text editor to correct the data file.

A.5.6 Formatted Data File

Files created using the "Save data set as" item in the data set menu have a human-readable format. You can also create files using an editor in this format; be sure that files are saved as plain text files, not in a word processor's special format. Table A.1 shows a formatted data file.

The first few lines of the file contain information *Arc* uses to assign names, labels, and so on. Statements are of two types: assignment statements, like `dataset = hald`, which assigns `hald` to be the name of the data set, and constructions that start with a `begin` and continue until an `end` is reached.

The assignment `columns = 5` specifies that the file has five columns. Specifying the number of columns is required only if some of the columns on the file are not used. The assignment `missing = ?` sets the missing value place holder that is filled-in for missing values. If the `delete-missing` assignment is equal to `t`, then cases that have missing values for any variable will be removed from the data set. If `delete-missing` is `nil`, then all the cases are used and cases with missing data are deleted as necessary in the computations. The first option guarantees that all plots/models are based on exactly the same observations, while the latter option allows each graph/model to use the maximum possible number of observations.

The `begin ... end` constructions are used to define blocks of information. The `begin description` construction allows typing in documentation for the data set. The documentation should be fairly short, 10 lines or less, although there is no formal limit. The construction `begin variables` is followed by one line for each column of data to be used that describes the data. These lines are in the format

column number = variable name = variable description

The variable name is a short label. You should avoid the use of the mathematical symbols +, -, =, and / in variable names. The variable description is

TABLE A.1   A Formatted Data File

```
dataset = hald  ...  The name of the data set is hald.
begin description  ...  The following lines describe the data:
The hald data were originally presented in the
statistics literature in A. Hald (1960),
Statistical Theory with Engineering Applications, Wiley.
end description  ...  End of the description
; a comment to skip  ...  Any line beginning with a ; is skipped
missing = ?  ...  Sets the missing value place holder.
delete-missing = nil  ...  If t, cases with missing data are deleted.
columns = 5  ...  Specifies the number of columns of data.
begin variables  ...  Begin defining variables
Col 0 = X1 = Chemical 1
Col 1 = X2 = Chemical 2 ; another comment
Col 2 = X3 = Chemical 3
Col 3 = X4 = Chemical 4
Col 4 = Y  = Heat of hardness
end variables  ...  End of defined variables
begin lisp  ...  Begin lisp instructions
(send hald :make-interaction '("X1" "X2"))
end lisp
begin transformation
X1SQ = X1^2
Z1 = (cut x1 2)
end transformation
begin data  ...  The data follow this statement.
(7 26 6 60 78.5
1 29 15 52 74.3
11 56 8 20 104.3
11 31 8 47 87.6
7 52 6 33 95.9
11 55 9 22 109.2
3 71 17 6 102.7
1 31 22 44 72.5
2 54 18 22 93.1
21 47 4 26 115.9
1 40 23 34 83.8
11 66 9 12 113.3
10 68 8 12 109.4)
```

a longer description of a variable that may not extend over more than one line.

All lines between `begin lisp` and `end lisp` are executed. As shown in Table A.1, an interaction between X1 and X2 will be created and added to the data set. You can put any valid lisp commands here that you like. *Arc* saves factors and interactions by adding a *lisp* statement to the file rather than actually saving the values created.

Similar to a block of *lisp* statements is a block of transformations. These are of the form "label = expression," where the expression can use variable names that already exist. The expression can be in usual mathematical notation, or a *lisp* expression. This block starts with begin transformations and ends with end transformations. Any statement that would be valid on the "Add a variate" menu item can be put here (see Section A.7.2).

The data follows the line begin data. The data should be preceded by a left parenthesis "(" and terminated with a right parenthesis ")." Although the parentheses are not required, including them speeds up reading the data by a substantial amount.

A.5.7   Creating a Data Set from the Text Window

The typed command

```
>(make-dataset)
```

is equivalent to selecting "Load" from the Arc menu. Assuming that the variables *CanRate*, *CigCon*, and *Country* already exist, the command

```
>(make-dataset
  :data (list CanRate CigCon Country))
```

will create a new data set using the data specified in place of a data file. You will be prompted to supply variable names and to name the data set. You can skip all the prompting by typing

```
>(make-dataset
      :data (list CanRate CigCon Country)
      :data-names (list "CanRate" "CigCon" "Country")
      :name "Cancer")
```

A.5.8   Old-Style Data Files

Data files that worked with the *R-code* version 1 that was distributed with Cook and Weisberg (1994b) can also be used with *Arc*. You can convert old-style data sets to new-style data sets by reading the file into *Arc* and using the "Save data set as" item in the data set menu. New-style data sets can be converted to old-style data sets using the command

```
>(arc-new-to-old "newstyle" "oldstyle")
```

where "newstyle" is the name of the data file to be converted, and "oldstyle" is the name of the old-style file that will be created.

A.5.9   Missing Values

Missing values in the data file are indicated by a ? unless you change the setting `*default-missing-indicator*` to some other value using the "Settings" item in the Arc menu, or you set the `missing` = assignment on the data file. The missing value indicator can be almost anything, but *it cannot be blank, period ".", or any other special character.* As described above in Section A.5.6, if you have missing data, the program will use as many complete cases as are available in any given calculation. If you include the command `delete-missing` = t in the data file (see Table A.1), then instead cases with missing data on any variate are removed from the data set for all calculations.

A.6   THE *Arc* MENU

When you start *Arc*, you will get a menu called Arc. This menu allows you to read data files, get help, view and change settings, and do calculations with standard distributions. Also, the names of all data sets currently in use are listed in the Arc menu and you make a data set active by selecting it from this menu. Here are descriptions of the items in the Arc menu.

Load. This item is used to read data files (see Section A.5) or to load files of *lisp* statements. With Mac OS and Windows OS, the "Load" item on the File menu and the "Load" item on the Arc menu are identical.

Dribble. With Mac OS this item also appears in the Xlisp-Stat menu, and with Windows OS it also appears on the File menu. Its use is described in Section A.3.1.

Calculate Probability. This item is used to obtain p-values for test statistics. See Section 1.7.1, page 20.

Calculate Quantile. This item is used to obtain quantiles of standard reference distributions for use in computing confidence intervals. See Section 1.6.3, page 16.

Help. This item accesses the help system for *Arc* and for *Xlisp-Stat*. See Section A.12.1.

Settings. Selecting this item will open a window that lists the names of many settings that the user can use to customize *Arc*. You can view and change information about a setting by clicking on the name of the setting, and then selecting "Update selection" from the "Settings" menu. Changes you make will be for the current session only unless you save them either from the dialog when updating a setting or from the Settings menu. If you save, a file called `settings.lsp` will be created and it will be read every time you start *Arc*.

Quit. Quits *Arc* and *Xlisp-Stat*.

A.7   THE DATA SET MENU

When you load a data file, two new menus are created: a *data set* menu and a menu named *Graph&Fit*. The data set menu will have the same name as the data set; if your data set is called Accidents, then the name of the data set menu will also be Accidents. Use the data set menu to modify and view data; use the Graph&Fit menu to obtain plots and fit models. Although you can load many data sets simultaneously, only one data set is active at a given time. You can change between data sets by selecting the name of the data set in the Arc menu.

A.7.1   Description of Data

Several items in the data set menu permit displaying numeric summaries of the data. These menu items are:

Description. Display information about the data set in the text window.

Display Summaries. Display in the text window basic summary statistics on the variables you specify. These include means, standard deviations, minimum, median, and maximum of each variable, as well as a correlation matrix of all pairwise correlations. See Section 1.4.1.

Table Data. This item is used to obtain tables of counts, or of summary statistics. Tables of counts are called *cross-tabulations*, or *contingency tables*, and they give the number of observations for each combination of possible values for the conditioning variables. Tables of summary statistics present means or other summaries for a variate for each combination of the conditioning variables. An example is given in Section 2.1, page 28.

Tables can be created with up to seven conditioning variables, and as many variates as you like. Several options are available in the table dialog (see Figure 2.3, page 30). If you check the item "Make new data set," then the tabled data will be saved as a new data set; *Arc* chooses the name. You can use the tabled data in plots, to fit models, and so on. You can choose to have the data displayed in a list rather than a table, and you can choose up to six summary statistics to be computed for each combination of the conditioning variables.

The default is to create tables using all possible cases, but if you select the button "Table included cases," then any cases that have been deleted using a plot control will not be used. The final item has to do with the way cells in a cross-tabulation with observed zeroes are to be treated when creating a new data set. They can be treated either as zero counts or as missing. This is generally important only if the table is to be saved for fitting Poisson regression models, as it allows the user to distinguish between random zeroes and structural zeroes.

The setting `*print-columns*` determines the number of columns on a page used in formatting a table. You can change this setting with the "Settings" item in the Arc menu.

Creating large tables with many conditioning variables can be very slow.

Display Data. Lists the values of all the variables you specify in the text window. See Section 1.4.3.

Display Case Names. Show case names in a window. This window will be linked to all other plots, and so this provides an alternative way of seeing case names of selected points without cluttering the plot.

A.7.2   Modifying Data

The data set menu has several items that can be used to modify existing data or create new data.

Add a Variate. This item is used to add new variables to the data set. If you type the expression `new=(a+b)^2`, then a variable called *new* will be added to the data set, and its values will be equal to $(a + b)^2$, where both a and b are existing variables. The expression to the right of the equal sign must obey the rules for an expression in the language C. In particular, the term `(1-a)(1-b)` is incorrect and must be replaced by `(1-a)*(1-b)`.

The quantity to the right of the equal sign can also be any expression in the language *lisp* that when evaluated returns a list of n numbers. For example, typing `new=(^ (+ a b) 2)` will create a new variable *new* with values again equal to the values of $(a + b)^2$. As a more complex example, you can type

```
new1 = (recode-values x '(1 2 3 4) '(0 0 0 1))
```

which will create a new variable *new1* by taking the existing variable x with values 1, 2, 3, and 4 and recoding so that if x equals one, two, or three, then *new1* = 0, and if $x = 4$, then *new1* = 1. The `recode-values` function and several other useful functions for manipulating data are described in Section A.12.

Delete Variable. Delete an existing variable. This needs to be done with care; if you delete X1, for example, you can have problems with interactions or factors derived from X1.

Rename Variable. Change the name of an existing variable. This can also cause problems if the name appears in previously defined models.

Transform. Create new variables using power and log transformations. See Section 6.3, page 107.

Make Factors. Make factors from existing variables. See Sections 12.1 and 12.8.1.

Make Interactions. Make interactions between existing factors and variables. An interaction is a product of terms, so, for example, the interaction of a term S and another term R has values given by $S \times R$. An interaction between a factor $\{F\}S$ and another variable R is defined by multiplying R by each of the indicator variables that make up the factor. The interaction of $\{F\}S$ and $\{F\}R$ consists of the products of all indicator variables in one factor with all indicator variables in the other factor. Interactions can involve more than two terms, again by multiplying values.

In the dialog produced by selecting this item the user specifies a set of terms and the program will adds all possible two-factor and/or three-factor interactions among these terms to the data set, depending on the buttons selected at the bottom of the dialog. Higher-order interactions requires using the dialog more than once.

Set Case Names. *Arc* attaches a label to every case, and it uses these labels in plots to identify points. If you have one or more text variables, *Arc* uses the first text variable it encounters to be the case labels. If you do not have any text variables, then case numbers are used as the labels. The dialog produced by this item is used to change the variable used to define case labels.

Save Data Set As. Saves the data set in a formatted file, as described in Section A.5.6, page 556.

Remove Data Set. Deletes the current active data set and all its menus, models, and plots.

A.8   GRAPHICS FROM THE GRAPH&FIT MENU

The Graph&Fit menu has six graphics commands.

Plot of. The plotting method for histograms, 2D and 3D scatterplots.

Scatterplot Matrix of. Use this item to create scatterplot matrices. The use of this item is described in Section 7.1, page 140.

Boxplot of. Use this item to create boxplots. The use of this item is described in Section 2.1, page 28.

Multipanel Plot of. Multipanel plots allow the user to view a sequence of plots in one window. For 2D plots, the user specifies one or more variables to be assigned to a changing axis, as well as one variable to be assigned to a fixed axis. The fixed axis can be either the vertical or the horizontal. When the plot is drawn, the user can change between the multiple panels of this plot using a slidebar plot control. Multipanel plots are used in added-variable plots, Section 10.5, and in *Ceres* plots, Section 16.2.

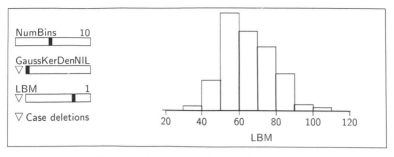

FIGURE A.2   A histogram showing the plot controls.

If a variable for the fixed axis is not selected, then histograms are drawn. Once again, a slidebar allows switching between plots.

Probability Plot of. Creates probability plots, Section 5.10.

Set Marks. This menu item allows setting, changing, or removing a marking variable. This variable is used to determine the color, symbol, or by default both, of plotted points. The settings menu item can be used to use only color or symbols rather than both. Marks can also be changed using the "Mark by" box on any dialog for creating a plot. The same marking variable will be used for all plots in a data set.

A.8.1   Histograms and Plot Controls

Histograms have four plot controls. A sample histogram is shown in Figure A.2. The first three plot controls are slidebars, which are used to select a value from a set of values. The slidebar for "NumBins" is used to select the number of bins in the histogram.

The second slidebar is called the *density estimate* slidebar. As the slider is moved to the right, the value of the *bandwidth* on the slider increases from zero. A *kernel density estimate*, using a Gaussian kernel, is superimposed on the plot. The smoothness of the density estimate depends on the bandwidth with larger values giving smoother estimates. If the bandwidth is too large, the estimate is oversmooth and biased; and if it is too small, it is undersmooth and too variable. The optimal choice of the bandwidth depends on the true density and the way that optimality is defined. One way to get an initial guess at a good bandwidth is to use a *normal reference rule*, which is the optimal bandwidth when the density being estimated is a normal density. *Arc* computes this value (given by the formula $h = 0.79\tilde{s}n^{1/5}$, where $\tilde{s}$ is the minimum of the sample standard deviation and the interquartile range divided by 1.34) and assigns it the value of 1 on the slidebar. This choice of bandwidth will generally oversmooth, so values in the range 0.6 to 0.8 are likely to be useful in most problems. If you use 0.6, for example, the bandwidth is then $0.6h$, where h is the normal reference rule.

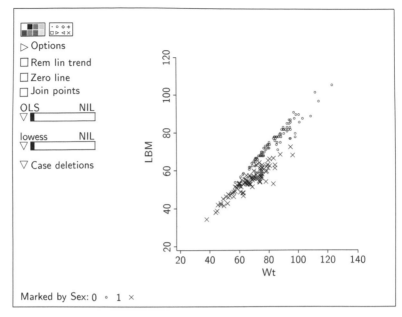

FIGURE A.3   A 2D plot showing the plot controls.

The density smoother slidebar has a *pop-up menu*, obtained by holding down the mouse button on the triangle to the left of the slidebar. This menu has four options. The item "Fit by marks," permits fitting a separate density estimate for each marked group, with the normal reference rule computed separately for each group. The next entry allows adding a different bandwidth multiplier to the slidebar. You can display the value of the bandwidth in the text window using the item "Display actual bandwidth" from the slidebar's pop-up menu. The final entry "Display summary statistics" displays basic summaries for each marked group, or for all the data with no marking, in the text window.

The next control is a transformation slidebar, as discussed in Section 5.2. The final plot control called "Case deletions" is described in the next section.

A.8.2   Two-Dimensional Plots and Plot Controls

Figure A.3 shows all the standard 2D plot controls. At the top left of the plot are the "Color palette" and the "Symbol palette." These are used to change a color and/or symbol assigned to a point in the plot. Simply select the points you want to change by clicking the mouse on them or by dragging over them with the mouse button down, and then click on the color or symbol you want. Colors and symbols are *linked* between plots, so changing one plot changes all plots (you can turn linking off by selecting "Unlink view" from the plot's menu). A side effect of using these palettes is that any marking variables, if defined, are removed from the plot and all linked plots.

The "Options" plot control gives a dialog for changing the appearance of the plot by modifying the axis labels, tick marks, and other features of the plot. An example of the options dialog is shown in Figure 3.5, page 46. By selecting check boxes in the options dialog, you can add transformation slidebars to the plot and a slidebar to control jittering.

The options dialog can also be used to add an arbitrary function to a plot. In the text area near the bottom of the dialog, type the function you want plotted. For example, y=x will add a straight line of slope 1 to the plot, while y=3+2*x+3*x^2 will add this quadratic function to the plot.

The "Rem lin trend" plot control is selected by clicking the mouse in the small rectangle, and unselected by clicking again. It computes the OLS regression of the vertical variable on the horizontal variable (WLS if weights were specified in the plot's dialog) and then replaces the quantity on the vertical axis with the residuals from this regression. Linear trends cannot be removed if transformation slidebars are on the plot. "Zero line" adds a horizontal line at zero to the plot, and "Join points" joins the points together with straight lines.

In the parametric smoother slidebar, which is initially marked "OLS," you can use the pop-up menu to fit using OLS, M-estimation with fixed tuning parameter (Huber, 1981), or using logistic regression, Poisson regression, or gamma regression. If you have specified weights, they are used in the fitting. For logistic regression, a plot of the proportion of successes on the vertical axis with weights equal to the number of trials will fit the correct logistic regression. If the number of trials equals one for all cases, you do not need to specify weights. The use of the "Power curve" smoother is discussed in Sections 13.1.2 and 16.1.1.

The smoothers available in the nonparametric smoothing slidebar include *lowess*, described in Section 9.5.2, and a very primitive slice smoother described in Section 3.2. The "lowess+−SD" option in the pop-up menu uses lowess to compute an estimate of the mean function, then computes the squared residuals from the lowess smooth, and smoothes them to estimate the variance function. The lines shown on the plot are the estimated mean and the estimated mean function plus and minus one times the square root of the estimated variance function.

Both smoother slidebars allow fitting a separate curve to each group of marked points by selecting the "Fit by marked group" option. On the parametric smoother slidebar, there are three "Fit by marked group" options: fit a separate line to each group, fit with common slopes, and fit with common intercepts; see Section 12.4.

The final plot control is called "Case deletions," and this is used to remove or restore selected cases. If you select a point or set of points by clicking on them or dragging over them, you can delete them from calculations by selecting the item "Delete selection from data set" from the menu. These cases will then be excluded from all calculations, and all previous calculations and plots will be updated. Cases are restored using another item in this plot control's menu.

One source of confusion in using the case deletions plot control is the difference between deleting points with this control and removing them using the item "Remove selection" from a plot's menu. *Deleting a case* removes it from all plots and from all computation and fitted models. *Removing a case* makes the point invisible in all plots, but it does not delete it from computations that are separate from plots.

If a marking variable is used in the plot, a marking legend is included at the bottom left of the plot. If you click on one of the symbols, only the cases for that symbol will be displayed in this plot and in all linked plots. This is like using the "Remove selection" item in the plot's menu, as the points are not removed from any other computations. If you click on the name of the marking variable, all cases are displayed.

A.8.3   Three-Dimensional Plots

The plot controls for 3D plots are described in Chapter 8.

A.8.4   Boxplots

Boxplots are described in Section 1.5.2, page 12, and Section 2.1, page 28.

The "Show anova" plot control on boxplots with a conditioning variable allows displaying an abbreviated Analysis of Variance table at the bottom of the plot. Weights, if set, are used in computing the Analysis of Variance. Boxplots are not linked to other plots.

A.8.5   Scatterplot Matrices

Scatterplot matrices are initially described in Section 7.1. The plot controls for these plots include transformation slidebars, introduced in Section 5.2, the "Case deletions" pop-up menu and a control called "Transformations," described in Sections 13.2.2 and 22.3.

Scatterplot matrices have a hidden control that is used to copy any frame of the scatterplot matrix into its own window by *focusing*. To get the new plot, move the mouse over the frame of interest and simultaneously press option-shift-mouse-click on Mac OS, or control-shift-mouse-click on Windows or Unix. If the mouse is over a variable name, a histogram of that variable will be produced.

A.9   FITTING MODELS

Models are fit by selecting one of the fit items from the Graph&Fit menu. The choices are summarized below.

Fit Linear LS. This item is used to fit linear least squares regression. The regression dialog is illustrated in Figure 6.6 and described in Section 6.3, page 107, and Section 11.3.2, page 270.

Fit Binomial Response. See Chapter 21.

Fit Poisson Response. See Section 23.3.

Fit Gamma Response. See Section 23.4.

Inverse Regression. This item is used to fit inverse regression (Quadratic fit, Cubic fit, Sliced Inverse Regression (SIR), pHd, or SAVE). These methods are described on the Web site for this book.

Graphical Regression. This item is described in Chapter 20.

A.10   MODEL MENUS

When a model is specified in *Arc*, a menu is added to the menu bar. The name of the menu and the name of the model are the same. We call the menu the *regression menu*. Items in this menu that create plots are all discussed in the main text; look in the index under Regression menu.

A.11   ADDING STATISTICS TO A DATA SET

Most of the statistics computed from the fit of a model that have a value for each case in the data can be added to the data set using an item in the model's menu. For example, suppose the regression of *Time* on T_1 is named L1. From the "L1" menu, select the item "Add to dataset," and in the resulting menu select `L1:Residuals`. This will save the residuals as a variable called *L1.Residuals* in the dataset.

A.12   SOME USEFUL FUNCTIONS

Several functions are provided that will help manipulate data. All of these functions require typing, and they can be used with the "Add a variate" menu item.

Repeating a Value. The function `repeat` is used to create a list of values that are all identical. For example,

```
>(repeat 1 10)
(1 1 1 1 1 1 1 1 1 1)
```

Generating a Patterned List. The function `gl`, which is short for generate levels, can be used to generate a patterned list of numbers. The function takes two or three arguments, *a*, *b*, and *n*; if *n* is not supplied, it is set equal to *ab*. The function will return a list of length *n* of the numbers 0 to *a* − 1 in groups of size *b*. For example,

```
>(gl 2 3 12)
(0 0 0 1 1 1 0 0 0 1 1 1)
```

has generated a list of length 12 for 2 levels in groups of three.

Sequences. The functions `iseq` and `rseq` can be used to generate a list of equally spaced numbers. Here are three examples:

```
>(iseq 10)
(0 1 2 3 4 5 6 7 8 9)
>(iseq 1 10)
(1 2 3 4 5 6 7 8 9 10)
>(rseq 0 1 11)
(0 0.1 0.2 0.3 0.4 0.5 0.6 0.7 0.8 0.9 1)
```

Cut. The `cut` function is used to cut a continuous predictor into a few discrete values. For example, (cut x 4) returns a list the same length as *x*, with all values equal to 0 for values in the lower quartile of *x*, 1 for the next quartile, 2 for the third quartile, and 3 for the upper quartile. The "4" can be replaced by any positive integer bigger than 1. (cut x '(2.1 9)) returns a list the same length as *x*, with all values equal to 0 for values in *x* smaller than or equal to 2.1; all values equal to 1 for *x* values between 2.1 and 9; and all values equal to 2 for all other values. (cut x '(2.1 9) :values '("A" "B" "C")) returns a list the same length as *x*, with all values equal to "A" for *x* values smaller or equal to 2.1, "B" for all *x* values between 2.1 and 9, and "C" for all other values. (cut x 4 :values #'mean) returns a list the same length as *x*, with all values less than the first quartile equal to the mean of the values less than the first quartile, values between the first and second quartile equal to the mean of those numbers, and so on. Generally, if the keyword :values is the name of a function like #'mean or #'median, then that function is applied to all the values in the group and used as the value for the group.

Recode. The function `recode` is used to replace a character list by a numeric list. For example, if *x* has all values equal to "F" or "M," then (recode x) will return a list the same length as *x*, with all the "F" replaced by 0 and all the "M" replaced by 1. Values are assigned to the levels of *x* by alphabetizing the levels.

The function `recode-value` can be used to change a specific value in a list *x* to something else. For example, (recode-value x 3 300) will return a list like *x* with all values of *x* = 3 changed to 300 and all other values unchanged.

The function `recode-values` is used to recode many values at once. For example,

```
(recode-values x '(1 2 3 4) '(0 0 0 1))
```

will return a list like x, with all values of 1, 2, or 3 replaced by 0 and all values of 4 replaced by 1.

Truncation. The function `truncate-at-val` returns a list with all values that pass a test set equal to zero. For example, `(truncate-at-val x 0)` replaces all values of x less than 0 by 0. `(truncate-at-val x 0 :value -10)` replaces all values of x less than 0 by -10, while

```
(truncate-at-val x 0 :test 'greater :value 10)
```

replaces all values of x greater than 0 by 10. The `:test` keyword can have the values `'less`, `'lessorequal`, `'greater` or `'greaterorequal`. In addition, it can be the name of any *lisp* function of two arguments that evaluates to true or nil.

Paste. The function `paste` can be used to combine several variables into a single text variable. For example, suppose the variable *Variety* is a text variable giving the name of the variety of soybeans used in a particular observation, with values like "Heifeng," "Champion," and so on, and the variable *Year* gives the year in which that observation was taken, either 1994 or 1995. The command

```
>(paste Variety Year :sep "-")
```

would return a list with elements like "Heifeng-1994." The default separator is a blank space.

Case Indicator. This function requires two arguments, a case number j and length n. It returns a list of n zeros, except the jth element is 1. For example, remembering that case numbers always start at zero, we have

```
>(case-indicator 3 7)
(0 0 0 1 0 0 0)
```

Polar Coordinates. Four functions are available for working with polar coordinates. If x and y are two lists of rectangular coordinates, then the corresponding polar coordinates are the radii $r = (x^2 + y^2)^{1/2}$ and angles $\theta = \arctan(y/x)$. The command `(polar-radius x y)` returns r. The command `(polar-angle x y)` returns θ, and the command `(polar-coords x y)` returns (r, θ). Finally, the command `(rect-coords r theta)` returns (x, y).

A.12.1 Getting Help

Help is available for most dialogs from a button on the dialog. Help can also be obtained for any function and for most methods using the "Help" item in the Arc menu. For example, to get help about the :r-squared method, type :r-squared in the help dialog. If available, help will be displayed in another dialog.

To get help about all functions that include cdf in their name, type

```
>(apropos 'cdf)
CDF
BIVNORM-CDF
F-CDF
POISSON-CDF
BETA-CDF
BINOMIAL-CDF
T-CDF
CAUCHY-CDF
NORMAL-CDF
CHISQ-CDF
GAMMA-CDF
```

To get help with the CDF of a *t*-distribution, enter t-cdf in the help dialog.

References

Acton, F. S. (1959). *The Analysis of Straight-Line Data*. New York: Wiley.

Agresti, A. (1990). *Categorical Data Analysis*. New York: Wiley.

Agresti, A. (1996). *An Introduction to Categorical Data Analysis*. New York: Wiley.

Allison, T. and Cicchetti, D. (1976). Sleep in mammals: Ecological and constitutional correlates. *Science*, 194, 732–734.

Altman, N. S. (1992). An introduction to kernel and nearest-neighbor nonparametric regression. *American Statistician*, 46, 175–185.

Anscombe, F. (1973). Graphs in statistical analysis. *American Statistician*, 27, 17–21.

Atkinson, A. C. (1985). *Plots, Transformations and Regression*. Oxford: Oxford University Press.

Atkinson, A. C. (1994). Transforming both sides of a tree. *American Statistician*, 48, 307–312.

Barnett, V. (1978). The study of outliers: Purpose and model. *Applied Statistics*, 27, 242–250.

Bates, D. M. and Watts, D. G. (1988). *Nonlinear Regression Analysis and Its Applications*. New York: Wiley.

Becker, R. and Cleveland, W. (1987). Brushing Scatterplots. *Technometrics*, 29, 127–142, reprinted in Cleveland, W. and McGill, M. (1988), *op. cit.*, 201–224.

Beckman, R. J. and Cook, R. D. (1983) Outlier … s. *Technometrics*, 25, 119–147.

Bedrick, E. J. and Hill, J. R. (1990). Outlier tests for logistic regression: A conditional approach. *Biometrika*, 77, 815–827.

Berk, K. N. (1998). Regression diagnostic plots in 3-D. *Technometrics*, 40, 39–47.

Berk, K. N. and Booth, D. E. (1995). Seeing a curve in multiple regression. *Technometrics*, 37, 385–398.

Berk, R. and Freedman, D. (1995). Statistical assumptions as empirical commitments. In Blomberg, T. and Cohen, S. (eds.), *Law, Punishment, and Social Control: Essays in Honor of Sheldon Messinger*, Part V. Berlin: Aldine de Gruyter, pp. 245–248.

Birkes, D. and Dodge, Y. (1993). *Alternative Methods of Regression*. New York: Wiley.

Bland, J. (1978). A comparison of certain aspects of ontogeny in the long and short shoots of McIntosh apple during one annual growth cycle. Unpublished Ph.D. dissertation, University of Minnesota, St. Paul, Minnesota.

Bliss, C. I. (1970). *Statistics in Biology*, Vol. 2. New York: McGraw-Hill.

Bloomfield, P. and Watson, G. (1975). The inefficiency of least squares. *Biometrika*, 62, 121–128.

Bowman, A. and Azzalini, A. (1997). *Applied Smoothing Techniques for Data Analysis.* Oxford: Oxford University Press.

Box, G. E. P. and Cox, D. R. (1964). An analysis of transformations. *Journal of the Royal Statistical Society, Series B*, 26, 211–246.

Box, G. E. P. and Draper, N. R. (1987). *Empirical Model-Building and Response Surfaces.* New York: Wiley.

Breiman, L. and Friedman, J. (1985). Estimating optimal transformations for multiple regression and correlation. *Journal of the American Statistical Association*, 80, 580–597.

Breiman, L. and Spector, P. (1992). Submodel selection and evaluation in regression. The *X* random case. *International Statistical Review*, 60, 291–320.

Brillinger, D. (1983). A generalized linear model with "Gaussian" regression variables. In Bickel, P. J., Doksum, K. A. and Hodges Jr., J. L. (eds.), *A Festschrift for Erich L. Lehmann.* New York: Chapman & Hall. pp. 97–114.

Brown, P. J. (1993). *Measurement, Regression, and Calibration.* Oxford: Oxford University Press.

Burt, C. (1966). The genetic determination of differences in intelligence: A study of monozygotic twins reared together and apart. *British Journal of Psychology*, 57, 137–153.

Buse, A. (1982) The likelihood ratio, Wald, and Lagrange multiplier tests: An expository note. *American Statistician*, 36, 153–157.

Carroll, R. J. and Ruppert, D. (1988). *Transformations and Weighting in Regression.* New York: Chapman & Hall.

Casella, G. and Berger, R. (1990). *Statistical Inference.* Pacific Grove, CA: Wadsworth & Brooks-Cole.

Clark, R., Henderson, H. V., Hoggard, G. K., Ellison, R. and Young, B. (1987). The abiltiy of biochemical and haematolgical tests to predict recovery in periparturient recumbent cows. *New Zealand Veterinary Journal*, 35, 126–133.

Clausius, R. (1850). Über die bewegende Kraft der Wärme und die Gezetze welche sich daraus für die Wärmelehre selbst ableiten lassen. *Annalen der Physik*, 79, 368–397, 500–524.

Chen, C. F. (1983). Score tests for regression models. *Journal of the American Statistical Association*, 78, 158–161.

Cleveland, W. (1979). Robust locally weighted regression and smoothing scatterplots. *Journal of the American Statistical Association*, 74, 829–836.

Cleveland, W. and McGill, M. (1988). *Dynamic Graphics for Statistics.* New York: Chapman & Hall.

Cochran, W. G. and Cox, G. (1957). *Experimental Designs*, 2nd ed. New York: Wiley.

Collett, D. (1991). *Modelling Binary Data.* New York: Chapman & Hall.

Cook, R. D. (1977). Detection of influential observations in linear regression. *Technometrics*, 19, 15–18.

Cook, R. D. (1986a). Assessment of local influence (with discussion). *Journal of the Royal Statistical Society, Series B*, 48, 134–169.

Cook, R. D. (1986b). Discussion of "Influential observations, high leverage points and outliers in linear regression," by S. Chatterjee and A. Hadi, *Statistical Science*, 1, 379–416.

Cook, R. D. (1993). Exploring partial residual plots. *Technometrics*, 35, 351–362.

Cook, R. D. (1994). On the interpretation of regression plots. *Journal of the American Statistical Association*, 89, 177–189.

Cook, R. D. (1996). Graphics for regression with a binary response. *Journal of the American Statistical Association*, 91, 983–992.

Cook, R. D. (1998a). Principal Hessian directions revisited (with discussion). *Journal of the American Statistical Association*, 93, 84–100.

Cook, R. D. (1998b). *Regression Graphics: Ideas for Studying Regressions Through Graphics*. New York: Wiley.

Cook, R. D. and Croos-Dabrera, R. (1998). Partial residual plots in generalized linear models. *Journal of the American Statistical Association*, 93, 730–739.

Cook, R. D., Hawkins, D. and Weisberg, S. (1992). Comparison of model misspecification diagnostics using residuals from least mean of squares and least median of squares fits. *Journal of the American Statistical Association*, 87, 419–424.

Cook, R. D., Peña, D. and Weisberg, S. (1988). The likelihood displacement: A unifying principle for influence. *Communications in Statistics, Part A—Theory and Methods*, 17, 623–640.

Cook, R. D. and Prescott, P. (1981). On the accuracy of Bonferroni significance levels for detecting outliers in linear models. *Technometrics*, 23, 59–63.

Cook, R. D. and Weisberg, S. (1982). *Residuals and Influence in Regression*. London: Chapman & Hall.

Cook, R. D. and Weisberg, S. (1983). Diagnostics for heteroscedasticity in regression. *Biometrika*, 70, 1–10.

Cook, R. D. and Weisberg, S. (1989). Regression diagnostics with dynamic graphics (with discussion). *Technometrics*, 31, 277–311.

Cook, R. D. and Weisberg, S. (1990a). Three dimensional residual plots. In Berk, K. and Malone, L. (eds.), *Proceedings of the 21st Symposium on the Interface: Computing Science and Statistics*. Washington, DC: American Statistical Association, 1990, pp. 162–166.

Cook, R. D. and Weisberg, S. (1990b). Confidence curves for nonlinear regression. *Journal of the American Statistical Association*, 85, 544–551.

Cook, R. D. and Weisberg, S. (1991). Comment on Li (1991), *op. cit. Journal of the American Statistical Association*, 86, 328–332.

Cook, R. D. and Weisberg, S. (1994a). Transforming a response variable for linearity. *Biometrika*, 81, 731–737.

Cook, R. D. and Weisberg, S. (1994b). *An Introduction to Regression Graphics*. New York: Wiley.

Cook, R. D. and Weisberg, S. (1997). Graphics for assessing the adequacy of regression models. *Journal of the American Statistical Association*, 92, 490–499.

Cook, R. D. and Wetzel, N. (1993). Exploring regression structure with graphics (with discussion). *TEST*, 2, 33–100.

Copas, J. (1983). Regression, prediction and shrinkage. *Journal of the Royal Statistical Society, Series B*, 45, 311–354.

Copas, J. (1988). Binary regression models for contaminated data (with Discussion). *Journal of the Royal Statistical Society, Series B*, 50, 225–265.

Cox, D. R. (1958). *The Planning of Experiments*. New York: Wiley.

Dalal, S., Fowlkes, E. B. and Hoadley, B. (1989). Risk analysis of the space shuttle: Pre-challenger prediction of failure. *Journal of the American Statistical Association*, 84, 945–957.

Dawson, R. (1995). The "unusual episode" data revisited. *Journal of Statistical Education*, 3, an electronic journal available at www.stat.ncsu.edu/info/jse.

Duan, N. and Li, K. C. (1991). Slicing regression: A link-free regression method. *Annals of Statistics*, 19, 505–530.

Eaton, M. (1986). A characterization of spherical distributions. *Journal of Multivariate Analysis*, 20, 272–276.

Edgeworth, F. (1887). On discordant observations. *Philosophical Magazine*, 23 (5), 364–375.

Ehrenberg, A. S. C. (1981). The problem of numeracy. *American Statistician*, 35, 67–71.

Enz, R. (1991). *Prices and Earnings Around the Globe*. Zurich: Union Bank of Switzerland.

Ezekiel, M. (1924). A method of handling curvilinear correlation for any number of variables. *Journal of the American Statistical Association*, 19, 431–453.

Ezekiel, M. (1930). The sampling variability of linear and curvilinear regressions. *Annals of Mathematical Statistics*, 1, 275–333.

Ezekiel, M. (1941). *Methods of Correlation Analysis*, 2nd ed. New York: Wiley.

Ezekiel, M. and Fox, K. A. (1959). *Methods of Correlation and Regression Analysis*. New York: Wiley.

Federer, W. T. (1955). *Experimental Designs*. New York: Macmillan.

Federer, W. T. and Schlottfeldt, C. S. (1954). The use of covariance to control gradients in experiments. *Biometrics*, 10, 282–290.

Finney, D. (1947). *Probit Analysis*, 1st ed. Cambridge: Cambridge University Press.

Finney, D. (1971). *Probit Analysis*, 3rd ed. Cambridge: Cambridge University Press.

Fisher, R. A. (1921). On the probable error of a coefficient of correlation deduced from a small sample. *Metron*, i (4), 1–32.

Fisherkeller, M. A., Friedman, J. H. and Tukey, J. W. (1974). PRIM-9: An interactive multidimensional data display and analysis system. Reprinted in Cleveland, W. and McGill, M., *op. cit*, pp. 91–110.

Flury, B. and Riedwyl, H. (1988). *Multivariate Statistics: A Practical Approach*. New York: Chapman & Hall.

Forbes, J. (1857). Further experiments and remarks on the measurement of heights by boiling point of water. *Transactions of the Royal Society of Edinburgh*, 21, 235–243.

Fox, J. (1991). *Regression Diagnostics*. Newberry Park, CA: Sage.

Freedman, D. (1983). A note on screening regression equations. *American Statistician*, 37, 152–157.

Freund, R. J. (1979). Multicollinearity etc., some "new" examples. *Proceedings of ASA Statistical Computing Section*, Washington, DC: American Statistical Association. pp. 111–112.

Fuller, W. A. (1987). *Measurement Error Models.* New York: Wiley.

Furnival, G. and Wilson, R. (1974). Regression by leaps and bounds. *Technometrics,* 16, 499–511.

Gauss, C. F. (1821–1826). Theoria Combinationis Observationum Erroribus Minimis Obnoxiae (Theory of the combination of observations which leads to the smallest errors). *Werke,* 4, 1–93.

George, E. and McCulloch, R. (1993). Variable selection via Gibbs sampling. *Journal of the American Statistical Association,* 88, 881–889.

Gilstein, C. Z. and Leamer, E. E. (1983). The set of weighted regression estimates. *Journal of the American Statistical Association,* 78, 942–948.

Gnanadesikan, R. (1977). *Methods for Statistical Analysis of Multivariate Data.* New York: Wiley.

Graybill, F. (1992) Matrices with applications in statistics, 2nd ed. Belmont, CA: Wadsworth.

Green, P. J. and Silverman, B. (1994). *Nonparametric Regression and Generalized Linear Models.* London: Chapman & Hall.

Hald, A. (1960). *Statistical Theory with Engineering Applications.* New York: Wiley.

Hall, P. and Li, K. C. (1993). On almost linearity of low dimensional projections from high dimensional data. *Annals of Statistics,* 21, 867–889.

Härdle, W. (1990). *Applied Nonparametric Regression.* Cambridge: Cambridge University Press.

Härdle, W. and Stoker, T. (1989). Investigating smooth multiple regression by the method of average derivatives. *Journal of the American Statistical Association,* 84, 986–995.

Hastie, T. and Tibshirani, R. J. (1990). *Generalized Additive Models.* New York: Chapman & Hall.

Henderson, H. V. and Searle, S. R. (1981). On deriving the inverse of a sum of matrices. *SIAM Review,* 23, 53–60.

Hernandez, F. and Johnson, R. A. (1980). The large sample behavior of transformations to normality. *Journal of the American Statistical Association,* 75, 855–861.

Hinkley, D. (1975). On power transformations to symmetry. *Biometrika,* 62, 101–111.

Hinkley, D. (1985). Transformation diagnostics for linear models. *Biometrika,* 72, 487–496.

Hoaglin, D., Iglewicz, B. and Tukey, J. W. (1986). Performance of some resistant rules for outlier labeling. *Journal of the American Statistical Association,* 81, 991–999.

Hosmer, D. W. and Lemeshow, S. (1989). *Applied Logistic Regression.* New York: Wiley.

Huber, P. (1981). *Robust Statistics.* New York: Wiley.

John, J. A. and Draper, N. R. (1980). An alternative family of power transformations. *Applied Statistics,* 29, 190–197.

Johnson, K. (1996). *Unfortunate Emigrants: Narratives of the Donner Party.* Logan, UT: Utah State University Press.

Johnson, M. (1987). *Multivariate Statistical Simulation.* New York: Wiley.

Johnson, M. P. and Raven, P. H. (1973). Species number and endemism: The Galápagos Archipelago revisited. *Science,* 179, 893–895.

Kay, R. and Little, S. (1987). Transformations of the explanatory variables in the logistic regrsion model for binary data. *Biometrika*, 74, 495–501.

Kerr, R. A. (1982). Test fails to confirm cloud seeding effect. *Science*, 217, 234–236.

Kuehl, R. O. (1994). *Statistical Principles of Research Design and Analysis*. Belmont, CA: Duxbury.

Larsen, W. A. and McCleary, S. A. (1972). The use of partial residual plots in regression analysis. *Technometrics*, 14, 781–790.

Li, K. C. (1991) Sliced inverse regression for dimension reduction (with discussion). *Journal of the American Statistical Association*, 86, 316–342.

Li, K. C. (1992). On principal Hessian directions for data visualization and dimension reduction: Another application of Stein's lemma. *Journal of the American Statistical Association*, 87, 1025–1039.

Li, K. C. and Duan, N. (1989). Regression analysis under link violation. *Annals of Statistics*, 17, 1009–1052.

Lindenmayer, D. B., Cunningham, R., Tanton, M. T., Nix, H. A. and Smith, A. P. (1991). The conservation of arboreal marsupials in the mountain ashforests of central highlands of Victoria, South-East Australia: III. The habitat requirements of Leadbeater's possum *Gymnobelideus leadbeateri* and models of the diversity and abundance of arboreal marsupials. *Biological Conservation*, 56, 295–315.

Linhart, H. and Zucchini, W. (1986). *Model Selection*. New York: Wiley.

Little, R. J. A. and Rubin, D. B. (1987). *Statistical Analysis with Missing Data*. New York: Wiley.

Mallows, C. (1973). Some comments on C_p. *Technometrics*, 15, 661–676.

Mallows, C. (1986). Augmented partial residual plots. *Technometrics*, 28, 313–320.

Mantel, N. (1970). Why stepdown procedures in variable selection? *Technometrics*, 12, 621–625.

McCullagh, P. and Nelder, J. A. (1989). *Generalized Linear Models*, 2nd ed. New York: Chapman & Hall.

McQuarrie, A. D. R. and Tsai, C-L. (1998). *Regression and Time Series Model Selection*. Singapore: World Scientific.

Mickey, R., Dunn, O. J. and Clark, V. (1967). Note on the use of stepwise regression in detecting outliers. *Computers and Biomedical Research*, 1, 105–109.

Mielke, H. W., Anderson, J. C., Berry, K. J., Mielke, P. W., Chaney, R. L. and Leech, M. (1983). Lead concentrations in inner-city soils as a factor in the child lead problem. *American Journal of Public Health*, 73, 1366–1369.

Miller, A. J. (1990). *Subset Selection in Regression*. New York: Chapman & Hall.

Mosteller, F. and Tukey, J. W. (1977). *Data Analysis and Regression*. Reading, MA: Addison-Wesley.

Nelder, J. A. and Wedderburn, R. W. M. (1972). Generalized linear models. *Journal of the Royal Statistical Society, Series A*, 135, 370–384.

Noll, S., Weibel, P., Cook, R. D. and Witmer, J. (1984). Biopotency of methionine sources for young turkeys. *Poultry Science*, 63, 2458–2470.

Pearson, K. (1930) *Life and Letters and Labours of Francis Galton*, Vol IIIa. Cambridge: Cambridge University Press.

Poisson, S. D. (1837). *Recherches sur la Probabilité des Jugements en Matière Criminelle et en Matière Civile, Précédées des Regles Générales du Calcul des Probabilitiés.* Bachelier, Imprimeur-Libraire pour les Mathematiques, la Physiques, etc., Paris.

Rencher, A. C. and Pun, F. C. (1980). Inflation of R^2 in best subset regression. *Technometrics*, 22, 49–53.

Ruppert, D., Wand, M., Holst, U. and Hössjer, O. (1997). Local polynomial variance-function estimation. *Technometrics*, 39, 262–273.

Ryan, T., Joiner, B. and Ryan, B. (1976). *Minitab Student Handbook.* North Scituate, MA: Duxbury.

Sacher, C. and Staffeldt, E. (1974). Relation of gestation time to brain weight for placental mammals: Implications for the theory of vertebrate growth. *American Naturalist*, 18, 593–613.

Samaniego, F. J. and Watnik, M. R. (1997). The separation principle in linear regression. *Journal of Statistics Education*, Vol. 5, Number 3, available at http://www.stat.ncsu.edu:80/info/jse/v5n3/samaniego.html.

Scheffé, H. (1959). *The Analysis of Variance.* New York: Wiley.

Schott, J. (1996). *Matrix Analyis for Statistics.* New York: Wiley.

Searle, S. R. (1982) *Matrix Algebra Useful for Statistics.* New York: Wiley.

Simonoff, J. (1996). *Smoothing Methods in Statistics.* New York: Springer-Verlag.

St. Laurent, R. (1990). The equivalence of the Milliken–Graybill procedure and the score test. *American Statistician*, 44, 36–37, correction p. 328.

Tierney, L. (1990). *Lisp-Stat: An Object-Oriented Environment for Statistical Computing and Dynamic Graphics.* New York: Wiley.

Tuddenham, R. D. and Snyder, M. M. (1954). Physical growth of California boys and girls from birth to age 18. *California Publications on Child Development*, 1, 183–364.

Tukey, J. W. (1949). One degree of freedom for nonadditivity. *Biometrics*, 5, 232–242.

Tukey, J. W. (1977). *Exploratory Data Analysis.* Reading, MA: Addison-Wesley.

Velilla, S. (1993). A note on the multivariate Box–Cox transformation to normality. *Statistics and Probability Letters*, 17, 259–263.

Velleman, P. (1982). Applied nonlinear smoothing. In Leinhardt, S. (ed.), *Sociological Methodology 1982*. San Francisco: Jossey-Bass, pp. 141–177.

Verbyla, A. P. (1993). Modelling variance heterogeneity: Residual maximum likelihood and diagnostics. *Journal of the Royal Statistical Society, Series B*, 55, 493–508.

Weisberg, H., Beier, H., Brody, H., Patton, R., Raychaudhari, K., Takeda, H., Thern, R. and Van Berg, R. (1978). *s*-Dependence of proton fragmentation by hadrons. II. Incident laboratory momenta, 30–250 GeV/c. *Physics Review D*, 17, 2875–2887.

Weisberg, S. (1985). *Applied Linear Regression*, 2nd ed. New York: Wiley.

Weisberg, S. (1986). A linear model approach to back calculation of fish length. *Journal of the American Statistical Association*, 81, 922–929.

Williams, D. A. (1982). Extra-binomial variation in logistic linear models. *Applied Statistics*, 31, 144–148.

Williams, D. A. (1987). Generalized linear model diagnostics using the deviance and single case deletions. *Applied Statistics*, 36, 181–191.

Wood, F. S. (1973). The use of individual effects and residuals in fitting equations to data. *Technometrics*, 15, 677–695.

Woodley, W. L., Simpson, R., Bioindini, R. and Berkeley, J. (1977). Rainfall results 1970–1975: Florida area cumulus experiment. *Science*, 195, 735–742.

Author Index

579

Subject Index

WILEY SERIES IN PROBABILITY AND STATISTICS

ESTABLISHED BY WALTER A. SHEWHART AND SAMUEL S. WILKS

Editors
Vic Barnett, Noel A. C. Cressie, Nicholas I. Fisher,
Iain M. Johnstone, J. B. Kadane, David G. Kendall, David W. Scott,
Bernard W. Silverman, Adrian F. M. Smith, Jozef L. Teugels;
Ralph A. Bradley, Emeritus, J. Stuart Hunter, Emeritus

Probability and Statistics Section

*ANDERSON · The Statistical Analysis of Time Series
ARNOLD, BALAKRISHNAN, and NAGARAJA · A First Course in Order Statistics
ARNOLD, BALAKRISHNAN, and NAGARAJA · Records
BACCELLI, COHEN, OLSDER, and QUADRAT · Synchronization and Linearity:
  An Algebra for Discrete Event Systems
BASILEVSKY · Statistical Factor Analysis and Related Methods: Theory and
  Applications
BERNARDO and SMITH · Bayesian Statistical Concepts and Theory
BILLINGSLEY · Convergence of Probability Measures, *Second Edition*
BOROVKOV · Asymptotic Methods in Queuing Theory
BOROVKOV · Ergodicity and Stability of Stochastic Processes
BRANDT, FRANKEN, and LISEK · Stationary Stochastic Models
CAINES · Linear Stochastic Systems
CAIROLI and DALANG · Sequential Stochastic Optimization
CONSTANTINE · Combinatorial Theory and Statistical Design
COOK · Regression Graphics
COVER and THOMAS · Elements of Information Theory
CSÖRGŐ and HORVÁTH · Weighted Approximations in Probability Statistics
CSÖRGŐ and HORVÁTH · Limit Theorems in Change Point Analysis
DETTE and STUDDEN · The Theory of Canonical Moments with Applications in
  Statistics, Probability, and Analysis
DEY and MUKERJEE · Fractional Factorial Plans
*DOOB · Stochastic Processes
DRYDEN and MARDIA · Statistical Analysis of Shape
DUPUIS and ELLIS · A Weak Convergence Approach to the Theory of Large Deviations
ETHIER and KURTZ · Markov Processes: Characterization and Convergence
FELLER · An Introduction to Probability Theory and Its Applications, Volume 1,
  Third Edition, Revised; Volume II, *Second Edition*
FULLER · Introduction to Statistical Time Series, *Second Edition*
FULLER · Measurement Error Models
GHOSH, MUKHOPADHYAY, and SEN · Sequential Estimation
GIFI · Nonlinear Multivariate Analysis
GUTTORP · Statistical Inference for Branching Processes
HALL · Introduction to the Theory of Coverage Processes
HAMPEL · Robust Statistics: The Approach Based on Influence Functions
HANNAN and DEISTLER · The Statistical Theory of Linear Systems
HUBER · Robust Statistics
IMAN and CONOVER · A Modern Approach to Statistics
JUREK and MASON · Operator-Limit Distributions in Probability Theory
KASS and VOS · Geometrical Foundations of Asymptotic Inference

*Now available in a lower priced paperback edition in the Wiley Classics Library.

Applied Probability and Statistics Section

*Now available in a lower priced paperback edition in the Wiley Classics Library.

*Now available in a lower priced paperback edition in the Wiley Classics Library.

*Now available in a lower priced paperback edition in the Wiley Classics Library.

*Now available in a lower priced paperback edition in the Wiley Classics Library.

*Now available in a lower priced paperback edition in the Wiley Classics Library.

Texts and References (Continued)

KOTZ, REED, and BANKS (editors) · Encyclopedia of Statistical Sciences: Update Volume 2

LAMPERTI · Probability: A Survey of the Mathematical Theory, *Second Edition*

LARSON · Introduction to Probability Theory and Statistical Inference, *Third Edition*

LE · Applied Categorical Data Analysis

LE · Applied Survival Analysis

MALLOWS · Design, Data, and Analysis by Some Friends of Cuthbert Daniel

MARDIA · The Art of Statistical Science: A Tribute to G. S. Watson

MASON, GUNST, and HESS · Statistical Design and Analysis of Experiments with Applications to Engineering and Science

MURRAY · X-STAT 2.0 Statistical Experimentation, Design Data Analysis, and Nonlinear Optimization

PURI, VILAPLANA, and WERTZ · New Perspectives in Theoretical and Applied Statistics

RENCHER · Methods of Multivariate Analysis

RENCHER · Multivariate Statistical Inference with Applications

ROSS · Introduction to Probability and Statistics for Engineers and Scientists

ROHATGI · An Introduction to Probability Theory and Mathematical Statistics

RYAN · Modern Regression Methods

SCHOTT · Matrix Analysis for Statistics

SEARLE · Matrix Algebra Useful for Statistics

STYAN · The Collected Papers of T. W. Anderson: 1943–1985

TIERNEY · LISP-STAT: An Object-Oriented Environment for Statistical Computing and Dynamic Graphics

WONNACOTT and WONNACOTT · Econometrics, *Second Edition*

WILEY SERIES IN PROBABILITY AND STATISTICS

ESTABLISHED BY WALTER A. SHEWHART AND SAMUEL S. WILKS

Editors
Robert M. Groves, Graham Kalton, J. N. K. Rao, Norbert Schwarz, Christopher Skinner

Survey Methodology Section

BIEMER, GROVES, LYBERG, MATHIOWETZ, and SUDMAN · Measurement Errors in Surveys

COCHRAN · Sampling Techniques, *Third Edition*

COUPER, BAKER, BETHLEHEM, CLARK, MARTIN, NICHOLLS, and O'REILLY (editors) · Computer Assisted Survey Information Collection

COX, BINDER, CHINNAPPA, CHRISTIANSON, COLLEDGE, and KOTT (editors) · Business Survey Methods

*DEMING · Sample Design in Business Research

DILLMAN · Mail and Telephone Surveys: The Total Design Method

GROVES and COUPER · Nonresponse in Household Interview Surveys

GROVES · Survey Errors and Survey Costs

GROVES, BIEMER, LYBERG, MASSEY, NICHOLLS, and WAKSBERG · Telephone Survey Methodology

*Now available in a lower priced paperback edition in the Wiley Classics Library.